CHEMISTRY FOR THE CONSUMER

KU-258-030

This book is due for return on or before the date last stamped below unless an extension of time is granted

2196

CHEMISTRY FOR THE CONSUMER

WILLIAM R. STINE

Wilkes College

CONTRIBUTORS
Owen D. Faut, Chapters Nine and Twenty-One
Mary Rees, Chapter Sixteen
Edward B. Stockham, Chapter Twenty
James J. Bohning, Chapter Twenty-Two

Allyn and Bacon, Inc.
Boston   London   Sydney   Toronto

To H.D., C.R., and C.J.S.

Copyright © 1978 by Allyn and Bacon, Inc.,
470 Atlantic Avenue, Boston, Massachusetts 02210.

All rights reserved. Printed in the United States of America.
No part of the material protected by this copyright notice may be
reproduced or utilized in any form or by any means, electronic or
mechanical, including photocopying, recording, or by any
information storage and retrieval system, without written permission
from the copyright owner.

LIBRARY OF CONGRESS CATALOGING IN PUBLICATION DATA

Stine, William R.
   Chemistry for the consumer.

   Bibliography: p.
   Includes index.
   1. Chemistry — Popular works. I. Title.
QD37.S67   540   77-28525
ISBN 0-205-06005-6

Illustrations on pages 4, 5, 7, 8, 33, 39, 53, 71, 85, 102, 125, 137,
171, 205, 257, 291, 366, 486 copyright © 1978 by Jim Chiros.

Second printing . . . August, 1979

C O N T E N T S

The nonscientist, faced with needing or simply wanting to learn chemistry, is often treated like a second-class citizen. In some instances, the student is provided with a course that avoids chemistry as much as possible and takes either an historical or social approach. Somewhat more often, the nonscientist is treated to a traditional chemistry course in a less rigorous form than that offered to science students, but with emphasis on physical chemistry. The usual subjects, such as stoichiometry, quantum numbers, and colligative properties are treated superficially in a way that is less than desirable for both student and instructor.

Chemistry for the Consumer provides an alternative. It recognizes the special needs of the nonscientist, who does not require a solid foundation in the traditional areas of chemistry, but may benefit greatly from an understanding of some of the chemistry that affects us continually throughout life.

The availability of good food, solutions to the energy crisis, and maintenance of good health are matters of great concern to every person and it is these subjects that set the tone of the book. The concentrated coverage of nuclear chemistry acknowledges the likelihood that nuclear fission will become increasingly important as a source of energy. The subject is discussed in such a way that the reader should acquire an appreciation of the technical aspects as well as an understanding of the safety features and environmental consequences. The major designs of atomic bombs are discussed as an interesting subject in itself and to provide a clear understanding of why a nuclear power plant is not a potential nuclear bomb.

Nuclear medicine is another major application of nuclear chemistry that emphasizes how scientists ranging from nuclear physicists to physicians can collaborate so as to make advances that may have very beneficial consequences. It also serves to emphasize how to work on a very destructive aspect of nuclear science, i.e., atomic bombs, eventually helped in the development of peaceful uses of great significance.

In Part Two, certain fundamental principles are considered as a background for later sections. These include chemical bonding, with emphasis on the periodic table, some organic chemistry, acidity, solubility, and some key aspects of the chemistry of carbohydrates.

In Parts Three and Four, the emphasis is on food—how to grow it, how to protect it while it is growing, and how to process and preserve it in a useful and safe form. In the coverage of agricultural chemistry, the discussion of food traces the events all the way from the agricultural laboratory (where scientists have contributed greatly to the quality of many food crops) to the farm or home garden (where fertilizers, pesticides, and herbicides are routinely used).

The chemistry involved in producing and protecting food is focused on five areas—alcoholic beverages, baking, dairy products, general problems of food preservation (including the use of food additives), and fats.

In Part Five, the subject is drugs. The early emphasis of this coverage is directed toward certain products that are used on a daily basis in the home, i.e., soaps and toothpastes. Several aspects of dental chemistry are covered, including the chemistry of tooth decay, the role of fluoride as a preventative, and the composition of the materials most often used by dentists for filling teeth. This coverage is followed by a discussion of drugs that are used as preventatives and treatments for various diseases. Addictive drugs and the treatments used in dealing with this problem are also considered.

The subject of birth control concludes the coverage of drugs. The general subject of steroid chemistry, which goes well beyond birth control, is introduced, and the mode of operation of the various types of oral contraceptives and other techniques are considered.

A few environmental issues are singled out for separate coverage in Chapter Twenty-one, but several are integrated throughout the text. Among the latter are the coverage of insect control (Chapter Ten) and detergents (Chapter Sixteen). The problems of insect control are mostly environmental issues, and detergents have been plagued with some serious environmental problems. In addition, considerable emphasis is directed to matters of the inner environment as they relate to food and drugs.

Finally the author wishes the reader a pleasant journey through the book. It is hoped that you may gain a better understanding of subjects that will be of continual concern to you as you act as a consumer of such things as energy, food, and drugs, as well as the air you breath and the water you drink.

William R. Stine

ACKNOWLEDGMENTS

Many people have been instrumental in the development of this book. I am particularly indebted to each of the individuals whose names appear on the title page, but particularly to Dr. Owen Faut who is responsible for the direction of several parts of the book.

The response from all those students who have been subjected to the topics presented herein has been a major factor in the development of many parts of the book in its present form. Most notably, the enthusiasm and suggestions offered by the teachers in the Hazleton, Pennsylvania school district were particularly valuable.

I must also express my appreciation to the many who have contributed meaningful illustrations to the text, particularly Jim Kane for his careful photographic work.

To those who reviewed the book as it was developing, I wish to offer thanks for the many helpful suggestions. I am particularly indebted to James Bills, Robert Balahura, John Amend, Stanley Manahan, Conrad Trumbore, Mark Souto, R. P. Steiner, Dennis Scott, Rudolph Bottei, and Floyd Kelly.

I am also grateful to Mr. Robert Summersgill for his expression of confidence. His thoughtfulness and consideration were greatly appreciated.

I have also appreciated the persistence and enthusiasm of John Shine. The latter has been particularly helpful. I thank Jane Dahl for editing out all the witches, although it is impossible to express my full admiration and respect.

Finally, words are totally inadequate to express my respect and gratitude for Miss Ruth Jacob for her efforts in my behalf. Her pride in seeing a job well done is virtually unparalleled.

NUCLEAR AND RADIOCHEMISTRY

PART ONE

I N the four chapters in Part One, we will examine the story behind the controversial subject of nuclear science. It is a topic that spans all of the traditional scientific disciplines of physics, biology, and chemistry and one which has become a great challenge to many individuals in the fields of engineering and medicine as well.

In order to appreciate the many interesting and important applications and problems of nuclear chemistry, the coverage of Chapter One is meant to provide an understanding of atomic structure and radioactivity as a background for later coverage.

MASS AND WEIGHT

The terms *mass* and *weight* are often used interchangeably but, strictly speaking, they are not the same. The mass of an object is a constant property that does not depend on its location. Weight depends on the mass but also depends on the gravitational attraction. As an example, an astronaut has the same mass whether he is standing on earth or orbiting in space, but his weight is much greater on earth under the influence of gravity. To be strictly correct, we should say that an individual has a mass of 150 pounds or 68 kilograms in the metric system (Appendix B contains a discussion of the metric system) but the common practice is to say that this person has a weight of 150 pounds even though his weight will vary from place to place.

Even among scientists, weight is frequently used when mass is the more precise term. Sometimes, this is simple carelessness but, in some instances, it is done for good reason and helps to avoid confusion (1). For example, the term *mass* is usually used when discussing very small units of matter. In any case, the reader need not be concerned about the distinction.

ATOMIC STRUCTURE

The basic units of all matter are called **elements.** There are approximately 105 elements known at this time. A complete list of the elements and their symbols appears on the inside of the front and back covers.

ATOMIC STRUCTURE: SPONTANEOUS RADIOACTIVE DECAY

The Atom: The basic particle of matter

The elements may be found pure as in the case of helium gas or silver metal, or different elements may be combined as in the case of water (H_2O), which is a substance formed by combination of the elements hydrogen and oxygen.

The smallest unit of an element, which possesses all the properties of that element, is the **atom.** The basic stable units of the atom are **protons** and **electrons** (2). These two particles can combine to give a third unit called the **neutron.** Some important characteristics of each are given in Table 1.1.

A scale of atomic weights, expressed in atomic mass units (amu), has been developed to describe the extremely small weights of atoms, using numbers easy to handle in mathematical calculations. Carbon, with an atomic weight of 12.000, has been accepted as the standard for this scale. By comparison, the weight of a hydrogen atom is approximately 1 amu. In other words, one atom of carbon is 12 times heavier than one atom of hydrogen. It has also been determined that 6.023×10^{23}* atoms of hydrogen have a mass of approximately 1 gram. This number (6.023×10^{23}) is called *Avogadro's number.*

* In the event that the "exponential notation" is unfamiliar, refer to Appendix A for clarification.

One atom of carbon is 12 times one atom of hydrogen

As indicated in Table 1.1, protons and neutrons are located in the central portion of the atom known as the **nucleus** and they are surrounded by electrons in a variety of arrangements called *orbitals*. The arrangement of the electrons is of major importance in describing the bonds that hold atoms together, and will be considered in Chapter Five. For now, the simple picture of a spherical atom with the nucleus surrounded by electrons in circular orbits is sufficient. Such a model

TABLE 1.1   CHARACTERISTICS OF ATOMIC PARTICLES

PARTICLE	LOCATION	CHARGE	EXACT MASS (amu)*	APPROXIMATE MASS (amu)*
Proton	Nucleus	+1	1.0078252	1
Electron	Outside the nucleus	−1	0.0005468	0
Neutron	Nucleus	0	1.0086654	1

*amu = 1 atomic mass unit = 1.66×10^{-24} grams.

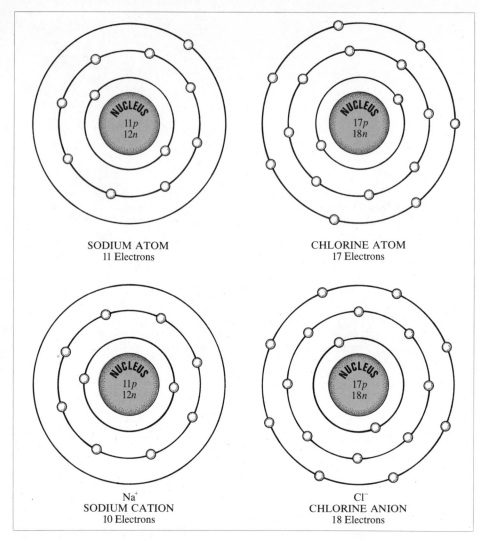

SODIUM ATOM
11 Electrons

CHLORINE ATOM
17 Electrons

Na^+
SODIUM CATION
10 Electrons

Cl^-
CHLORINE ANION
18 Electrons

was proposed by the Danish physicist Niels Bohr in 1913 and is commonly called "The Bohr Atom." The model is generally regarded as outdated, but it does account nicely for certain nuclear phenomena that will be encountered in later sections.

In a neutral atom, the number of protons equals the number of electrons. If they are unequal, the atom exists as a charged particle called an **ion.** If the number of protons exceeds the number of electrons, a positively charged **cation** results. In the reverse situation, a negatively charged particle called an **anion** exists when the number of electrons exceeds the number of protons. In short, most of the elements listed in the table on the inside front and back covers may exist as neutral atoms or as ions.

Elements may also be chemically combined to form **compounds.** The smallest unit of a compound, which possesses all the properties of that compound, is the **molecule.** In some cases, molecules contain only one element. For example, an

Normally-occurring hydrogen has a significant natural abundance of both ^{1}H and D, whereas T is unstable and undergoes a radioactive decay.

RADIOACTIVITY

We have already seen two isotopes, ^{14}C and ^{3}H (tritium), which are so unstable that they undergo radioactive decay. To understand this phenomenon, consider the role of the neutrons in an atomic nucleus. The number of neutrons present is very influential in determining the stability of the nucleus. A stable atom is a delicately-balanced combination of attractive and repulsive forces. Particles with opposite charges, e.g., protons and electrons, attract one another. On the other hand, particles with the same charge repel one another, i.e., electrons repel each other and protons behave in the same way toward other protons. The influence of neutrons on the balance is seemingly more subtle and yet plays a major role in determining whether an atomic nucleus is stable or will undergo some type of radioactive decay, thereby liberating energy and yielding a new, more stable atom.

The neutron/proton (n/p) ratio is often considered to be of prime importance in determining stability. This ratio reaches an upper limit of 1.5 for stable isotopes at ^{209}Bi. As the number of protons (atomic number) in the nucleus increases, the repulsive force between the protons also increases. To decrease this disruptive, destabilizing force, it is necessary to move the repelling protons farther apart by addition of extra neutrons. However, a point is reached where too many neutrons are present, the atom becomes unstable, and radioactive decay occurs. The atom will decay in a manner that will bring the neutron/proton ratio closer to a stable value for that element. However, even a more suitable neutron/proton ratio does not serve to stabilize any nuclei with a mass greater than that of ^{209}Bi. Consequently, all elements above ^{209}Bi are unstable and undergo some mode of radioactive decay. Figure 1.1 gives a plot of the number of neutrons versus protons for all known isotopes.

According to Figure 1.1, stability is observed for the nuclei represented by blackened squares. The lines for $n/p = 1.0$ and 1.5 are added for reference. Even some nuclei within these two extremes are unstable, but all nuclei lying outside the extremes are unstable.

If a particular nucleus is not a stable one, it will undergo a radioactive decay to produce a new nucleus. If the product nucleus is a stable one, no further decay occurs. If not, additional decay will occur until a stable nucleus is finally produced.

Radioactive decay can occur in several ways. Discussion of the three modes of spontaneous decay follows.

Alpha Decay

An alpha particle or helium nucleus, symbolized α, $^4_2\alpha$, or $^4_2$He, may be emitted by a nucleus, which gains greater stability in the process. This type of decay is particularly common to elements with a very high atomic number (above 83), since the loss of an alpha particle causes a decrease in the number of protons in the nucleus. Since the alpha particle consists of 2 protons and 2 neutrons, the neutron/proton ratio is 1. Its emission from a nucleus causes an atom to proceed to a higher n/p

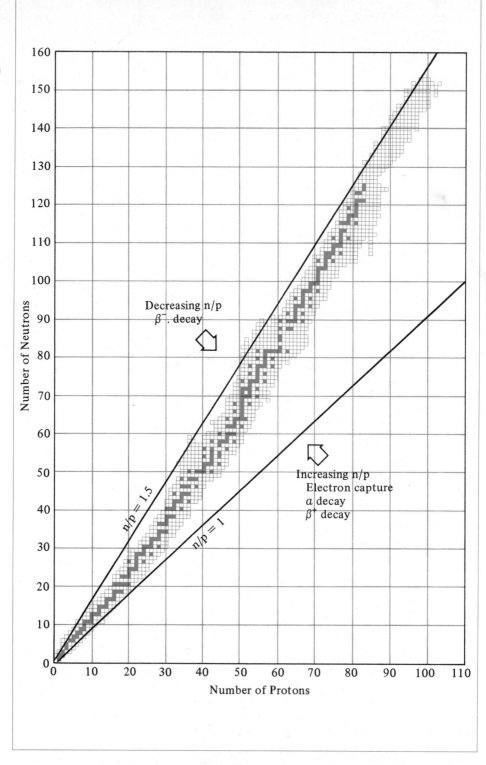

(assuming n/p is greater than 1 before decay). The following reaction is a typical example.

$$^{235}_{92}\text{U} \rightarrow {}^{4}_{2}\alpha + {}^{231}_{90}\text{Th}$$
$$n/p = \tfrac{143}{92} = 1.554 \qquad n/p = \tfrac{141}{90} = 1.567$$

Beta Decay

In order to lower the neutron/proton ratio, the conversion of a neutron to a proton can occur within the nucleus. Recalling the description of a neutron as a combination of a proton and an electron, the conversion of a neutron to a proton must be accompanied by the emission of a negative electron, described in this situation as a beta (β^-) particle to denote that the nucleus is the source of the electron. An example is the decay of indium-116 in which an indium atom changes into a tin atom due to the additional proton.

$$^{116}_{49}\text{In} \rightarrow {}^{0}_{-1}\beta + {}^{116}_{50}\text{Sn} \ (\beta^- \text{ decay, electron emission})$$
$$n/p = \tfrac{67}{49} = 1.367 \qquad n/p = \tfrac{66}{50} = 1.320$$

In addition to alpha decay, an increase in the n/p may be accomplished in two ways. The two modes of decay are beta-plus (β^+) decay, also called *positron emission,* and *electron capture*. These two processes are the opposite of β^- decay and occur when the n/p is too low. Examples follow.

$$^{116}_{51}\text{Sb} \rightarrow {}^{0}_{+1}\beta + {}^{116}_{50}\text{Sn} \ (\beta^+ \text{ decay, positron emission})$$
$$n/p = \tfrac{65}{51} = 1.275 \qquad n/p = \tfrac{66}{50} = 1.320$$

In this example, a proton appears to be splitting into a neutron and a positron, with the neutron being retained by the nucleus.

$$^{195}_{79}\text{Au} + {}^{0}_{-1}e \rightarrow {}^{195}_{78}\text{Pt} \ (\text{electron capture})$$
$$n/p = \tfrac{116}{79} = 1.468 \qquad n/p = \tfrac{117}{78} = 1.500$$

In the electron capture process, an electron enters the nucleus and combines with a proton to form a neutron.

Gamma Decay

There are many examples of nuclei that exist in unstable forms and can relieve some of the instability by emission of energy in the form of gamma rays. Gamma rays are different from alpha and beta particles since they are not particles at all. In fact, they are hard to visualize and even harder to explain. They are a type of

Figure 1.1   Neutrons versus protons for known isotopes. The arrows indicate how a nucleus may change in each of the ways described under α and β decay.

electromagnetic radiation, like light, which is also hard to describe and, in fact, is alternately described in two different ways:

1. as wavelike in character
2. as small bundles of energy called quanta or photons

In most cases, γ rays (or x-rays) are given off only when the other types of decay occur. One common example of γ emission is that observed for cobalt-60, which undergoes beta decay to nickel-60 and emits two γ rays in the process. This example will be mentioned later under radiation therapy.

$$^{60}_{27}\text{Co} \rightarrow {}^{60}_{28}\text{Ni} + {}^{0}_{-1}\beta + 2\gamma$$

In a few cases it is possible to observe the release of gamma radiation without any α or β decay occurring. One very important example of this occurs when technetium-99m undergoes the following conversion.

$$^{99m}\text{Tc} \rightarrow {}^{99}\text{Tc} + \gamma$$

This is a process commonly known as an **isomeric transition,** symbolized IT, in which an unstable form of the isotope becomes more stable by the emission of γ radiation. The symbol m (metastable) in ^{99m}Tc signifies that the nucleus is some-what unstable and will become stable by γ emission. Such a process, in which only γ emission occurs, is rarely observable because the isotopes that undergo such changes are very unstable. Even 99^mTc loses half of its radioactivity in only six hours. Some other metastable isotopes remain active for only a fraction of a second. Such a process presumably occurs even in the decay of ^{60}Co (see previous equation) but the ^{60m}Ni, which is initially formed, is so unstable that it immediately decays by emitting γ radiation and forms stable ^{60}Ni, which is the observed product.

OTHER TYPES OF IONIZING RADIATION

There are many types of radiation (particulate or electromagnetic) that are described as ionizing radiation. Table 1.2 lists the common types of ionizing radiation.

X-rays are a specific kind of γ rays given off following an electron capture process. During this process, energy is released in the form of x-rays when an outer-orbital electron drops into the space vacated by the electron that is captured by the nucleus. The following example illustrates the process. According to this example, electron capture by the radioactive beryllium nucleus leaves an unstable arrangement of the orbital electrons. The system is stabilized by moving an electron from the outer shell to the inner shell and emitting energy in the form of an x-ray.

Greater quantities of x-rays are normally generated by x-ray machines. The generation process is similar to throwing a rock against a wall. In the process, the

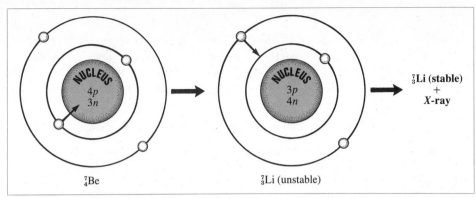

wall takes on the energy and emits it as sound waves and heat. Similarly, in an x-ray machine, a target material is bombarded by high speed (high energy) electrons. These high energy electrons interact strongly with electrons of the target, knocking out some of the target electrons and causing other orbital electrons to rearrange. The result is the release of energy in the form of x-rays.

Some of the characteristics of the other types of ionizing radiation listed in Table 1.2 will be considered in later chapters.

TABLE 1.2   TYPES OF IONIZING RADIATION

TYPE	SYMBOL	MASS	CHARGE	DESCRIPTION	SOURCE
Alpha	$_2^4\alpha$ or $_2^4$He	4	+2	helium nucleus	spontaneous radioactive decay (primarily from heavy atoms)
Beta (negatron)	$_{-1}^0 e$ or $_{-1}^0\beta$	0	−1	negative electron	spontaneous radioactive decay
Beta (positron)	$_{+1}^0 e$ or $_{+1}^0\beta$	0	+1	positive electron	spontaneous radioactive decay
Proton	$_1^1$H	1	+1	hydrogen nucleus	artificially produced by nuclear reactors (accelerators)
Neutron	$_0^1 n$	1	0	proton plus electron	artificially produced by nuclear reactors (accelerators)
Gamma	γ	0	0	electromagnetic radiation	spontaneous radioactive decay
X-ray		0	0	electromagnetic radiation	x-ray machines, rearrangement of orbital electrons

NUCLEAR REACTIONS

Beginning with the discussion of alpha decay, a nuclear reaction was written for each radioactive decay process that was described. A complete and balanced equation should be written for each nuclear reaction that is considered. In order to write these equations accurately, it is necessary to adhere to the following rules:

1. The sum of the atomic numbers on each side of the equation must be equal.
2. The sum of the mass numbers on each side of the equation must be equal.

For examples, refer to each of the nuclear reactions written thus far, but also consider the following problems. In each problem, part of a radioactive decay process is described, but each reaction must be completed by properly balancing the equation in order to identify the missing product.

Problem 1.1. Tritium (^{3}H) undergoes a radioactive decay with emission of an electron (negative beta particle). Write the complete reaction.

Discussion. The problem should be approached stepwise by first writing the portion of the equation which is known.

$$^3\text{H} \rightarrow {}_{-1}^{0}\beta + \text{?}$$

Next, include the atomic number for hydrogen. This gives the sum of the atomic numbers on the left side of the equation and allows us to write the atomic number and mass number for the unknown product nucleus.

$$^3_1\text{H} \rightarrow {}_{-1}^{0}\beta + {}_2^3\text{?}$$

Finally, consult a table of the elements to find the symbol and name of the element that has an atomic number of 2. It is helium and the complete and balanced equation for the beta decay of tritium is

$$^3_1\text{H} \rightarrow {}_{-1}^{0}\beta + {}_2^3\text{He}$$

Problem 1.2. Aluminum-25 undergoes radioactive decay by electron capture. Write the complete reaction.

Discussion. Once again, write as much of the equation as possible from the information given in the problem and in a table of the elements. This gives

$$^{25}_{13}\text{Al} + {}_{-1}^{0}e \rightarrow {}_{12}^{25}\text{?}$$

Finally, a table of the elements shows that magnesium is the element with atomic number 12. Therefore the final equation is

$$^{25}_{13}\text{Al} + {}_{-1}^{0}e \rightarrow {}_{12}^{25}\text{Mg}$$

Problem 1.3. Thorium-232 undergoes a radioactive decay to yield radium-228. Write the complete reaction.

Discussion. This time we know the product that is formed but not the mode of decay. However, the problem is approached in the same fashion to give

$$^{232}_{90}\text{Th} \rightarrow {}^{228}_{88}\text{Ra} + {}^{4}_{2}?$$
or
$$^{232}_{90}\text{Th} \rightarrow {}^{228}_{88}\text{Ra} + {}^{4}_{2}\alpha$$

THE PERIODIC TABLE

As a cross-check for the answers to the previous problems, let us take our first look at the chemist's most frequent source of information on the elements—the periodic table (Table 1.3 and inside front cover). This listing gives the elements in order of increasing atomic number, i.e., increasing number of protons in the nucleus. The atomic weights are averages that result from a contribution from all of the naturally-occurring isotopes of each element (see *Isotopes*). The symbols and sometimes the names of the elements are also given in the periodic table.

In most radioactive decay processes, there is a conversion from one ele-

TABLE 1.3   THE PERIODIC TABLE OF THE ELEMENTS

Period	I A	II A	III B	IV B	V B	VI B	VII B		VIII		I B	II B	III A	IV A	V A	VI A	VII A	VIII A
1																		2 He 4.00
2	3 Li 6.94	4 Be 9.01											5 B 10.81	6 C 12.01	7 N 14.01	8 O 16.00	9 F 19.00	10 Ne 20.18
3	11 Na 23.00	12 Mg 24.31											13 Al 26.98	14 Si 28.09	15 P 30.97	16 S 32.06	17 Cl 35.45	18 Ar 39.95
4	19 K 39.10	20 Ca 40.08	21 Sc 44.96	22 Ti 47.90	23 V 50.94	24 Cr 52.00	25 Mn 54.94	26 Fe 55.85	27 Co 58.93	28 Ni 58.71	29 Cu 63.55	30 Zn 65.37	31 Ga 69.72	32 Ge 72.59	33 As 74.92	34 Se 78.96	35 Br 79.90	36 Kr 83.80
5	37 Rb 85.47	38 Sr 87.62	39 Y 88.91	40 Zr 91.22	41 Nb 92.91	42 Mo 95.94	43 Tc* 98.91	44 Ru 101.07	45 Rh 102.91	46 Pd 106.4	47 Ag 107.87	48 Cd 112.40	49 In 114.82	50 Sn 118.69	51 Sb 121.75	52 Te 127.60	53 I 126.90	54 Xe 131.30
6	55 Cs 132.91	56 Ba 137.34	57 La 138.91	72 Hf 178.49	73 Ta 180.95	74 W 183.85	75 Re 186.2	76 Os 190.2	77 Ir 192.22	78 Pt 195.09	79 Au 196.97	80 Hg 200.59	81 Tl 204.37	82 Pb 207.2	83 Bi 208.98	84 Po* [210]	85 At* [210]	86 Rn* [222]
7	87 Fr* [223]	88 Ra* 226.02	89 Ac* [227]	104 Rf* [261]	105 Ha* [262]													

atomic number 1
symbol of the element H
atomic weight 1.01†

Lanthanides

58 Ce 140.12	59 Pr 140.91	60 Nd 144.24	61 Pm* [147]	62 Sm 150.4	63 Eu 151.96	64 Gd 157.25	65 Tb 158.93	66 Dy 162.50	67 Ho 164.93	68 Er 167.26	69 Tm 168.93	70 Yb 173.04	71 Lu 174.97

Actinides

90 Th* 232.03	91 Pa* 231.04	92 U* 238.03	93 Np* 237.05	94 Pu* [244]	95 Am* [243]	96 Cm* [247]	97 Bk* [247]	98 Cf* [251]	99 Es* [254]	100 Fm* [257]	101 Md* [258]	102 No* [255]	103 Lr* [256]

*All isotopes are radioactive.

† All atomic weights have been rounded to 0.01.

[ ] Indicates mass number of longest known half-life.

ment to another. Gamma decay is the exception. For the other common decay processes the principal nuclear changes are as follows.

1. For α decay: the number of protons decreases by 2.
2. For β decay: (a) electron emission—the number of protons increases by 1, (b) positron emission—the number of protons decreases by 1, and (c) electron capture—same as positron emission.

Therefore, the nuclear reactions in these problems can be checked for accuracy by consulting the periodic table. In Problem 1.1, electron emission increases the atomic number to yield the element immediately following the original in the periodic table. Thus, hydrogen (atomic number 1) was converted to helium (atomic number 2). In Problem 1.2, the opposite occurred, aluminum (atomic number 13) was converted to magnesium (atomic number 12), because a proton is converted to a neutron when the electron is captured by the nucleus.

In alpha decay, two protons (and two neutrons) are emitted, so the atomic number decreases by two. An example appeared in Problem 1.3, in which thorium (atomic number 90) was converted to radium (atomic number 88).

Using the periodic table in this way should help in understanding the radioactive decay processes. At the same time, it is important to become familiar with the table, which will be considered in more detail in Part Two and used throughout the text.

DETECTION OF RADIATION

Radioactive emissions are often described as **ionizing radiation** because they interact with molecules or atoms to produce ions. As noted previously, cations are species with fewer electrons than protons, whereas anions have an excess of electrons. Later, we will consider the biological effects of such radiation. When ionization is made to proceed in a suitable chamber, such as a Geiger-Muller counter, the amount of radiation causing the ionization can be measured. In order to do this, a chamber filled with a suitable gas is exposed to the radiation, causing some of the gas molecules to ionize. Figure 1.2 gives a schematic diagram of a typical radiation counter.

The counter also contains a positively-charged wire (collecting electrode) and a negatively-charged outer wall (electrode). The ions formed from the decay are attracted to the electrodes and produce a signal when they make contact. Such a system can easily be calibrated so that the signal is a direct measure of the intensity of the radiation being monitored.

Radiation of very low energy is described as *soft*. Radiation of this sort has great difficulty in penetrating any matter including the window of a counter (Figure 1.3). Such radiation is often counted by the technique of *liquid scintillation*. In this case, a special compound is added directly to the radiating material in order to absorb the radiation energy. This compound then reemits a portion of the energy as light that can be measured quantitatively and translated to describe the intensity of the radioactivity. The isotope ^{14}C is commonly counted by the liquid scintillation technique and is one of the most widely used isotopes in various types of re-

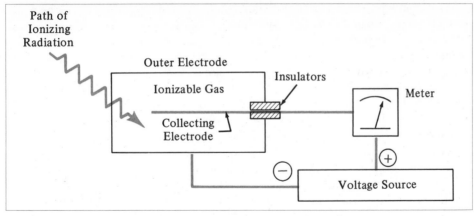

Figure 1.2   Schematic diagram of a typical radiation counter (3).

Figure 1.3   A radiation counter (courtesy of The Nucleus).

search. It decays by emission of a β^- particle with a maximum energy of only 0.155 meV.* The reaction is

$$^{14}_{6}C \rightarrow {}^{14}_{7}N + {}^{0}_{-1}\beta$$

The hazards of such radiation and others will be considered in Chapter 2 but, for now, the radiation from ^{14}C is very soft, which means:

1. The isotope is safe to handle, although care must be taken to avoid any ingestion,
2. the radiation is stopped before it can pass through the window of a radiation counter.

* meV = Million electron volts, a unit of energy commonly used in describing radioactive decay.

Although the amount of radiation being given off by a source may be monitored by sophisticated detection equipment, it is also very important to determine the amount of radiation to which a person has been exposed. Radiation detection badges or dosimeters are worn at all times by researchers who handle radioactive materials and by workers at nuclear power plants (Figure 1.4). The construction of the radiation badges varies from very simple to quite complex, depending on the type of radiation to be monitored. For example, a worker exposed to beta radiation only, need not wear a badge that can distinguish between beta and gamma radiation. But if a worker is exposed to more than one kind of radiation, or to radiation of varying intensities, he must wear a more sophisticated badge.

Many badges are made of photographic film which is shielded to prevent exposure by light rays but is exposed (blackened) by the more penetrating types of ionizing radiation. Badges made to detect different types of radiation consist of several layers of photographic film separated by filters (absorbers) of increasing absorbing power. Thus, the top layers of film are exposed by all radiation, whereas the bottom layers will be exposed only by the more penetrating types of radiation. The type of radiation can be determined by examination of the different layers of film. The amount of radiation can be calculated from the degree of darkening of each layer of film.

Figure 1.4   Radiation dosimeters (courtesy of Dosimeter Corporation).

SUMMARY

The atom is composed of protons and neutrons, which are located in the nucleus, plus electrons, which are found outside the nucleus. When the number of protons does not equal the number of electrons, atoms exist as ions.

Most elements exist in multiple forms, called isotopes, which differ in the number of neutrons found in the nucleus. Some of these isotopes are stable, whereas others are radioactive and undergo alpha, beta, or gamma decay in order to achieve stability. X-rays may be emitted as a result of electron capture processes or by x-ray machines.

The products of nuclear decay processes may be predicted by properly balancing the equations that describe these processes. The periodic table may serve as a handy reference source to assist in balancing nuclear equations.

Emitted radiation is accurately detected by the use of sophisticated counting devices or by the technique of liquid scintillation in the case of very soft radiation. Film badges can be conveniently worn by persons working in the vicinity of radiation so that the level of radiation exposure may be monitored.

PROBLEMS

1. Give a definition or example of each of the following:
   - a) element
   - b) atom
   - c) neutron
   - d) cation
   - e) anion
   - f) compound
   - g) molecule
   - h) isotopes
   - i) alpha particle
   - j) beta particle
   - k) n/p
   - l) x-rays

2. Give the number of protons, neutrons, and electrons in each of the following atoms or ions. Also give the name of each and indicate if the symbol describes a cation, anion, or neutral atom. Refer to the inside front or back cover for a table of atomic symbols, numbers, and weights.

   Example: $^{59}Fe^{+3}$

   Answer: $^{59}_{26}Fe^{+3}$ has

	26 protons
	33 neutrons
	23 electrons
and is a	cation of iron

   - a) $^{64}Cu^{+2}$
   - b) ^{198}Au
   - c) ^{235}U
   - d) $^{131}I^{-}$
   - e) ^{113}Sn
   - f) $^{203}Hg^{+2}$
   - g) ^{74}As
   - h) $^{68}Ga^{+3}$
   - i) $^{90}Sr^{+2}$
   - j) $^{19}F^{-}$

3. Write complete nuclear equations to describe each of the processes listed.

   Example: β^{-} decay of ^{203}Hg

   Answer: $^{203}_{80}Hg \rightarrow ^{203}_{81}Tl + ^{0}_{-1}\beta$

   - a) electron capture by tin-113 accompanied by gamma emission
   - b) alpha decay of bismuth-214
   - c) positron emission from indium-106
   - d) gamma emission by technetium-99m

e) β^- decay of krypton-87 accompanied by gamma emission

f) alpha decay of cerium-142

g) positron emission by cesium-125

h) gamma emission by yttrium-91m

i) electron capture by platinum-186

j) β^- decay of europium-158

4. Write a complete nuclear equation to describe each of the processes listed. See footnote *after* solving the problems.*

a) decay of plutonium-241 to produce americium-241

b) decay of oxygen-15 to produce nitrogen-15

c) decay of thorium-232 to produce radium-228

d) decay of nickel-59 to produce cobalt-59

5. Determine whether the n/p ratio increases, decreases, or remains the same for the isotopes in each of the following processes.

a) problem 3(e)   d) problem 3(a)

b) problem 3(c)   e) problem 3(b)

c) problem 3(h)

6. Why is it that most elements do not have whole-number atomic weights?

7. Explain the advantage of liquid scintillation counting compared to other techniques.

REFERENCES

1. Sienko, M.J., and Plane, R.A. *Chemistry.* 5th ed. New York: McGraw-Hill, 1976, p. 21.

2. Feynman, R.P. "Structure of the Proton." *Science 183* (1974): 601.

3. Kastner, J. *The Natural Radiation Environment.* Washington, D.C.: U.S. Atomic Energy Commission, 1968, p. 28.

* Parts (b) and (d) appear to be the same process. However, oxygen-15 decays by positron emission, whereas nickel-59 decays by electron capture.

I N order to consider the potential hazards of radioactive emissions, it is necessary to be familiar with some of the ways of describing radiation. The conventional ways of expressing the amount of radiation are by the following units:

Curie: the unit that describes the amount of radioactive material in a sample in terms of the number of radioactive disintegrations per second.

Roentgen: the unit that describes the quantity of radioactivity given off by a sample (only applicable to x- or gamma radiation).

Rad: the unit of absorbed dose.

Rem: the unit of absorbed dose expressed in terms of the biological effect of radiation that varies from one type of tissue to another and from one type of radiation to another.

RADIATION LEVELS

It is useful to put certain statistics into their proper perspective in order to work toward a full understanding of the hazards of radiation. The data in Tables 2.1 and 2.2 provide a striking contrast between normal radiation exposures and the levels of radiation dose that bring about immediate and sometimes very acute symptoms.

Several comments to the data seem appropriate. Most obvious is the low level of normal radiation exposure, as shown in Table 2.2, compared to the doses given in Table 2.1. The data in Table 2.1 are crude but the details are unimportant under normal circumstances. The data given in Table 2.2 are variable and even the various average values are cited higher and lower by other reference sources. The recommended maximum annual exposure from man-made sources is 0.170 rem (170 millirems) so that the normal exposure is well within the suggested limits. By comparison to the levels in Table 2.1 there would appear to be no cause for concern even if the 170 millirem level is exceeded somewhat, although this implies that it is appropriate to ignore the recommended maximum level. It is generally agreed that the 0.170 rem level is an arbitrary one and yet its existence alone points up the fact that even low levels of radiation are cause for concern. The reasons will be explained in the following paragraphs.

DATA ON RADIOACTIVE ISOTOPES

Since type and energy of radiation vary greatly, it is desirable to have sufficient knowledge to assess the dangers of any particular isotope. Tables of necessary

RADIATION HAZARDS

TABLE 2.1   PROBABLE EFFECTS OF WHOLE-BODY SHORT-TERM
RADIATION DOSES (1)

DOSE (rems)	PROBABLE CLINICAL EFFECT
0 to 25	No observable effects.
25 to 100	Slight blood changes but no other observable effects.
100 to 200	Vomiting in 5 to 50 percent within three hours, with fatigue and loss of appetite. Moderate blood changes. Except for the blood-forming system, recovery will occur in all cases within a few weeks.
200 to 600	For doses of 300 rems and more, all exposed individuals will exhibit vomiting within two hours or less. Severe blood changes, accompanied by hemorrhage and infection. Loss of hair after two weeks for doses over 300 rems. Recovery in 20 to 100 percent of cases within one month to a year.
600 to 1000	Vomiting within one hour, severe blood changes, hemorrhage, infection, and loss of hair. From 80 to 100 percent of exposed individuals will succumb within two months; those who survive will be convalescent over a long period.

TABLE 2.2   ANNUAL RADIATION EXPOSURES IN THE UNITED STATES* (2)

SOURCE	MILLIREMS†
Natural sources	
A. External to the body	
1. From cosmic radiation	50.0
2. From the earth	47.0
3. From building materials	3.0
B. Inside the body	
1. Inhalation of air	5.0
2. Elements found naturally in human tissues	21.0
Total, natural sources	126.0
Man-made sources	
A. Medical procedures	
1. Diagnostic x-rays	50.0
2. Radiotherapy x-rays, radioisotopes	10.0
3. Internal diagnosis, therapy	1.0
Subtotal	61.0
B. Atomic energy industry, laboratories	0.2
C. Luminous watch dials, television tubes, radioactive industrial wastes, etc.	2.0
D. Radioactive fallout	4.0
Subtotal	6.2
Total, man-made sources	67.2
Overall total	193.2

* Estimated average exposures to the gonads, based on 1963 report of the Federal Radiation Council.
† One thousandth of a rem.

data are readily available in many sources including handbooks of physics and chemistry. Such tables generally list the kind of information given in Table 2.3.

Consider the significance of each piece of information. The only data requiring explanation are half-life and decay energy.

Half-Life

The **half-life** is the length of time required for the radioactivity to diminish by one-half. This may best be understood by considering a particular example such as the ^{203}Hg as indicated in Table 2.4 and Figure 2.1. It can be seen from Table 2.4 that one-half of the radioactivity is lost every 47 days. The graphical representation in Figure 2.1 shows the same behavior, but emphasizes even more clearly that decay is most rapid when the amount of radioactive material is greatest, while the half-life is the same regardless of the level of radioactivity. In fact, one can determine the half-life of an isotope at any level of activity by making a series of measurements of the activity over a suitable period of time. The point at which the time equals zero is the time when the first measurement is taken. The activity level is arbitrarily taken as 100% relative to all of the later measurements but this does not suggest that no decay has occurred prior to the first measurement. In fact, the isotope may have only a small fraction of its original activity although its half-life is unaffected.

Half-lives vary greatly, ranging from fractions of a second to thousands of

TABLE 2.3   ISOTOPE INFORMATION FOR ^{14}C AND ^{203}Hg

INFORMATION	EXAMPLE 1	EXAMPLE 2
isotope	^{14}C	^{203}Hg
percent natural abundance	near zero	zero
atomic mass	14	203
half-life	5568 years	47 days
mode of decay	β^-	β^-, γ
energy of decay	0.155 meV	0.21, 0.279 meV

TABLE 2.4   TIME AND PERCENT ACTIVITY FOR ^{203}Hg

TIME (NUMBER OF HALF-LIVES)	PERCENT ACTIVITY
0	100
1 × 47 days (1)	50
2 × 47 days (2)	25
3 × 47 days (3)	12.5
4 × 47 days (4)	6.25
5 × 47 days (5)	3.125
10 × 47 days (10)	0.00098

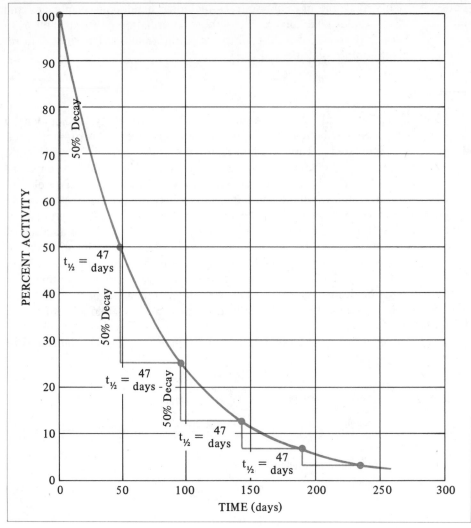

Figure 2.1 Radioactive decay of mercury-203.

years. Carbon-14 dating is a major application of radioactive decay and is dependent on half-life. This subject will be discussed at length in the next chapter.

An even more practical aspect of half-life is the shelf-life or useful life of an isotope. Iodine-131 is sometimes used in the treatment of an overactive thyroid gland. The length of time that radioactive emissions will continue to occur from this isotope can be calculated quite accurately, and this piece of information is needed in order to know how long a patient will be exposed to the radioactivity from a given dose of ^{131}I.

The knowledge of half-lives of radioactive isotopes is also very important when determining how to safely dispose of them (see discussion in Chapter Four). An isotope with a short half-life (e.g., ^{203}Hg, $t_{1/2} = 47$ days) may simply be al-

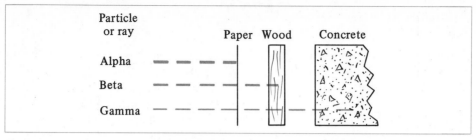

Figure 2.2   Relative penetration of alpha, beta, and gamma radiation (4).

lowed to decay to a very low level of activity on standing in a suitably shielded container. On the other hand, isotopes with very long half-lives may require burial in a suitable location under the direction of the Nuclear Regulatory Commission (NRC). The problem was previously dealt with under the direction of the Atomic Energy Commission (AEC).

Decay Energy

Another important piece of information about isotopes is the decay energy (see Table 2.3). This is the energy of the emitted particle or ray. The situation is complicated for β particle emission since β particles are emitted with a range of energies so that the energy value listed describes the maximum energy observed. A mysterious particle called a *neutrino** is said to be emitted along with the β particle and it also carries energy. Therefore, the tabulated decay energy for β particles is the maximum energy for β particles from any particular isotope. In other words, the total amount of energy released in beta decay is distributed between the beta particle and the neutrino and the distribution is variable. Therefore, most beta particles are emitted with energies lower than the tabulated values.

     The decay energy is very important for predicting the ability of a certain type of radiation to penetrate matter. Higher energy radiation has greater penetrating power. Practical consequences include knowing that the radiation will penetrate the body if therapeutic radiation treatments are planned. (See discussion in Chapter Three.) Safe storage conditions can only be determined by knowing how effectively the radiation will penetrate a potential container. Some isotopes can be handled very safely in glass containers and may not even penetrate air to any great extent. Other isotopes must be stored in thick-walled lead containers to prevent escape of the radiation. Figure 2.2 provides a good indication of the relative penetrating ability of α, β, and γ radiation.

     Knowing the range of alpha and beta particles not only assists in determination of safety hazards but also provides a very accurate means of determining the thickness of certain materials, with precision in the range of a millionth of an inch (5).

---

* At one time the neutrino served as a convenient idea to explain the variation in beta decay energies. Recently, more direct evidence has been reported (3) to confirm the existence of neutrinos. The neutrino has very little practical importance since its interaction with matter is negligible.

BIOCHEMICAL BASIS OF
RADIATION DAMAGE

In order to understand the hazards of ionizing radiation, it is necessary to digress in order to consider the basic concepts, which are categorized under the heading of molecular biology. Molecular biology is an outgrowth of the overlap of work done in the more established areas of biochemistry and genetics. The convergence of efforts in the two areas is nicely illustrated by the properties of a microorganism called *Neurospora crassa,* better known as bread mold.

Traditionally the biochemist has addressed problems relating to chemical pathways of metabolism in living systems. For example, he might study the details of how common sugar (sucrose) is ultimately broken down to CO_2 and H_2O with resultant benefit to the living organism, including simple microorganisms such as bacteria and molds. It turns out that metabolic pathways of bacteria often parallel those used by sophisticated organisms (i.e., you). This allows the chemist ready access to these biochemical systems and also accounts for the ability of many bacteria to thrive in other living systems.

The geneticist has traditionally studied phenomena relating to reproduction noting how certain traits are passed on from one generation to another, whereas others are not. The familiar occurrence of so-called dominant (e.g., brown eyes) and recessive (e.g., blue eyes) traits illustrates the complexity of the subject. In addition, there are many common examples of specific genetic breeding of both plants and animals.

As for the bread mold, the classical geneticist and biochemist might have categorized particular hybrid forms according to some of the descriptions listed in Table 2.5.

As indicated in Table 2.5, the classical geneticist might have described the hybrid mold by its appearance, whereas the biochemist might have determined that a particular hybrid is best described in terms of a chemical compound that is an essential nutrient for its survival. This nutrient is not required for the normal nonhybrid mold that has the chemical systems to synthesize the nutrient for itself.

TABLE 2.5   PROPERTIES OF BREAD MOLD HYBRIDS

APPEARANCE* (GENETICIST DESCRIPTION)	EXTERNAL GROWTH REQUIREMENT* (BIOCHEMIST DESCRIPTION)
albino	alanine
colonial	arginine
fluffy	histidine
pale	leucine
peach	tryptophan
snowflake	thiamine
yellow	pantothenic acid

* These two columns of characteristics do *not* relate to one another horizontally, i.e., the albino hybrid does not necessarily have a requirement for alanine.

To illustrate the significance of such a finding, consider the following general scheme of reactions, which is intended to simulate some metabolic sequence.

$$I \rightarrow II \rightarrow III \rightarrow IV \rightarrow V \rightarrow VI \rightarrow VII \text{ etc.}$$

This kind of sequence is typical for living systems. As an example, the complete breakdown of common table sugar (sucrose) to give carbon dioxide, water, and energy, which can be used to power a living system, involves more than 20 steps. Thus, if the organism is supplied with any compound signified by I, II, etc., it can utilize this material as a fuel for the general pathway previously described. However, suppose a hybrid form of bread mold is unable to utilize compounds I, II, and III and requires that compounds IV or beyond be supplied in order to sustain its growth. This suggests the inability to carry out the reaction III $\rightarrow$ IV. Such an observation generally indicates the absence of the catalyst for the conversion of III to IV and the system is normally described as having a requirement for IV. A *catalyst* is a substance that increases the speed of a reaction. In some instances, a catalyzed reaction may occur so slowly in the absence of a catalyst that it may not seem to go at all. Catalysts in biochemical systems are proteins functioning as enzymes. Thus, it becomes necessary to probe how a defect such as a missing enzyme may develop and cause a hybrid to result. In order to do this, a knowledge of how proteins are normally synthesized by living systems is required and this knowledge is the basic material that comes under the heading of molecular biology.

First, let us consider the chemical structure of proteins. They are very large molecules, called *polymers*, which are composed of many repeating units called *monomers*. In the case of proteins, the monomers are *amino acids*, which are molecules characterized as having one group of atoms called an amino group ($-NH_2$) and another group of atoms called an acid group ($-COOH$). A very simple example is the amino acid known as glycine.

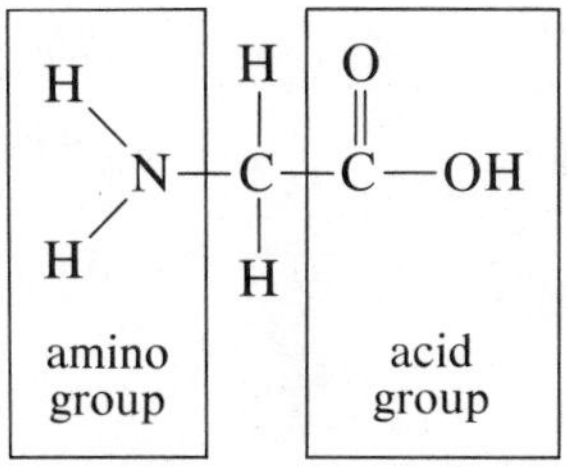

Glycine (an amino acid)

There are 20 different amino acids that are commonly used as building blocks of proteins. The hormone, insulin, is an example of a relatively small protein consisting of two chains of 21 (A-chain) and 30 (B-chain) amino acid units joined by sulfur bridges as shown in Figure 2.3. The names of the 20 common amino acids are given in the figure and the position of each is indicated by a numbered ball. Other proteins vary from just a few units long to more than 10,000 units and may exist as single chains or clusters of many polymer chains.

However, in addition to the number of amino acid units and the number of chains, the sequence in which the amino acid units appear is also of major impor-

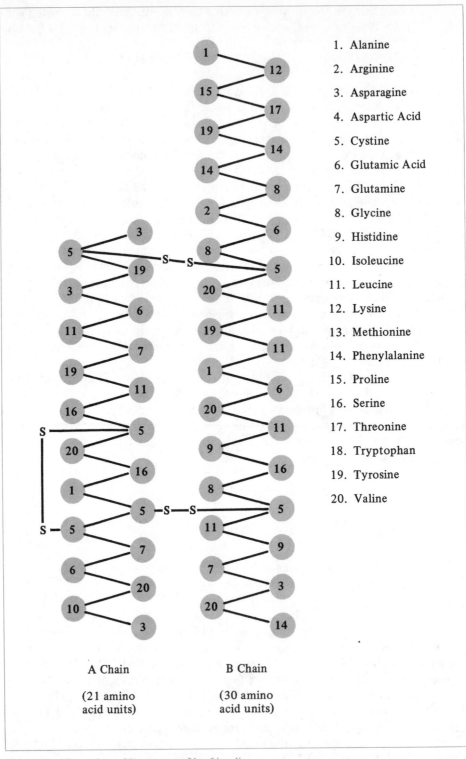

1. Alanine
2. Arginine
3. Asparagine
4. Aspartic Acid
5. Cystine
6. Glutamic Acid
7. Glutamine
8. Glycine
9. Histidine
10. Isoleucine
11. Leucine
12. Lysine
13. Methionine
14. Phenylalanine
15. Proline
16. Serine
17. Threonine
18. Tryptophan
19. Tyrosine
20. Valine

A Chain

(21 amino
acid units)

B Chain

(30 amino
acid units)

Figure 2.3   The amino acid sequence of beef insulin.

tance. If the amino acid sequence of insulin is altered, the molecule may be unable to carry out its normal function, which is to assist in the metabolism of carbohydrates.

If a particular protein normally serves as an enzyme to catalyze a chemical reaction that occurs in some living organism, a change in the sequence of amino acids in that protein may cause the appearance of a hybrid form of the organism. Such a change is called a *mutation*. In many instances, the mutation is lethal.

The Telegraph Code Versus The Genetic Code

The next major piece of the puzzle is how the synthesis of proteins is guided so as to assure that the end product is biochemically active, e.g., as an enzyme to catalyze some reaction involved in metabolism.

We now come to the portion of the molecular biology story that is generally regarded as genetics. The so-called genetic material is the chromosome consisting of smaller segments known as **genes.** It is often stated that each gene contains the information to guide the synthesis of one protein. Thus, different organisms, with different genes, have different collections of proteins and exhibit different properties.

The next question is: what is the chemical structure of the chromosome and the gene and what kind of information is stored in each gene that allows it to direct protein synthesis and pass the information on to future generations?

Chromosomes are another type of polymer made up of the repeating units symbolized A, C, G, and T. These four symbols are the common shorthand notation used to describe a set of four complex compounds which also may be linked together to form polymeric chains. Thus, a typical gene (or chromosome) is merely a long chain of the four repeating units, which may occur in a wide variety of sequences such as ATGGCTACAAGTC. . . . This type of molecule is called deoxyribonucleic acid or simply DNA.

To summarize before proceeding, chromosomes are DNA molecules. The DNA is made of very long chains of the units A, C, G, and T. Smaller segments of the DNA chains are called genes and each gene contains information to direct the synthesis of a protein.

The information is in the form of a code, known as the *genetic code*. Thus, for protein synthesis, there must be information in the genetic code to direct the placement of the amino acids into their proper sequence in a protein so that the protein will be normal and functional. If mistakes are made in the synthesis of a protein, a defective product may be made. An example might be a nonfunctional form of an enzyme which is necessary for catalysis of the conversion of III to IV mentioned earlier.

The genetic code is an intriguing system for transmitting information, much like the international telegraph code.* It is interesting to consider why the telegraph code developed in its present form. The dot and dash signals correspond to short and longer signals with a telegraph key or light source. Combinations of dots and dashes are necessary to provide enough signals to communicate a 26-letter al-

---

* The international telegraph code is often referred to as Morse Code, although the latter is different and seldom used. Both codes appear in Table 2.7.

phabet, numbers, punctuation, parentheses, etc. The notation 2^n allows prediction of the number of symbols available using n digit signals. The 2 describes the number of symbols available. If n equals 3, 2^n equals eight, which means that there are eight different, three-digit signals that can be constructed using only two symbols (dot and dash). Table 2.6 explains why as few as one and as many as 5-digit signals are required to provide a sufficient number of signals to transmit a 26-letter alphabet and other necessary information.

Similarly, the genetic code must be sufficiently complex to contain all the required information. Recall that the code is found in the sequence of A, C, G, and T in DNA, and since four symbols appear in the code compared to the two in the telegraph code, the term 4^n (n = the number of digits in a signal) can be used to determine the minimum number of digits required. Two facts apply. First of all, the code must direct the placement of the 20 common amino acids into proteins. Secondly, the system must be automatic so that it is logical to expect uniformly-sized signals, i.e., the same number of digits in every signal. Since 4^n equals only 16 for a two-digit signal, it is obvious that a three-digit signal is necessary to properly place 20 amino acids. Thus, a portion of DNA such as TACGCACTGAACTGT . . . is actually a series of coded signals TAC, GCA, CTG, AAC, TGT, etc., each of which specifies one of the 20 amino acids. Some amino acids have more than one three-letter code.

One last piece of the puzzle remains to be considered and that is the source of amino acids. Certain vitamin supplements contain one or two amino acids, but most come from another source of protein (e.g., meat) in the diet, which can be broken down into the twenty amino acids by the digestive system.

It is interesting to consider the validity of the advertising of several hair sprays and shampoos that reportedly contain protein to replace protein lost from the hair by washing, curling, combing, etc. Some basic facts tend to refute the claim. First of all, hair is a particular type of protein. Unless the product actually has hair suspended in it, it does not contain anything that faintly resembles the properly assembled amino acids found in hair protein. The production of hair occurs only internally so that the only possible way to gain any advantage from such a product is to drink it, which is not recommended.

The General Picture

Earlier in this chapter the subject of molecular biology was developed to emphasize the role of DNA (chromosomes) in directing the synthesis of other biochemi-

TABLE 2.6   REQUIREMENTS OF THE INTERNATIONAL TELEGRAPH CODE

n (NUMBER OF DIGITS)	2^n	TOTAL NUMBER OF SIGNALS
1	2	2 (using only 1 digit)
2	4	6 (using 1 and 2 digits)
3	8	14 (using 1, 2, and 3 digits)
4	16	30 (using 1, 2, 3, and 4 digits)
5	32	62 (using 1, 2, 3, 4, and 5 digits)

INTERNATIONAL	MORSE CODE
ALPHABET	

INTERNATIONAL	MORSE CODE
A ·—	A ·—
B —···	B —···
C —·—·	C ·· ·
D —··	D —··
E ·	E ·
F ··—·	F ·—·
G ——·	G ——·
H ····	H ····
I ··	I ··
J ·———	J —·—·
K —·—	K —·—
L ·—··	L ——
M ——	M ——
N —·	N —·
O ———	O · ·
P ·——·	P ·····
Q ——·—	Q ··—·
R ·—·	R · ··
S ···	S ···
T —	T —
U ··—	U ··—
V ···—	V ···—
W ·——	W ·——
X —··—	X ·—··
Y —·——	Y ·· ··
Z ——··	Z ··· ·

NUMERALS	
1 ·————	1 ·——·
2 ··———	2 ··—··
3 ···——	3 ·······
4 ····—	4 ····—
5 ·····	5 ———
6 —····	6 ······
7 ——···	7 ——··
8 ———··	8 —····
9 ————·	9 —··—
0 —————	0 ————

PUNCTUATION	
Period (.) ·—·—·—	Period (.) ··—··—··
Comma (,) ——··——	Comma (,) ·—·—·
Interrogation (?) ··——··	Interrogation (?) —··—·
Colon (:) ———···	Colon (:) —·— ··
Semicolon (;) —·—·—·	Semicolon (;) ··· ··
Hyphen (-) —····—	Hyphen (-) ···· ·—··
Slash (/) —··—·	Slash (/) ·——
Quotation marks (") ·—··—·	Quotation marks (") ··—· —··

Table 2.7   Telegraph Codes.

cals including proteins. Some of these proteins are enzymes (biochemical catalysts), which facilitate the conduct of all biochemical reactions including those involved in metabolism. As the molecular biology story suggests, radiation damage
can be tolerated by many chemical components such as carbohydrates, fats, and

proteins but not by DNA. The DNA contains the information to direct the formation, utilization, and storage of the other components. Therefore, damage to DNA can be lethal, whereas damage to other biochemicals may only cause temporary symptoms due to a deficiency of some requirement. If the radiation dose is high enough to destroy much of the DNA content of the cells of a certain type of tissue, recovery from the damage is questionable. Radiation effects have been studied in many living systems including many types of microorganisms such as bread mold. Depending on dose, results range from no effect, to formation of hybrids, to death.

It was noted earlier (Table 2.2) that humans are continuously subjected to many types of radiation. One of the most prevalent is cosmic radiation, which results from nuclear chemistry in the upper atmosphere. The interesting question that arises is how are humans able to tolerate such radiation over a period of many years without apparent adverse effects? Certainly cosmic rays are very penetrating (gamma radiation) and have no difficulty interacting with cellular DNA. However, no ill effects are evident (6). This is most easily appreciated by considering the curve in Figure 2.4 and recalling that DNA is the key target for the effects of radiation. According to this description, there is a linear increase in the frequency of mutation when only a single hit by radiation is required to cause mutation. In contrast, when two or more hits are required to cause a mutation, there is a low frequency of mutation when the dose is low, but mutation becomes significant when the dose is high. The latter behavior is the pattern observed for humans and other organisms that have the ability to carry out chemical self-repair of damaged DNA.

Elegant chemical mechanisms have been discovered for repair of damage to DNA caused by ultraviolet radiation (7). These findings clearly demonstrate the ability of many living systems to resist significant damage from low-level radiation and to recover from some higher doses.

Even when chemical repair is possible, it may not be adequate to combat higher doses of radiation for a number of reasons. For one thing, when considerable damage is done, the amount of repair required may be too great. One can

Figure 2.4   Frequency of mutation versus dose for single and multihit mutations.

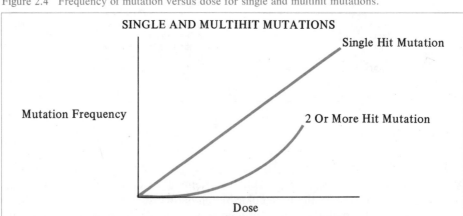

draw an analogy to an automobile after an accident. If the damage is minor, it may only be necessary to replace a fender or a bumper to restore the car to its original form. But if the damage is too great, what is left may not even resemble an automobile and repair may be impossible. In a similar way, if damage to DNA is too extensive, it may not be possible to restore the damaged DNA to its original form. To put it another way, if two hits by radiation are required to produce a mutation, the damage caused by the first hit may be repaired before a second hit occurs, if radiation levels are low.

In addition, very high levels of radiation may even knock out the chemical systems required for repair. Finally, there is also the possibility of damage to the immunological systems, which provide resistance to infection. In such cases, permanent damage and death may occur.

The theory of single and multiple-hit mutations is the logical conclusion drawn from experiments on organisms ranging from very primitive to very sophisticated. The difference in the effects of long, low-intensity doses and short,

The Mutant: A potential hazard of radioactive emissions

high-intensity doses can be readily observed. Some organisms show a greater tendency to be affected equally by either type of radiation exposure, whereas other organisms are better able to resist low radiation levels due to better developed repair mechanisms.

In any case, repair mechanisms are not always error-free so that even low levels of radiation may cause minor mutations. A simple example may be the slow but progressive aging process, which may be partly due to the constant exposure to cosmic and certain types of man-made radiation. In addition, damage to those cells involved in reproduction may not be observable until future generations, which may end up using a genetic code that has been altered by radiation damage long before.

Specific Concerns

After one has established the hazards that might be expected in handling a particular radioisotope, he should also acknowledge the possibility that errors could be made during the processing of an order for an isotope, so that he may end up handling something much different, and possibly much more dangerous, than he had expected (8).

The idea of soft radiation, e.g., the weakly penetrating low energy beta radiation emitted by carbon-14, was introduced earlier. Similarly, other isotopes that emit low energy alpha and beta particles are very safe to work with unless they are taken internally. It seems a rather simple problem to avoid the intake of any radioactive materials and yet this is not quite as simple as it appears. There is the classic case of the workers who painted radium coatings on watch dials to make them glow in the dark. Radium-226 emits weakly penetrating alpha particles and is therefore quite safe to handle. Unfortunately, many of these painters were found to have repeatedly placed their brushes between their lips in order to get a fine point on the brushes. Thus, they acquired a large amount of radioactive material. The long half-life of ^{226}Ra (1622 years) plus its tendency to accumulate in bone make it a very serious health hazard (9). This tendency is expected since calcium, which is a major component of bones, and radium have very similar properties. This is suggested by the fact that both of these elements appear in the same group in the periodic table. This subject is discussed further in Chapter Five.

More subtle, but possibly just as important, is the radioactive fallout from nuclear bomb testing or from processing of ores, fuels, and radioactive wastes. If radioactive materials get into water or food, there is the chance they may be ingested with serious consequences. For example, fallout on land used for grazing cattle may cause radioactive material to be taken in by the cow. The cow later gives milk or is slaughtered and the radioactive isotopes may be consumed by humans. The total exposure to radioactivity from fallout seems insignificant (about 4 millirems per year), as shown in Table 2.2, but the effects may be significant due to the possibility that certain isotopes may concentrate in specific areas of the body.

Radioactive isotopes of so-called noble or inert gases (see later discussion), such as radon, argon, and krypton, may be released into the atmosphere at various stages in the handling of radioactive materials, such as those involved in electric power production. For example, krypton-85 is a product from the break-down (fission) of uranium that occurs in nuclear reactors. One source has suggested that "eventually, it is likely that all of the presently allowable exposure of the general population to radioactivity from whatever source will be exhausted by exposure to ^{85}Kr in the air (10)."

This is not to say that nothing can be done about these gases. A process for removing radioactive noble gases does exist, and this or some other means could be utilized in order to trap most of the gases before they can escape into the atmosphere (11).

To some, the noble gases may not seem to be much of a problem since they are normally chemically inert, and yet the possibility exists that they could be inhaled and dissolve in the blood. If that happens, they might simply be returned to the lungs and exhaled or they might be retained to some extent. It has even been suggested that they might become incorporated to some extent into fatty tissue and thus become a long term, internal load of radioactivity (12).

The effect of low-level radiation is quite subtle. With the previous discussion of molecular biology as a background, it is possible to appreciate how genetic damage might occur by preventing proper synthesis of certain essential enzymes. In addition, there is the possibility of preventing formation of proteins that may be an important part of a mechanism that determines the rate of synthesis of certain other proteins. Not only is it desirable to have an adequate supply of essential proteins, but it is also necessary to have control over the rate of synthesis of all cellular components, so as to have control over the growth of the entire cell. Major characteristics of cancer cells are rapid growth and unregulated reproduction. Thus, radiation may only affect a small portion of the total DNA content of an organism and still cause the tremendous proliferation of cells typical of cancer. One source has, thus, described ionizing radiation as an "invisible bullet (12)."

SUMMARY

A comparison of the radiation dose levels that cause dramatic clinical symptoms and the average dose levels, to which the population is exposed, suggests that there is no cause for concern over current or somewhat increased dose levels. Unfortunately, this conclusion ignores the possibility that ionizing radiation can also cause certain subtle but permanent chemical changes in living organisms.

In order to determine the hazards and proper techniques for handling a particular radioisotope, it is necessary to be able to interpret published data on these isotopes. The half-life can be used to assess the useful lifetime of an isotope as well as the time period over which it is a potential hazard. By knowing the mode of decay and the decay energy, it is possible to determine the ability of the decay particle or γ-ray to penetrate matter. This information is also necessary for determining the proper method for detection of the isotope.

DNA is the genetic material that directs all body functions by controlling the synthesis of proteins, some of which serve as catalysts for metabolism and other chemical events in the body. The information in the DNA is in the form of the genetic code, which is much like the international telegraph code. The genetic code contains the information to direct the placement of the 20 common amino acids into protein chains that are synthesized in the body. An alteration of the coded information, by radiation damage or other means, may have no effect, or may cause mutations, some of which are lethal. Most cells have the capacity to repair minor damage to DNA so that low levels of radiation can be tolerated for the most part, although subtle mutations may occur and have serious long-term consequences. The latter fact explains the description of ionizing radiation as an "invisible bullet."

PROBLEMS

1. Give a definition or example of each of the following.
   a) half-life
   b) neutrino
   c) molecular biology
   d) hybrid
   e) catalyst
   f) enzyme
   g) chromosomes
   h) genes
   i) DNA
   j) protein
   k) genetic code
   l) monomer
   m) polymer

2. What is the largest single source of radiation from
   (a) natural sources?
   (b) man-made sources?

3. If a human received 10 rems of radiation each year for 60 years, no effects might be observed. On the other hand, if he received the same total of 600 rems in one dose, it would likely be fatal. Explain.

4. What are the differences between the international telegraph code and the genetic code?

5. Suppose the genetic code contained five code letters, A, C, G, T, and X. How many digits would be required to provide all the information necessary to direct protein synthesis?

6. What is the source of the monomers used for protein synthesis in the body?

7. Explain how exposure of normal bread mold to radiation might cause a mutation.

8. Why is the advertising for protein-containing shampoos (and other hair products) misleading?

9. Translate the following sentence, which is given in the international telegraph code:

```
—    · · · ·   ·   — — ·   ·   — ·   ·   —   · ·   — · — ·   — · — ·   — — —   — · ·   ·   · ·
· ·   · · —   · · ·   ·   — · ·   —   — — —   — · ·   · ·   · — ·   ·   — · — ·   —   —
· · · ·   ·   · — — ·   · — · ·   · —   — · — ·   ·   — —   ·   — ·   —   — — —   · · — ·   —
· · · ·   ·   · · — —   — — — — —   — · — ·   — — —   — —   — —   — — —   — ·   · ·
— —   · ·   — ·   — — —   · —   — · — ·   · ·   — · ·   · · ·   · ·   — ·   —   — — —
—   · · · ·   ·   · ·   · — ·   · — — ·   · — ·   — — —   · — — ·   ·   · — ·   · · ·   ·
— — · —   · · —   ·   — ·   — · — ·   ·   · ·   — ·   · — — ·   · — ·   — — —   —   ·   · ·
— ·   · · ·   · — · — · —
```

REFERENCES

1. Glasstone, S., and Sesonske, A. *Nuclear Reactor Engineering*. Princeton, N.J.: Van Nostrand, 1963, p. 532.

2. Asimov, I., and Dobzhansky, T. *The Genetic Effects of Radiation*. Washington, D.C.: U.S. Atomic Energy Commission, 1966, p, 37.

3. Barish, B.C. "Experiments With Neutrino Beams." *Scientific American 229* (1973): 30.

4. Phelan, E. W. *Radioisotopes in Medicine*. Washington, D.C.: U.S. Atomic Energy Commission, 1966, p.8.

5. Baker, P.S., Fuccillo, Jr., D.A., Gerrard, M.A., and Lafferty, Jr., R.H. *Radioisotopes in Industry*. Washington, D.C.: U.S. Atomic Energy Commission, 1965, p.11.

6. Gillette, R. "Radiation Standards: The Last Word or at Least a Definitive One." *Science 178* (1972): 966.

7. Setlow, J.K., and Setlow, R.B. "Nature of the Photoreactivable Ultraviolet Lesion in DNA." *Nature 197* (1963): 560.

8. "Suppliers' Errors Beset Radiochemicals Users." *Chemical and Engineering News*, May 3, 1971, p. 24.

9. Aub, J.G., Evans, R.D., Hempelmann, L.H., and Martland, H.S. "The Late Effects of Internally-Deposited Radioactive Materials in Man." *Medicine 31* (1952): 221.

10. Novick, S. *The Careless Atom*. Boston, Mass.: Houghton, Mifflin Co., 1969, p. 142.

11. "Process Removes Radioactive Noble Gases." *Chemical and Engineering News*, May 28, 1973, p. 15.

12. Lewis, R.S. *The Nuclear Power Rebellion*. New York: Viking Press, 1972, p. 61.

I N this chapter we will consider many of the common applications of radio-chemistry spanning the areas of archaeology, scientific research, and medicine. The processes of nuclear fission, breeding, and nuclear fusion are also discussed. Nuclear fission is the primary process employed as a source of energy for generating electricity in nuclear power plants. This subject is pursued in more detail in Chapter Four.

CARBON-14 ARCHAEOLOGICAL DATING

In Table 2.2 it was noted that the general population is continually subject to cosmic radiation. The radiation is mostly γ radiation at the surface of the earth, but is a mixture of many types of radiation in space. For example, the solar wind contains a high level of protons from the sun. Cosmic rays can interact with gases in the upper atmosphere to produce neutrons. These neutrons can then react with nitrogen found in N_2 gas in the atmosphere according to the following reaction.

$$^{14}_{7}N + ^{1}_{0}n \rightarrow ^{14}_{6}C + ^{1}_{1}H$$

The shorthand notation ^{14}N (n, p) ^{14}C describes the process in which ^{14}N is converted to ^{14}C by neutron bombardment accompanied by proton emission. Thus, the overall sequence proceeds from cosmic rays to ^{14}C. The ^{14}C then finds its way into the earth's atmosphere and becomes a part of the biological carbon pool (biosphere) as it is incorporated into live plants and animals. Any living material continually takes up and gives off ^{14}C until an equilibrium is reached and the level remains constant.

On the average, approximately two ^{14}C atoms are deposited every second on every square centimeter of earth surface as determined by experiments using balloons. The ^{14}C is readily detectable because of the beta emission that accompanies its decay.

$$^{14}_{6}C \rightarrow ^{14}_{7}N + ^{0}_{-1}\beta$$

More important, this reaction represents a process for loss of ^{14}C from the biosphere. Thus, an equation for formation and another for destruction of ^{14}C describe a situation in which the total concentration of ^{14}C reaches a so-called "steady state" or constant concentration.

APPLICATIONS
OF RADIOCHEMISTRY

The steady state is easily understood by noting that a greater buildup of ^{14}C would lead to faster decay since the rate of decay is proportional to the concentration of material (^{14}C) that is decaying. This point was already made in the discussion of Figure 2.1.

Any living material, which is continually exchanging ^{14}C with the biosphere, will have this same steady-state level. However, when death occurs, the intake of ^{14}C ceases, whereas radioactive decay continues at a rate governed by the half-life of ^{14}C, which is 5568 years. After 5568 years, one-half of the normal level of ^{14}C found in the organic material will be lost. Similarly, any other ^{14}C level allows a calculation of the age of the sample from which it is obtained (1). The normal steady-state level of ^{14}C in living organisms undergoes approximately 15 radioactive disintegrations per gram of carbon per minute. This is normally described as 15 counts per minute (cpm) per gram of carbon. Any carbon-containing sample that displays only 7.5 cpm per gram of carbon must be about 5568 years old.

Carbon-14 dating is not always successful

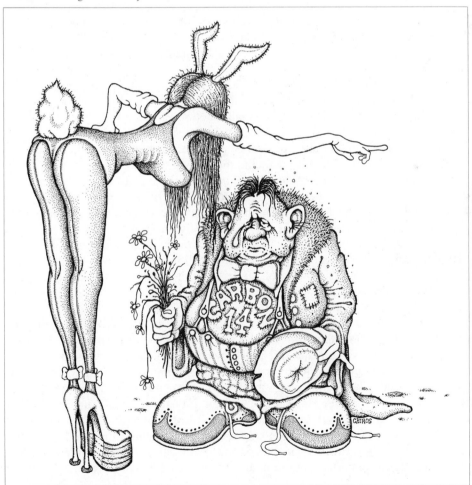

In order to determine the ^{14}C level, the carbon content of a sample is normally converted to CO_2 by burning and the CO_2 is monitored to determine the amount of ^{14}C. In all cases, the ^{12}C isotope is the one present in greatest amount but it is not radioactive.

Coal is an example of a material that has been out of contact with the biosphere for so long (more than 50,000 years) that virtually all of its ^{14}C has decayed. Thus, we have a substance for which the very long half-life of ^{14}C is much too short to allow any accurate determination of its age. The dead sea scrolls and wood found in Egyptian pyramids have ages within the range of ^{14}C work. Workers in this field normally place the upper limit for use of the technique at about 50,000 years. See Problem 4(d).

The balance of ^{14}C has potentially been upset by many factors over the centuries. The most dramatic example has been the burning of fossil fuels. Like coal, the other fossil fuels are very old and have been out of contact with the biosphere for long periods of time. Thus, the ^{14}C level is unusually low (near zero), so that, when these fuels are burned, it is predominantly ^{12}C changing to $^{12}CO_2$. This release of nonradioactive carbon causes a dilution of the normal steady state carbon pool. This phenomenon has led some to predict as much as a 12% dilution effect, which would then increase the "apparent" age of any materials exposed. However, some plants that died in 1950 showed only a 2% dilution. The difference between predicted and observed has been attributed to absorption of the CO_2 by the oceans.

Perfection of the technique for counting old ^{14}C-containing samples was a major technological advance in itself. The problem lies in the detection of very low radiation levels that may be a small fraction of the normally-encountered background radiation (mostly cosmic rays). Cosmic rays are very penetrating (mostly gamma radiation) and are thus easily detected by any radiation counters whether this is desired or not. A technique known as *anticoincidence counting* was developed by W.F. Libby to counteract the problem. In this technique, the sample is placed in a container that is surrounded by efficient shielding in order to filter out all radiation other than cosmic. The container is also located in a system that consists of two counters. One counter is exposed to both the cosmic radiation and the ^{14}C (beta emitter) sample. The other counter is shielded from the weak beta emission and detects only the cosmic radiation. One can determine the difference between the observation of the first counter (cosmic plus beta) and the second counter (cosmic only) and get the intensity of the beta emission, which indicates the ^{14}C content.

There has also been an interesting description of the use of ^{14}C dating for estimating the age of certain iron samples (2). Three major forms of processed iron are wrought iron, cast iron, and steel, all of which are alloys (mixtures) of iron and carbon. Thus, it is possible to determine the age of many iron samples provided the carbon, which was combined with the iron, was obtained from a source that had not been dead for very long. For example, if coal were the source of carbon used, then the iron sample could not be dated. Fortunately, wood and charcoal were common fuels used in the iron smelting process in early times and, thus, became the usual source of the carbon content of the carbon alloys. Detailed information on the age of certain iron samples has been useful in piecing together many aspects of the historical development of the iron age from the time of its beginning in about 6000 B.C.

Archaeological Dating Using the Rubidium-Strontium Clock

There are several other examples of archaeological dating based on slow (long half-life) radioactive decay. One of the more widely used methods is the rubidium-strontium clock. This method of dating depends on the following radioactive decay.

$$^{87}Rb \rightarrow \ ^{87}Sr + \ _{-1}^{0}\beta \ (t_{1/2} = 4.7 \times 10^{10} \ \text{years})$$

Since ^{87}Sr is not radioactive, it remains unchanged once it is formed from radioactive Rb. It is possible to examine a mineral sample of rubidium and determine the amount of ^{87}Rb and ^{87}Sr present. With this information the age of the mineral can be calculated (3).

Since the rubidium-strontium clock method of dating is more complicated (requires confirming evidence) than the ^{14}C method, the ^{14}C method is the technique of choice. However, as noted earlier, the ^{14}C method is limited in usage since it has an upper limit of 50,000 years. Since ^{87}Rb has a half-life of 47 billion years, the rubidium-strontium clock can be used even in determining the age of the earth, which is estimated at 4.5 billion years.

One application of the Rb-Sr clock has been its use in estimating the age of some lunar samples brought back during the Apollo 15 mission. The experimenters determined an age of 3.3×10^9 years for the samples (4).

ISOTOPIC TRACERS

There are a great many useful applications of radioisotopes as tracers. These techniques allow determination of the amount of any atom located in any particular portion of a system, living or dead. In later discussion the use of tracers for clinical diagnosis will be illustrated. In addition, much biochemical research utilizes various isotopically-labeled metabolites to monitor chemical processes in laboratory animals.

To illustrate the kinds of things that are involved in tracer work, consider the use of tracers to study the extent of binding of mercury ions to human serum albumin using the technique of *equilibrium dialysis* (5). Such an experiment is particularly relevant in relation to the mass of publicity dealing with mercury poisons. This experiment may be used to describe the extent to which Hg^{+2} ions are able to bind to the protein, albumin, which is a major component of the blood. A schematic diagram of the apparatus is pictured in Figure 3.1, in which radioactive $^{203}Hg^{+2}$ ions are present.

The key to success of such a dialysis is the semipermeable membrane that separates the two compartments of the system. This membrane is permeable to small molecules like H_2O and Hg^{+2} salts, but it is impermeable to large protein molecules such as albumin. Thus, all the chemical components, including Hg^{+2}

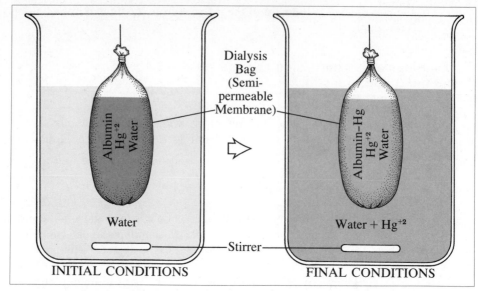

Figure 3.1   An equilibrium dialysis study of the binding of mercury to serum albumin.

ions, can move freely throughout the system except for the protein, which remains fixed.*

After passage of sufficient time to allow mixing, a sample can be removed
from each compartment and the radioactive ^{203}Hg (β and γ emission) can be
counted. The mercury will exist as follows:

outside the dialysis bag:	free Hg^{+2} ions only
inside the dialysis bag:	free Hg^{+2} ions
	Hg^{+2} bound to albumin

Since the membrane allows free passage of the unbound mercury ions in
both directions, the concentration of free Hg^{+2} will become the same both inside
and outside. Therefore, one can readily determine the excess count inside, which
can be attributed to the mercury-203 that is bound to the albumin.

Osmosis is a phenomenon very similar to dialysis except that only solvent
molecules are able to pass through the membrane. This requires that the membrane be impermeable to many small molecules and ions but permeable to solvent.
Solvent molecules will then move in the direction from low concentration of dissolved material to high concentration in an attempt to equalize the concentration
on both sides of the membrane.

Nonradioactive Isotopic Tracers

In addition to the many radioactive isotopes that are useful for monitoring a wide
range of phenomena, a few nonradioactive isotopes have some useful applica-

* Dialysis is also the basic phenomenon involved in kidney function.

tions (6). Among these isotopes are ^2H, ^{13}C, ^{15}N, and ^{18}O, none of which is the most abundant isotopic form of the four elements. Their failure to undergo radioactive decay may seem to be a deterrent to easy analysis of these isotopes, but other sophisticated analytical techniques, e.g., nuclear magnetic resonance (NMR) and mass spectroscopy, can be used for detection of these isotopes. In fact, these two techniques have the potential for providing much more detailed information concerning the location of the isotope in a molecule or a mixture of molecules. In some cases, the radioactive and nonradioactive isotopes of any given atom, e.g., ^{14}C and ^{13}C respectively, may nicely complement one another. If an experimenter is trying to monitor some complex chemical process, he may find it very advantageous to use the much simpler, but cruder, radioactive assay to assist him in developing a technique for isolation of the isotopically-labeled component of his system. Then he may revert to using the nonradioactive isotope in the same experiment in order to gain the more detailed information available from the more sophisticated techniques.

Two additional practical advantages result from the use of stable isotopes. For one thing, the danger of exposure to radiation is eliminated. This is a particular advantage in avoiding environmental pollution since the nonradioactive isotope will not become a dangerous pollutant no matter what its fate. For another, not only is the isotope stable, but so is the compound which contains it. For example, when CO_2 labeled with ^{14}C undergoes β decay, the carbon converts to ^{14}N so that the original CO_2 becomes contaminated with NO_2. Such an event may not normally be a serious problem but it can be avoided altogether by the use of the stable ^{13}C isotope, which is also readily distinguished from the common ^{12}C isotope by the techniques mentioned above.

$$^{14}CO_2 \rightarrow {}_{-1}^{0}\beta + {}^{14}NO_2$$

The incorporation of isotopes (both radioactive and nonradioactive) into chemical compounds is nicely illustrated by observing the results of photosynthesis by certain plants in the presence of $^{14}CO_2$. Photosynthesis (discussed in Chapter Nine) is the process whereby chlorophyll-containing plants utilize CO_2 and water to produce oxygen (O_2) and sugar with energy from the sun being used to drive the conversion. The sugar found in the plant (as cellulose, starch, or simple sugars) is found to be labeled with the ^{14}C. Such an experiment serves to confirm other experiments that have been done in an attempt to understand photosynthesis, but it also suggests a way of incorporating either ^{13}C or ^{14}C into the sugars, which may then be used to study other processes. In effect, the biochemist allows a plant to do the chemistry for him or her and then separates the sugar from the plant system. In a similar way, many other isotopically-labeled compounds can be synthesized by purely chemical or biochemical means.

CLINICAL APPLICATIONS OF
X-RAYS AND RADIOISOTOPES

Nuclear medicine is a rapidly growing field. Approximately 12 million nuclear medicine procedures are now carried out each year in the United States. In fact, about 10% of all patients admitted to hospitals are evaluated with a single radioisotope, technetium-99m (7). Several problems and examples will be described in the following sections to illustrate how radiation can be applied in both therapy and diagnosis. The major advantage of the nuclear methods is their noninvasive nature, which makes them preferable to more radical surgical techniques.

Therapy

Therapeutic applications of radiochemistry can be subdivided into two types. One type, called *teletherapy,* involves exposure to an external source of radiation, whereas the second type is internal radiation therapy.

EXTERNAL RADIATION THERAPY

Recall that alpha and beta particles interact quite strongly with all matter, resulting in very short-range penetration of most materials. For alpha particles, the range is less than 0.01 millimeter in body tissues, so that penetration from an external source is not possible. Beta particles seem better suited due to greater penetration (e.g., 10.50 millimeters for a 2.18 meV beta particle in water), but are still not very useful, unless they are accelerated to energies in excess of a billion electron volts (beV) (8). Similarly, high energy electromagnetic radiation (x and γ) can readily penetrate body tissues when externally directed on a particular organ or other tissue. The most common examples are machine-generated x-rays and γ rays from a cobalt-60 or cesium-137 source, which may be directed on a known tumor site. The hope is that tumor cells will be preferentially destroyed, whereas normal cells will be left unaffected. While such a hope is much too optimistic, there is clear evidence that rapidly-multiplying cells are much more sensitive to radiation than relatively dormant cells. Rapid multiplication is an unfortunate characteristic of many tumor cells, so that they are often affected more by radiation than are normal cells.

A specific example of teletherapy has been the treatment of Cushing's disease, which is often due to increased corticotropin (a hormone) production by the pituitary gland. Direct irradiation of the pituitary has led to a decreased output of the hormone (8).

INTERNAL RADIATION THERAPY

Internal radiation therapy, using radioactive isotopes, is less common because it is necessary to have the isotope localize almost exclusively at the site of the tumor growth. Unfortunately, there are few examples of radiopharmaceuticals (radioactive drugs) that show this kind of selectivity or, if they do, the particular site may be too sensitive to radiation to permit such treatment.

Iodine-131. One example of such internal irradiation involves the use of ^{131}I for treatment of hyperthyroidism. Overactive thyroid may be the result of many things, including tumor activity. Even in the absence of tumor growth, the killing of specific numbers of thyroid cells can serve to cut the thyroid hormone production down to a normal level. The isotope ^{131}I works well for this purpose for two important reasons. First of all, it binds quite selectively to the thyroid due to the involvement of iodine in production of the thyroid hormones. Secondly, the half-life of ^{131}I is 8.06 days, so that its radioactive lifetime is of short duration, which contributes to controlling the radiation dose. In other words, a large dose can be administered to achieve the desired effect but with subsequent loss of radioactivity following rather quickly. A long half-life would require much lower doses to avoid adverse long-term effects of large amounts of radiation. The iodine is normally administered in the form of a simple sodium iodide (NaI) pill.

Phosphorus-32.   A rather nonspecific example of internal radiation therapy has involved the use of radioactive sodium phosphate labeled with radioactive phosphorus-32 (^{32}P). Phosphate is a common chemical component of all living systems, so that any phosphate taken internally freely circulates throughout the body in the bloodstream. In fact, this seemingly undesirable feature has been turned into an advantage in the sometimes successful treatment of leukemia (over-production of white cells) and polycythemia vera (overproduction of red blood cells) (8).

Chromium-51.   Chromium wire enriched in ^{51}Cr (electron capture decay plus gamma emission, $t_{1/2} = 27.8$ days) has been subjected to testing at Argonne Cancer Research Hospital in a technique known as *brachytherapy* (9). By this technique, a partially radioactive chromium wire is implanted at a disease site (tumor) in the body. The isotope ^{51}Cr has a conveniently short half-life. Other radioactive isotopes that have been used in brachytherapy are radium and radon gas in needles or seeds. However, both of these have reportedly been found to suffer problems of leakage and loss.

Other Isotopes.   Still other isotopes have been used in brachytherapy. Among these are yttrium-90 (^{90}Y) and gold-198 (^{198}Au), which have been used for pituitary implantation in the treatment of Cushing's disease, mentioned earlier (8). Palladium-109 (^{109}Pd) wire has been implanted in the brain to relieve symptoms of Parkinson's disease. The ^{90}Sr-^{90}Y parent-daughter mixture (see later section on radiochemical cows) has been incorporated into the tip of a needle and implanted so as to relieve pain by destroying pain fibers in the spinal cord (8).

EXTRACORPOREAL IRRADIATION

Another technique that has been used to treat disease is extracorporeal irradiation (9). By this technique a physician splices an external plastic tube into the blood flow. It is then possible to irradiate the blood with gamma radiation as it flows through the plastic tube. Certain types of leukemia have reportedly responded to this treatment.

BORON-10 NEUTRON-CAPTURE THERAPY

Perhaps the most intriguing approach to radiation therapy used in the treatment of cancer is boron-10 neutron-capture therapy, which involves both internal and external procedures. The nuclear reaction for this method is shown below.

$$^{10}_{5}\text{B} + ^{1}_{0}n \rightarrow ^{7}_{3}\text{Li} + ^{4}_{2}\alpha$$

The procedure consists of administering a dose of some boron-containing sub-stance, followed by external irradiation with neutrons. Because they are un-charged, neutrons readily penetrate tissue and interact quite specifically with boron atoms. When this happens, alpha particles are produced at the site of the boron. Since they are the most massive of the common forms of radiation, alpha particles are biologically the most destructive. In spite of this, as was previously noted (see Chapter Two), alpha particles are neither dangerous nor useful when supplied from outside the body. However, in neutron-capture therapy, the alpha particles can theoretically be produced directly at the site of the tumor. As usual, radiation does not damage tumor cells exclusively but the chances may be greatly increased by incorporating boron-10 atoms into certain substances that will local-ize very specifically in tumors. One of the positive features of this approach is the absence of any dose of radioactive isotopes. Both boron-10 and lithium-7 are stable, nontoxic isotopes in the quantities used, so that even if some of the boron tends to go to normal tissues, these tissues will be unaffected unless neutrons are focused directly on them. In addition, it is possible to exert much greater control over the radiation dose, since alpha particles are only produced when the external radiation source is in use.

Diagnosis

Diagnostic applications of radiochemistry are a combination of external and in-ternal methods. In a typical procedure, an isotope is taken orally or by injec-tion. Once again, the isotope must concentrate to some extent in the desired tissue. After passage of a suitable period of time to allow the concentration to occur, the gamma radiation is checked (scanned) by placing a suitable detector near the skin at a location closest to the section of tissue being investigated.

The *scintillation camera* is a relatively new device that can be used for measuring the radioactive emissions from inside the body. The technique is called *imaging,* since a photographic image of an entire organ can be obtained if an iso-tope concentrates uniformly in that organ. This permits detection of various abnormalities such as a blockage or leakage. If an isotope is known to avoid tumor sites in an organ, one looks for "cold spots," or areas of unusually low concentra-tion of the isotope, as an indication of tumor activity. On the other hand, some isotopes preferentially localize in tumors, in which case "hot spots" indicate tumors. In some instances the radioactivity of body wastes may be monitored. Specific examples follow.

Thyroid Function. One of the more common tests is one which monitors thyroid function (10). Ingested iodine is rapidly absorbed from the intestine as iodide ion (I^-) and is circulated through the bloodstream. The iodide, which reaches the thyroid gland, is concentrated and incorporated into two thyroid hormones that regulate the general rate of metabolism.

The test studies uptake of radioactive ^{131}I by the thyroid. The isotope ^{131}I is commercially available in capsules of sodium iodide (NaI). The patient swallows the pill with water and measurements of the radioactivity level of the thyroid are made at 6 and 24 hour periods by placing an external radiation detector at the neck.

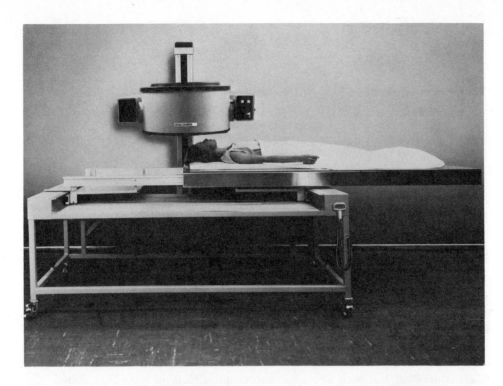

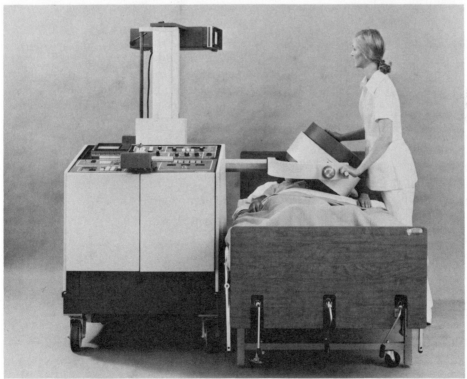

Figure 3.2 Scintillation cameras. (Courtesy of Picker Nuclear.)

Normal thyroid accumulates 7.5 to 25% of the oral dose in 6 hours and 12 to 46.5% in 24 hours. A significantly low absorption (about half of normal) is typical of congestive heart failure. A high reading suggests hyperthyroidism.

Vitamin B_{12}.   Another test measures the absorption of vitamin B_{12} (Schilling Test) (10). Because each molecule of vitamin B_{12} contains an atom of cobalt, it can be labeled with radioactive ^{57}Co. Radioactive ^{57}Co has a half-life of 270 days, which makes it preferable to the Co^{60}, which is normally used for external radiation therapy and has a half-life of 5.3 years.

After the patient has been fasting for 12 hours, a capsule of radioactive B_{12} is given orally. Two hours later, one milligram of nonradioactive B_{12} is injected intramuscularly in order to facilitate excretion of some excess B_{12} in the urine. All urine output is collected for 24 hours. The amount of B_{12} excreted is determined by counting the ^{57}Co in the urine and determining the percentage of the original dose that is excreted. If excretion is greater than 8% of the dose, the result is considered normal. If it is less than 8%, more radioactive B_{12} is given plus a capsule of "intrinsic factor," which is required for normal absorption of B_{12} out of the digestive tract into the bloodstream. The urine is again monitored for a 24-hour period. If the excretion of radioactive B_{12} now exceeds 8%, a deficiency known as "pernicious anemia," which is due to an absence of intrinsic factor, is the likely diagnosis. If the result is still below 8%, other problems are indicated.

Technetium-99m.   Among the most successful efforts in nuclear medicine has been the development of methods that employ ^{99m}Tc. As noted previously, ^{99m}Tc decays by isomeric transition—emission of gamma radiation. Since there are no charged particles emitted, there is little tissue damage, which is very desirable during diagnosis. Equally important, the amount of detectable radiation per rad (absorbed dose) is unusually high, since most of the emitted radiation has an energy that is ideal for normal detection cameras. With many other isotopes, much of the radiation is emitted with energies that are too high to permit efficient detection, so that the dosage must be increased to get enough detectable radiation. The result is greater damage to the tissues.

Heart Function.   A number of isotopes have been used to monitor the flow of blood through the heart and in other parts of the body. Heart function can be monitored by the use of ^{131}I or ^{99m}Tc-labeled albumin, which can be used to scan the blood-filled chambers of the heart. Tumors and pericardial effusion (leakage) can be detected by imaging.

A relative newcomer to this type of study is thallium, as $^{201}Tl^+$, but it is an excellent choice. It has a short half-life (74 hours) and the gamma rays emitted as a result of the electron capture process can be detected very efficiently. This makes it possible to use very low doses of the isotope and, therefore, minimizes the amount of damage done to body tissues.

Bone Scanning.   It is also possible to examine certain characteristics of bone by using radiopharmaceuticals. To do this, intravenous injection of ^{85}Sr or ^{99m}Tc is given and scans are taken at a specified later time (e.g., 72 hours). The ^{85}Sr is taken up by bone (primarily calcium phosphate) since strontium is chemically similar to calcium. Hot spots of radioactivity, observed by scanning, are indicative of

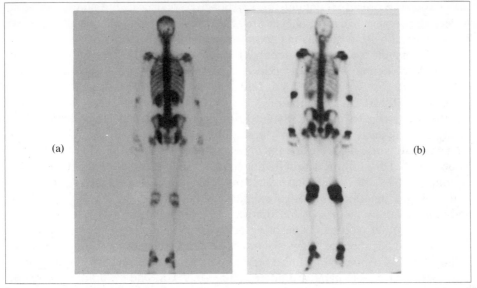

(a)

(b)

Figure 3.3   The Technetium-99m bone scan. (From Early, Paul J., Razzak, Muhammed Abdel, and Sodee, D. Bruce. *Textbook Of Nuclear Medicine Technology*, 2d ed. St. Louis: The C. V. Mosby Co., 1975.) See text for discussion.

rapid bone cell growth, which may indicate the presence of tumor cells, since tumor cells are often characterized by rapid growth. Normal bone development in children can also be observed as shown in Figure 3.3, in which ^{99m}Tc is rapidly taken up in the new bone growth. The scan in Figure 3.3 (a) shows a normal uptake of a salt of ^{99m}Tc. In (b), we see the unusual but also normal uptake of ^{99m}Tc in a 14-year-old patient in which the increased concentration of the radioactive isotope indicates new bone growth.

DETECTION OF TUMORS

Radiopharmaceuticals can be used in a number of ways to obtain a wide variety of information, but none is more important than diagnosis of tumors. A few examples have already been mentioned. Some other methods, which are used for diagnosis of tumors in some specific parts of the body, are considered in the following paragraphs.

The Liver.   The liver can be monitored by using various radioisotopes attached to individual molecules or to groups of molecules, called *colloids*. Rose bengal is a complex compound that concentrates in the liver. It can be labeled with ^{131}I to permit imaging of the liver.

One of the most common nuclear medicine procedures utilizes colloids, labeled with ^{99m}Tc, in order to image the liver and locate abnormalities.

The Brain.   A relatively new radiopharmaceutical with great diagnostic potential is ^{111}In-bleomycin. It can be used to detect the presence of a variety of tumors (11) and is a good radioisotope to give to humans. It has a half-life of 67 hours, which is long enough to prepare a patient for a radioactive scan test, but short enough to cause little damage to the body. The energy of this isotope's gamma ray emissions is also ideal for use with existing scintillation cameras.

Bleomycin is a naturally-occurring material that is selectively absorbed and broken down by tumors. It is produced by a strain of bacteria known as *Streptomyces verticullus*. When ^{111}In is coupled to bleomycin, the resulting molecule becomes very important. Since the bleomycin concentrates in tumors, these tumors can then be "seen" by gamma ray detection equipment since ^{111}In undergoes beta decay accompanied by gamma ray emission. Using this compound, a diagnostician can distinguish between an area of dead tissue (e.g., due to a blood clot) and a brain tumor. This differentiation had been impossible using other available radioisotopes. It has also been found that ^{111}In-bleomycin is absorbed to a greater extent by some tumors than is other existing radioisotopes, thereby giving a better and more accurate picture of the tumor.

Four other isotopes have often been used to provide information on the presence and location of brain tumors. The four are mercury-203, arsenic-74, gallium-68, and copper-64. All four tend to concentrate to some extent in tumors. The ^{203}Hg is a typical gamma emitter. The ^{74}As, ^{68}Ga, and ^{64}Cu are all positron emitters and, thus, offer a special advantage in pinpointing the exact location of the tumor. Positrons have the unusual property of decaying immediately after emission into 2 gamma rays, which are found to travel in opposite directions. The gamma rays are formed according to the following reaction. The electron may be supplied by any atom in the path of the positron.

$$_{+1}^{0}\beta + {}_{-1}^{0}\beta \rightarrow 2\gamma$$

By using a detection system much like the anticoincidence counting system mentioned earlier, it is possible to obtain a record of the gamma radiation that is simultaneously detected by two radiation counters that are positioned on opposite sides of the head.

The Pancreas.   A very ingenious strategy is sometimes employed in the detection of tumors of the pancreas. A major function of the pancreas is the synthesis of several proteins (e.g., insulin), which are required for digestion. This makes it possible to deliver a radioactive isotope to the pancreas by replacing the sulfur atom of the amino acid, methionine, with a radioactive selenium atom, ^{75}Se. The substitution of Se for S does not alter the uptake of the amino acid by the pancreas because of the great similarity of two atoms as indicated by their relative locations in the periodic table.

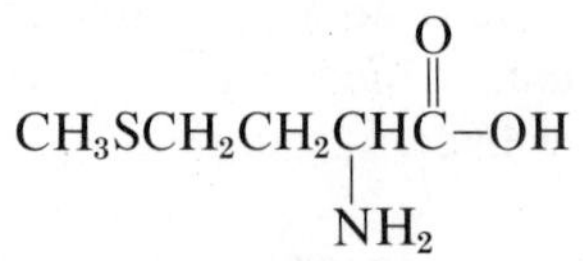

$$
\underset{\text{methionine}}{CH_3SCH_2CH_2\underset{\underset{NH_2}{|}}{C}HC\overset{\overset{O}{\|}}{-}OH}
\qquad
\underset{^{75}\text{selenomethionine}}{CH_3SeCH_2CH_2\underset{\underset{NH_2}{|}}{C}HC\overset{\overset{O}{\|}}{-}OH}
$$

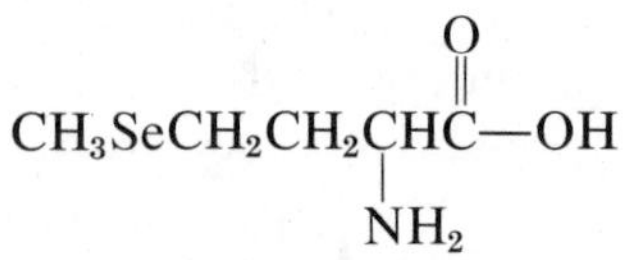

In examining an image of the pancreas, a cold spot suggests an abnormality, since it indicates that a portion of the pancreas is not utilizing amino acids normally, possibly due to tumor activity.

X-rays. Diagnostic x-rays are normally machine-generated x-rays that pass through tissue in different amounts depending on the path. As noted previously (see Figure 2.3), different materials absorb radiation to a different extent. Similarly, high-density areas of the body, such as bone, absorb x-rays more efficiently and cause a lower exposure of a photographic plate as a result. This causes a light area on the plate.

Radiochemical Cows

Cows produce milk that can be removed by milking. Similarly, radioactive isotopes decay to produce other isotopes that can be separated by milking. In this situation, milking involves a separation of two elements that differ in some physical property, such as the solubility of their ions in water. One example is the following tin-indium system.

$$-_{-1}^{0}e + {}_{50}^{113}Sn \xrightarrow[t_{1/2} = 118 \text{ days}]{\text{electron capture}} {}_{49}^{113m}In$$

$$\downarrow \begin{array}{l} \gamma \text{ emission} \\ t_{1/2} = 104 \text{ min} \end{array}$$

$${}_{49}^{113}In$$

According to these reactions, the tin-113 decays to indium-113m by electron (beta particle) capture. This decay is followed by emission of gamma radiation during which the metastable nucleus decays to form ^{113}In. This system is a typical cow that can be used as a source of ^{113m}In. The ^{113m}In cannot be purchased and shipped because of its very short half-life. However, the parent, ^{113}Sn, can be handled at leisure and then milked to remove the short half-life daughter (^{113m}In).

Descriptions of other examples of useful cow systems follow.

$${}_{42}^{99}Mo \xrightarrow{\beta^- \text{ decay}} {}_{43}^{99m}Tc$$
$$t_{1/2} = 67 \text{ hr} \qquad\qquad t_{1/2} = 6 \text{ hr}$$

Technetium-99m has been used extensively for diagnosis of tumors. Its short half-life makes transport and storage impractical but the molybdenum parent can be made, shipped, and used at a convenient pace as a source of ^{99m}Tc.

Much the same situation is true for each of the following examples. Important applications of each have been described earlier.

$${}_{32}^{68}Ge \xrightarrow{\text{electron capture}} {}_{31}^{68}Ga$$
$$t_{1/2} = 280 \text{ days} \qquad\qquad t_{1/2} = 68 \text{ min}$$

$${}_{39}^{87}Y \xrightarrow{\text{electron capture}} {}_{38}^{87}Sr$$
$$t_{1/2} = 80 \text{ hr} \qquad\qquad t_{1/2} = 2.8 \text{ hr}$$

A typical cow system has the parent isotope embedded into some kind of medium, e.g., an ion exchange resin. As decay occurs, some of the daughter isotope takes the place of the parent. At some point, the chemist washes the resin in

Figure 3.4 A "cow." Photo of a generator used to produce samples of technetium-99m. A charge vial A containing 20 ml of isotonic saline solution is placed onto the double needle B and vented through tube C fitted with a cotton pledget filter D. Elution begins automatically when a shielded evacuated collection vial E is placed onto needle F. As the saline is removed from charge vial A, it passes through shielded column G loaded with the parent, molybdenum-99. Technetium-99m is selectively eluted and the eluate then passes through filter H and needle F into shielded collection vial E. (Photo and line drawing courtesy of New England Nuclear.)

such a way that the daughter washes off the resin and the parent remains embedded. An example is shown in Figure 3.4. Cows can be divided into two types:

1. *Converse Cows:*

   The use of ^{113}Sn as a source of ^{113m}In is an example of a converse cow in which the parent has a longer half-life than the daughter.

2. *Reverse Cows:*

   Many decay processes occur with the formation of longer half-life products. Such a system in which the parent has a shorter half-life than the daughter would seem less useful since it would be much easier to purchase the daughter and thus avoid the problem of milking (separation). However, such a reverse cow system may be useful when both the parent and the daughter have relatively short

A Reverse Cow

half-lives. In such a case it may be possible to produce the parent (in a nuclear reactor) and ship it to a user. In some cases, the parent may even be completely decayed before delivery, or shortly thereafter, so that the sample is pure daughter at a relatively high level of activity.

PRODUCTION OF OTHER PARTICLES: INDUCED RADIOACTIVE DECAY

The phenomena known as *transmutation, nuclear fission,* and *nuclear fusion* are the next items to be considered. They are related since they occur when certain nuclei are bombarded with various high energy radiation such as alpha, beta, and others. The other particles are available from nuclear reactions such as the two that follow. Here, the nuclei are bombarded with alpha particles causing production of neutrons and protons.

$$^{9}_{4}\text{Be} + ^{4}_{2}\alpha \rightarrow ^{12}_{6}\text{C} + ^{1}_{0}n$$
shorthand notation: $^{9}_{4}\text{Be}(\alpha, n)^{12}_{6}\text{C}$

$$^{14}_{7}\text{N} + ^{4}_{2}\alpha \rightarrow ^{17}_{8}\text{O} + ^{1}_{1}\text{H (or } ^{1}_{1}p)$$
shorthand notation: $^{14}_{7}\text{N}(\alpha, p)^{17}_{8}\text{O}$

This now gives us a total of five different types of radiation (α, β, γ, n, and p) that can be used to bombard other nuclei to cause nuclear reactions to occur. However, the availability of five different types of radiation brings up a practical problem that must be dealt with experimentally in order to allow the radiation to induce certain desired chemical changes. The problem is that not all particles are emitted with energies that will permit them to attack other nuclei productively. Some particles are emitted with energies that are too low to permit them to penetrate other nuclei, whereas other particles have such high energy that they pass right through the nuclei without causing any change. In general, it is charged particles, both positive and negative, that have difficulty penetrating other nuclei due to repulsions by electrons surrounding the nuclei or by the nuclei themselves, depending on the charge of the particle. Thus it is common to increase the kinetic energy (the energy of motion) of the charged particles by acceleration, using devices known as **cyclic** or **linear accelerators.** These devices merely speed up charged particles by suitable interactions with a series of charged plates. The well-known cyclotron is an example of a cyclic acceleration device.

In contrast, neutrons, which have a zero charge, have a tendency to penetrate completely through a nucleus without any effect. Thus carbon in the form of graphite is often used as a **moderator** to slow the neutrons down so that they can come under the influence of nuclear forces and cause reactions to occur. Other moderators will be discussed in a later section.

Transmutation

Now that we have seen how to effectively utilize different types of radiation, the formation of new elements by *transmutation* can be considered. The following are four examples of reactions that have been used to produce isotopes that were previously unknown.

$$^{96}_{42}\text{Mo} + {}^{2}_{1}\text{H} \rightarrow {}^{1}_{0}n + {}^{97}_{43}\text{Tc} \text{ (Technetium-97)}$$

$$^{142}_{60}\text{Nd} + {}^{1}_{0}n \rightarrow {}^{0}_{-1}\beta + {}^{143}_{61}\text{Pm} \text{ (Promethium-143)}$$

$$^{209}_{83}\text{Bi} + {}^{4}_{2}\alpha \rightarrow 3{}^{1}_{0}n + {}^{210}_{85}\text{At} \text{ (Astatine-210)}$$

$$^{230}_{90}\text{Th} + {}^{1}_{1}p \rightarrow 2{}^{4}_{2}\alpha + {}^{223}_{87}\text{Fr} \text{ (Francium-223)}$$

Among the more ambitious examples of transmutation has been the synthesis of the **transuranium elements.** These are the elements following uranium (atomic number 92) in the periodic table. Some of these elements may have been present in the earth's crust at the time of its formation about 4.5 billion years ago but the half-lives of the isotopes are too short to permit survival to present times.

The first successful synthesis was reported in 1940 when elements 93 and 94, called neptunium and plutonium, were produced according to the following reactions.

$$^{238}_{92}\text{U} + {}^{2}_{1}\text{H} \rightarrow {}^{239}_{92}\text{U} + {}^{1}_{1}\text{H} \text{ (Deuteron Bombardment)}$$

$$^{239}_{92}\text{U} \rightarrow {}^{239}_{93}\text{Np} + {}^{0}_{-1}\beta \text{ (Spontaneous Decay } t_{1/2} = 23.5 \text{ min)}$$

$$^{239}_{93}\text{Np} \rightarrow {}^{239}_{94}\text{Pu} + {}^{0}_{-1}\beta \text{ (Spontaneous Decay } t_{1/2} = 2.35 \text{ days)}$$

Nuclear Fission

Still another phenomenon made possible using available moderated or accelerated radiation is nuclear **fission**. Atomic bombs used in World War II were operated by nuclear fission. Nuclear fission is also the basis of current nuclear power plants.

The fission process may be described as the splitting of certain large atoms into smaller atoms due to the absorption of energy in the form of a bombarding neutron. Due to the difference in magnitude of the nuclear forces of the parent nucleus and the smaller fragments, the entire process is accompanied by the liberation of large amounts of energy.

One common example is the fission of uranium-235 that results from bombardment with neutrons as illustrated by the following.

$$^{235}_{92}U + ^{1}_{0}n \rightarrow ^{135}_{53}I + ^{97}_{39}Y + 4^{1}_{0}n$$
$$\rightarrow ^{139}_{56}Ba + ^{94}_{36}Kr + 3^{1}_{0}n$$
$$\rightarrow ^{103}_{42}Mo + ^{131}_{50}Sn + 2^{1}_{0}n$$
$$\rightarrow ^{139}_{54}Xe + ^{95}_{38}Sr + 2^{1}_{0}n$$

As shown, there are several possible induced decay routes (and many more not shown), but all of the routes are characterized by production of more neutrons than are consumed, so that the process can become a self-perpetuating chain reaction in which the product neutrons can turn around and become the bombarding particles for fission of more ^{235}U. Fortunately, it is possible to carry out such a reaction under so-called **critical** or **subcritical** conditions in which an average of one or less product neutrons become bombarding neutrons for continuation of the chain. Such conditions can be readily achieved by addition of an efficient neutron absorbing material, such as cadmium or boron, which can be added in any amount desired to maintain the critical or subcritical condition. The amount of fissionable material is also a decisive factor in controlling the criticality of particular situations. The extreme condition in which an average of more than one product neutron becomes a bombarding particle is termed **supercritical** and is one of the conditions applicable to atomic bombs in which the level of neutron bombardment, fission, and production of more neutrons is allowed to accelerate tremendously, ultimately resulting in an explosion (see further discussion in Chapter Four).

Breeder Reactors

Nuclear fission breeder reactors provide another very important method of carrying out the fission process on more readily available starting materials. The neutron-activated fission of ^{235}U would seem to be quite sufficient to deliver all the necessary energy except that the supply of ^{235}U is limited and does not constitute a long-range energy source. The isotope, ^{238}U, is much more prevalent in nature, but it is not fissionable. Therefore, the following sequence of reactions is useful for conversion of nonfissionable ^{238}U to fissionable plutonium-239 (^{239}Pu).

$$^{238}_{92}U + ^{1}_{0}n \rightarrow ^{239}_{92}U \text{ (Neutron Bombardment)}$$
$$^{239}_{92}U \rightarrow ^{239}_{93}Np + ^{0}_{-1}\beta \text{ (Spontaneous Decay, } t_{1/2} = 23.5 \text{ min)}$$
$$^{239}_{93}Np \rightarrow ^{239}_{94}Pu + ^{0}_{-1}\beta \text{ (Spontaneous Decay, } t_{1/2} = 2.35 \text{ days)}$$

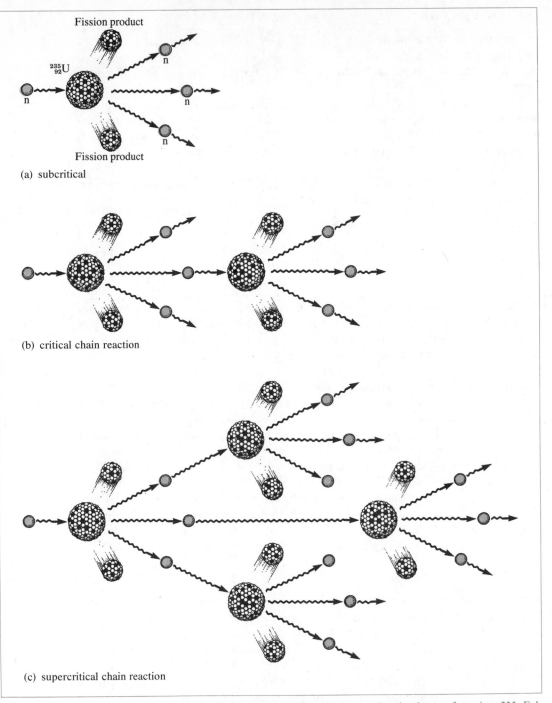

(a) subcritical

(b) critical chain reaction

(c) supercritical chain reaction

Figure 3.5 A nuclear fission chain reaction initiated by neutron bombardment of uranium-235. Following fission, the neutrons which are released may be lost or captured by other nuclei which do not undergo fission. (a) If less than one of the neutrons released by fission is effective in causing further fission, a chain reaction will be subcritical and cannot be sustained. (b) If an average of one neutron from each fission event goes on to cause another fission, the chain reaction will be critical and self-sustaining. (c) If more than one neutron from each fission causes additional fissions, the process becomes supercritical and can lead to an explosion.

Thus, a potential fission fuel, ^{238}U, can be utilized to *breed* ^{239}Pu, which is directly fissionable. The reactions almost duplicate those first used to synthesize Np and Pu.

An analogous situation exists with respect to naturally-occurring thorium-232. The isotope ^{232}Th is also nonfissionable but it too can be converted to a fissionable isotope, ^{233}U, by the following conversion.

$$^{232}_{90}Th + ^{1}_{0}n \rightarrow ^{233}_{90}Th + \gamma \text{ (Neutron Bombardment)}$$
$$^{233}_{90}Th \rightarrow ^{233}_{91}Pa + ^{0}_{-1}\beta \text{ (Spontaneous Decay } t_{1/2} = 22 \text{ min)}$$
$$^{233}_{91}Pa \rightarrow ^{233}_{92}U + ^{0}_{-1}\beta \text{ (Spontaneous Decay } t_{1/2} = 27 \text{ days)}$$

Thus, the overall conversion

$$^{238}_{92}U \rightarrow ^{239}_{94}Pu$$

and

$$^{232}_{90}Th \rightarrow ^{233}_{92}U$$

can be used to breed the fissionable product isotopes from the nonfissionable, but *fertile* isotopes.

Nuclear Fusion

Many solid materials may be made to mix together thoroughly by melting (fusion) the solids into liquids, which then freely mix. Similarly, nuclei may be made to combine when the conditions are right for it. This process is also called *fusion* and is the reverse of fission. In other words, nuclear fission is the process in which large nuclei split into smaller fragments, whereas small nuclei are combined to form heavier nuclei by fusion. Surprisingly, just as fission was seen to liberate large amounts of energy, so also does fusion. It seems an impossible contradiction that two opposite processes can both give off energy. However, just as stability of isotopes disappears for all isotopes above ^{209}Bi, it is known that the opposite, i.e., maximum stability, is found at about mass number 60. The stability of atoms in that range of atomic weights is consistent with the observed high abundance in nature of such things as iron and nickel. Therefore, in theory, fusion of small nuclei, below mass number 50, to build larger, more stable nuclei will release energy, just as very large nuclei can undergo fission to form more stable smaller nuclei and release energy.

Fission is initiated by neutron bombardment. Similarly, fusion requires combination of nuclei that exhibit strong repulsions for one another, so that extremely high energy must be supplied to initiate fusion. Once initiated, fusion is also self-sustaining. The so-called hydrogen bomb proceeds by fusion that is initiated by a fission device. Fusion holds great promise as an energy source for the future but great technological problems lie along the way. The most obvious is the extremely high temperature (75 to 100 million degrees Celsius) required for fusion, which makes containment impossible in any known vessel. In fact, much effort has gone into development of some sort of magnetic bottle. Even reaching such temperatures is a major obstacle although, among other things, it is hoped that lasers may offer a solution to the problem (12).

Fusion processes, represented by the following reactions, are thought to be responsible for the energy output of the sun.

$$2{}_1^1\text{H} \rightarrow {}_1^2\text{H} + {}_1^0\beta$$
$${}_1^1\text{H} + {}_1^2\text{H} \rightarrow {}_2^3\text{He}$$

The fusion reactions of hydrogen bombs are:

$${}_1^2\text{H} + {}_1^3\text{H} \rightarrow {}_2^4\text{He} + {}_0^1n$$
$${}_3^6\text{Li} + {}_0^1n \rightarrow {}_2^4\text{He} + {}_1^3\text{H}$$
$${}_1^2\text{H} + {}_1^3\text{H} \rightarrow {}_2^4\text{He} + {}_0^1n \text{ etc. (chain continues)}$$

Source of starting materials:

deuterium: from D_2O of sea water
tritium: from step 2 (above)
neutrons: from nuclear reactors and steps 1 and 3 (above)
lithium: from ores

SUMMARY

Carbon-14 dating is an important technique for determining the age of samples derived from sources that were once alive. Cosmic radiation interacts with nitrogen-14 in the atmosphere to produce carbon-14. The carbon can then be incorporated into living organisms and the supply is continuous. The radioactive decay of carbon-14 is likewise continuous and the two processes tend to balance out in a condition called a "steady state." When an organism dies, incorporation ceases and only the decay process remains, so that the level of carbon-14 steadily decreases at a rate determined by its half-life (5568 years). Thus, it is theoretically possible to determine the age of a sample by measuring the level of carbon-14 that remains. Fifty thousand years is the upper limit of the range of application of carbon-14 dating techniques since the carbon-14 content becomes too low for accurate measurement. Fossil fuels are too old to be dated by this technique. The rubidium-strontium clock is another dating technique that has permitted the dating of much older samples, e.g., from the moon, due to the very long half-life of rubidium-87.

Radioactive isotopes are of major importance as tracers in many areas of research. They make it possible to monitor the movement of isotopes in systems undergoing changes. Other unusual, but nonradioactive isotopes, e.g., carbon-13, provide similar information. The techniques used to monitor these isotopes have both advantages and disadvantages compared to those used to monitor radioactive decay.

One of the more rapidly advancing fields of science is nuclear medicine. Both internal and external radiation therapy have been used for many years. Diagnostic applications are the subject of much current research activity. Noninvasive imaging techniques are often much preferred to more radical surgical methods.

Transmutation is a major application of radioactive decay. It has led to the synthesis of many new elements and is commonly used to produce radioactive isotopes for use in medicine and research. Transmutation involves bombarding a target nucleus with some sort of particle that is usually either accelerated or moderated in order to achieve reaction.

Nuclear fission is another important process. It is the basic mechanism that occurs in nuclear power plants. Breeder reactors function by producing fissionable fuel from fertile, nonfissionable isotopes.

Nuclear fusion is the ultimate goal for nuclear power production but many technological problems are yet to be solved. The energy output of the sun and the energy released by hydrogen bombs are the result of nuclear fusion processes.

PROBLEMS

1. Give a definition or example of each of the following terms.
   a) anticoincidence counting
   b) Rb-Sr clock
   c) cosmic radiation
   d) cast iron
   e) dialysis
   f) teletherapy
   g) accelerators
   h) radiopharmaceuticals
   i) brachytherapy
   j) extracorporeal irradiation
   k) pernicious anemia
   l) radiochemical cow
   m) m (e.g., ^{99m}Tc)
   n) transmutation
   o) fission
   p) fusion
   q) breeders
   r) critical
   s) supercritical
   t) tracer

2. Write the completed equation for each of the following nuclear processes.
   a) $^2H + {}^3H \rightarrow \alpha + ?$
   b) $^6Li(n, \alpha)?$
   c) $^{235}U + n \rightarrow {}^{94}_{37}Rb + {}^{140}_{55}Cs + ?$
   d) $^{27}_{13}Al(^2H, \alpha)?$
   e) $^{233}U + n \rightarrow ? + {}^{95}Kr + 3n$
   f) $^{238}U + n \rightarrow$
   g) isomeric transition of ^{99m}Tc
   h) $^{14}N(n, ?)?$
   i) decay of $^{14}CO_2$
   j) $^{108}In(e, \gamma)?$

3. A mixture of one gram of ^{226}Ra (an alpha emitter) and a few grams of ^9Be makes an excellent source of neutrons (ca. 10^7 neutrons/sec). Write the nuclear reactions which account for this fact.

4. Each of the following questions pertains to the radioactive decay of carbon-14 labeled CO_2 obtained from living systems. Consult Figure 3.6 when needed.
   a) What is meant by "the steady state?"
   b) How are the readings obtained for comparison with the data plotted along the vertical axis?
   c) Why is there a value of 15 cpm plotted on the vertical axis?
   d) Why is 50,000 years considered to be the upper limit for use of the ^{14}C dating method?
   e) What is the apparent age of a wood sample, found in an Egyptian pyramid, that

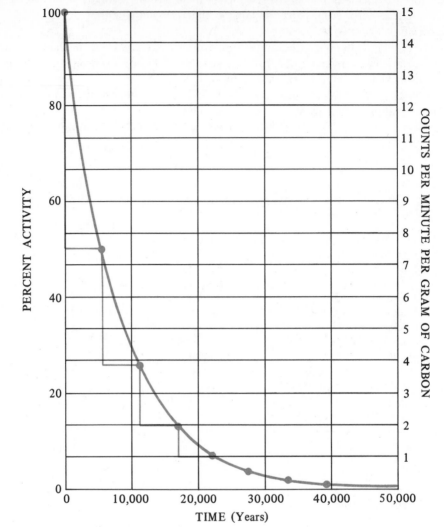

Figure 3.6   Radioactive decay of carbon-14.

was burned to produce CO_2 with a radioactivity level of 6.7 cpm/gram of carbon?

  f) What is the apparent age of a cast iron sample that was similarly treated to pro-
     duce CO_2, which has a radioactivity level of 11 cpm/gram of carbon?

5. Distinguish between dialysis and osmosis.

6. Why is ^{57}Co, rather than ^{60}Co, used to monitor vitamin B_{12} absorption?

7. What type of radiation is most effective (i.e., most damaging) in treating tumors
   when used from (a) outside the body and (b) inside the body? Explain.

8. What is the reaction for boron-10 neutron-capture therapy?

9. Give a diagnostic application of radiochemistry.

10. What is the value of a converse cow?

REFERENCES

1. Protsch, R., and Berger, R. "Earliest Radiocarbon Dates for Domesticated Animals." *Science 179* (1973): 235.
2. Van der Merwe, N.J. *The Carbon-14 Dating of Iron.* Chicago: University of Chicago Press, 1969.
3. Faul, H. *Nuclear Clocks,* Washington, D.C.: U.S. Atomic Energy Commission, 1966, pp. 20–23.
4. Murphy, V.R., Evensen, N.M., Jahn, B., Coscio, Jr., M.R., Dragon, J.C., and Pepin, R.O. "Rubidium-Strontium and Potassium-Argon Age of Lunar Sample 15555." *Science 175* (1972): 419.
5. Katz, S.A., Parfitt, C., and Purdy, R. "Equilibrium Dialysis." *Journal of Chemical Education 47* (1970): 721.
6. Maywiyoff, N.A., and Ott, D.G. "Stable Isotope Tracers in the Life Sciences and Medicine." *Science 181* (1973): 1125.
7. Heindel, N.D., ed. *An Introduction to the Chemistry of Radiopharmaceuticals.* New York: Masson, Inc., 1978, in press.
8. Lawrence, J.H., Manowitz, B., and Loeb, B.S. *Radioisotopes and Radiation.* New York: McGraw-Hill, 1966, Chapter 2.
9. Shore, B., and Hatch, F. *Biological Implications of the Nuclear Age.* U.S. Atomic Energy Commission Symposium Series No. 16, Washington, D.C., 1969, pp. 1–20.
10. Davidson, L., and Henry, J.B. *Clinical Diagnosis.* Philadelphia: W.B. Saunders, 1969, Chapter 8.
11. "Indium Radionuclide Helps Detect Cancers." *Chemical and Engineering News,* April 2, 1973, p. 12.
12. Metz, W.D. "Laser Fusion: A New Approach to Thermonuclear Power." *Science 177* (1972): 1180.

E LECTRICAL energy derived from nuclear sources appears to have unlimited short and long-range potential. In the following discussion we will consider some of the features and problems of the various types of nuclear power plants currently in use or planned for future use.

TYPES OF NUCLEAR POWER PLANTS

The three possible categories of power plants identified by the type of nuclear process are: (1) ^{235}U-induced fission, (2) ^{239}Pu or ^{233}U breeding, followed by induced fission, and (3) fusion.

The plutonium breeder system, in the form of the *liquid metal fast breeder reactor* (LMFBR), seems to offer the most likely long-range answer to energy production from nuclear sources. Widespread development of breeder reactors for electric power production may become a necessity in the future as supplies of rare ^{235}U dwindle. To replace ^{235}U, the all but inexhaustible nonfissionable ^{238}U may become prominent in energy production. Even long before the ^{235}U supplies are depleted, the supply and demand picture may well price the ^{235}U up to levels at which it will not be economical to employ it for power production (1).

The so-called *light water reactor* (LWR) employs ^{235}U fission in a system in which H_2O serves as a coolant.* Surprisingly, there is a heavy use of fossil fuels to power the facilities needed to concentrate or enrich the ^{235}U from its normal ore concentration of 0.7% to about 3%. Nuclear weapons require uranium that is about 97% ^{235}U and requires several thousand more stages of enrichment (2).

Nuclear reactors for electric power production take many forms. So far, the LWR and LMFBR have been mentioned. A more complete list appears in Table 4.1.

Each of the types listed in Table 4.1 has a name that is meant to indicate something about its mode of operation. There are many possible variables. These include the type and design of the coolant system, the moderator, whether the process is simple fission or breeding, and whether it is a fast or thermal neutron reactor system.

First, consider the concept of a **fast reactor.** The words *fast* and *thermal*

* Heavy water, ^2H_2O or D_2O, is water in which the element hydrogen is present as the isotope with one neutron in the nucleus—called *deuterium*. Light water contains only the isotope ^1H.

NUCLEAR POWER
GENERATION

TABLE 4.1   TYPES OF NUCLEAR POWER PLANTS

ABBREVIATION	REACTOR TYPE
LWR	Light Water Reactor
HWR	Heavy Water Reactor
PWR	Pressurized Water Reactor
BWR	Boiling Water Reactor
HTGCR	High Temperature Gas-Cooled Reactor
LMFBR	Liquid Metal Fast Breeder Reactor

(slow) are used to describe the energy of the neutrons used to bring about the fission process or to breed a fissionable isotope. Fast neutrons are high-energy neutrons (generally 1 meV or higher), whereas thermal or slow neutrons are low energy neutrons (<1 meV). It was mentioned in Chapter Three that some isotopes will not capture fast neutrons. In such cases, some sort of moderator is required to decrease their energy. In the jargon of the nuclear scientist, it is common to refer to the **cross section** (σ) of a nucleus, which describes the probability that it will capture a neutron. A large cross-section signifies a high probability of capture (and

Figure 4.1   Pennsylvania Power and Light Company's Susquehanna Steam Electric Station, with twin boiling water reactor generating units planned for operation in 1980 and 1982. (Courtesy of Pennsylvania Power and Light Company.)

possibly fission). An example of an isotope with a high fission cross-section for thermal neutrons but a low cross-section for fast neutrons is ^{235}U. Therefore, the fission of ^{235}U can only be initiated with neutrons that have been properly moderated.

On the other hand, ^{238}U requires fast bombarding neutrons in order for capture to occur and begin the series of reactions involved in the breeding of ^{239}Pu. In other words, ^{238}U has a zero cross-section for capture of thermal neutrons but a reasonably high cross-section for capture of fast neutrons.

There are many consequences of this difference between ^{235}U and ^{238}U. It was previously noted that ^{235}U makes up about 0.715% of naturally-occurring uranium. Put another way, there is one ^{235}U atom for every 139 ^{238}U atoms. In these proportions, the ^{235}U cannot be made to undergo sustained fission to power a nuclear reactor for, while the ^{238}U nucleus cannot capture a thermal neutron, it can cause a so-called inelastic scattering of the neutron, which serves to lower the energy of the neutron to a level below that required even for ^{235}U fission. However, when the uranium ore is enriched up to about 3% ^{235}U, the fission process becomes a significant event and can be built up to and maintained at critical self-sustaining levels. The fission reactions produce more neutrons than they consume so that even though the ^{238}U does capture some, the process can proceed in such a way that at least one product neutron from each fission can be moderated by the system and used again as a bombarding particle.

Quite logically, there has been concern over the possibility that a runaway

Figure 4.2   The measurement of a uranium oxide fuel element is a routine quality control procedure. In a nuclear power plant each pellet can provide about as much energy as a ton of coal. (Courtesy of Westinghouse Electric Corp.)

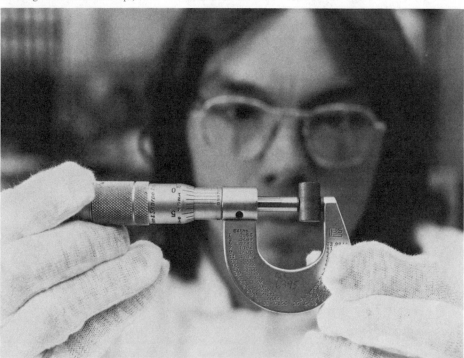

nuclear power plant could be a potential nuclear bomb. Such an event is not possible for a number of reasons. The fuel composition is a major factor. Just as uranium ore has too little ^{235}U to give the critical fission condition required to sustain a reactor, even the uranium enriched to 3% ^{235}U is not sufficient to produce the conditions necessary for a nuclear explosion. Samples enriched to about 97% ^{235}U are common for nuclear weapons.

In addition, two other circumstances prevent bomb formation. In the first place, the reactor design is such that the nuclear reactions cannot be contained to allow the system to build up to bomb intensity. If control were lost, the heat would cause the core of the reactor to come apart (melt). Finally, a bomb requires that the components be combined in about 10^{-6} seconds, which is impossible within the design of a power generating plant.

Referring to Table 4.1, consider the other distinguishing features of the nuclear power plants. Coolants and moderators are two of the major variables. Table 4.2 lists the most important examples of each. The discussion following gives some of the advantages and disadvantages of each type of system.

Quite clearly, all of the alternatives for coolants and moderators have both advantages and disadvantages. Rather logically, this has led to the use of almost

TABLE 4.2   ADVANTAGES AND DISADVANTAGES OF VARIOUS COOLANTS AND MODERATORS

COOLANT	ADVANTAGES	DISADVANTAGES
Gas (He, CO_2)	Efficient heat transfer Chemically inert	Expensive
Light Water (H_2O)	Inexpensive Can also serve as moderator	Captures many neutrons[1] Requires high pressure containment
Heavy Water (D_2O)	Captures very few neutrons Can also serve as moderator	Expensive Requires high pressure containment
Liquid Metal (sodium)	Can tolerate high temperature without pressure buildup Efficient heat transfer High specific heat[3]	Reacts violently with air and water Poor moderator[2]

MODERATOR	ADVANTAGES	DISADVANTAGES
Light Water	Inexpensive Good moderator	Captures many neutrons[1]
Heavy Water	Good moderator Captures very few neutrons	Expensive
Graphite (carbon)	Inexpensive Captures a few neutrons	Fair moderator
Beryllium (Be)	Captures a few neutrons	Toxic Expensive Fair moderator

[1] Requires enriched fuel.
[2] An advantage for fast reactors.
[3] Shows small temperature increase while absorbing large quantities of heat.

all of the alternatives in the design of at least one nuclear power plant somewhere in the world. Of these, most of the distinguishing features appearing in Table 4.2 and the reactor types listed in Table 4.1 are quite evident from the names, but additional comments seem appropriate on some of them.

The *boiling water reactor* (BWR) is a system in which water is used as both coolant and moderator. A schematic diagram of the BWR appears in Figure 4.3. The system performs in the following way:

1. Heat is generated by nuclear fission in the core of the reactor.
2. Water flows through the core and acts as moderator and coolant.
3. As a coolant, the water extracts heat from the core and is converted to steam.
4. The steam is used to drive a turbine, which generates electricity.

Figure 4.3   A boiling water reactor (BWR) nuclear power plant.

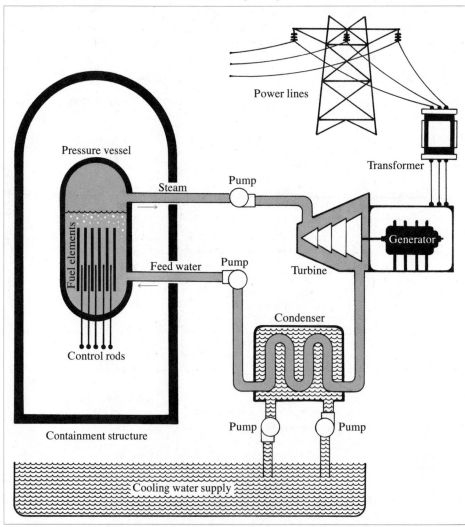

Figure 4.4   Pressure vessel being lowered into the chamber at Brunswick Nuclear Power Plant (Carolina Power & Light Company), Unit #2, located in Southport, North Carolina. This 821 megawatt plant went on the line in 1975. (Courtesy of the General Electric Company.)

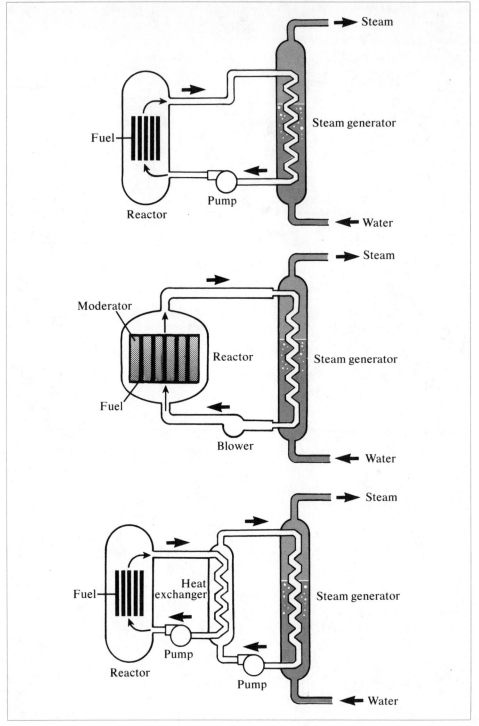

Figure 4.5 Nuclear steam-supply components in a pressurized-water reactor (PWR); Figure 4.6 Nuclear steam-supply components in a gas-cooled reactor (GCR); Figure 4.7 Nuclear steam-supply components in a liquid-metal-cooled fast breeder reactor (LMFBR).

Figures 4.5 through 4.7 are schematic diagrams of the steam-supply components of three other types of nuclear power plants (3). In each type of system, the steam is used to drive turbines and generate electricity as outlined in Figure 4.3. In fact, once the steam has been generated, it makes no difference what the source is. Power plants run on coal, oil, or gas all use the steam in essentially the same way.

In the *pressurized water reactor* (PWR), the water again acts as coolant and moderator, but it is kept under pressure so that it does not boil as it circulates past the core. Instead, the steam generator is separated from the core. By keeping the coolant water pressurized, the operating temperatures can be much greater and the efficiency of the setup is improved. This advantage is largely offset by the expense of the additional component of the system. A major application of the PWR has been the nuclear submarine, which benefits from the compact size available with the PWR. Equally important to the submarine is the absence of a demand for oxygen, which is required to power all fossil fuel generators. Conventional submarines must surface frequently to take on a supply of air.

In a *gas cooled reactor* (GCR), the setup is very similar to a PWR except that the gas acts only as a coolant. Therefore, a moderator (usually graphite) must be included in the core.

The LMFBR is the most sophisticated design to be considered here. The coolant is liquid sodium. Recalling that fast neutrons are required for productive

Figure 4.8   The interior of the pressurized water reactor nuclear power plant at the Tennessee Valley Authority's Sequoia nuclear station is shown. The reactor itself, located at the bottom of the cavity in the center of the photograph, is bordered by four steam generators and a pressurizer. Each steam generator is 67 feet high and has a diameter of 14 feet. The twin reactor system at this site is rated at 2.3 million kilowatts. (Courtesy of Westinghouse Electric Corp.)

Figure 4.9   A nuclear submarine at sea. (Courtesy of the General Electric Company.)

bombardment of ^{238}U, this design does not include a moderator. Liquid sodium is pumped through the left-hand loop (heat exchanger) and the intermediate loop. The latter transfers heat to the steam generator. Since sodium reacts violently with water, the sodium flowing through the core is isolated from the water in the steam generator. In this way, even if the system does spring a leak, the nuclear core is protected from the site of the problem. In addition, the liquid sodium, which flows through the core, becomes intensely radioactive via neutron bombardment. The second loop isolates the radioactivity from the water supply passing through the steam generator.

SAFETY PROBLEMS

One concern over the safety of a nuclear power plant is the danger of meltdown of the reactor core if the coolant system fails. This is an important and very controversial subject. The following comments are a synopsis of an excellent article written on the subject by R. Philip Hammond (4).

> There would seem to be two major problems that could develop in the event of a meltdown. Although a nuclear explosion is not possible, for reasons cited earlier, the reactor core could attain temperatures too hot to contain it. Under such circumstances, the core would also melt and become free to sink. One possible consequence of this has been described as the "China Principle" in which it is imagined that the nuclear core would go clear through the earth and come out in China. More realistically, it has been suggested that the core would sink into the earth perhaps 30 or 40 feet. In this location, the radioactive sample is well shielded and recovery operations can be conducted safely. During the meltdown some more easily vaporized radioactive materials might be released but they would represent only a small fraction of the fission products (4).

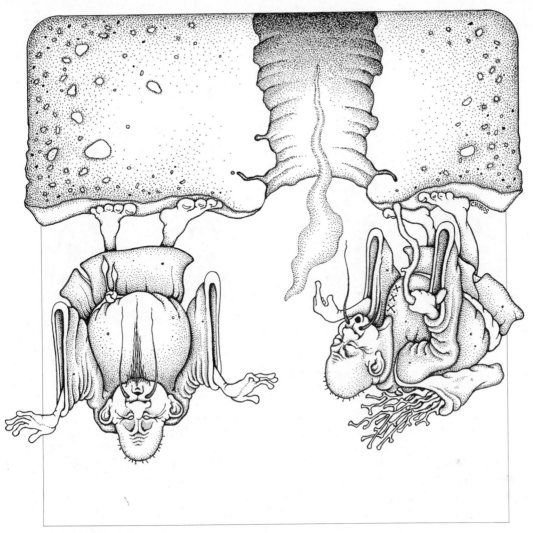

The China Principle

There is also concern over the possibility of a steam explosion when the molten core comes into contact with water in the system. Such an explosion might be expected to disperse radioactive starting materials and fission products over a very wide area.

According to Hammond, an explosion of this type seems an "incredible" contradiction. The contradiction is simply that a runaway reactor is not possible as long as the water (coolant) remains available, so that a meltdown could only occur if the coolant system is not available and in this situation, the water would not be available for a steam explosion.

Another major concern is the problem of disposal of radioactive waste products formed during the fission process (5). The fact that disposal problems have been divided into two groups serves to emphasize the magnitude of the problem (6). Relatively short-term radioactivity arises from medium atomic

Figure 4.10   Steam turbine for generating electricity at the Carolina Power & Light Company's Brunswick plant. The 821 megawatt turbine uses steam from a boiling water reactor to drive a generator that is located beyond the wall at the top of the photograph. The turbine section shown is 117 feet long. The generator is on the same shaft and extends for another 59 feet. The S-shaped structures on either side of the turbine are insulated steam pipes and valves. (Courtesy of the General Electric Company.)

weight fission products of uranium and plutonium. Of these, strontium-90, cesium-137, and krypton-85 present the major problem. They have half-lives of 29 years, 30 years, and 10.6 years respectively, so that in 700 years less than one ten-millionth of the radioactivity remains.

More important is the other group of radioactive waste products formed by neutron absorption by the original fuel. That is, not all of the ^{235}U or other fissionable fuel does actually undergo fission. The products formed by this neutron absorption are called *actinides* (actinium, thorium, uranium, neptunium, plutonium, etc.). They are all very toxic and have extremely long half-lives. For example, plutonium-239, which is formed in both light water reactors and breeder reactors, has a half-life of 25,000 years, so that adequate waste disposal is of great importance.

About 75% of the radioactive waste is reportedly stored at Hanford, Washington, and in June 1973, a leak, later found to have been occurring for 51 days prior to its discovery, was found in one of the storage tanks (7). The accident and the failure in observing the 115,000 gallon leak for so long was later attributed to aging tanks, primitive monitoring technology, managerial laxity, and human error on the part of the privately owned company that was contracted to attend to the waste material. Even more unfortunate is the 422,000 gallons estimated loss between August 1958 and June 1973 from several other tanks. No matter where the blame lies, these incidents serve to emphasize the magnitude of the problem of handling radioactive waste materials.

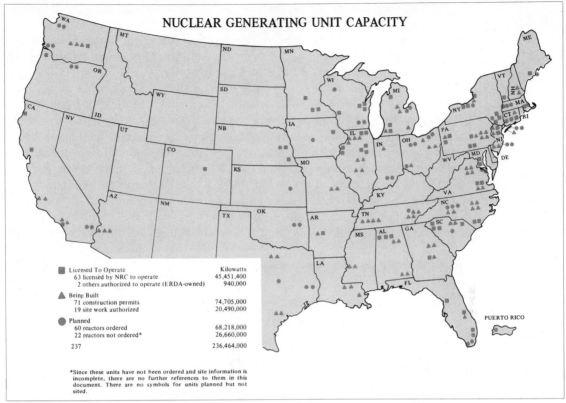

Figure 4.11 Locations of existing and planned nuclear power plants in the United States as of December 1976. (Courtesy of the Energy Research and Development Administration.)

On the brighter side is the opinion that the burden of radioactive waste material will actually become an asset in about 500 years. It has been noted that the gamma ray emissions from fission products will have mostly decayed away and the remaining hazard will come from easily shielded alpha particle emissions from the actinides, which may then be separated out and used as reactor fuels or for other applications (8).

The use of fusion processes for power production is far in the future. By comparison with standard production techniques, it has the obvious advantages of neither consuming fossil fuels nor releasing atmospheric pollutants. In addition, there is no danger of explosion. However, there is the need to deal with high energy neutrons and radioactive tritium (a weak β emitter).

FURTHER OBSERVATIONS

In the period between 1974 and 1985, it is expected that the demand for electric power will increase by about 50 percent. Although coal is becoming increasingly important, there are severe pollution problems that must be dealt with in using many types of coal. At the same time, the use of oil and gas is decreasing due to shortages and rising costs.

Consequently, at the present time, an electric power company planning construction of facilities to generate additional electric power, or to replace existing facilities, has only two options—coal and nuclear. The United States is fortunate to have some very large reserves of coal, but many countries around the world do not and for them, nuclear is the only option.

There are possibilities for the use of solar energy, geothermal energy, and for the use of gas and liquid fuels obtained from coal, but none of these will be available in the near future or, if they are, the costs may be prohibitive. On the other hand, the technology for using nuclear power is here now and unless a particular power company has a large supply of inexpensive coal, it will find that nuclear is the most economical. Therefore, it is often the only real option, particularly in poorer countries of the world. But even in this country, economic pressures cannot be overlooked. Although the government should and does oversee the construction and operation of nuclear power facilities, the investment and operation of the facilities is in the hands of private industry, and most people

"Of course it's perfectly safe. Any accident would be in complete violation of the guidelines established by the Nuclear Regulatory Commission."

Copyright © 1976. Reprinted by permission of S. Harris.

agree that is where it belongs for a number of reasons, including the economics.
However, as long as private industry has the responsibility, it is going to be
bound by the economics and we have to face the fact that nuclear power is here
now and the development of nuclear facilities is advancing to an even greater extent. It was recently estimated that some 38 countries outside the United States
have a total of 260 nuclear power reactors either operating, under construction,
or on order (9). This is not to condone the trend or to overlook the serious problems associated with nuclear power. It is merely to acknowledge the realities
of the times in which we live.

THE ATOMIC BOMB

Last, and hopefully least important, is the story of the nuclear bomb. If nothing
else, an understanding of the nuclear bomb serves to help appreciate the relative
safety that is potentially available in commercial applications of nuclear power.

Perhaps two characteristics of a nuclear weapon best illustrate the unique
design that is capable of such violent energy release. One such property is the
purity of the fissionable material (^{235}U or ^{239}Pu).* Highly concentrated (approximately 97%) ^{235}U is necessary to avoid the absorption of neutrons by ^{238}U so that
repeated bombardment of the fissionable isotope is an efficient and rapidly self-
generating process.

The other major feature of a nuclear bomb is achievement of a **critical mass**
of fissionable material. The critical mass is simply the weight of isotope required
for the reactor to achieve the supercritical condition in which more than one product neutron becomes a bombarding neutron.

The concept of critical mass can easily be understood by considering a
hypothetical spherically-shaped sample of fissionable material. Using the well-
known relationships,

$$\text{volume of a sphere} = 4/3\pi r^3$$
$$\text{surface area of a sphere} = 4\pi r^2,$$

it can readily be seen that as r increases (corresponding to a greater mass), the volume will increase faster than the surface area. Therefore the ratio, volume/surface
area, increases as the sample size increases. When the surface area is large, product neutrons can escape quite easily from the surface of the sphere and not again
become involved in the reaction. However, as the volume of the sphere becomes
larger, a greater percentage of the product neutrons cannot escape and therefore
become bombarding neutrons. It is easy to see that when a certain critical mass is
reached, the reaction can become rapidly self-perpetuating and lead to a nuclear
explosion.

In order to achieve a critical mass at exactly the right instant but not before, it is necessary to design a bomb that has the capability of forming a critical
mass but does not have such a mass up to the moment of detonation. For this purpose there are two designs, illustrated in Figure 4.12.

* The bomb dropped on Hiroshima was a uranium bomb, whereas a plutonium bomb was used at
Nagasaki.

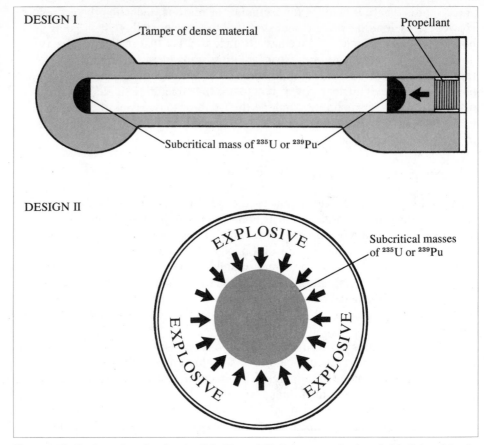

Figure 4.12 Designs of nuclear bombs (10). From C.F. Behrens et al., *Atomic Medicine,* 5th ed., (Baltimore: The Williams and Wilkins Company, 1969).

In the first design, a propellant (small explosive) displaces the subcritical mass at one end of the system very rapidly into the subcritical mass on the other end, producing a larger critical mass that may explode.

In the second design, a nonnuclear explosive is detonated causing an implosion of the subcritical mass in the fissionable core (^{235}U or ^{239}Pu). Although the mass is not changed, the decrease in the surface area of the core decreases the probability that a neutron may escape from the core while, at the same time, increasing the probability that each product neutron will become a bombarding neutron. The volume also decreases. In fact the volume/surface area (V/SA) decreases. This suggests that the sample is less likely to become critical since it appears that the amount of sample is decreased. However, while the size is decreased, the amount of fissionable material is unchanged and the success of the design results from lower surface area plus closer packing of the fuel. In effect, the density of the core is increased by the implosion so that the product neutrons are more likely to bombard other atoms. Thus, the implosion of the fissionable core creates the conditions required for a supercritical process.

The so-called hydrogen bomb is a device that operates by a fusion process as outlined in Chapter Three. The extreme conditions required for the fusion process are commonly achieved by using a fission device to trigger the bomb.

OTHER APPLICATIONS

Other aspects of nuclear chemistry will be presented in later chapters. These include: agriculture (Chapter Nine), insect control (Chapter Ten), food preservation (Chapter Fourteen), and antibiotic production (Chapter Eighteen).

SUMMARY

There are many variations in the design of nuclear power plants. Of these, the light water reactor (LWR) and the liquid metal fast breeder reactor (LMFBR) are the most popular. The probability (cross section) that a nucleus will capture a neutron determines whether an unmoderated (fast) or a moderated (thermal) reactor design is used. Among the other major variables are coolant and moderator.

The problems of safety to the population and the environment are a major concern in nuclear power generation. A nuclear power plant is clearly not a potential nuclear bomb but the release of radioactivity, e.g., from a meltdown, is an ever-present danger, although the disposal of radioactive wastes may be the most nagging problem of all.

PROBLEMS

1. Give a definition or example of each of the following:
   a) coolant
   b) moderator
   c) specific heat
   d) critical mass
   e) boiling water reactor
   f) pressurized water reactor
   g) heavy water reactor
   h) cross section

2. Compare the relative advantages of light and heavy water
   a) as a coolant
   b) as a moderator

3. In order to achieve sustained fission of ^{235}U, a moderator must be present. Why?

4. The letters LMFBR may very well become increasingly important anytime you turn on lights, applicances, etc. What do the letters stand for and how does each component of the system, described in the name, differ from other alternate systems?

5. Explain the two designs of atomic bombs shown in Figure 4.12.

6. Two samples of ^{239}Pu have radii of 5 centimeters and 25 centimeters. Calculate volume/surface area for each sample and explain why one sample is more likely to be suitable for use in an atom bomb.

REFERENCES

1. Krieger, J.H. "All Parties Must Get Moving On Fast-Breeder Development." *Chemical and Engineering News,* May 3, 1971, p. 29.
2. Metz, W.D. "Uranium Enrichment: U.S. One Ups European Centrifuge Effort." *Science 183* (1974): 1270.

3. Lyerly, R.L., and Mitchell, W. *Nuclear Power Plants*. Washington, D.C.: U.S. Atomic Energy Commission, 1969, pp. 9–13.

4. Hammond, R.P. "Nuclear Power Risks." *American Scientist 62* (1974): 155.

5. Farney, D. "Ominous Problem: What to do with Radioactive Waste." *Smithsonian 5* (1974): 20.

6. Kube, A.S., and Rose, D.J. "Disposal of Nuclear Wastes." *Science 182* (1973): 1205.

7. Gillette, R. "Radiation Spill at Hanford: The Anatomy of an Accident." *Science 181* (1973): 728.

8. Cohen, B.L. "Radioactive Waste Disposal." *Science 184* (1974): 746.

9. Abelson, P.E. "A Global Rush Toward Nuclear Energy." *Science 191* (1976): 901.

10. Behrens, C.H., King, E.R., and Carpender, J.W.J. *Atomic Medicine,* 5th ed. Baltimore, Md.: Williams and Wilkins, 1969, Chapter 2.

CHEMICAL PRINCIPLES: REVIEW AND PREVIEW

P A R T   T W O

The material presented in this section is intended for frequent reference as it relates to coverage in later chapters. The coverage is aimed toward those who have had no prior exposure to chemistry other than the material presented in Part One, although even the more experienced student may want to consult this material to check the structure of some complex carbohydrate or other compound. The names and symbols of some simple and combination cations and anions, many of which have been mentioned in Part One or appear in later chapters, are presented in the following tables and on the inside back cover for easy reference. Two additional tabulations appear on the inside front and back covers. The first is called the periodic table and lists the elements in order of increasing atomic number (the number of protons). The second tabulation is alphabetical but gives the same in-

COMMON SIMPLE IONS

ELEMENT	COMMON SIMPLE ION	NAME OF ELEMENT	NAME OF ION (IF DIFFERENT FROM NEUTRAL ATOM)
H	H^+	hydrogen	proton
Li	Li^+	lithium	
O	O^{-2}	oxygen	oxide
F	F^-	fluorine	fluoride
Na	Na^+	sodium	
Mg	Mg^{+2}	magnesium	
Al	Al^{+3}	aluminum	
S	S^{-2}	sulfur	sulfide
Cl	Cl^-	chlorine	chloride
K	K^+	potassium	
Ca	Ca^{+2}	calcium	
Fe	Fe^{+2} Fe^{+3}	iron	ferrous or iron (II) ferric or iron (III)
Cu	Cu^{+1} Cu^{+2}	copper	cuprous or copper (I) cupric or copper (II)
Zn	Zn^{+2}	zinc	
Br	Br^-	bromine	bromide
Rb	Rb^+	rubidium	
Sr	Sr^{+2}	strontium	
Ag	Ag^+	silver	
Cd	Cd^{+2}	cadmium	
Sn	Sn^{+2} Sn^{+4}	tin	stannous or tin (II) stannic or tin (IV)
I	I^-	iodine	iodide
Cs	Cs^+	cesium	
Ba	Ba^{+2}	barium	
Pb	Pb^{+2}	lead	
Ra	Ra^{+2}	radium	

formation as the periodic table. The ions listed in the following tables are listed in the order in which the elements (the central atom of the combination ions) appear in the periodic table.

COMMON COMBINATION IONS

ELEMENT	COMBINATION ION	NAME OF ION
H	OH^-	hydroxide
C	HCO_3^-	hydrogen carbonate or bicarbonate
	CO_3^{-2}	carbonate
N	NH_4^+	ammonium
	NO_3^-	nitrate
	NO_2^-	nitrite
P	$H_2PO_4^-$	dihydrogenphosphate
	HPO_4^{-2}	monohydrogenphosphate
	PO_4^{-3}	phosphate
S	HSO_4^-	hydrogen sulfate or bisulfate
	SO_4^{-2}	sulfate
	HSO_3^-	hydrogen sulfite or bisulfite
	SO_3^{-2}	sulfite
Cl	OCl^-	hypochlorite
Cr	CrO_4^-	chromate
	$Cr_2O_7^{-2}$	dichromate
Mn	MnO_4^-	permanganate
As	AsO_4^{-3}	arsenate
Br	OBr^-	hypobromite
	BrO_3^-	bromate
	BrO_4^-	perbromate

S ALTS. The cations and anions in the tables on pages 80 and 81 may be combined to form uncharged salts, some of which were mentioned in Part One (e.g., sodium iodide and sodium phosphate). The exact formula of a salt can be determined by consulting the tables. For example, aluminum chloride would form by combination of Al^{+3} and 3 Cl^- to give $AlCl_3$, which has a net charge of zero. Additional examples are given below, including salts of combination ions, which are handled by putting the entire combination ion in parentheses when it occurs more than once in a compound (e.g., calcium nitrate). In Part One it was not necessary to be concerned about detailed structures of the salts that we encountered since we were dealing only with changes in the nuclei of the atoms and were able to ignore the surrounding atoms and electrons. In spite of this, you will find on reviewing Part I, that almost every isotope described was a part of a complex molecule or ion. Although we sometimes encounter pure elements, we will generally find elements combined with other elements in compounds or ions.

$H^+ + Cl^- = HCl$	hydrogen chloride
$2\,K^+ + S^{-2} = K_2S$	potassium sulfide
$2\,Na^+ + CO_3^{-2} = Na_2CO_3$	sodium carbonate
$Ba^{+2} + SO_4^{-2} = BaSO_4$	barium sulfate
$Zn^{+2} + HPO_4^{-2} = ZnHPO_4$	zinc monohydrogenphosphate
$Ca^{+2} + 2\,NO_3^- = Ca(NO_3)_2$	calcium nitrate
$Cd^{+2} + 2\,HCO_3^- = Cd(HCO_3)_2$	cadmium bicarbonate
$2\,Al^{+3} + 3\,SO_4^{-2} = Al_2(SO_4)_3$	aluminum sulfate
$3\,NH_4^+ + PO_4^{-3} = (NH_4)_3PO_4$	ammonium phosphate
$2\,NH_4^+ + HPO_4^{-2} = (NH_4)_2HPO_4$	ammonium monohydrogenphosphate
$NH_4^+ + H_2PO_4^- = NH_4H_2PO_4$	ammonium dihydrogenphosphate
$Sn^{+4} + 4\,NO_3^- = Sn(NO_3)_4$	stannic nitrate or tin (IV) nitrate

THE PERIODIC TABLE

In Chapter One, the periodic table of the elements was introduced as a source of information for checking or predicting the products of certain nuclear reactions. In the present coverage, the periodic table will be considered in greater detail with

CHEMICAL BONDING:
THE PERIODIC TABLE

two goals in mind. The first is to provide sufficient background for an understanding of some of the information presented in the tables of simple ions and combination ions. Without this understanding, it would be necessary to have such tables constantly available for reference or to memorize all of the information. After repeated exposure to many of the common ions, one automatically commits some of the information to memory and this is very helpful. However, if one encounters some of the less common elements, there must be some means for determining the kind of information that is given in these tables in order to understand the chemistry involved.

The second reason for considering the periodic table is to understand bonding, which is the major goal of this chapter.

The history of the periodic table dates back to 1869, when the Russian chemist Dmitri Mendeleev and the German chemist Lothar Meyer recognized that certain elements exhibit similar physical and chemical properties. Furthermore, they recognized that when the elements are arranged in order of increasing atomic weight, those elements with very similar properties showed up periodically at reg-

TABLE 5.1   THE PERIODIC TABLE

Key:
- atomic number: 1
- symbol of the element: H
- atomic weight: 1.01†

Period	I A	II A	III B	IV B	V B	VI B	VII B	VIII			I B	II B	III A	IV A	V A	VI A	VII A	VIII A
1																		2 He 4.00
2	3 Li 6.94	4 Be 9.01											5 B 10.81	6 C 12.01	7 N 14.01	8 O 16.00	9 F 19.00	10 Ne 20.18
3	11 Na 23.00	12 Mg 24.31											13 Al 26.98	14 Si 28.09	15 P 30.97	16 S 32.06	17 Cl 35.45	18 Ar 39.95
4	19 K 39.10	20 Ca 40.08	21 Sc 44.96	22 Ti 47.90	23 V 50.94	24 Cr 52.00	25 Mn 54.94	26 Fe 55.85	27 Co 58.93	28 Ni 58.71	29 Cu 63.55	30 Zn 65.37	31 Ga 69.72	32 Ge 72.59	33 As 74.92	34 Se 78.96	35 Br 79.90	36 Kr 83.80
5	37 Rb 85.47	38 Sr 87.62	39 Y 88.91	40 Zr 91.22	41 Nb 92.91	42 Mo 95.94	43 Tc* 98.91	44 Ru 101.07	45 Rh 102.91	46 Pd 106.4	47 Ag 107.87	48 Cd 112.40	49 In 114.82	50 Sn 118.69	51 Sb 121.75	52 Te 127.60	53 I 126.90	54 Xe 131.30
6	55 Cs 132.91	56 Ba 137.34	57 La 138.91	72 Hf 178.49	73 Ta 180.95	74 W 183.85	75 Re 186.2	76 Os 190.2	77 Ir 192.22	78 Pt 195.09	79 Au 196.97	80 Hg 200.59	81 Tl 204.37	82 Pb 207.2	83 Bi 208.98	84 Po* [210]	85 At* [210]	86 Rn* [222]
7	87 Fr* [223]	88 Ra* 226.02	89 Ac* [227]	104 Rf* [261]	105 Ha* [262]													

Period 1: 1 H 1.01†

Lanthanides	58 Ce 140.12	59 Pr 140.91	60 Nd 144.24	61 Pm* [147]	62 Sm 150.4	63 Eu 151.96	64 Gd 157.25	65 Tb 158.93	66 Dy 162.50	67 Ho 164.93	68 Er 167.26	69 Tm 168.93	70 Yb 173.04	71 Lu 174.97

Actinides	90 Th* 232.03	91 Pa* 231.04	92 U* 238.03	93 Np* 237.05	94 Pu* [244]	95 Am* [243]	96 Cm* [247]	97 Bk* [247]	98 Cf* [251]	99 Es* [254]	100 Fm* [257]	101 Md* [258]	102 No* [255]	103 Lr* [256]

*All isotopes are radioactive.

†All atomic weights have been rounded to 0.01.

[] Indicates mass number of longest known half-life.

ular intervals. This periodic behavior, or *periodicity,* is the basis for the periodic table of the elements.

This kind of thinking seems rather trivial at this point in time, but in 1869 only about 60 elements were known and the concept of atomic number was not yet understood. Nevertheless, the observations made by Mendeleev and Meyer are the basis for the modern version of the periodic table in which one acknowledges the dramatic similarities among certain groups of elements when they are listed in order of increasing atomic number as shown in Figure 5.1. To cite just two examples, the elements in group IA are all very reactive, metallic substances, which exist as ions with a plus one charge (e.g., Na^+, K^+) when combined with other elements. At the other extreme, the group VIIIA elements are normally inert substances that exist as gases and are rarely found in combination with other elements.

Within the periodic table, the vertical columns are called **groups**; the horizontal rows are called **periods**. It is the elements within each of the groups that exhibit similar characteristics, but it is the length of each of the periods that allows us to predict when one element may be similar to another. This periodicity is summarized in Table 5.2 and can be confirmed by simply counting across each of the periods. Elements 57–71 and 89–103 are regarded as parts of the sixth and seventh periods, respectively. They are tabulated separately for convenience and because of certain structural similarities within each series. In our consideration of the elements, we will concentrate on the elements in groups IA-VIIIA. The transition elements and the elements in the lanthanide (atomic number 57–71) and actinide (atomic number 89–103) series are more complex and will not be considered here.

Hydrogen exhibits certain features of both the group IA and VIIA elements, and yet its properties are very different from both of these groups. In fact, many sources include a periodic table with hydrogen appearing in group IA or group VIIA, but it is shown separately here.

In order to appreciate the periodic behavior of the elements, we must now consider the role of electrons in an atom since it is the arrangement of the electrons that is the primary factor in determining the properties of each element. It is the properties of the elements in group VIIIA that open up the entire picture. As previously noted, these elements have little tendency to combine with any other elements. In fact, they are commonly known as the *noble* or *inert gases*. The current interpretation of this behavior is that each of these elements has a stable

TABLE 5.2   THE PERIODICITY OF ELEMENTS IN THE PERIODIC TABLE

PERIOD	NUMBER OF ELEMENTS
1	2
2	8
3	8
4	18
5	18
6	32
7	32

A noble gas has little tendency to combine with others

arrangement of electrons and, therefore, has little tendency to give up electrons (to form a cation) or accept electrons (to form an anion). These inert gases are said to have a closed-shell structure. At one time, it was popular to describe these shells as circular orbits symmetrically arranged about the nucleus (see Chapter One for examples). However, the electrons are not arranged in circular orbits. The three-dimensional arrangements are actually quite complex and yet the circular orbits and other similar representations, such as the "partial circle designation," which follows, for the sodium atom and ion and for the chlorine atom and

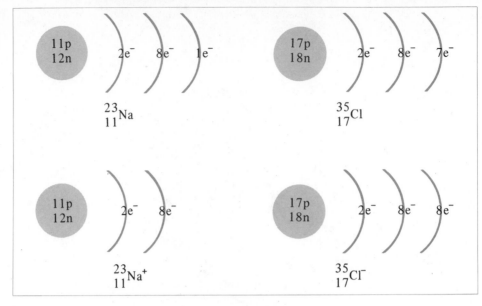

chloride ion, are still in use. Each representation describes a series of electronic levels arranged in order of increasing energy, with the first two electrons in the lowest energy (most stable) level, and the remaining electrons in higher energy levels. In addition, each level has a certain maximum capacity for electrons and the order in which these levels are filled is quite complex. Fortunately, when dealing with elements in the groups numbered IA-VIIIA, the arrangement of the electrons is quite simple since all of the elements have an electronic arrangement in which each of the inner levels is filled, while the outer shell contains the number of electrons indicated by the group number. For example, sodium (Na) appears in group IA and has one electron in its outer shell. Chlorine (Cl) is in group VIIA and has 7 electrons in its outer shell. These outer-shell electrons are commonly called **valence electrons**.

The noble gases are in group VIIIA. Except for helium, the atoms of the noble gases all have 8 valence electrons and, in view of the inert (very stable) nature of the noble gases, it is clear that the magic number is eight—an *octet* of electrons. In fact, this finding is the basis of the well-known **octet rule**, which says that any element is most stable when it has a set of eight valence electrons. In this state, an element has a structure that is electronically equivalent to a noble gas.

Let us take a closer look at the sodium and chlorine atoms as they combine to form sodium chloride, which consists of Na^+ and Cl^-. As sodium becomes a cation, it loses its one valence electron and attains an electronic arrangement equivalent to that found in the noble gas, neon (Ne). Similarly, the chlorine atom achieves the electronic structure of argon (Ar) by gaining one electron and forming an ion—the chloride ion.

Since both of the examples given in the previous paragraph constitute chemical changes in the elements, it is easy to see how important the valence electrons are in determining the chemical properties, i.e., the chemical reactions, of the various elements.

Since the noble gases neither take on nor give off valence electrons under normal circumstances, it is rather common to say that they have no valence electrons. Either way, zero valence electrons or eight valence electrons, both represent the same stable structure of the atom.

Helium is an exception to the octet rule since the first shell of electrons has a maximum capacity of only two electrons. Nevertheless, the noble gas structure is very stable as evidenced by the fact that Li^+ is the stable ion of lithium, because it has the same electronic structure as He.

Ions, such as Li^+, Na^+, and K^+, are termed **monovalent cations** to signify the loss of one valence electron. The elements in group IA are termed the **alkali metals** and all readily form monovalent cations. Ions such as Li^{+2}, Na^{+2}, and K^{+2} do not exist since none conform to a noble-gas structure.

The elements in group IIA are called the **alkaline earth metals**. When ionized, they exist as divalent cations. Examples are Mg^{+2}, Ca^{+2}, and Ba^{+2}.

The group VIIA elements are called the **halogens** and they are commonly found as **monovalent anions** in compounds such as sodium chloride (NaCl), potassium bromide (KBr), and calcium chloride ($CaCl_2$).

THE ELECTRON DOT SYMBOLISM

Perhaps the simplest way to predict what sort of ion is most likely to form is to draw an *electron dot formula*. This is illustrated for the elements in the second

$$Li\cdot \qquad Be\cdot \qquad \cdot\overset{\cdot}{B}\cdot \qquad \cdot\overset{\cdot}{\underset{\cdot}{C}}\cdot \qquad \cdot\overset{\cdot}{\underset{\cdot}{N}}\cdot \qquad :\overset{\cdot\cdot}{\underset{\cdot}{O}}\cdot \qquad :\overset{\cdot\cdot}{\underset{\cdot\cdot}{F}}\cdot \qquad :\overset{\cdot\cdot}{\underset{\cdot\cdot}{Ne}}:$$

period. In this representation, only the valence electrons are included and we see once again that the number of valence electrons corresponds to the group number of each of these elements. It should also be noted that there are four distinct subshells available for placement of the valence electrons. The electrons remain separated when possible, but must pair up for the elements in groups VA-VIIIA.

Thus, we can readily predict how many electrons must be lost or gained by each element in order to achieve an inert gas structure.

COVALENT BONDING

Up to this point, we have focused attention on the elements on the left and right sides of the periodic table. We shall continue to restrict attention to the elements in groups IA-VIIIA, but let us consider elements such as carbon and nitrogen, which illustrate another type of behavior when they are found in combination with other atoms. The electron dot structures are as follows:

$$\cdot\overset{\cdot}{\underset{\cdot}{C}}\cdot \qquad\qquad \cdot\overset{\cdot\cdot}{N}\cdot$$
$$\text{group IVA} \qquad\quad \text{group VA}$$

In the case of carbon, it would seem that the carbon atom must either gain or lose 4 electrons in order to achieve an inert gas structure. Both possibilities are highly unlikely. If an atom loses electrons to form a cation, it is more difficult to remove each successive electron because the increasing positive charge causes the atom to have a steadily greater attraction for electrons. The same sort of problem arises as an atom takes on electrons to become an anion and becomes increasingly more negative. This negative charge tends to repel additional electrons. The gain or loss of one or two electrons occurs fairly readily to form monovalent and divalent ions, but ions with a higher valence are less common.

Instead of forming ions, some atoms achieve an inert gas structure in an alternate way—by formation of **covalent bonds**, as illustrated for methane (CH_4). In

$$\cdot \overset{\displaystyle\cdot}{\underset{\displaystyle\cdot}{C}} \cdot + 4H \cdot \longrightarrow H : \overset{\displaystyle\cdot\cdot}{\underset{\displaystyle\cdot\cdot}{C}} : H = H - \overset{\displaystyle H}{\underset{\displaystyle H}{C}} - H$$

methane

this process, the carbon atom effectively gains four electrons and, thus, achieves the electronic structure like that in neon. The lines between the atoms in the second representation of methane each represent two electrons and depict covalent bonds holding the atoms together. Thus, the carbon atom in methane is tetravalent and the carbon and hydrogen atoms are found to *share* the bonding electrons.

In the case of the nitrogen atom, there is a need for three additional electrons to achieve an inert gas structure. Consequently, three hydrogen atoms can enter to form the ammonia molecule, in which the nitrogen atom is trivalent. Here again, the nitrogen and hydrogen atoms *share* valence electrons so that each atom achieves an inert gas structure.

$$\cdot \overset{\displaystyle\cdot\cdot}{N} \cdot + 3H \cdot \longrightarrow H : \overset{\displaystyle\cdot\cdot}{\underset{\displaystyle H}{N}} : H \; = \; H - \overset{\displaystyle\cdot\cdot}{\underset{\displaystyle H}{N}} - H$$

VALENCE

In previous sections, the term *valence* has been used several times and explained by example. There are actually two ways to look at valence. For those compounds that form ions, valence describes the charge on the cation or anion. Thus, for calcium chloride ($CaCl_2$), a divalent cation combines with a monovalent anion in a one to two ratio. For covalent compounds, valence refers to the number of atoms or groups of atoms attached to a particular atom. However, in both instances, valence is the combining capacity of an atom, and this more general definition is applicable to both ionic and covalent compounds.

One can also refer to the valence of the various atoms in a salt such as sodium nitrate, $NaNO_3$, which is made up of a sodium cation and a nitrate anion. The combination anion consists of a nitrogen atom surrounded by three oxygen atoms, as represented by

Each of the bonds between the nitrogen and oxygen atoms is a pair of electrons that is being shared by both atoms. In the nitrate anion, the combination of atoms has an excess of one electron and is negatively charged, but in addition to the ionic attraction between Na^+ and NO_3^- in sodium nitrate, there are the strong covalent bonds between the nitrogen and the oxygen atoms, which result from the sharing of electrons.

The same kind of covalent bonding is present in many other compounds such as water (H_2O) and hydrogen fluoride (HF). These are not ionic compounds. They contain covalent O—H and F—H bonds as shown below.

$$\overset{\cdot\cdot}{\underset{H \quad H}{O}} \qquad :\overset{\cdot\cdot}{\underset{\cdot\cdot}{F}}-H$$

In Chapter Six, we will begin to look at organic compounds, in which covalent bonds predominate.

SUMMARY

Formulas can be drawn for ionic compounds by knowing the charges on the isolated cations and anions. The periodic table may be used to assist in determining the charges on the ions. The periodic recurrence of properties of elements can be traced to their electronic structures, with the great stability of the inert gases serving as the point of reference in the arrangement of the electrons. The partial circle designation, the electron dot symbolism, and the octet rule can be used to emphasize the importance of the valence electrons in determining the ability of individual atoms to form ions or to bond covalently to other atoms.

PROBLEMS

1. Write formulas for each of the following compounds. Consult the table of simple ions and combination ions when necessary (pp. 80 and 81).

   a) hydrogen chloride
   b) sodium hydroxide
   c) potassium sulfide
   d) lithium oxide
   e) sodium carbonate
   f) potassium sulfate
   g) barium sulfate
   h) ammonium nitrate
   i) zinc monohydrogenphosphate
   j) strontium nitrate
   k) calcium nitrate
   l) magnesium iodide
   m) cadmium bicarbonate

   n) sodium bisulfite
   o) aluminum sulfate
   p) potassium hydroxide
   q) ammonium phosphate
   r) ammonium monohydrogenphosphate
   s) ammonium dihydrogen phosphate
   t) cupric sulfate
   u) tin (IV) nitrate
   v) potassium bromate
   w) sodium nitrite
   x) lead chloride
   y) stannous fluoride
   z) silver chloride

2. Give definitions or examples of each of the following:
   a) valence
   b) covalent bond
   c) period
   d) group
   e) octet rule
   f) inert gas
   g) alkali metal
   h) electron dot formula

3. Give an example of each of the following:
   a) monovalent cation
   b) divalent cation
   c) trivalent cation
   d) monovalent anion
   e) divalent anion
   f) trivalent anion

4. The inertness of the group VIIIA elements was a major key to the electronic arrangement of atoms. Explain.

5. What is the charge on the ions formed by the group IIA elements?

6. Why do the group VIIA elements all have similar properties?

7. What inert gas configuration is found in each of the ions in the following compounds?
   a) LiI
   b) $CaBr_2$
   c) Na_2Te

8. Explain the relationship between the group number and the valence of elements in groups IA-VIIIA.

9. Answer each of the following questions for the second and fourth periods in the periodic table.
   a) How many elements appear in the period?
   b) How many electrons are found in the group VIA element in the period?
   c) How many valence electrons are found in the group VIA element?
   d) What is the group VIA element?
   e) Give the formula for the sodium salt* of the group VIA element.
   f) Draw the electron dot symbol for the group VIA element.
   g) Draw the electron dot symbols for the sodium salt of the group VIA element.
   h) What is the valence of the ion of the group VIA element?
   i) Is the octet rule adhered to in the group VIA element or ion?
   j) Give the formula for the carbon compound† of the group VIIA element.
   k) Draw the electron dot symbol for the carbon compound of the group VIIA element.

* The sodium salts of the group VIA elements are ionic.
† The carbon compounds of the group VIIA elements are covalent.

C OVALENT bonds are the major feature of compounds containing carbon. There is no simple ion of carbon but it does exist in some important combination ions such as bicarbonate and carbonate, both of which are derived from carbonic acid (H_2CO_3).

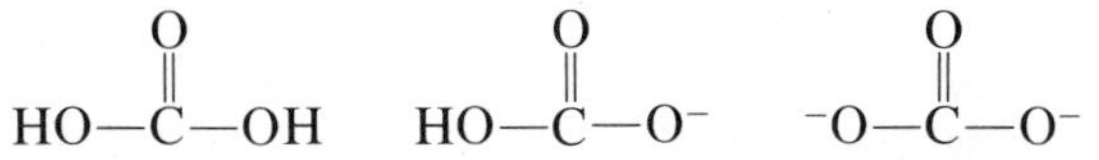

$$\underset{\substack{\text{carbonic}\\\text{acid}}}{HO-\overset{\displaystyle O}{\overset{\|}{C}}-OH} \qquad \underset{\substack{\text{bicarbonate}\\\text{ion}}}{HO-\overset{\displaystyle O}{\overset{\|}{C}}-O^-} \qquad \underset{\substack{\text{carbonate}\\\text{ion}}}{{}^-O-\overset{\displaystyle O}{\overset{\|}{C}}-O^-}$$

In each of the species shown above, the carbon-oxygen bonds are covalent. All are regarded as inorganic compounds, as are all of the other salts mentioned previously. More often, carbon appears in so-called organic compounds, which are characterized by having covalent carbon-hydrogen bonds. Even more important, the vast majority of organic compounds contain carbon-carbon bonds, which may be single, double, or even triple bonds as shown in the following three examples.

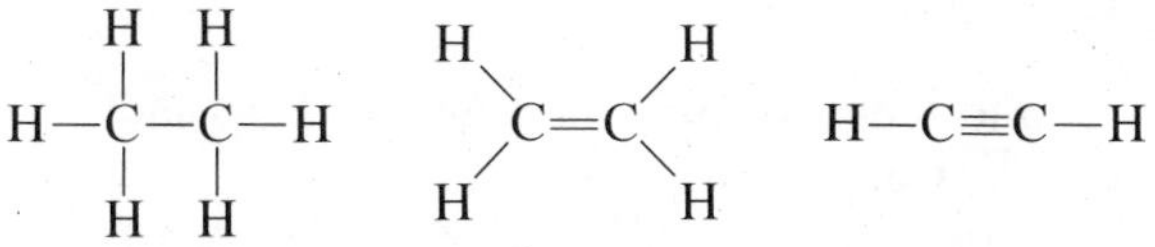

$$\underset{\text{ethane}}{H-\overset{\overset{\displaystyle H}{|}}{\underset{\underset{\displaystyle H}{|}}{C}}-\overset{\overset{\displaystyle H}{|}}{\underset{\underset{\displaystyle H}{|}}{C}}-H} \qquad \underset{\text{ethylene}}{\overset{\displaystyle H}{\underset{\displaystyle H}{}}C=C\overset{\displaystyle H}{\underset{\displaystyle H}{}}} \qquad \underset{\text{acetylene}}{H-C\equiv C-H}$$

In these and all other stable compounds of carbon, *the magic number is four,* which means that carbon almost always appears with four covalent bonds and is said to be tetravalent. In this arrangement every carbon atom obeys the octet rule. Even in carbon dioxide, there are four covalent bonds.

$$O=C=O$$
carbon dioxide

Organic chemistry is a major area of chemistry. Proteins and nucleic acids (DNA) are compounds held together by a wide variety of covalent bonds, including carbon-carbon bonds. The early meaning of the distinction between inorganic and organic compounds was that the latter were produced by living organisms. In current usage, the term *organic* refers more generally to all compounds containing carbon bonded to hydrogen and other elements, whether or not the compound is associated with any living organism.

ORGANIC CHEMISTRY

THE HYDROCARBONS

The simplest group of organic compounds are those containing only carbon and hydrogen, known as **hydrocarbons.** The simplest hydrocarbon contains only a single carbon, and recalling the magic number 4, the formula must be CH_4. The formula also takes into account that hydrogen is a **monovalent atom,** i.e., forms only one covalent bond at any time. Some of the other major atoms found in

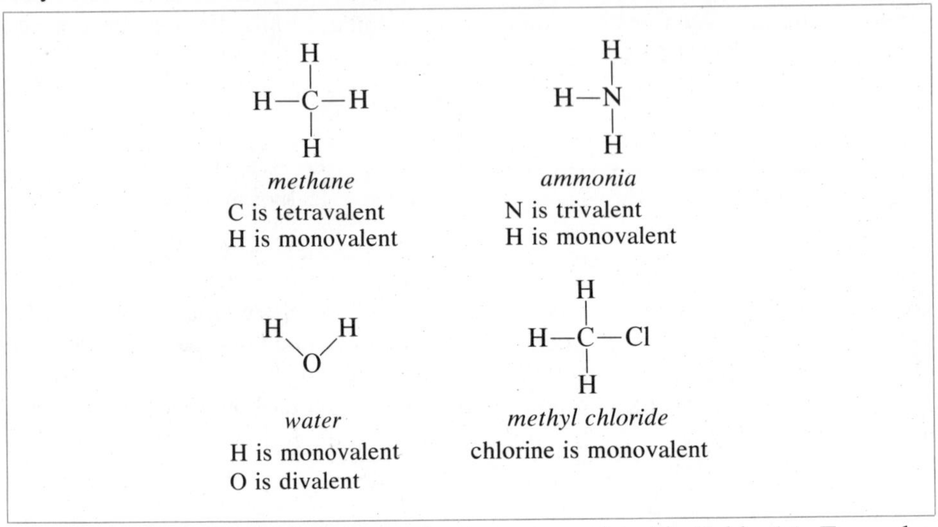

methane
C is tetravalent
H is monovalent

ammonia
N is trivalent
H is monovalent

water
H is monovalent
O is divalent

methyl chloride
chlorine is monovalent

organic compounds and their normal valences are listed in Table 6.1. Examples are methane, ammonia, water, and methyl chloride.

Noting the proper valence, the series of hydrocarbons continues with **ethane, propane, butane,** etc. which have the structures shown, and since a se-

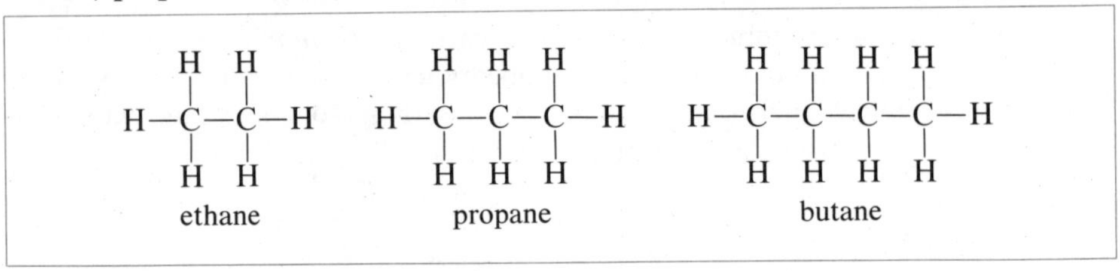

ethane

propane

butane

TABLE 6.1   THE NORMAL VALENCE OF ATOMS COMMONLY FOUND IN ORGANIC COMPOUNDS

ELEMENT	NORMAL VALENCE
H	1
C	4
O	2
N	3
X (F, Cl, Br, I)	1
S	2
P	3 or 5

quence such as C—H—C would make hydrogen divalent, the structures with carbon chains are the only ones possible for the two-, three-, and four-carbon compounds. The only variation possible among these three compounds is isobutane. This compound is an **isomer** (another form) of butane, which has the same formula (C_4H_{10}) as the form with a linear chain of carbon atoms.

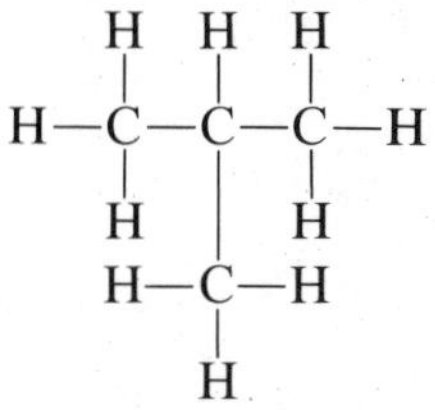

isobutane

As the number of carbon atoms increases to five, the number of possible isomers becomes three. The five-carbon hydrocarbons are called **pentanes.**

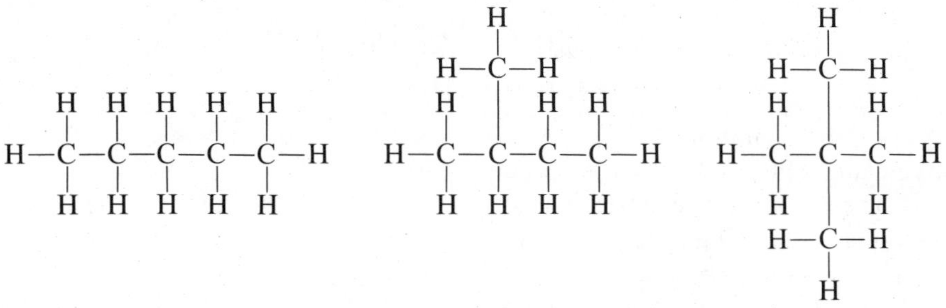

The six-carbon hydrocarbons are called **hexanes** and they exist in five isomeric forms (see below and page 94).

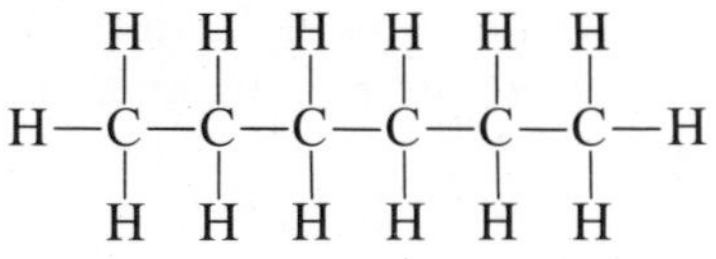

I

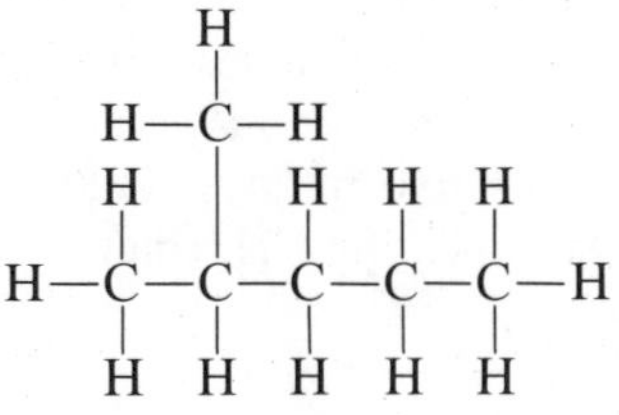

II

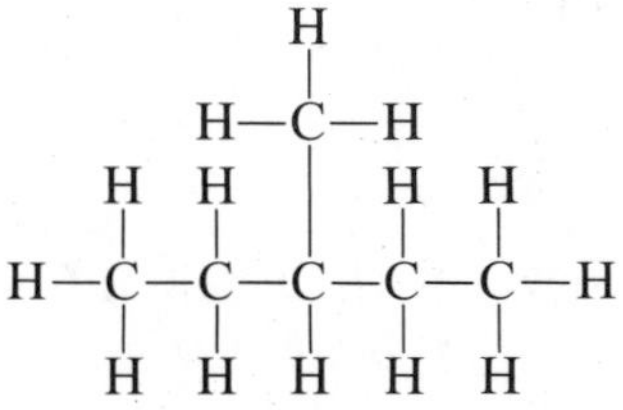

III

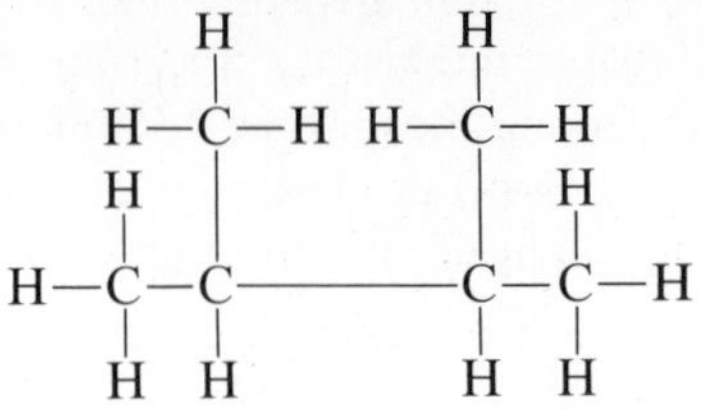

Since these formulas become rather unwieldy, several shorthand notations are in use. These are illustrated for compounds I-V.

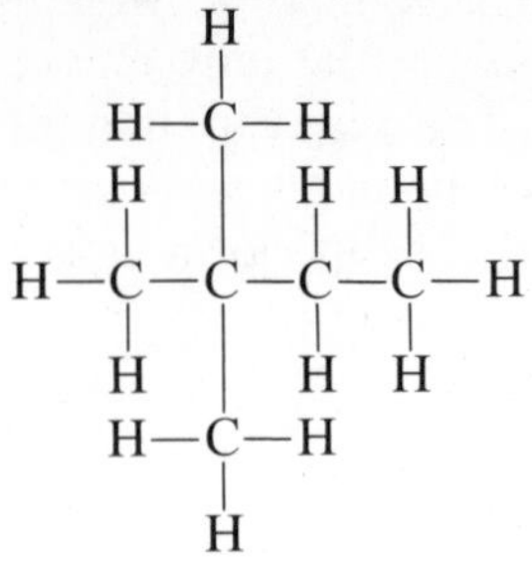

COMPOUND I:

(a) $C-C-C-C-C-C$

(b)

(c) $CH_3CH_2CH_2CH_2CH_2CH_3$

(d) $CH_3(CH_2)_4CH_3$

The first representation shows only the carbon skeleton and while it is the most convenient, it is quite incomplete and will not be used.

The second abbreviation is also very convenient and is frequently used for compounds containing long carbon chains. The end of each line in this abbreviation represents a carbon atom. Recalling that carbon is tetravalent, it should be noted that this abbreviation assumes that the two terminal carbon atoms each have three hydrogens, and the four carbon atoms within the chain each have two hydrogens.

The third abbreviation is called a *single-line formula* and is more cumbersome, but it is complete with all the hydrogens. Actually it is not really that difficult if one recognizes that it is merely the carbon skeleton with hydrogens added to make each carbon tetravalent.

The fourth formula is actually an abbreviated form of the third abbreviation in which the repeating CH_2 (methylene) groups are lumped together and counted.

COMPOUND II:

(a) 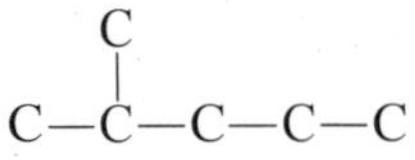 $\begin{array}{c} \quad\;\; C \\ \quad\;\; | \\ C-C-C-C-C \end{array}$

(b)

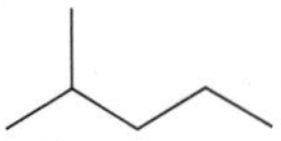

(c) $\begin{array}{c} \quad\;\; CH_3 \\ \quad\;\; | \\ CH_3CHCH_2CH_2CH_3 \end{array} = CH_3CH(CH_3)CH_2CH_2CH_3 = (CH_3)_2CHCH_2CH_2CH_3$

(d) $\begin{array}{c} \quad\;\; CH_3 \\ \quad\;\; | \\ CH_3CH(CH_2)_2CH_3 \end{array}$

For this compound, the pattern is similar except that it has a CH_3 (methyl) group as a branch off the main chain. The only thing new here is the alternate forms of the third abbreviation. The branched methyl group may be placed in parentheses following the carbon to which it is attached, or it may be lumped together with the methyl group at the end of the chain, since both of these groups are attached to the same carbon in the chain. It should be noted that

pentavalent carbon

$(CH_3)_2CHCH_2CH_2CH_3 \neq CH_3CH_3CHCH_2CH_2CH_3$

trivalent carbon

since the structure on the right has two impossible arrangements of carbon atoms.

Quite obviously there is potential for confusion in the use of parentheses but the confusion can be avoided altogether if one remembers to "think four" for carbon. However, to avoid all confusion, parentheses will only be used to abbreviate chains of repeating methylene groups. Any methyl or larger branches will be shown as they appear in the first abbreviation under (c) in Compound I and below.

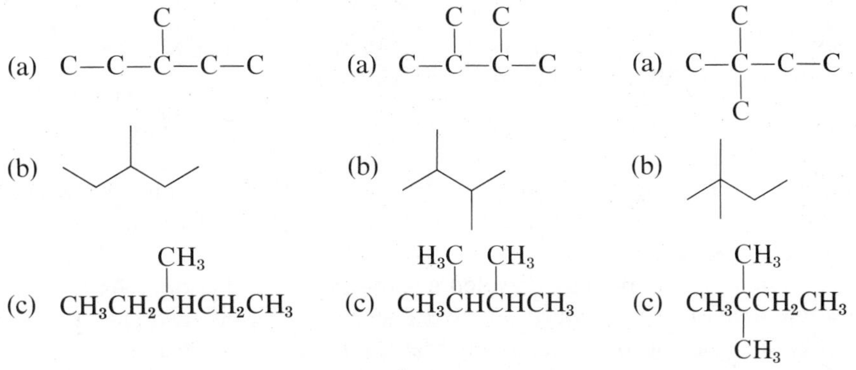

COMPOUND III:	COMPOUND IV:	COMPOUND V:

COMPOUND III:

(a)
$$\begin{array}{c} C \\ | \\ C-C-C-C-C \end{array}$$

(b)

(c) $CH_3CH_2CHCH_2CH_3$ with CH_3 branch

(d) none possible

COMPOUND IV:

(a)
$$\begin{array}{c} C \quad C \\ | \quad | \\ C-C-C-C \end{array}$$

(b)

(c) $CH_3CHCHCH_3$ with H_3C and CH_3 branches

(d) none possible

COMPOUND V:

(a)
$$\begin{array}{c} C \\ | \\ C-C-C-C \\ | \\ C \end{array}$$

(b)

(c) $CH_3CCH_2CH_3$ with CH_3 above and CH_3 below

(d) none possible

Obviously as the number of carbon atoms increases, the number of isomers increases dramatically, when all the patterns of branching are taken into account.

As for the straight-chain isomers, consult Table 6.2 for the names and boiling points. It can be seen that an unbranched isomer such as $CH_3(CH_2)_3CH_3$ is named as *normal* pentane or simply *n*-pentane. The branched isomers have more complex names, which are discussed later.

It can be seen from Table 6.2 (page 96) that there is a regular increase in the boiling point as the chain length increases. There is also a steady increase in melting point with a few irregularities. Nevertheless, it can be seen that the early members of the series are gases even far below room temperature,* whereas the last few entries in the table are solids at room temperature.

Branching causes a lowering of both the boiling and melting points. For ex-

* 20°C = 68°F.

TABLE 6.2   LINEAR HYDROCARBONS

FORMULA	NAME	BOILING POINT(°C)	MELTING POINT(°C)
CH_4	methane	−162	−183
CH_3CH_3	ethane	−88	−172
$CH_3CH_2CH_3$	propane	−42	−187
$CH_3(CH_2)_2CH_3$	n-butane	0	−138
$CH_3(CH_2)_3CH_3$	n-pentane	36	−130
$CH_3(CH_2)_4CH_3$	n-hexane	69	−95
$CH_3(CH_2)_5CH_3$	n-heptane	98	−91
$CH_3(CH_2)_6CH_3$	n-octane	126	−57
$CH_3(CH_2)_7CH_3$	n-nonane	151	−54
$CH_3(CH_2)_8CH_3$	n-decane	174	−30
$CH_3(CH_2)_{10}CH_3$	n-dodecane	216	−10
$CH_3(CH_2)_{12}CH_3$	n-tetradecane	252	6
$CH_3(CH_2)_{14}CH_3$	n-hexadecane	280	18
$CH_3(CH_2)_{16}CH_3$	n-octadecane	308	28
$CH_3(CH_2)_{18}CH_3$	n-eicosane	343	37
$CH_3(CH_2)_{20}CH_3$	n-docosane	369	44
$CH_3(CH_2)_{22}CH_3$	n-tetracosane	391	54
$CH_3(CH_2)_{28}CH_3$	n-triacontane	450	66

ample, branched butane (isobutane) has a B.P. of −12°* and a M.P. of −159° compared to 0° and −138°, respectively, for n-butane.

Petroleum

Crude oil is a complicated mixture of hydrocarbons, including the linear, branched, and certain other variations such as cyclic and unsaturated hydrocarbons (discussed later).

A major step in refining petroleum is fractional distillation, which separates the components of the mixture according to boiling point with the low boiling (volatile) components being removed first. Not everything distills, but the high boiling residue is used in making paraffin, lubricating greases (petroleum jelly), and asphalt.

The major fractions and their uses are shown in Table 6.3.

It can be seen from Table 6.3 that there is an overlap among the various fractions. This gives the refiner flexibility in deciding how much of each fraction to separate, so that he can maximize his profits in response to demand for various fractions. Even more flexibility is available due to **cracking.** This is a process whereby the refiner can heat a fraction, such as the kerosene fraction, using a suitable catalyst and cause the relatively long chains to break into smaller fragments with lower boiling points. In this way, gasoline can be produced from kerosene. The cracking process is illustrated by the cracking of butane, which produces ethylene and hydrogen. The ethylene is not used as a fuel but it is in demand as a starting material for the production of polyethylene.

$$CH_3CH_2CH_2CH_3 \xrightarrow[\text{high temperature}]{\text{catalyst}} 2\,CH_2{=}CH_2 + H_2$$

* All degrees Celsius.

TABLE 6.3 PETROLEUM FRACTIONS

FRACTION	USUAL RANGE OF HYDROCARBONS	APPROXIMATE BOILING RANGE (°C)	USES
Gas	$C_1 - C_4$	below 20°	fuel, plastics
Petroleum ether	$C_5 - C_6$	30 − 60°	solvents
Ligroin	$C_6 - C_8$	60 − 120°	solvents
Gasoline	$C_5 - C_{12}$	40 − 205°	fuel
Kerosene	$C_{12} - C_{18}$	175 − 325°	diesel fuel, jet fuel, home heating
Lubricating oil	C_{17}-higher	above 300°	lubricants (greases)
Residue	C_{20}-higher	above 350°	asphalt

In addition to simple cracking, a process known as *catalytic reforming* may be carried out. This process tends to convert some long-chain hydrocarbons into branched isomers, which improves their efficiencies for use as gasoline. The efficiency is described quantitatively by the *octane rating* scale, which arbitrarily gives *n*-heptane a rating of zero and isooctane* a rating of 100. In other words, *n*-heptane and other straight-chain hydrocarbons knock very badly when used alone as gasoline, whereas the highly-branched isomers burn very smoothly. Consequently, catalytic reforming can improve the octane rating dramatically since some very highly branched isomers may have octane ratings greater than 100. On the other hand, *n*-octane has an octane rating of − 19. Tetraethyllead,

* The word *isooctane* is an unsystematic name used only in this context.

Figure 6.1   An important consumer of diesel fuel. (Courtesy of J. Bohning.)

Figure 6.2   A catalytic cracker. (Courtesy of Mobil Oil Corp.)

Figure 6.3 An instrument for measuring octane rating. (Courtesy of Waukesha Engine Division, Dresser Industries, and ASTM).

$Pb(CH_2CH_3)_4$, is traditionally the most popular additive that can be used to accomplish the same purpose, although automobile engines are presently being engineered to use gasolines without lead in order to minimize pollution.

$$CH_3CCH_2CHCH_3$$

isooctane
octane rating = 100

$$CH_3CH_2CH_2CH_2CH_2CH_2CH_3$$

n-heptane
octane rating = 0

Natural Gas

The major component (ca. 95%) of natural gas is methane. Lesser amounts of the C_2—C_4 hydrocarbons are also present. Because of the difference in boiling point, a propane-butane fraction is removed by cooling under pressure, which causes the propane and butane to liquify. The propane-butane fraction is then compressed into cylinders and sold as *bottled gas*.

100 Alkanes, Alkenes, and Alkynes

Thus far, we have concentrated on hydrocarbons with only carbon-carbon single bonds. These compounds are called **alkanes.** Alternative forms in which hydrocarbons can exist are called **alkenes** and **alkynes,** which signify the presence of carbon-carbon double and triple bonds, respectively. Compounds with double or triple bonds are termed **unsaturated,** whereas compounds with only single bonds are termed **saturated.** The heavily advertised polyunsaturated cooking oils contain many carbon-carbon double bonds.

The simplest compounds in the alkene and alkyne series are ethylene and acetylene—both of which are very important commercially.

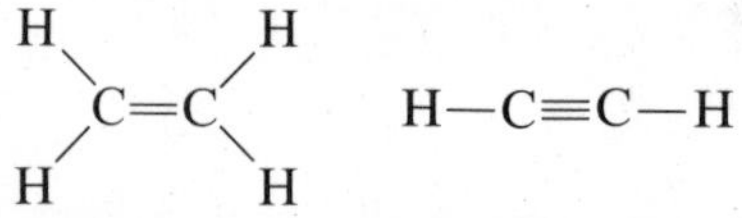

ethylene   acetylene

As the formula indicates, each of the carbons in ethylene is sharing four valence electrons with the other carbon atom, so that each atom is tetravalent. The conversion from ethane (an alkane) to ethylene (an alkene) is accompanied by a loss of two hydrogen atoms from adjacent carbon atoms. The same is true for the conversion from ethylene to acetylene. In acetylene, each carbon is sharing six valence electrons with the other carbon atom.

The alkene structure is a particularly common one, which appears many times in later chapters and exhibits a special type of isomerism called *geometrical isomerism.* A compound such as dichloroethylene illustrates the phenomenon since there are three different forms which are known to exist, as shown below.

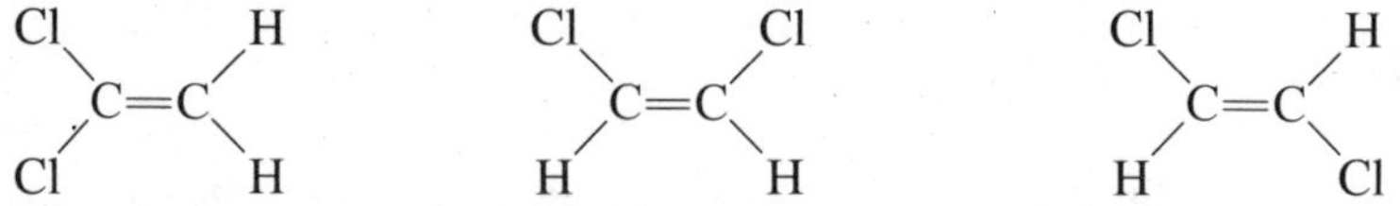

1,1-dichloroethylene   *cis*-1,2-dichloroethylene   *trans*-1,2-dichloroethylene

First there is the distinction between the 1,1 and 1,2 isomers in which the two chlorines are on the same or adjacent carbon atoms, and the carbons are numbered from left to right. This type of isomerism is the same as what has been observed for simple saturated hydrocarbons where branching can appear from any point in the chain. The two isomers termed *cis* and *trans* are **geometric isomers,** which result from the rigidity of a carbon-carbon double bond. The common ball and stick models for 1,2-dichloroethane and 1,2-dichloroethylene illustrate the important difference between alkanes and alkenes, which accounts for the appearance of geometric isomers of alkenes. The two arrangements of 1,2-dichloroethane are not regarded as isomers* because they may be interconverted by simple rotation about the carbon-carbon single bond. In the alkene, the second bond between the carbons restricts the rotation and prevents the interconversion. In other

* They are described as different *conformations* of the same isomer.

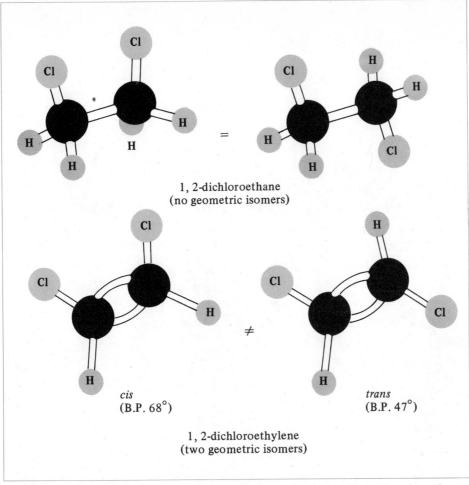

1, 2-dichloroethane
(no geometric isomers)

cis
(B.P. 68°)

trans
(B.P. 47°)

1, 2-dichloroethylene
(two geometric isomers)

words, one of the bonds between the carbons would have to be broken in order to permit rotation. Consequently, the *cis* and *trans* isomers are distinctly different compounds with different physical properties (e.g., boiling point). The term *cis* signifies that the two identical atoms or groups are on the same side of the double bond, whereas in the *trans* isomer, the two groups are on opposite sides.

Cyclic Hydrocarbons

Compounds such as cyclohexane and benzene are also found in petroleum. In fact, the cycloalkanes (called naphthenes) are particularly abundant in California

cyclohexane

benzene

The Cycloalkanes are particularly abundant in California petroleum.

petroleum. These compounds differ from the open-chain analogs in that the ends
are joined, but the physical and chemical properties are very similar.

FUNCTIONAL GROUPS

While hydrocarbons are important as fuels and in other direct applications, they
are also important starting materials for the synthesis of other organic chemicals
that differ from hydrocarbons due to the presence of atoms or groups of atoms
called **functional groups.** These groups give compounds physical and chemical
properties that are very different from the simple hydrocarbons. The list of possible compounds is large, but some of the oxygen- and nitrogen-containing functional groups are shown in Table 6.4. The symbol R is the common shorthand notation used to designate a radical, which may be obtained from any molecule by
removing a hydrogen. For example, if a hydrogen atom is removed from methane
(CH_4), the result is a methyl radical. This species has a trivalent carbon and must

combine with some atom, such as chlorine (Cl), to produce methyl chloride (CH_3Cl), in which the carbon is in the normal tetravalent form. The chlorine atom then acts as a functional group and gives the compound properties that are very different from simple methane. Once the methyl radical is attached to a chlorine or other functional group, it is usually referred to as a methyl group, although it is not a functional group.

An alternative is to combine the methyl radical with a functional group such as hydroxyl (OH) to form methyl alcohol.

For the sake of convenience, the organic chemist uses the symbol R in writing compounds ranging from very simple ones like methyl alcohol to very complex compounds such as cholesterol.

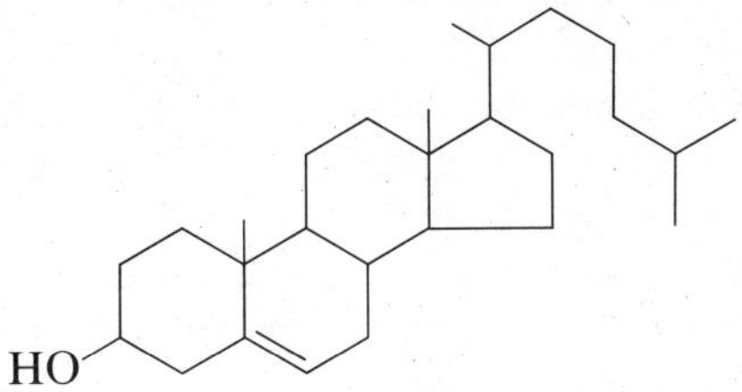

= ROH

cholesterol

TABLE 6.4   COMMON FUNCTIONAL GROUPS CONTAINING OXYGEN AND NITROGEN

GROUP	GENERAL FORMULA	TYPE OF COMPOUND
—OH	ROH	alcohol
—OR'	ROR'	ether
$-\overset{\text{O}}{\overset{\|}{\text{C}}}-\text{H}$	$\text{R}-\overset{\text{O}}{\overset{\|}{\text{C}}}-\text{H}$	aldehyde
$-\overset{\text{O}}{\overset{\|}{\text{C}}}-\text{R}'$	$\text{R}-\overset{\text{O}}{\overset{\|}{\text{C}}}-\text{R}'$	ketone
$-\overset{\text{O}}{\overset{\|}{\text{C}}}-\text{OH}$ (—COOH)	$\text{R}-\overset{\text{O}}{\overset{\|}{\text{C}}}-\text{OH}$ (RCOOH)	acid
$-\overset{\text{O}}{\overset{\|}{\text{C}}}-\text{OR}'$	$\text{R}-\overset{\text{O}}{\overset{\|}{\text{C}}}-\text{OR}'$	ester
$-\overset{\text{O}}{\overset{\|}{\text{C}}}-\text{NH}_2$	$\text{R}-\overset{\text{O}}{\overset{\|}{\text{C}}}-\text{NH}_2$	amide
—O—O—H	R—O—O—H	hydroperoxide
$-\text{NH}_2$	$\text{R}-\text{NH}_2$	amine
$-\text{NH}_2 + -\text{COOH}$	$\underset{\underset{\text{NH}_2}{\|}}{\text{RCH}}-\text{COOH}$	amino acid
$-\text{C}\equiv\text{N}$	$\text{R}-\text{C}\equiv\text{N}$	nitrile
$-\text{N}=\text{C}=\text{O}$	$\text{R}-\text{N}=\text{C}=\text{O}$	isocyanate

NOMENCLATURE

The systems used for naming organic compounds are quite complex. Since a variety of organic compounds will be encountered in later chapters, let us consider a few aspects of the most systematic approach used for naming—the IUPAC system. The initials IUPAC stand for the International Union of Pure and Applied Chemistry, which is an organization that has formulated rules for naming organic compounds. Typical IUPAC names are illustrated and discussed below.

$$CH_3CH_2CH_2CH_2CH_2CH_3 \qquad \text{IUPAC name: hexane}$$

The compound above may be called either *n*-hexane or simply hexane. According to the IUPAC rules, the first step is to locate and identify the longest continuous carbon chain and name it. The second step is to locate and identify any branches or functional groups. Since the compound above has neither, the name is simply hexane under the IUPAC system. The name *n*-hexane is the common name.

The second step in formulating a IUPAC name is illustrated in the following compounds:

$$\underset{\displaystyle CH_3CH_2CH_2CH_2\overset{\textstyle \overset{\textstyle CH_3}{|}}{C}HCH_3}{} \qquad \text{2-methylhexane}$$

$$\underset{\displaystyle CH_3CH=CHCH_2\overset{\textstyle \overset{\textstyle CH_3}{|}}{C}HCH_3}{} \qquad \text{5-methyl-2-hexene}$$

$$CH_3CH=CHCH_2\overset{\overset{\textstyle CH_3}{|}}{C}HCH_2\overset{\overset{\textstyle O}{\|}}{C}-OH \qquad \text{3-methyl-5-heptenoic acid}$$

In the first compound, a methyl branch appears. It is identified and its position is located by numbering the atoms in the six-carbon chain. Thus, the methyl is located on the second carbon (C—2), which is identified in the name. The numbering is done from right to left in this example in order to arrive at the smallest number for the branch. In other words, if the numbering were done from left to right, the name would be 5-methylhexane, which would be incorrect.

In the second compound, there are three features to be described by the name—the longest chain, the double bond, and the branch. The name 2-hexene covers the first two. The double bond lies between C—2 and C—3 and is located by identifying the first carbon in the chain that is part of the double bond. Once again the numbering is done from the end that gives the smaller number, i.e., the end closer to the double bond. The compound is also identified as a hexene rather than a hexane because the "ene" ending signifies a carbon-carbon double bond. Finally, the methyl group at carbon 5 is identified and located. It would be logical, but incorrect, to name this compound as 2-methyl-4-hexene, because the IUPAC rules require numbering from the end closer to the double bond. This may seem silly and even contradictory, but it must be remembered that the system must be uniform and has to work for a wide variety of compounds and, while there may appear to be certain inconsistencies, the IUPAC system works very well.

The third compound has an additional complication, which is the acid group. In this case, the numbering is started from the acid group.

Many other compounds will be given and named in later chapters. In some cases, the IUPAC name will be given and in others a common name will be used, but for the most part, we need not be too concerned about the origin of either type of name, since the goal of this section is to provide you with enough information to be able to understand how organic compounds are named, rather than to be able to formulate names for all of the very complex molecules that we may encounter.

POLYMERS

Of all of the types of known organic compounds, polymers are among the most important to the consumer. Within the last 40 years, we have been bombarded with an ever-increasing list of products, including plastics, packaging materials, and synthetic fibers, in which organic polymers are the basic ingredient. Chemical terms, such as polyesters, polyurethanes, acrylics, and vinyls, and brand names, such as Orlon, Acrilan, Dacron, Plexiglas, Lucite, Teflon, Saran, and Formica, are well known.

We have already considered two important types of polymers (DNA and proteins), and in Chapter Eight, we will consider the polymers (starch, cellulose, and glycogen) that are used for storage of carbohydrates in plants and animals.

Polymers are very large molecules, i.e., they have very high molecular weights, and are made up of long chains of repeating units, called *monomers*. The term *polymer* is derived from Greek: *poly* = many; *meros* = parts. In the case of proteins, the monomers are the 20 common amino acids. In DNA, there are four monomers, symbolized A, C, G, and T.

Polyethylene

Unsaturated compounds are also useful monomers. For example, ethylene can be polymerized to form polyethylene by treatment with the proper catalyst. The

ethylene   polyethylene

polymer consists of the units $-CH_2CH_2-$ repeating over and over to give giant molecules with weights in the hundreds, thousands, or even millions of atomic mass units.

A closer look at the polymerization process reveals the fact that only the double bond is participating in the reaction. In fact, it is only the second bond of the double bond that is affected. This is not surprising in view of the nature of a carbon-carbon double bond for, although the two bonds are usually pictured as being equivalent, they are actually very different. One important difference between the bonds is their strength. One of the bonds, called a sigma (σ) bond, is very strong. The other bond is called a pi (π) bond and is much weaker. It is the π bond that comes into play in polymerization.

If we recall that a covalent bond is formed by combining a pair of electrons, we can picture the ethylene molecule as follows:

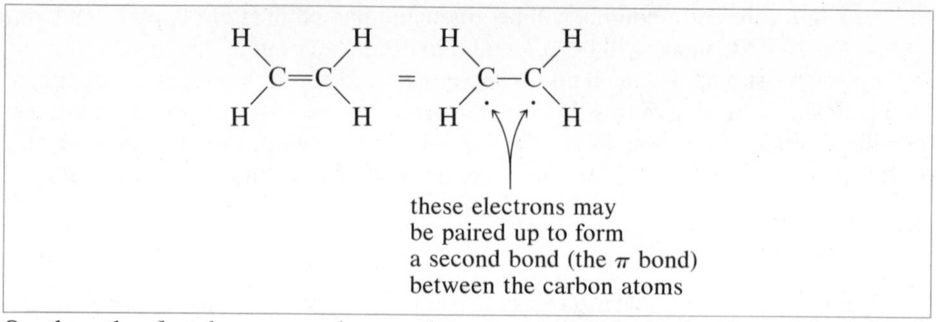

$$H_2C=CH_2 \quad = \quad H_2C-CH_2$$

these electrons may
be paired up to form
a second bond (the π bond)
between the carbon atoms

On the other hand, we can also see how two ethylene molecules might be persuaded to combine under the influence of a suitable catalyst. The same process

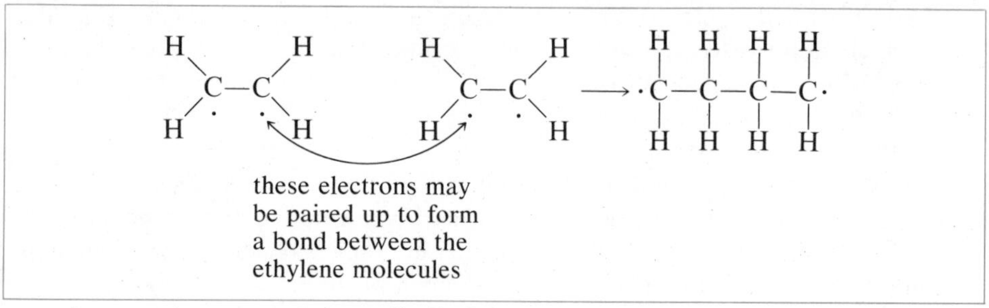

these electrons may
be paired up to form
a bond between the
ethylene molecules

may then continue on and on until all of the monomer is used up or the chain stops growing in one of several ways.

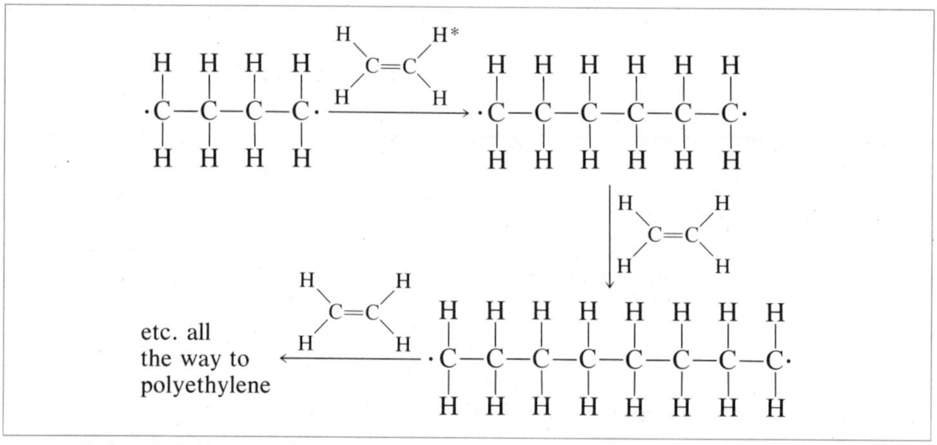

etc. all
the way to
polyethylene

Polyethylene is an important polymer, which sees considerable use in bags, bottles, pails, pipe, electrical insulation, and toys, depending on the method of processing.

Vinyl Polymers

Now that we have seen that only the double bond is involved in polymerization, it takes very little imagination to surmise what would happen if one or more of the

* This is a common shorthand notation that indicates a reaction between the compound to the left of the arrow and the compound shown alongside the arrow.

hydrogens in the ethylene molecule is replaced by a methyl group or a functional group, such as a chlorine atom.

If a hydrogen of ethylene is replaced by a methyl group, the monomer becomes propylene and the polymer becomes polypropylene.

propylene   polypropylene

Polypropylene is used in fibers, steering wheels, pipe, indoor-outdoor carpeting, plastic bottles (which can tolerate higher temperatures than polyethylene), and kitchenware.

If a hydrogen of ethylene is replaced by a chlorine atom, the monomer becomes chloroethylene, better known as vinyl* chloride, and the polymer becomes polyvinyl chloride (PVC). PVC is also used as plastic pipe, floor coverings, packaging, plastic raincoats, garden hose, auto seat covers, phonograph records, and in upholstery and luggage as a replacement for leather.

Another well-known polymer is Teflon, in which all of the hydrogens of ethylene are replaced by fluorine atoms, which render the polymer quite inert and stable, even at high temperatures.

tetrafluoroethylene   polytetrafluoroethylene
Teflon

Some of the other well-known vinyl polymers follow.

styrene   polystyrene
styrofoam insulation

vinylidene
chloride   poly(vinylidene chloride)
Saran

* The radical that forms when a hydrogen is removed from ethylene is called a *vinyl radical*.

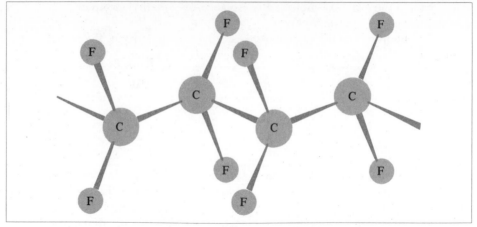

Figure 6.4   The Teflon polymer chain.

Figure 6.5   A familiar product made of Plexiglas. (Photo by James J. Kane, Jr.)

acrylonitrile → polyacrylonitrile fibers for rugs and clothing (Acrylan, Orlon)

methyl methacrylate* → poly(methyl methacrylate) hard contact lenses (Lucite, Plexiglas)

2-hydroxyethyl methacrylate* → poly(2-hydroxyethyl methacrylate) soft contact lenses

The last two of the vinyl polymers described provide an interesting contrast. Both are transparent plastics, but the physical properties of the two are quite different. As the name suggests, hard contact lenses are a hard, brittle plastic, which is often irritating to the eyes. On the other hand, the polymeric material found in the soft contact lenses absorbs a large quantity of water due to the addition of the OH group, which closely resembles the water molecule. Consequently, the soft contact lens acts like a film of water resting in the eye and usually causes very little irritation. In Chapter Seven, we will look at the reason for the hydrophilic (water-loving) property of the polymer used in soft contact lenses.

* The well-known acrylics are polymers formed from monomers that are are slight variations of the compound called *acrylic acid*.

Condensation Polymers

Polymers formed from ethylene or substituted ethylenes, i.e., the vinyls, are often called **addition polymers** because the monomer units simply add together to form long chains without any side-products. Furthermore, the functional groups that are sometimes present in the monomers do not normally participate in the polymerization reactions, although they do influence the properties of the polymers.

An entirely different kind of polymerization process occurs during the production of **condensation polymers,** in which functional groups react with one another to link the monomer molecules together to form polymeric chains. In addition, a small molecule (often water) is released when condensation polymers form.

Polyesters

A major class of condensation polymer is the *polyester.* A simple ester can be formed by combination of an alcohol (R—OH) and an acid $(R'—\overset{\overset{O}{\|}}{C}—OH)$. *

$$R—O\,\boxed{H} + \boxed{HO}\overset{\overset{O}{\|}}{—C}—R' \longrightarrow R'—\overset{\overset{O}{\|}}{C}—OR + H_2O$$

alcohol   acid   ester

A polyester will form if both the alcohol and the acid are *bifunctional* monomers as in the case of ethylene glycol and terephthalic acid. The fact that a

ethylene   terephthalic   dimer
glycol   acid

polymer forms is most easily appreciated by looking at the dimer (combination of two monomers), which forms in the first step of the polymerization. At the dimer stage, free alcohol and acid groups are still available for continued growth of the polymer chain. Due to the bifunctional nature of the monomers, the growing chain always has a functional group available at each end. The major applications of this particular polyester are as the fiber known as Dacron and in tire cords.

Thermoplastic and Thermosetting Resins

Polymers that are processed into useful end products are usually mixed with other ingredients, which impart special properties. The basic polymer component is re-

---

* The prime notation is used to distinguish between two different radicals, R and R'.

ferred to as a **resin.** The polymers that we have considered up to this point are called **thermoplastic** resins to signify that they can be heated and melted without being destroyed. For this reason, they can be molded into many different shapes and processed into many useful products. As we have seen, the major feature of thermoplastic polymers is the long linear chains of repeating units.

Another type of resin is the **thermosetting** polymer. This type of polymer is characterized by a complex three-dimensional, weblike structure, in which the individual polymer chains are interconnected. A thermosetting resin cannot be melted to allow molding into various shapes. Instead, it must be polymerized directly into its final shape.

A polyester can be a thermosetting resin if at least one of the monomers is trifunctional as in the example on page 112, in which two chains are shown growing, each consisting of five monomer units. The extra OH functional groups, which are located on each of the chains, make it possible for the diacid monomer to form a cross-link between the chains as shown.

The result is a thermosetting polyester. Thermosetting resins are very hard materials. This characteristic, coupled with their excellent insulating properties, makes them ideal for use as cases for radios and other electrical circuits.

Rubber

The **vulcanization** of rubber is another process that introduces cross-links between polymer chains. Natural rubber is an addition polymer with the following repeating unit:

$$\left(CH_2-\underset{\underset{CH_3}{|}}{C}=CH-CH_2\right)_n \qquad \textit{cis}\text{-polyisoprene (natural rubber)}$$

It is an elastic substance that becomes very sticky when hot. In 1839, Charles Goodyear perfected the process, which he called *vulcanization,* in which natural

$$-CH_2-\underset{\underset{}{\overset{\overset{}{CH_3}}{|}}}{C}=CH-CH_2-CH_2-\underset{\overset{\overset{}{CH_3}}{|}}{C}=CH-CH_2-$$

natural rubber

$$-CH_2-\underset{\underset{CH_3}{|}}{C}=CH-CH_2-CH_2-\underset{\underset{CH_3}{|}}{C}=CH-CH_2-$$

sulfur
heat, catalyst

$$-\underset{\underset{S}{|}}{CH}-\underset{\overset{\overset{}{CH_3}}{|}}{C}=CH-CH_2-CH_2-\underset{\overset{\overset{}{CH_3}}{|}}{C}=CH-\underset{\overset{}{\underset{S}{|}}}{CH}-$$

vulcanized rubber

$$-\underset{\underset{CH_3}{|}}{CH}-C=CH-CH_2-CH_2-\underset{\underset{CH_3}{|}}{C}=CH-CH-$$

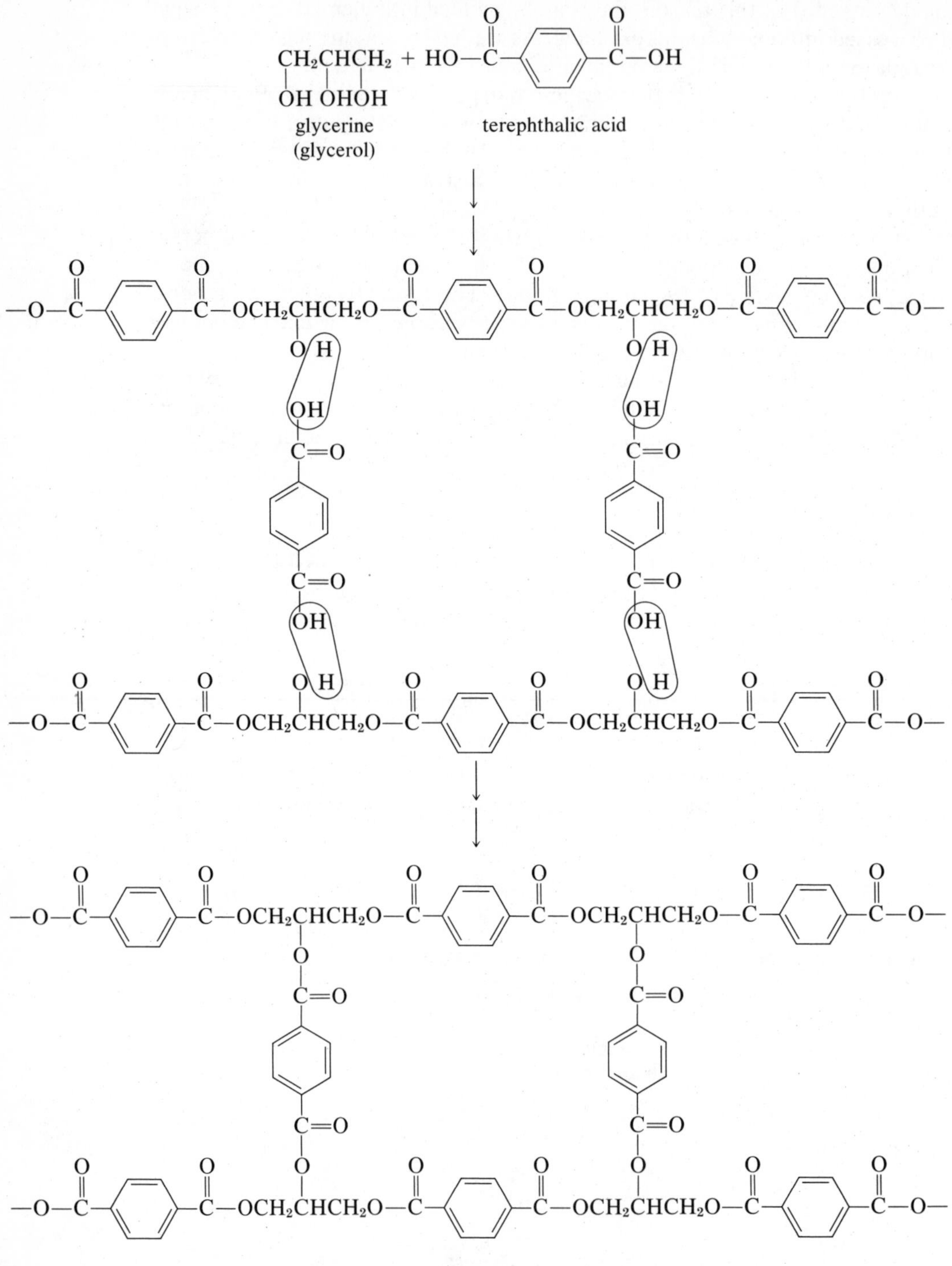
glycerine
(glycerol)
terephthalic acid

Figure 6.6 Charles Goodyear discovers the miracle of vulcanization. (Courtesy of Goodyear.)

rubber is treated with sulfur to form a network of cross-links. This makes the rubber harder and stronger and eliminates the tackiness of untreated rubber.

OTHER IMPORTANT POLYMERS

Although the list of possible polymeric materials is almost endless, let us conclude our look at polymer chemistry by considering three additional classes of polymers that are among the best-known and most widely used of all.

Figure 6.7 Plastic and vulcanized rubber bowling balls. (Courtesy of Brunswick Corporation; photo by James J. Kane, Jr.)

Figure 6.8 Steps In Tire Production. (a) Worker measures sulfur, which is required for vulcanization. (b) The uncured tire is placed in a curing mold and subjected to pressure and heat for a specified period of time. This causes the soft, gummy "green" tire to be transformed into a tough, long-wearing passenger tire. Simultaneously, the tire tread is impressed and the tire emerges ready for the final inspection. (c) Open molds show indentations that determine the design of the tread. (Courtesy of General Tire.)

Polyamides

One of the most important polymeric resins is *nylon,* which has seen a wide range of applications, including such things as fibers for clothing, stockings, and carpets and in tire cords. To the chemist, nylon is not a single polymer but is any one of many polymers, called **polyamides,** which are formed by condensation polymerization of a diacid and a diamine.

The most successful nylon is known as nylon 66, which is formed from two six-carbon monomers as follows:

$$HO-\overset{\overset{O}{\|}}{C}(CH_2)_4\overset{\overset{O}{\|}}{C}\boxed{-OH} + \overset{H}{\underset{H}{\diagdown}}N(CH_2)_6N\overset{H}{\underset{H}{\diagup}}$$

adipic
acid   hexamethylene diamine

$$\downarrow$$

$$HO-\overset{\overset{O}{\|}}{C}(CH_2)_4\overset{\overset{O}{\|}}{C}-\underset{\underset{H}{|}}{N}(CH_2)_6\underset{\underset{H}{|}}{N}-\overset{\overset{O}{\|}}{C}(CH_2)_4\overset{\overset{O}{\|}}{C}-\underset{\underset{H}{|}}{N}(CH_2)_6\underset{\underset{H}{|}}{N}- + n\,H_2O$$

nylon 66

Proteins are also polyamides, usually called *polypeptides,* in which the monomers are amino acids. Amino acids are unusual monomers for condensation

$$\overset{H}{\underset{H}{\diagdown}}N-\underset{\underset{R}{|}}{C}H-\overset{\overset{O}{\|}}{C}-OH \qquad \text{an amino acid}$$

polymers, because both functional groups are present in the same compound. This would seem to simplify matters except for the presence of the group represented by R, which may be any of 20 different groups. In other words, there are 20 different amino acids that are commonly found in proteins. They combine as shown

$$\overset{H}{\underset{H}{\diagdown}}N-\underset{\underset{R_1}{|}}{C}H-\overset{\overset{O}{\|}}{C}\boxed{-OH} + \overset{H}{\underset{H}{\diagdown}}N-\underset{\underset{R_2}{|}}{C}H-\overset{\overset{O}{\|}}{C}\boxed{-OH} + \overset{H}{\underset{H}{\diagdown}}N-\underset{\underset{R_3}{|}}{C}H-\overset{\overset{O}{\|}}{C}-OH + \text{etc.}$$

$$\downarrow$$

$$\overset{H}{\underset{H}{\diagdown}}N-\underset{\underset{R_1}{|}}{C}H-\overset{\overset{O}{\|}}{C}-\underset{\underset{H}{|}}{N}-\underset{\underset{R_2}{|}}{C}H-\overset{\overset{O}{\|}}{C}-\underset{\underset{H}{|}}{N}-\underset{\underset{R_3}{|}}{C}H-\overset{\overset{O}{\|}}{C}- \qquad \text{a protein}$$

above under the same conditions used to make synthetic polyamides, but proteins cannot be readily synthesized because it is necessary to assemble the amino acid units in exactly the right order in order to produce a biochemically functional pro-

tein. If one were to mix all of the necessary amino acids and allow the polymerization to occur, the polymer would have a random sequence of amino acid
units. R. Merrifield has developed an automated technique for synthesizing proteins, but it is a very tedious procedure in which the polymer chain is lengthened
by reaction with one amino acid at a time. Many days are required to complete the
synthesis of even very simple proteins. It was truly a monumental research accomplishment, but hardly one which suggests that it is practical to synthesize proteins on a commercial scale. When a protein is needed, it is usually isolated from
an animal or microorganism. For example, the protein, insulin, is routinely isolated from the pancreas of cattle.

Polyurethanes

The polyurethanes have become an extremely important class of polymers, particularly as foams in furniture, automobile safety padding, soles for shoes, pillows,
insulation, life preservers, and as fiber for carpets.

A polyurethane is a sort of combination polyester and polyamide. It is like
most condensation polymers since it results from the reaction of the functional
groups in a diisocyanate and a diol (or triol) but, unlike other condensation
polymers, no water or other small molecule is liberated in the process.

$$CH_2CH_2 + O{=}C{=}N{-}(CH_2)_4{-}N{=}C{=}O$$
$$\underset{OH\ \ OH}{|\ \ \ |}$$

ethylene glycol   tetramethylene diisocyanate

$$\downarrow$$

$$CH_2CH_2O{-}\overset{O}{\overset{||}{C}}{-}\underset{H}{\overset{}{N}}(CH_2)_4\underset{H}{\overset{}{N}}{-}\overset{O}{\overset{||}{C}}{-}OCH_2CH_2{-}$$
$$\underset{OH}{|}$$

a polyurethane

Formaldehyde Polymers

The simplest aldehyde is known as formaldehyde, which has the formula, CH_2O.
It is capable of acting as a monomer by forming one-carbon bridges between units
of other monomers, such as phenol and melamine to form the well-known thermosetting resins, Bakelite and Formica. Both polymers are very hard and durable.

Leo H. Baekeland was a Belgian-born chemist who came to the United
States in 1899. In 1909, he received a patent for this process for producing a thermosetting resin, which he called Bakelite, from the combination of phenol and
formaldehyde. It was the first commercial synthetic resin and dominated the plastics field until the 1930s.

The monomer, phenol, has three sites (numbered 2,4, and 6) that can react
with formaldehyde to form a very complex web-like structure. Baekeland perfected the process for interrupting the polymerization before the polymer was
fully cross-linked, so that it can be molded into various shapes before the final

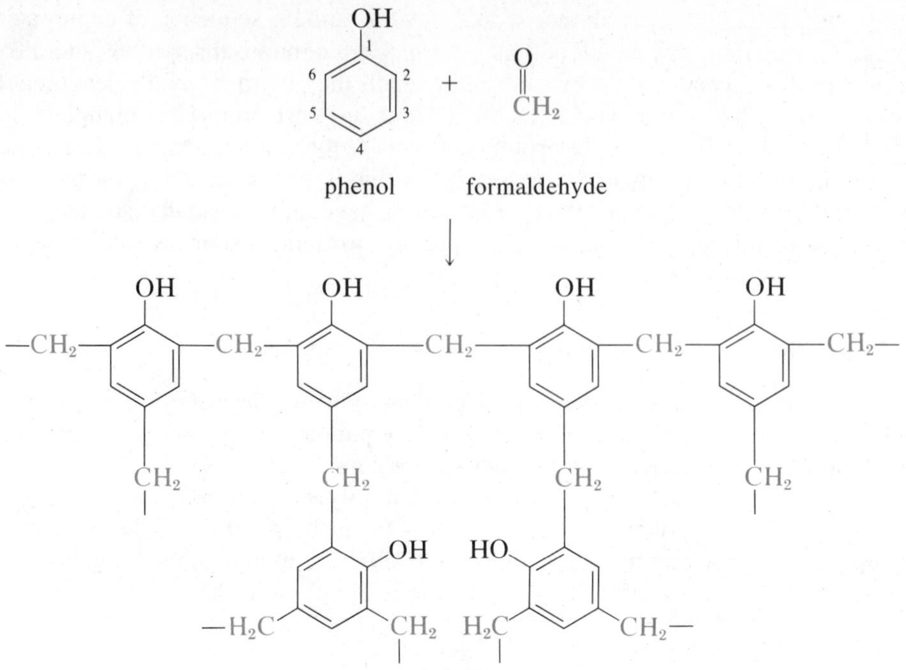

phenol   formaldehyde

a phenol-formaldehyde resin
Bakelite

cure (complete cross-linking) to form a hard thermosetting plastic. The Bakelite resin structure can also be shown using an abbreviated representation with the symbol P to signify the phenol ring.

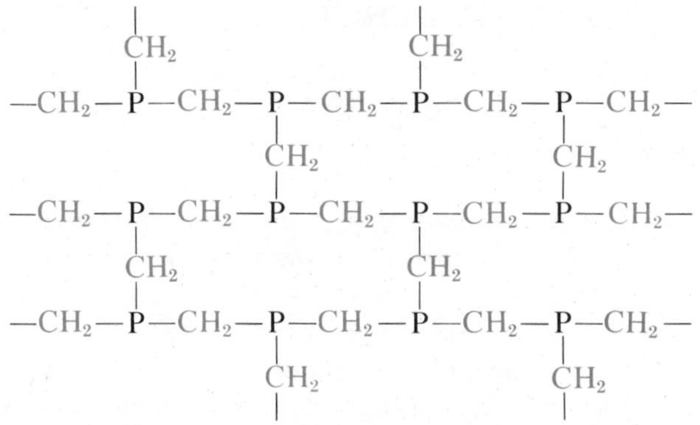

Bakelite
(P = phenol)

Bakelite resins are extremely resistant to heat and chemicals. They are used as electrical insulation, brake linings, pulleys, handles for cooking utensils, adhesives, telephone headsets, radio cases, buttons, and many more applications in which a hard durable plastic is required.

In a similar fashion, the trifunctional melamine molecule can form a rigid thermosetting resin with formaldehyde forming a one-carbon bridge between the melamine monomers. The fully cross-linked structure of the melamine-

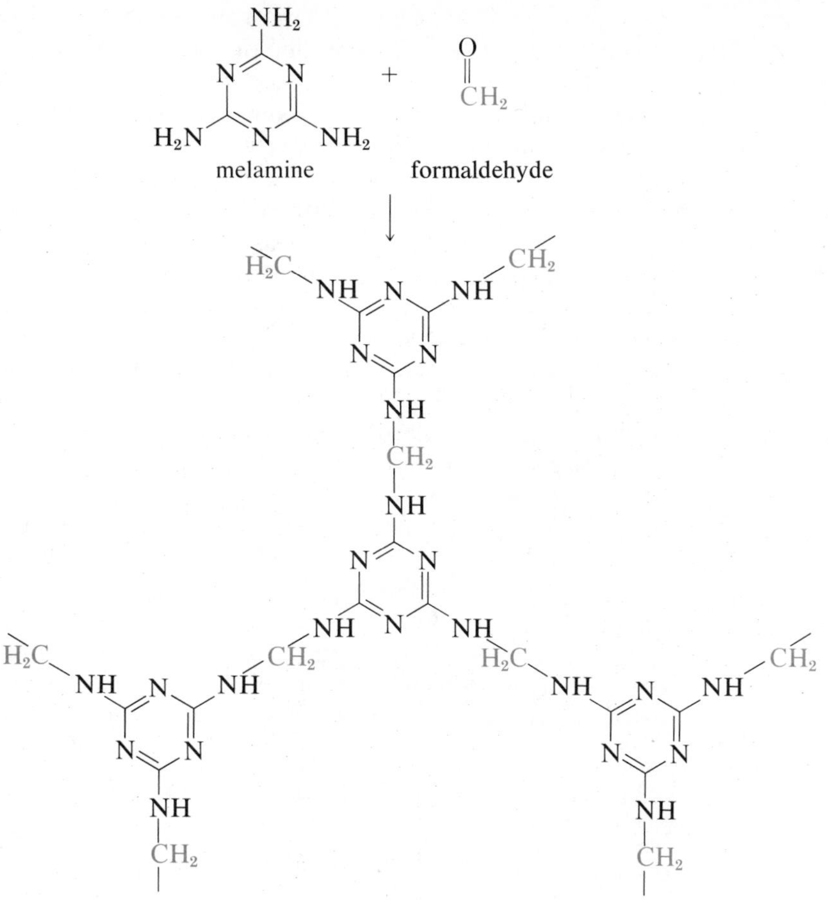

NH_2

melamine   formaldehyde

a melamine-formaldehyde resin

formaldehyde resin can be represented in the same way as the phenol-formaldehyde resin (previously shown), except that each phenol is replaced by melamine.

Melamine-formaldehyde resins are used in making dinnerware and decorative surface coatings for counter tops, tables, and wall coverings under the brand name Formica.

SUMMARY

Organic chemistry is the chemistry of carbon. The carbon atom shows considerable versatility in its bonding to other atoms and to other carbon atoms, including single, double, and triple bonds, but in all cases, the octet rule is obeyed and carbon is tetravalent.

Among the hydrocarbons, many isomeric forms result from the branching of the carbon chains. The chains may be symbolized by several convenient abbreviations. As the chains become longer, the boiling temperatures steadily increase, with exceptions due to branching, which lowers the boiling point. The boiling point is the basis for classifying and separating the common hydrocarbons obtained from petroleum. These include gasoline, kerosene, and asphalt. Catalytic cracking and reforming are processes for increasing the utility of natural hydrocarbons.

Unsaturation introduces the potential for geometrical isomerism. In some cases, *cis-trans* isomers exhibit very different properties. Ring structures are also of great importance among organic compounds. A wide variety of functional groups greatly increases the versatility in the structure and properties of organic compounds. Organic compounds are commonly named by the IUPAC system, although some compounds are routinely identified by a less systematic common name.

A major application of organic chemistry is polymer chemistry. Addition polymers are formed by polymerization of unsaturated monomers. A variety of functional groups may be present in the monomer but do not participate in the polymerization, although they do exert a strong influence over the properties of the finished product. Addition polymers are generally thermoplastic polymers, which can be molded into a variety of shapes.

Condensation polymers are formed when monomers contain two or more functional groups and the groups react with one another. Polyesters, polyamides, etc. may be thermoplastic, if each monomer has only two functional groups, or thermosetting, if additional functional groups are available for cross-linking of the polymer chains. Vulcanized rubber is harder and sturdier than natural rubber, due to cross-linking of the polymer chains. Formica and other melamine-based polymers are very durable, rigid, thermosetting plastics, which find many useful applications.

PROBLEMS

1. Give a definition or example for each of the following.
   a) valence
   b) covalent bond
   c) methane
   d) bottled gas
   e) cracking
   f) *cis*
   g) *trans*
   h) isomers
   i) single-line formula
   j) petroleum fraction
   k) alkane
   l) alkene
   m) alkyne
   n) unsaturated
   o) saturated
   p) geometrical isomerism
   q) cyclic hydrocarbon
   r) functional group
   s) monomers
   t) polymers
   u) bifunctional monomer
   v) resin
   w) thermoplastic resin
   x) thermosetting resin
   y) vulcanization
   z) Orlon
   aa) Acrilan
   bb) Dacron
   cc) Plexiglas
   dd) Lucite
   ee) Teflon
   ff) Saran
   gg) Formica

2. Draw two formulas for each of the following. In one show all of the hydrogens. In the second use an abbreviation like that used in Problem 3.
   a) 2-hexene (show both *cis* and *trans* isomers)
   b) 3-chlorooctane
   c) 2-bromo-1-pentene
   d) 2-bromo-2-pentene (show both *cis* and *trans* isomers)

3. Name each of the following compounds.

   a)

   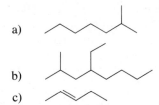

   b)

   c)

4. What is meant by the following: "The magic number for carbon is 4."

5. List all of the linear hydrocarbons from Table 6.2 that might be found in the kerosene fraction of crude oil. What other hydrocarbons might be present?

6. What are the major differences between addition polymers and condensation polymers?

7. What is the repeating unit in polyvinyl chloride?

8. What is the repeating unit in a common acrylic polymer?

9. What is the monomer used to make soft contact lenses?

10. What are some of the uses of polyurethanes?

11. What are the monomers used in making nylon 66? Explain the name.

12. Why is it difficult to synthesize proteins?

THE coverage of the chemistry of solutions is intended to serve as background material to permit an understanding of the behavior of ions and molecules in solution plus the reasons why substances do or do not mix to form solutions. We will focus attention on two aspects of solution chemistry, *acidity* and *solubility,* since these two properties are important and often interrelated.

In order to consider solutions, it is first necessary to consider the common terminology involved. In short, a *solute* dissolves in a *solvent* to yield a *solution*. In other words, a solute is a solid, liquid, or gas that can be dissolved in another substance, which is called a solvent. A simple example of a solution is salt in water, where salt is the solute and water is the solvent. The distinction between solute and solvent is not so clear when alcohol and water are mixed because both compounds are liquids, but neither is the distinction particularly important in this situation.

The most common examples of solutions are mixtures of gases (air), solutions of gases dissolved in liquids (carbonated soda, ammonia water), liquids in liquids (alcohol-water), solids in liquids (salt water), and solids in solids (alloys). Solutions in which water is the solvent are commonly referred to as *aqueous solutions*.

Most of the solutions mentioned are familiar and straightforward, with the possible exception of *alloys*. Alloys are solutions of two or more metals formed by heating to cause melting, mixing, and then cooling to form a homogeneous solid. Common examples are yellow gold (90% gold, 10% copper), sterling silver (92% silver, 8% copper), brass (copper, zinc, variable composition), bronze (copper, tin, variable composition), stainless steel (minimum 10% chromium, iron), and solder (tin, lead, variable composition).

Dentists use complex alloys called dental *amalgams*, for filling teeth. The alloys are called amalgams because mercury is one of the components of the mixture. Mercury is a very unusual metal because it is a liquid at room temperature. Silver (ca. 70%) and tin (ca. 25%) are the other major components of dental amalgams, along with copper and zinc in smaller amounts. Dental amalgams are discussed further in Chapter Sixteen.

Some of the other types of solutions are also discussed in later sections.

ACIDS AND BASES

Acidic and basic substances are familiar to everyone. The sour taste of vinegar or lemon juice is due to the presence of acetic and citric acid, respectively. Common bases are lye, lime, milk of magnesia, and baking soda. Bases are commonly

SOLUTION CHEMISTRY

described as bitter tasting and as slippery feeling when dissolved in water. More often, the chemist uses the following criteria to determine the presence of acids and bases.

Acids: Acids turn litmus paper from blue to red, neutralize bases, and form solutions with pH less than 7.

Bases: Bases turn litmus paper from red to blue, neutralize acids, and form solutions with pH greater than 7.

Before considering the origin of each of these properties, it is necessary to consider the definition of acids and bases. For simplicity, we shall consider only one definition, which will carry us through all of the acids and bases that we will encounter. In 1923, J.N. Brønsted proposed the definition. According to his description, an acid is a substance that will donate hydrogen ions (H$^+$), whereas a base is a substance that will accept hydrogen ions. In place of the phrase *hydrogen ions,* the chemist routinely uses the term *proton* since the hydrogen ion is merely a hydrogen atom minus the valence electron. Therefore, an acid may be described as a proton donor and a base as a proton acceptor.

The simplest of the common acids is hydrochloric acid, which is a solution of hydrogen chloride (HCl) in water. Pure hydrogen chloride is a gas but when dissolved in water, it undergoes the following reaction, which illustrates the proton-donating ability.

$$HCl \rightarrow H^+ + Cl^-$$

In more complete form, the ionization is generally described by the following reaction that shows the involvement of the water, which acts as a base (proton acceptor).

$$\underset{\text{acid}}{HCl} + \underset{\text{base}}{H_2O} \rightarrow H_3O^+ + Cl^-$$

The species, H$_3$O$^+$, is called the hydronium ion. The ability of the water molecule to accept a proton is easily understood from the electron dot structure as shown in the following equation.

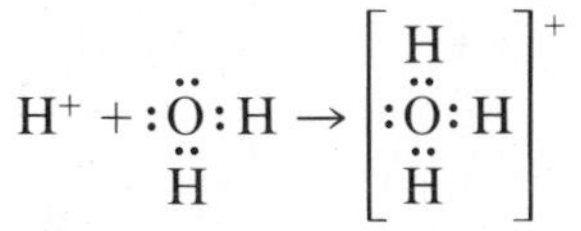

$$H^+ + \overset{..}{\underset{..}{:O}}:H \rightarrow \left[\begin{array}{c} H \\ .. \\ :\overset{..}{O}:H \\ .. \\ H \end{array} \right]^+$$

Bases dissolve in water to form basic or alkaline solutions. The term *alkali* is sometimes used in place of the word *base*.

Perhaps the most common base is sodium hydroxide, NaOH, sometimes called *lye*. In this case, the hydroxide ion (OH$^-$) functions as a base by accepting an available proton to form water.

Ammonia (NH$_3$) is another common base. Like HCl, ammonia is a gas that dissolves in water. The resulting solution is alternately known as *ammonia water* or ammonium hydroxide. The latter name is attributable to the following reaction. This reaction is an example of an equilibrium or reversible reaction. The contribution of each species depends on whether the reaction proceeds more efficiently to the left or to the right. In this case, the right-to-left reaction is favored.

$$: NH_3 + H_2O \longleftrightarrow NH_4^+ + OH^-$$

In other words, *ammonia water* is largely a solution of NH_3 in water that contains a trace of ammonium hydroxide (NH_4OH). This suggests that NH_3 is the basic substance, although the hydroxide ion can also function as a base.

Before considering additional examples, let us return to the properties of acids and bases cited earlier. Acids and bases were described as substances that turn litmus from blue to red and red to blue, respectively. Litmus paper is a porous paper that has been impregnated with an indicator dye called *litmus*. It can exist in two forms—red in acid and blue in base. Using the symbol In (indicator) for the litmus, the following equation symbolizes the process involved.

$$In + H^+ \rightarrow InH^+$$

blue   red

Another obvious characteristic of acids and bases is their ability to react with one another in a process known as *neutralization*, which can usually be described as the reaction of an acid and a base to form a salt plus water. For example:

$$HCl + NaOH \rightarrow H_2O + NaCl$$

or, in ionic form

$$H^+ + \cancel{Cl^-} + Na^+ + OH^- \rightarrow H_2O + Na^+ + \cancel{Cl^-}$$

TABLE 7.1   pH

pH	CONCENTRATION OF HYDROGEN ION (IN MOLES PER LITER)*		CONCENTRATION OF HYDROXIDE ION (IN MOLES PER LITER)*
1	0.1	or 10^{-1}	10^{-13}
2	0.01	or 10^{-2}	10^{-12}
3	0.001	or 10^{-3}	10^{-11}
4	0.0001	or 10^{-4}	10^{-10}
5	0.00001	or 10^{-5}	10^{-9}
6	0.000001	or 10^{-6}	10^{-8}
7	0.0000001	or 10^{-7}	10^{-7}
8	0.00000001	or 10^{-8}	10^{-6}
9		10^{-9}	10^{-5}
10		10^{-10}	10^{-4}
11		10^{-11}	10^{-3}
12		10^{-12}	10^{-2}
13		10^{-13}	10^{-1}
14		10^{-14}	10^{0}

* A mole is the amount of any compound equal to its molecular weight. The molecular weight is the sum of the atomic weights of all the atoms in the molecule, expressed in grams. For example, 18 grams of H_2O is one mole.

An acid may be described as a proton donor and a base as a proton acceptor.

or, eliminating the ions which do not participate

$$H^+ + OH^- \rightarrow H_2O$$

The third property of acids and bases is pH, which is a numerical description of the acidity of a solution. The origin of this scale is given in Table 7.1.

Several items need to be emphasized in Table 7.1. The pH value is numerically equal to the exponent of the hydrogen ion concentration. Thus, the pH decreases as the acidity increases. At the same time, as the pH decreases, the concentration of hydroxide ion decreases. At pH 7, the concentrations of H^+ and OH^- are equal and the solution is neutral. Below pH 7, the hydrogen ion concentration exceeds the hydroxide ion concentration and the solution is acidic. Above pH 7, the hydroxide ion predominates and the solution is basic (alkaline).

It can also be seen from Table 7.1 that each change of one pH unit represents a tenfold change in hydrogen ion (and hydroxide ion) concentration.

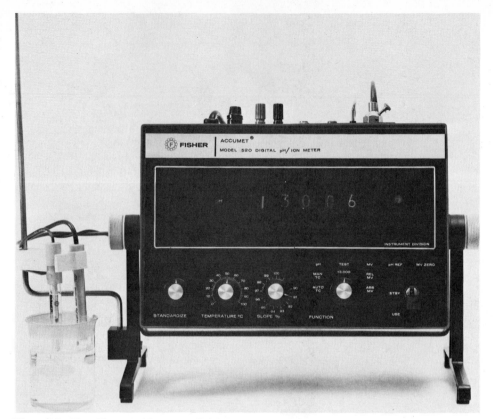

Figure 7.1 The pH meter. (Courtesy of Fisher Scientific.)

Copyright by Chemtech. Reprinted by permission.

"Try altering the pH a bit."

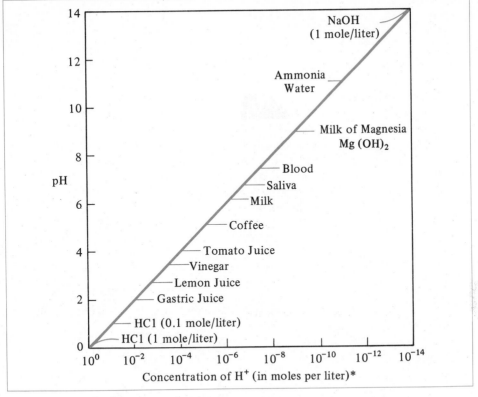

Figure 7.2   The pH of some common substances.

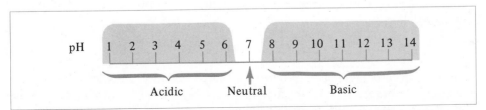

The pH of a solution can be measured directly by the use of a pH meter (see Figure 7.1) or some type of indicator paper. The pH values of some common substances are given in Figure 7.2.

Concentration Versus Strength

Acid *concentration* and acid *strength* are two distinctly different properties. The concentration of an acidic (or basic) solution merely describes the amount of acid (or base) dissolved in a specified quantity of solution. The concept of acid strength is somewhat more complex. It describes the relative ability of an acid to donate a proton (or a base to accept one). In equation form, the distinction between hydrogen chloride, a strong acid, and acetic acid, a weak acid, is easily understood. The strong acid ionizes completely in water, whereas the acetic ionizes only slightly, typically less than 5 percent.

$$HCl + H_2O \longrightarrow H_3O^+ + Cl^-$$

$$CH_3\overset{\overset{\displaystyle O}{\|}}{C}-OH + H_2O \xrightleftharpoons{} H_3O^+ + CH_3\overset{\overset{\displaystyle O}{\|}}{C}-O^-$$

The most common strong acids are hydrochloric, nitric (HNO_3) and sulfuric (H_2SO_4). Some important weak acids and common sources are listed in Table 7.2 and the structures are given in Figure 7.3.

The common strong bases are substances that contain hydroxide ion or produce it in large amounts when dissolved in water. Thus, sodium hydroxide (NaOH), potassium hydroxide (KOH), and calcium hydroxide ($Ca(OH)_2$) are strong. Sodium carbonate (washing soda) is also rather strong. It releases hydroxide ion according to the following equation.

$$CO_3^{-2} + H_2O \rightleftharpoons HCO_3^- + OH^-$$

carbonate   bicarbonate
ion   ion
(from Na_2CO_3)

Sodium bicarbonate (baking soda) is a much weaker base because of the following reaction, which proceeds more readily from right to left.

$$HCO_3^- + H_2O \rightleftharpoons H_2CO_3 + OH^-$$

bicarbonate   carbonic
ion   acid
(from $NaHCO_3$)

Carbonic acid is an unstable compound, which breaks down according to the following reaction.

$$H_2CO_3 \rightleftharpoons H_2O + CO_2$$

Thus, when carbonic acid is produced in large quantity, bubbles of CO_2 (carbonation) are observed. Looking at it another way, carbonic acid is merely a solution of CO_2 gas in water, much like ammonia water (NH_3 gas in water). In both

TABLE 7.2 COMMON WEAK ACIDS

WEAK ACID	COMMON OCCURRENCE
carbonic	soft drinks
acetic	vinegar
citric	citrus fruits
tartaric	grapes
malic	apples
lactic	sour milk, cheese
boric	eyewash solutions

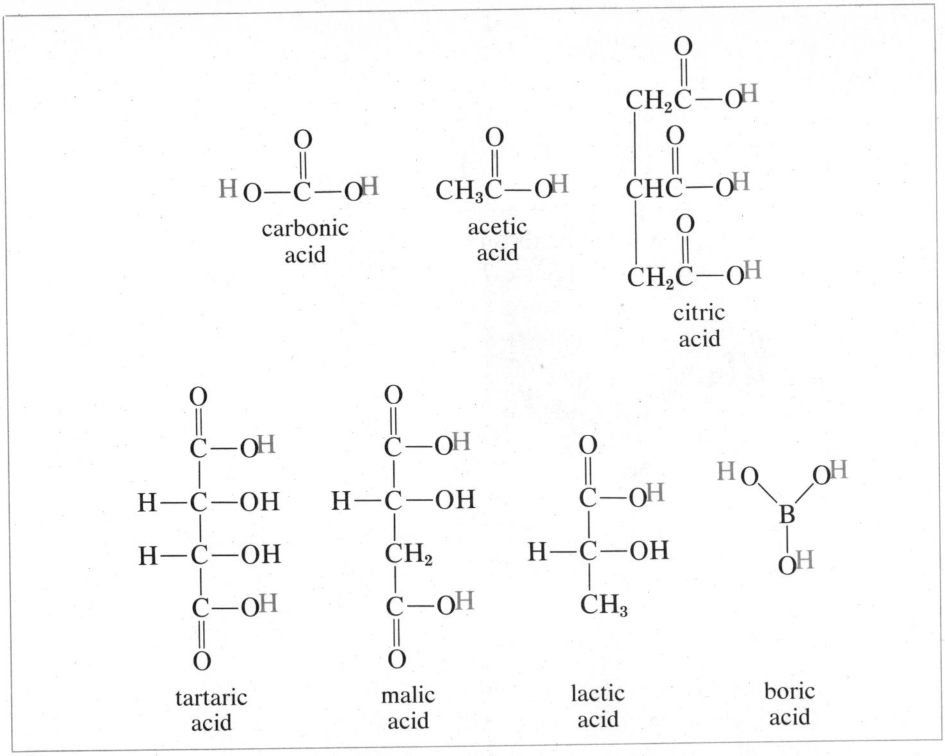

Figure 7.3   Some common weak acids. (The acidic hydrogens are shown in color.)

cases, only a limited amount of the gas will dissolve. When more is present, the gas will come out of solution. Such is the case when a container of carbonated beverage is opened, since extra CO_2 is often dissolved under pressure.

When sodium bicarbonate ($NaHCO_3$) is used as an antacid (e.g., Alka Seltzer) to neutralize stomach acids (e.g., HCl), a burp may result due to the CO_2, which is produced as follows.

$$HCO_3^- \ + \ HCl \rightarrow Cl^- \ + \ H_2CO_3$$
(as $NaHCO_3$)
$$\downarrow$$
$$H_2O \ + \ CO_2 \text{ (gas)}$$

Lime, also known as *quicklime,* is calcium oxide, which reacts with water to form *slaked lime,* which functions as a base by releasing hydroxide ions.

$$CaO \ + \ H_2O \rightarrow \ Ca(OH)_2$$
lime   slaked lime

Limestone is calcium carbonate, $CaCO_3$. It acts as a base in the same way as sodium carbonate, although calcium carbonate is practically insoluble in water in the absence of acid. Ammonium hydroxide is only weakly basic due to the reaction described earlier.

Copyright by Chemtech. Reprinted by permission.

SOLUBILITY

Like dissolves like. Oil and water do not mix. Salts and sugars dissolve in water. Oil paints dissolve in turpentine. Don't put beer in the freezer. These statements are simple examples of solubility phenomena, which are explained in this section.

Once again, we must consider the terminology used to describe solubility. An arbitrary value of 3 grams of solute per 100 grams of solvent is taken as the point of distinction. If the 3 grams or more goes into solution, the solute is regarded as soluble; if less goes in, the solute is said to be insoluble. Only rarely is the solubility close to the 3 gram level. Most of the time, solutes are either much more or much less soluble, so the distinction is very clear. The term *miscible* is generally used to describe liquids that are mutually soluble.

The term *concentration* was mentioned earlier. A concentrated solution is one with a large amount of solute dissolved. A *dilute* solution contains very little solute. The term *saturated* is used to describe a solution with a concentration equal to the capacity for the solute.

The Concept of Solubility

In order to understand why some substances are soluble in some solvents, consider the case of sodium chloride in water versus silver chloride in water. The former is, of course, very soluble. The latter is quite insoluble. When one mixes solutions of silver nitrate (soluble) and sodium chloride, a precipitate of silver chloride forms.

$$Ag^+ + NO_3^- + Na^+ + Cl^- \rightarrow \underset{\substack{\text{white} \\ \text{precipitate}}}{AgCl} + Na^+ + NO_3^-$$

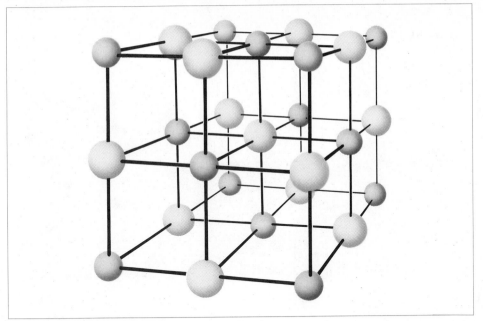

Figure 7.4   The crystal lattice of sodium chloride. The larger circles (color) represent Cl^- and the smaller gray ones Na^+. From Russell S. Drago, *Principles of Chemistry with Practical Perspectives*, 2d ed. (Boston: Allyn and Bacon, Inc., 1977), p. 31.

What is happening in these two cases? In the case of silver chloride, crystals will sit in water and be unaffected by the water, whereas crystals of sodium chloride will break down into individual ions and dissolve into the water.

A crystal of sodium chloride is typical of most salts. It has a three-dimensional structure (see Figure 7.4), called a lattice structure, of sodium and chloride ions, each surrounded by the other. The attractions between the cations and anions are quite strong, but vary from one compound to the next. In the case of NaCl, the attractions within the crystals are relatively weak so they are easily disrupted by interaction with water. In the case of AgCl, the forces holding the crystal together are too strong.

Let us consider what sort of attraction exists between the cations (or anions) and water molecules, which sometimes allows the water to break up the crystal structure. The fact is that there is an attraction between the polar salt and polar water molecules. It is obvious from the charges that the ions are polar, but what sort of polarity exists in a water molecule?

In order to answer this question, it is necessary to consider the phenomenon of *electronegativity*. This property may simply be described as the relative attraction of an atom for electrons in a molecule. An atom with a relatively strong attraction for electrons has a high electronegativity. As might be expected, the elements in group VIIA exhibit a high affinity for electrons as they attempt to simulate an inert gas structure. At the other end of the scale, the group IA elements have very low electronegativities in line with their tendency to form cations. Electronegativity values for all the elements are shown in Figure 7.5. The numbers are based on a value of 4.0 for fluorine. There is a general trend of in-

EN increases →

1 H 2.1																	
3 Li 1.0	4 Be 1.5											5 B 2.0	6 C 2.5	7 N 3.0	8 O 3.5	9 F 4.0	
11 Na 0.9	12 Mg 1.2											13 Al 1.5	14 Si 1.8	15 P 2.1	16 S 2.5	17 Cl 3.0	
19 K 0.8	20 Ca 1.0	21 Sc 1.3	22 Ti 1.5	23 V 1.6	24 Cr 1.6	25 Mn 1.5	26 Fe 1.7	27 Co 1.8	28 Ni 1.8	29 Cu 1.9	30 Zn 1.6	31 Ga 1.6	32 Ge 1.9	33 As 2.0	34 Se 2.4	35 Br 2.8	
37 Rb 0.8	38 Sr 1.0	39 Y 1.2	40 Zr 1.4	41 Nb 1.6	42 Mo 1.8	43 Tc 1.9	44 Ru 2.2	45 Rh 2.2	46 Pd 2.2	47 Ag 1.9	48 Cd 1.7	49 In 1.7	50 Sn 1.8	51 Sb 1.9	52 Te 2.1	53 I 2.5	
55 Cs 0.7	56 Ba 0.9	57–71 La–Lu 1.1–1.2	72 Hf 1.3	73 Ta 1.5	74 W 1.7	75 Re 1.9	76 Os 2.2	77 Ir 2.2	78 Pt 2.2	79 Au 2.4	80 Hg 1.9	81 Tl 1.8	82 Pb 1.8	83 Bi 1.9	84 Po 2.0	85 At 2.2	
87 Fr 0.7	88 Ra 0.9	89 Ac– 1.1–1.7															

EN decreases ↓

Figure 7.5 Electronegativity values of the elements. There is a general trend toward decreasing electronegativity proceeding from top to bottom of the periodic table and a trend toward increasing electronegativity from left to right.

creased electronegativity toward the top right side of the periodic table, although there are some important exceptions.

Considering the electronegativity values of oxygen and hydrogen, it is possible to account for the polarity of the water molecule by noting the polarity of the O-H bonds. The important factor is the vast difference in electronegativity

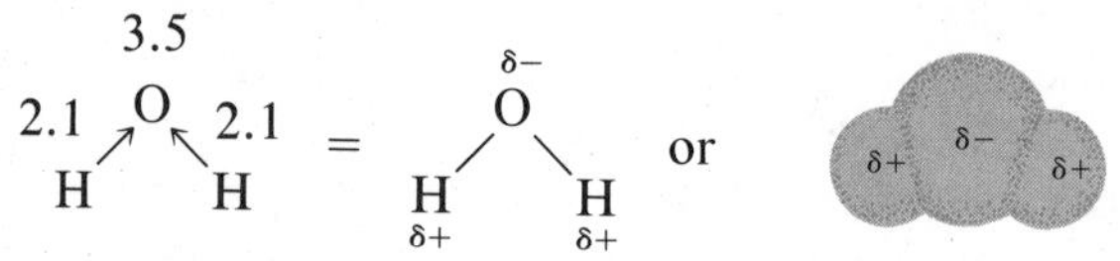

$$
2.1 \nearrow \overset{3.5}{O} \nwarrow 2.1 \quad = \quad \overset{\delta-}{O} \quad \text{or}
$$
$$
\underset{H \quad H}{} \qquad \underset{H \atop \delta+ \quad H \atop \delta+}{}
$$

between hydrogen and oxygen, which results in a significant distortion of the bonding electrons in the direction of the oxygen atoms (see arrows). As a result, the oxygen and hydrogen atoms carry a small amount of charge, sometimes called a *partial charge* and symbolized $\delta+$ or $\delta-$. It is not a full unit of charge, but there is sufficient distortion of the electrons so that the molecule is quite polarized and, therefore, able to interact strongly with polar solutes. (See Figure 7.6.)

The same sort of description applies to ethyl alcohol (C_2H_5OH) in which the polar O-H group interacts strongly with the polar water molecules and allows alcohol and water to mix completely. Similarly, ethylene glycol molecules interact very strongly with water and with themselves. The former leads to high solubility in water. The attraction between molecules of ethylene glycol makes them

$$CH_2 - CH_2$$
$$| \quad \quad |$$
$$OH \quad OH$$

ethylene glycol

difficult to separate and, since they must be separated in order for ethylene glycol to boil and vaporize, the boiling point is very high (198°C), which partly explains why ethylene glycol makes an excellent, permanent antifreeze.

Oil and water do not mix but will do so in the presence of soap. The chemical structure and action of soap is described in Chapter Sixteen, but consider the lack of solubility in the absence of soap. The constituents of petroleum are hydrocarbons, which means that these compounds consist only of carbon-carbon and carbon-hydrogen bonds. The C-C bonds are, of course, nonpolar; the C-H bonds are only very slightly polar (see Figure 7.5 for electronegativity values). Therefore neither allows a strong enough interaction with water to permit solubility. In effect, polar water molecules interact among themselves and exclude the hydrocarbon molecules. The same is true of food oils (see Chapter Fifteen).

In other words, the phrase "like dissolves like," has a simple meaning. Polar molecules tend to attract one another and dissolve. Similarly, nonpolar molecules tend to be mutually soluble. In the latter case, there are no strong attractions between the molecules but neither are there any significant forces that interfere with the mixing.

Temperature Effects

In general, solubility increases when the temperature is increased. The greater energy, in the form of increased motion of the molecules (i.e., increased kinetic energy), can often be used to disrupt the attractive forces of a crystal.

Figure 7.6   The dissolution of an ionic compound in water. (Adapted from Conrad L. Stanitski and Curtis T. Sears, *Chemistry for Health-Related Sciences: Concepts and Correlations* New York: Prentice-Hall, 1976, p. 107, by permission of Prentice-Hall.)

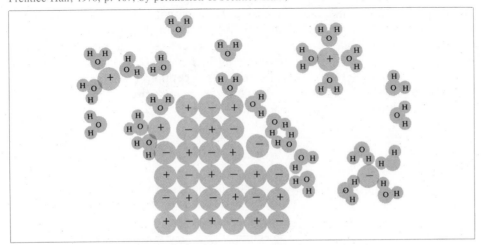

Gases dissolved in liquids provide a striking contrast. In this instance, solubility decreases when the temperature is increased. Gas molecules are free to escape from a liquid and the tendency to do so is increased by increasing the speed (i.e., the temperature) of the molecules. It is really quite amazing how much gas can be dissolved into most liquids, particularly when the gas is supplied under pressure as in the case of the carbon dioxide in beer and other carbonated beverages.

Even air dissolves in water to a considerable extent. A cold glass of water, allowed to warm to room temperature, will release bubbles of air. If a pan of water is heated on a stove, much bubbling will occur long before boiling begins, due to the escape of dissolved air.

The notable exception to the temperature effect on the solubility of gases in liquids is the dramatic decrease in solubility when a liquid is cooled to freezing. This phenomenon makes it quite dangerous to store beer or other carbonated beverages in a freezer due to the tremendous buildup of pressure inside the container when the liquid freezes and the gas comes out of solution.

SUMMARY

A solution is a homogeneous mixture in which a solute is dissolved in a solvent. Both solute and solvent may be gases, liquids, or solids. Air, carbonated beverages, salt water, and metal alloys are common solutions. Examples of alloys are yellow gold, sterling silver, and dental amalgams.

Certain properties can be used to characterize and define acids and bases, including the ability to neutralize one another. An acid, defined by the Brønsted theory, is a proton donor. A base is a proton acceptor. The pH scale can be used to designate the exact acidity of any solution, including the level found in many common fluids.

Acid concentration and acid strength are concepts that must be differentiated. Concentration describes the amount of acid in solution. Strength describes the extent to which it acts like an acid.

The solubility of most salts in water and the inability of oil and water to mix are typical solubility phenomena that are easily explainable on the basis of the relative polarity of solvents and solutes. Electronegativity assists in explaining or predicting the polarity of different compounds. When the electronegativity difference between two atoms is great, a bond between the atoms is very polar.

An increase in temperature generally increases solubility, although solutions in which gases act as solutes are a notable exception. Even the latter trend is dramatically reversed when the solution freezes and may present a definite hazard in the handling of many common substances.

PROBLEMS

1. Give a definition or example of each of the following.
   a) solute
   b) solvent
   c) solution
   d) carbonation

e) alloy
f) amalgam
g) litmus paper
h) acid
i) base
j) Brønsted acid
k) Brønsted base
l) hydrochloric acid
m) alkaline
n) lye

o) lime
p) slaked lime
q) neutralization
r) concentration
s) baking soda
t) washing soda
u) miscible
v) polar
w) electronegativity

2. What is ammonia water?

3. Write a balanced equation for the neutralization of each of the following.
   a) calcium hydroxide by nitric acid
   b) sulfuric acid by ammonium hydroxide
   c) hydrochloric acid by sodium bicarbonate

4. What color would a strip of litmus paper be after it had been moistened with lemon juice, milk of magnesia, vinegar, ammonia water, a solution of baking soda or a solution of washing soda? What acid or base is present in each case?

5. a) Give two examples of strong acids and explain why they are strong.
   b) Give two examples of strong bases and explain why they are strong.

6. What is the change in the amount of acid present when the pH changes from 2 to 4?

7. What is meant by "like dissolves like?"

8. Why do oil and water not mix?

9. Under what circumstances would a salt fail to dissolve in water?

10. Why does a carbonated drink fizz very vigorously when it comes in contact with ice?

A MAJOR class of compounds that will be encountered repeatedly in later chapters is the *carbohydrates*. Carbohydrates are very important biochemicals, which appear in many forms, both edible and inedible, some of which are much too complex for coverage here.

MONOSACCHARIDES AND DISACCHARIDES

The simplest of the carbohydrates are called *sugars* or *monosaccharides*. Of these, the most important ones contain six carbons and have the formula $C_6H_{12}O_6$ or $C_6(H_2O)_6$, which accounts for the use of the word carbohydrate (hydrated carbon).

Two of the most important monosaccharides are glucose (Glu) and fructose (Fru), which are combined in a disaccharide called sucrose (Glu-Fru), or common table sugar. The notation used for carbohydrates is illustrated for glucose in detail and in the common abbreviated form, which will be used throughout the text. In both forms, there is a six-membered ring consisting of five carbons and one oxygen. The reader should imagine that the ring is lying in the plane of the page. In

α-glucose   β-glucose

α-glucose (abbreviated form)   β-glucose (abbreviated form)

CARBOHYDRATES

Fructose is the sweetest of the sugars.

that way, the lines pointing from the ring toward the top of the page represent bonds to groups or atoms lying above the ring and lines pointing toward the bottom of the page represent bonds below the ring. In the abbreviation, all of the hydrogens attached directly to the ring are omitted.

It can be seen that glucose may exist in two slightly different isomeric forms, symbolized α and β, which rapidly interconvert and differ only in the position of the OH at C-1.* In all of the monosaccharides that we shall encounter, the α isomer will have the OH at C-1 below the ring, whereas, the β isomer will have the OH at C-1 above the ring.

No interconversion of the other OH groups can occur although other isomers do exist. For example, in galactose (discussed later), the OH at C-4 points up above the ring rather than below as in glucose, but glucose and galactose do not normally interconvert. It is the α form of glucose that participates in the formation of sucrose by combining with fructose. The fructose also exists in α and β forms but the β is incorporated into sucrose. Since fructose is the only sugar we shall encounter with a five-membered ring, we need not worry about the difference

* The α and β isomers also exist in two forms each. They are designated α-D, α-L, β-D and β-L but the L forms are biochemically unimportant and will not be considered.

between the α and β forms of this sugar. In fact, in considering sucrose, some other disaccharides, and some polysaccharides, we need only be concerned about

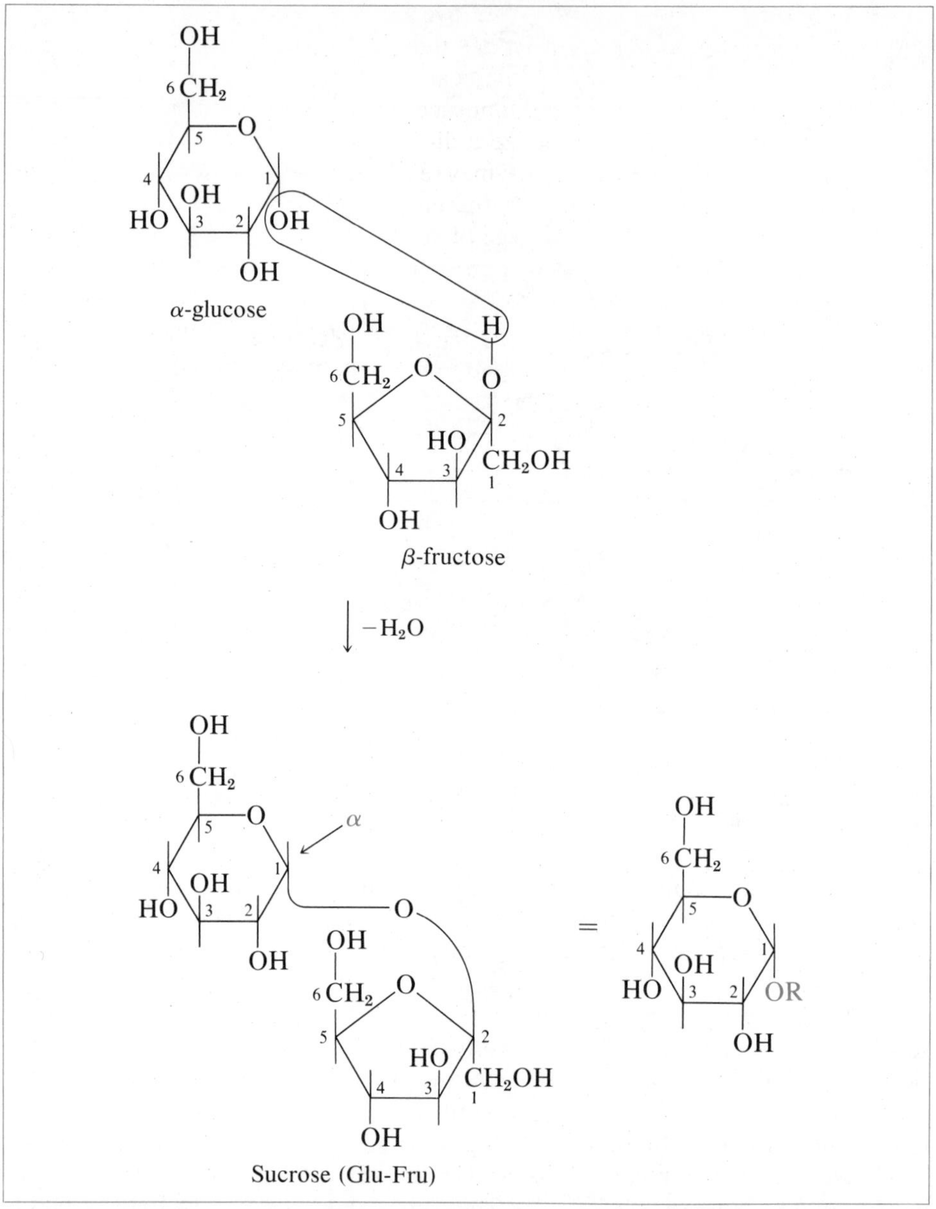

α-glucose

β-fructose

$-H_2O$

Sucrose (Glu-Fru)

the α-β isomerism of the monosaccharide ring on the left. The arrangement of atoms at the point of attachment to the ring on the left is a major factor in determining whether a carbohydrate is a useful food source. Consequently, the representation of sucrose with the fructose ring abbreviated as R is sufficient to show that the OR (R = fructose) is attached to the glucose ring in the same way as the OH at C-1 in α-glucose. Thus, the link between the two monosaccharide units in sucrose is α. Of course, we already know it is α from the reaction that showed α-glucose combining with fructose, but here and in other di- or polysaccharides it is important to be able to determine whether the linkage is α or β by inspection rather than by knowing some complicated reaction.

As a result of the α linkage, sucrose can be digested by all animals, including man. The reason is that man has the necessary enzyme to split sucrose into glucose and fructose, each of which can be used directly as a source of fuel. In effect, man can metabolize any monosaccharide either directly or by conversion to glucose but he can only utilize a di- or polysaccharide if he can break it down into monosaccharides. The enzymes required to catalyze the cleavage of an α linkage are normally available, but the enzymes required for cleavage of a β linkage are often missing. For cleavage of sucrose, the reaction is merely the reverse of that shown for formation of sucrose from glucose and fructose. The equal mixture of glucose and fructose that is produced in this reaction is called *invert sugar.* It is the main sugar found in honey and is produced by the honeybee, which carries out the cleavage of sucrose that it obtains from the nectar of flowers. Glucose is abundant in nature. Fructose occurs primarily in honey and fruit juices. Fructose is the sweetest of the carbohydrates followed by sucrose, glucose, maltose, and lactose, in that order.

Sucrose, usually known simply as *sugar,* is obtained for commercial use from sugarcane and sugar beets. The raw sugar is about 96-98% sucrose and is brown due to the presence of molasses, which is removed when the sugar is refined. Brown sugar is a variable product containing some molasses, and ranges from light brown to dark brown with the lighter, more purified brands having a milder flavor.

Two other important disaccharides are maltose (malt sugar) and lactose (milk sugar). Maltose is comprised of two glucose units, i.e., Glu-Glu, whereas,

Maltose (Glu-Glu)

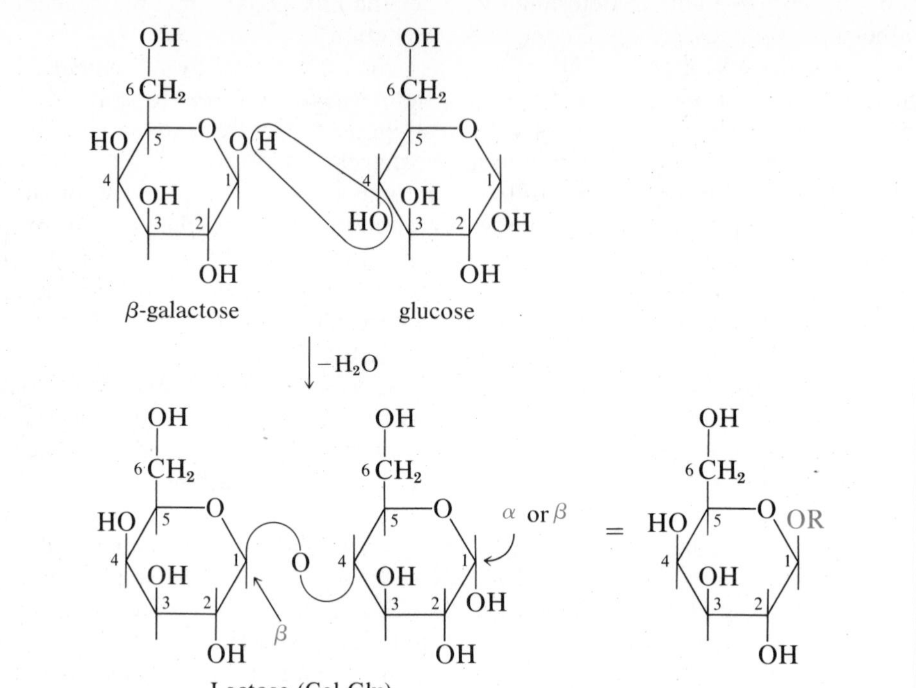

β-galactose   glucose

$-H_2O$

α or β

OR

β

Lactose (Gal-Glu)

lactose is a combination of a galactose and a glucose unit, i.e., Gal-Glu. Once again, we must take account of α and β isomeric forms, but first consider the carbons that are linked together in the disaccharides listed so far. In sucrose, the situation is again somewhat unique because the oxygen bridge between the two rings links C-1 of glucose to C-2 of fructose. In maltose, lactose, and most of the other carbohydrates to be considered, the linkages are 1,4, which means C-1 of the ring on the left is linked to C-4 of the ring on the right by an oxygen bridge.

In the coverage of brewing in Chapter Eleven, we will consider the breakdown of starch to form maltose and the subsequent conversion of maltose into alcohol. Maltose is readily metabolized by man and by yeast enzymes used for production of alcohol by fermentation. Once again, it is the α linkage to the left-hand ring that makes the disaccharide susceptible to attack by enzymes that man and yeasts have available for use.

Lactose differs from maltose in two ways. In maltose, the two units are glucose, whereas in lactose, there is one glucose and one galactose joined by a β linkage. Galactose differs from glucose only in the location of the OH at C-4. This is of no real consequence, since man is able to metabolize galactose once it is cleaved from the disaccharide. However, the β linkage is of very great significance. The importance of this point is discussed in detail in Chapter Thirteen (Dairy Products). The problem is that lactose is the carbohydrate found in milk and, because of the β linkage, some people are unable to metabolize the sugar and as a result suffer from what is known as *lactose intolerance*. This is attributed to the lack of an enzyme known as β-galactosidase (lactase), which catalyzes the cleavage of the β linkage in lactose. About 70% of the world's population—including blacks, Asians, and South American Indians—cannot digest lactose (1).

Mono- and disaccharides are major forms of dietary carbohydrate. When animals consume carbohydrates, they burn some and store the rest. Plants are able to manufacture some carbohydrate (glucose) by a process known as *photosynthesis* (Chapter Nine). Like man, once a plant has the carbohydrate, it may burn it or store it. The two major storage forms of carbohydrate in plants are *cellulose* and *starch,* although the common commercial sources of sucrose (sugarcane and sugar beets) do store a large amount of their carbohydrate as sucrose. Anyone who appreciates the flavor of fresh sweet corn enjoys the flavor of the sucrose that is converted to starch as the corn ages.

Cellulose is a polysaccharide consisting of glucose units (usually more than 1500) joined by β linkages. In combination with several other polymeric substances, cellulose is a major structural component of most plants. Wood and cotton have a high cellulose content that contributes to their rigid structures. Vegetables with a coarse texture (lettuce, cabbage, peas, beans, broccoli, etc.), fruits, and cereal bran have a high fiber content that is partly due to cellulose.

Cellulose $(Glu)_n$ β linkages

Because of the β linkages between the glucose units, cellulose is indigestible to man, since the necessary enzymes, called *cellulases,* are lacking. Cows and other ruminants (cud-chewing animals) are able to digest grasses and other feed that are high in cellulose, even though these animals also lack the necessary enzymes. The difference is that cows have a stomach consisting of several compartments, one of which is known as the *rumen.* A large number of bacteria that thrive in the rumen have the necessary enzymes for the breakdown of cellulose. The bacterial attack does not simply cleave the cellulose to glucose, but rather breaks it down into 2-, 3-, and 4-carbon fragments, which are stored as carbohydrate and fat by the cow. Methane and hydrogen are also formed in the breakdown of cellulose. Vegetables with a high cellulose content are said to increase the bulk in the diet and may aid digestion by stimulating contractions of the intestines, but they have little caloric value, in spite of the high content of carbohydrate, because humans cannot metabolize the cellulose.

The other major storage form of carbohydrates in plants is *starch*. It is also a polysaccharide of repeating glucose units but it differs from cellulose in two ways. For one thing the linkages between the monosaccharide units are α, so starch is readily digested, since the enzymes required for cleavage of α linkages are ubiquitous. The other major difference between cellulose and starch is the occurrence of two forms of starch. One form is linear and is called *amylose.* The

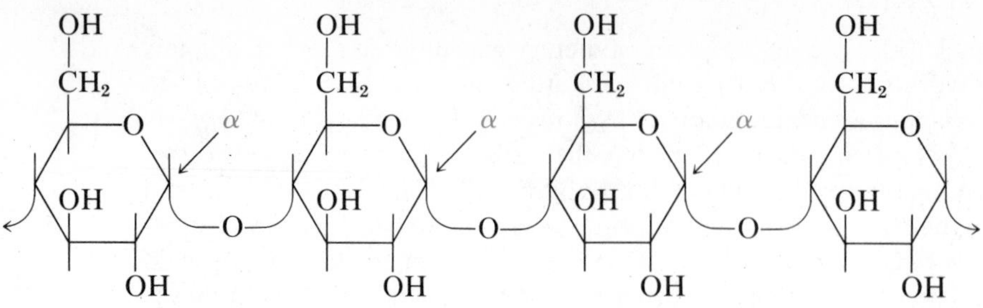

amylose
(Glu)$_n$
α linkages

other is branched and is called *amylopectin*. Amylose chains are commonly 1000-4000 units long, whereas, amylopectin often has more than 5000 glucose units. Natural starch (corn, potato) consists of 20-30% amylose and 70-80% amylopectin but some hybrids (primarily corn) may have different compositions. A hybrid known as waxy corn has 100% amylopectin. The proportion of the two components is very critical in determining the properties that decide the applicability of different starches. When heated and later cooled, amylose forms an opaque, rigid gel that makes it applicable for use in puddings and gum candies. On the other hand, amylopectin cools to a clear nongelling paste with a waxy consistency that makes it particularly suitable for thickening gravies and sauces. Prior to World War II, tapioca from the East Indies was used as a source of starch in much the same way that waxy (high amylopectin) starches are now used (2).

Note the structures of amylose and amylopectin. Amylose is identical to

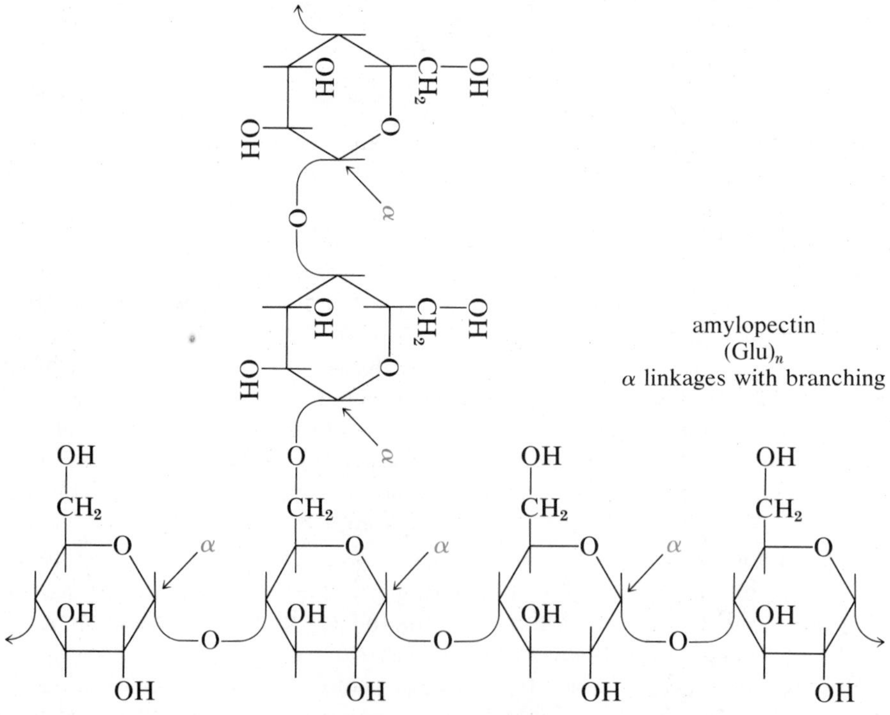

amylopectin
(Glu)$_n$
α linkages with branching

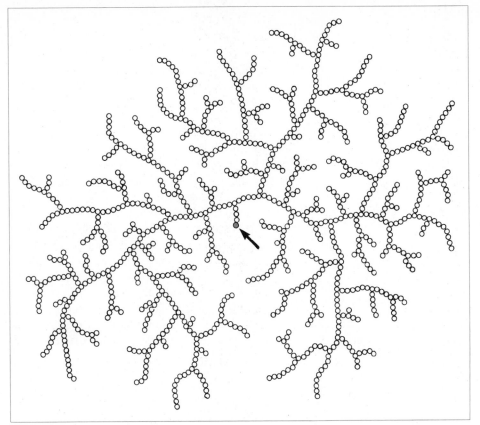

Figure 8.1   The structure of glycogen. See text for discussion. (From McGilvery, *Biochemistry: A Functional Approach* (Philadelphia: W.B. Saunders, 1970).

cellulose except for the α linkages between C-1 and C-4 of adjacent glucose units. In amylopectin, the linkages are again α, but in addition to the C-1 to C-4 linkages, there is an occasional link between C-1 and C-6 that gives the amylopectin its branched structure and different behavior when heated. Another consequence of the difference in structure is described in Chapter Eleven (see brewing).

The final carbohydrate to be described is *glycogen,* which is the polysaccharide used as the storage form of carbohydrate in animals and man. The glycogen structure is very similar to amylopectin, except that it has shorter chains—typically 12–18 units long. As in starch there are α linkages between the glucose units so that glycogen is readily metabolized.

A common representation of glycogen is given in Figure 8.1. The drawing is a cross section through glycogen showing the tree-like structure created by branched amylose chains. The short inner segments are actually a part of branches extending above and below the cross section. The circles represent glucose units; only one residue (arrow) does not have C-1 attached to another monosaccharide unit.

SUMMARY

Monosaccharides, disaccharides, and polysaccharides are the common forms of carbohydrates. Glucose is the most abundant monosaccharide. It may occur alone as a monosaccharide, as the disaccharide, maltose, or as the repeating unit in the polysaccharides, starch, glycogen, and cellulose. Fructose and galactose are the other major monosaccharides. Sucrose and lactose are the other major disaccharides.

Alpha and beta configurations of the monosaccharide units are of major importance in determining the digestibility of di- and polysaccharides. The alpha arrangement, found in sucrose, maltose, and starch, makes these carbohydrates readily digestible. A large portion of the world's adult population lacks the enzyme (lactase) necessary to break the beta linkage found in lactose. These people are unable to tolerate milk because they cannot metabolize the milk sugar.

Only ruminants (cows, termites) are able to metabloize cellulose due to the beta linkages between the glucose units. Even in ruminants, it is actually bacteria, thriving in the intestinal tract, that provide the enzymes required to break down the polysaccharide.

Amylose and amylopectin are the components of starch. They are linear and branched, respectively. The branching greatly affects the properties of different starches and determines their suitability for use in various food products. Glycogen is the storage form of carbohydrate in animals. It has a structure very much like amylopectin.

PROBLEMS

1. Give definitions or examples of each of the following.
   a) α and β linkages
   b) monosaccharide
   c) disaccharide
   d) polysaccharide
   e) starch
   f) glycogen
   g) cellulose
   h) ruminant
   i) brown sugar
2. Identify the following di- or polysaccharides.
   a) Gal-Glu
   b) Glu-Fru
   c) Glu-Glu
   d) -Glu-Glu-Glu-Glu-Glu-Glu-Glu-Glu-
   e) -Glu-Glu-Glu-Glu-Glu-Glu-Glu-Glu-
   $$\qquad\qquad\;\; Glu$$
   $$\qquad\qquad\;\; Glu$$
3. Draw a complete formula for maltose. How would you modify your drawing to symbolize the amylose component of starch? How would you modify your drawing of amylose to symbolize cellulose?

4. Why are humans unable to utilize cotton as a dietary source of carbohydrate? Are any animals able to do so?
5. Maltose and lactose differ in structure in two ways. What are they?

REFERENCES

1. Majtenyi, J.Z. "Food Additives—Food For Thought." *Chemistry* 47 (5) (1974): 6–13.
2. American Chemical Society. *Chemistry In The Economy*. Washington, D.C., 1973, pp. 257–59.

AGRICULTURAL CHEMISTRY

PART THREE

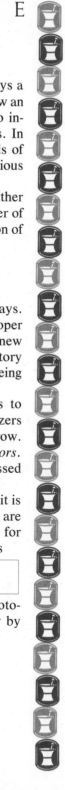

I N this section, we will consider some of the ways in which chemistry plays a role in agriculture. In this chapter, we will begin with a consideration of how an understanding of some of the steps in plant development can be used to increase productivity and influence the development of plants in useful ways. In Chapter Ten, we will focus attention on some of the problems and methods of dealing with insects, which are often a health hazard in addition to causing serious destruction to some of the major agricultural crops.

The topics discussed in Part Three are of concern to the consumer either directly, in the event he or she is a home gardener, or indirectly as a consumer of foods produced by others but for which he or she spends a considerable portion of income.

GROWTH REGULATION

Scientists can be credited with improving plant productivity in a number of ways. One is through genetic breeding. Another is through the development and proper use of fertilizers (discussed in a later section). A third approach is a relatively new one that many plant biochemists are actively researching, and while the full story lies in the future, it is at least possible to understand the questions that are being asked and tested (1, 2, 3, 4).

This new approach will likely involve the application of chemicals to plants. This may seem like fertilization, and yet it is distinctly different. Fertilizers provide the nutrients (e.g., nitrogen and phosphorus) that plants need to grow. The newer class of chemicals is more properly described as *growth regulators*. Weed killers (herbicides) are a type of growth regulator and these are discussed later.

In order to understand a possible mode of action of growth regulators, it is necessary to consider two metabolic processes common to all plants. These are *photosynthesis* and *respiration*. Photosynthesis is a very complex process for which many details remain obscure. The overall process of photosynthesis is

$$CO_2 + H_2O \xrightarrow{\text{light}} O_2 + \text{carbohydrate}$$

The carbohydrate appears in plants as starch, cellulose, simple sugars, etc. Photosynthesis can be described as *carbon fixation* since it is the major pathway by

HOW DOES YOUR GARDEN GROW?

Copyright © by Chemtech. Reprinted with permission of Chemtech.

which plants acquire carbon. The reverse of photosynthesis is respiration and two types of respiration are observed in plants. One type is called *photorespiration* and the other is known as *dark respiration*. As the name indicates, photorespiration requires light, whereas dark respiration occurs all the time—both in the light and in the dark.

Quite obviously, photosynthesis is required for the growth of the plant since carbohydrate is an essential component. It has been estimated that from 15 to 35 chlorophyll-containing leaves are required to supply the sugar for a mature apple (5). During the photosynthesis, chlorophyll-containing cells absorb energy from light, and act as chemical pumps by pumping the energy from the light into low energy compounds (CO_2 and water) to form high energy fuels (carbohydrate plus oxygen).

Dark respiration is also necessary in plants, just as it is in man, because it is by this process that chemical utilization (oxidation) of fuels (foods such as carbohydrates and fats) occurs in order to provide energy in aerobic* organisms. Anaerobic organisms derive useful energy from the same sources but by slightly different pathways—sometimes known as fermentation. This subject is discussed in Chapter Eleven.

Photorespiration is very much a mystery. The process does not provide plants with any useful energy. In fact, it seems to be totally wasteful since it is simply the reverse of photosynthesis. However, such a conclusion is a very dramatic contradiction in an evolutionary sense. Long before man ever recognized the potential or even the possibility of breeding, mother nature had a tremendous head start. The intentional breeding attempted by man may in some ways be more efficient, but the evolutionary, unintentional breeding that occurs in nature has been occurring for billions of years and is generally based on the survival of the fittest. In spite of man's ignorance of the value of photorespiration, it is hard to understand how a wasteful process, such as photorespiration, could have evolved, since its absence would seem to allow plants to grow more efficiently. One useful

* An aerobic organism is one that uses oxygen.

purpose that has been suggested (but not proven) is that photorespiration may be a defense mechanism for draining off excess energy in times of high light intensity when photosynthesis is too efficient and may actually be damaging. A second suggested purpose is the possibility that it may serve as a source of the amino acids glycine and serine (2).

In any case, the hope of many plant biochemists is to find methods, perhaps by treatment with chemicals, to enhance *net photosynthesis,* which is photosynthesis minus respiration, by either enhancing the photosynthesis or suppressing respiration.

Encouragement can be derived from two phenomena. For one thing, shade-tolerant plants (e.g., pachysandra) apparently have a better balance between photosynthesis and respiration, although they seem to be characterized by an overall slowdown of both processes so that the rate of growth is generally slow. The second and more encouraging fact is a chemical difference known to exist between normally fast-growing plants and slow-growing plants. Table 9.1 lists several species and their net photosynthesis rate under the conditions specified. Quite obviously, there is a clear distinction among the plants listed in the table. A fair amount of information is available about the chemical differences between plants in the two categories, which suggests that the chemistry involved in both photosynthesis and respiration may differ and be subject to alteration by man.

In any case, if the rates of photosynthesis and/or respiration can be modified, either chemically or genetically, it is conceivable that the rate of plant growth may be increased. The fact that the dry weight of most plants is 60–95% carbohydrate is a very convincing indication of the potential importance of research in this

TABLE 9.1 TYPICAL RATES OF NET PHOTOSYNTHESIS IN SINGLE LEAVES OF VARIOUS SPECIES AT HIGH ILLUMINANCE IN AIR CONTAINING 0.03% CO_2 AT 25° TO 30°.*

HIGH EFFICIENCY PLANTS	
SPECIES	NET PHOTOSYNTHESIS
Maize	46–63
Sugarcane	42–49
Sorghum	55
Pigweed	58
Bermuda grass	35–43
LOW EFFICIENCY PLANTS	
Spinach	16
Tobacco	16–21
Wheat	17–31
Rice	12–30
Orchard grass	13–24
Bean	12–17

* Net photosynthesis is given as milligrams of CO_2 fixed per square decimeter of leaf surface per hour (2).

area. Presumably, much of the increased productivity already achieved, e.g., by breeding, is based on changes in net photosynthesis and has taken the form of both higher yields per plant and shorter growing seasons.

FERTILIZATION

The most elementary aspect of agricultural chemistry is proper fertilization. Like animals, plants must have an adequate supply of nutrients in order to grow. In the following pages, we will turn our attention to fertilization, which requires an understanding of two closely related subjects, soils and fertilizers, since soil structure has a great influence on fertilizer requirements. The emphasis will be on some of the conditions that are necessary for both the commerical farmer and the home gardener to achieve maximum success with their crops.

Soils

Oxides (compounds containing oxygen) of silicon (Si), aluminum (Al), iron (Fe), and combinations of these elements are the most abundant components found in most soils. These species are found combined into very complex units of variable size. It is common to refer to soils as clay, silt, sand, or a combination, depending on the size of the particles. Among these three designations, the particle sizes in clay soils are smallest, ranging below 0.002 mm in diameter. Silt particles range from 0.002 to 0.05 mm and sand from 0.05 to 2 mm in diameter. In some cases, the classifications are further subdivided (e.g., coarse and fine sand).

Another common term is *loam*. Loam describes a mixture of clay, silt, and sand, which exhibits some of the characteristics of each. It is a normal practice to describe the soil texture in terms of the relative percentages of clay, silt, and sand. The more common designations of soil texture are listed below in order of increasing fineness.

sand	silt loam	silty clay loam
loamy sand	silt	sandy clay
sandy loam	sandy clay loam	silty clay
loam	clay loam	clay

A loamy soil is usually considered to be the best for growing crops. The two extremes, clay and sand, have properties that do not lend themselves to easy use, although, if one recognizes the problems associated with each type, it is possible to use them successfully.

Due to the small particle size of clay, it has a high surface area. It also has charges on the surface and these charges give the clay an unusually high capacity for holding nutrients (i.e., fertilizers). Thus, the loss of nutrients from clay soils is negligible in comparison to that in sandy soils. This may seem to be only an advantage until one realizes that the nutrients must be released for plant utilization.

Clay soils also hold much more water than sand so that leaching of nutrients is much less of a problem. However, too much water may cause a deficiency of air. Good aeration is required for root growth and to provide conditions that are suitable for a high level of soil microorganisms. Clay soils are also very heavy and sticky. They tend to crack when dry, although, some expansion and contraction, as the clay takes on and gives up water, respectively, is helpful in assisting aeration, since air is alternately forced out and drawn into the soil.

Sandy soils have some characteristics opposite to those found in clay. Sand has little capacity for water, or attraction for nutrients, so that water tends to pass through very freely often leaching nutrients out with it. Of course, as noted earlier, the failure to tightly bind nutrients is also an advantage since the nutrients become more available to plants as long as they remain in the soil. Sandy soils, therefore, require more frequent fertilization.

Figure 9.1   A characteristic common to nearly all soils is the accumulation of organic matter in and near the surface. (Photo courtesy of USDA, Soil Conservation Service.)

Loamy soils are well aerated due to looser packing, lower capacity for water retention, and because of the change in water content held by the clay particles. Loamy soils have the ability to bind nutrients (due to the clay component) but not excessively (due to the sand).

Another frequently mentioned component of good soil is organic matter. Organic matter consists of plant and animal residues plus microbes (microorganisms). These microbes include bacteria (often more than a billion per gram of soil), fungi, and actinomycetes. All three types participate in complex processes that increase the fertility of the soil. Some bacteria fix nitrogen,* which is one of the major nutrients required for plant growth (see *fertilizers* for detailed discussion), but most microbes simply feed on plant and animal residues and later release the nutrients from these residues back into the soil for use by growing plants. The process whereby organic matter is decomposed and inorganic ions are released is called **mineralization.** In mineral form these nutrients are more available and useful to growing plants.

Organic matter also tends to improve soil texture by countering the too great capacity for water retention in clay soils and by moderating the opposite effect in very sandy soils. A high level of organic material also indicates an adequate supply of nutrients to support the life of the microbes and crops.

This discussion points up one example of why man must be aware of the action of microbes. Later we will consider some of the positive contributions (e.g., fermentation and antibiotic production) of microbes as well as the potential hazards that must be dealt with (e.g., in food preservation and treatment of disease).

Very sandy soils have been found superior in one particular application. Salt water is generally considered useless for growing plants because the high salt concentration outside the plant tends to cause water to dialyze (see Chapter Three) out through the root membranes and causes the plant to dehydrate. In sandy soils, this is less of a problem since water tends to run through more freely, carrying salts, and preventing a buildup of salt, which could cause injury to the plant.

Finally, a brief mention of *lateritic soils* is in order. Such soils are common in the tropics where heavy rainfall and warmer climate can adversely affect the properties. Both rainfall and warm climate would seem to be desirable but it is a matter of too much of a good thing. Heavy rainfall tends to contribute to leaching both nutrients and organic matter out of the soil. The warmer climate assists rapid growth of microbes, insects, earthworms, etc. that may cause plant and animal residues to be mineralized too rapidly and, thus, contribute to the leaching.

Fertilizers

In this section, an attempt will be made to cover many of the steps that must be taken to assure proper plant growth. Part of the relationship between soils and fertilizers was mentioned in the previous section.

The basic problem is one of supplying all of the necessary nutrients to the plant. The situation is no different than that which confronts man. The human diet

---

* *Nitrogen fixation* is the process whereby nitrogen as N_2 is removed (fixed) from the air and converted to a useful form. See later discussion.

TABLE 9.2   PLANT NUTRIENTS

MACRONUTRIENTS	SECONDARY NUTRIENTS	MICRONUTRIENTS
Carbon (C)	Calcium (Ca)	Boron (B)
Hydrogen (H)	Magnesium (Mg)	Copper (Cu)
Oxygen (O)	Sulfur (S)	Iron (Fe)
Nitrogen (N)		Manganese (Mn)
Phosphorus (P)		Zinc (Zn)
Potassium (K)		Molybdenum (Mo)
		Chlorine (Cl)

must also contain certain essentials (amino acids from proteins, vitamins, minerals, etc.). Both plants and animals have very sophisticated chemical reaction pathways (metabolism) for utilizing nutrients either directly or by conversion into forms suitable for both function (e.g., proteins to serve as enzymes, or carbohydrates as a source of energy) and structure (e.g., fat or proteins of skin, hair, or nails).

One interesting phenomenon is the way in which plants and animals complement one another. Each can do things that the other cannot. For example, man can produce, transport, and apply fertilizer nutrients to plants. On the other hand, plants can utilize certain mineral nutrients, e.g., nitrogen as nitrate ion (NO_3^-), to produce amino acids, which the plant incorporates into protein. Man cannot utilize the nitrate and is actually dependent on plants and animals (which feed on plants) to obtain protein with a full complement of the nine essential amino acids, plus other dietary requirements like carbohydrates, fats, vitamins, and minerals.

This discussion will deal only with nutrient requirements of plants. Some sixteen nutrients are commonly considered essential for proper plant growth. For emphasis, these sixteen are often subdivided into three groups known as macronutrients, secondary nutrients, and micronutrients. The last group is slowly enlarging as researchers find new micronutrient requirements. The three categories are listed in Table 9.2.

Table 9.3 provides data to justify the three levels of nutrients given in Table 9.2, except for carbon, hydrogen, and oxygen, which are not tabulated since they

TABLE 9.3   NUTRIENT CONTENT OF VARIOUS FOODS (7)

CROP	YIELD/ACRE	NUTRIENTS IN CROP (POUNDS/ACRE)									
		N	P	K	Ca	Mg	S	Cu	Mn	Zn	B
Corn	100 bu	90	15	21	6	6	7	0.04	0.06	0.10	—
Apples	500 bu	30	5	37	8	5	10	0.03	0.03	0.03	0.01
Potatoes	400 bu	80	13	120	3	6	6	0.04	0.09	0.05	0.05
Tomatoes	15 ton	90	13	11	5	8	10	0.05	0.10	0.12	0.14
Soybeans	40 bu	150	15	43	7	7	4	0.04	0.05	0.04	0.01

TABLE 9.4   NUTRIENTS PRESENT IN THE UPPER SEVEN INCHES OF
ONE ACRE OF SOIL (7)

NUTRIENTS	POUNDS/ACRE
nitrogen	1000–6000
phosphorus	800–2000
potassium	up to 49,000
calcium	up to 500,000
magnesium	8000–26,000
sulfur	up to 3000
boron	40–400
chlorine	200–18,000
copper	4–400
iron	400–over 200,000
manganese	up to 200,000
molybdenum	about 4
zinc	20–600

are absorbed by the plant as water and via photosynthesis. This makes H_2O and
CO_2 the two direct nutrients. Water is provided by rain and irrigation. Abundant
CO_2 is available in the air and the supply is renewed in overabundance by burning
carbon compounds such as coal, oil, gasoline, and garbage.

The remaining nutrients are provided via the soil. An exception is nitrogen,
which can be obtained directly from the air by certain plants, called *legumes*.
Table 9.4 gives the content of the remaining thirteen nutrients found in the soil
layer in which most plants grow. In effect, Table 9.3 shows what plants take from
the soil, whereas Table 9.4 shows what is present in the soil. A quick comparison
suggests that the soil is in very good shape and there seems to be no real need for
fertilizer at all. Unfortunately, while the nutrients may well be in the soil, they
may be totally unavailable to the plants. The role of fertilizer is to provide the nutrient required and to do so in a form which the plant can use.

pH

In some cases, adequate plant nutrition may simply be a matter of preparing the
soil to release the nutrients available. One approach is to control the acidity of the
soil. Different types of soil have different acidities. Clay soils are usually acid,
limestone soils are basic or alkaline, and sandy soils are usually neutral. Acidity is
measured most commonly by the pH. If the pH is less than seven, the soil is acid.
The lower the pH, the more acid the soil. If the pH is greater than seven, the soil is
alkaline or basic. The subject of acidity and pH was introduced in Chapter Seven.

At one time farmers checked the pH by simply tasting the soil. If the soil
tasted sweet, the ground was alkaline. The idea of sweet and sour tastes of soil is
not unreasonable. Various foods have either sweet or sour tastes. Tomatoes can
be sour due to excess acid in the fruit. Sweet tomatoes are raised in sweeter soil,
although tomatoes cannot be raised in alkaline soil as easily as in acid soil. Fresh

fruit tastes sour when green, due to excess acid, but is sweet when ripe because the acid is converted to less acidic materials, usually sugars.

The generally desirable pH of the soil is slightly alkaline. If the soil is very alkaline, the required metal nutrients become very insoluble in water and unavailable to plants. If the soil is acid, the nutrients become too soluble and wash away with the rain. The preference for slight alkalinity is only a general rule. There are dramatic violations of this. Blueberry plants will grow well in many types of soil, but they will bear fruit only if the soil is quite acid. The desirable pH for blueberry production is about 4.5. Rhododendron will flower best in a somewhat sour ground. In some cases, the situation is not so clear-cut. United States government publications suggest that, for the growth of spinach, the soil should not be alkaline. It is suggested that beets do not grow well in acid soils. These recommendations imply that these two crops should be difficult to grow in the same soil. However, many organic gardeners find little difficulty in growing the two side by side. They argue that pH requirements for some crops are overrated and the only requirement is a properly prepared soil rich in organic matter. Such ground will normally have a pH close to 7.

In any case, it is often necessary to adjust the pH of the soil. We must also keep in mind that to change an acid soil to a sweet soil requires that we neutralize all the acid and add excess alkaline material. Acids and bases neutralize one another so it is merely a matter of adding one or the other until the proper pH is reached.

WHEN ALKALINITY IS TOO HIGH

Acidity may be increased by addition of many things including the following:

(1) **Alum.** Alum is $KAl(SO_4)_2 \cdot 12\ H_2O$.* It undergoes a reaction in water to produce acid. It is soluble in water but does not wash away easily in rain because the aluminum undergoes a complex series of reactions that keep it as a slightly soluble material at the point of application. Since the aluminum is the acid-producing component, this is desirable. Also, aluminum is not harmful to plants. There is abundant aluminum widespread throughout most soils.

The reactions of alum occur as follows, assuming the reaction is only with OH^- and not with any other component of the soil. $Al(H_2O)_6^{+++}$ is the reacting species.

$$Al(H_2O)_6^{+++} + OH^- \rightarrow Al(H_2O)_5(OH)^{++} + H_2O$$
$$Al(H_2O)_5(OH)^{++} + OH^- \rightarrow Al(H_2O)_4(OH)_2^+ + H_2O$$
$$Al(H_2O)_4(OH)_2^+ + OH^- \rightarrow Al(H_2O)_3(OH)_3 + H_2O$$
$$Al(H_2O)_3(OH)_3 + OH^- \rightarrow Al(H_2O)_2(OH)_4^- + H_2O$$

If the medium is strongly basic, the additional reactions can occur.

$$Al(H_2O)_2(OH)_4^- + 2\ OH^- \rightarrow Al(OH)_6^{---} + 2\ H_2O$$

* The formula $KAl(SO_4)_2 \cdot 12\ H_2O$ indicates that there are 12 molecules of water present for every one potassium, one aluminum, and two sulfate ions in a crystal of alum.

According to the previous reactions, the water molecules are able to release protons to neutralize base (OH^-), which is present in the soil. This may be rewritten as follows:

$$Al(H_2O)_6^{+3} \rightarrow H^+ + Al(H_2O)_5(OH)^{+2}$$

(2) Aluminum Sulfate, $Al_2(SO_4)_3$. The reactions for aluminum sulfate are the same as those given for alum.

(3) Sulfur. Many compounds of sulfur are acidic. Examples are sulfuric acid (H_2SO_4), sulfurous acid (H_2SO_3), and hydrogen sulfide (H_2S). In fact, in some areas near certain industries, there is concern over air pollutants including such gases as sulfur dioxide (SO_2) and sulfur trioxide (SO_3). These oxides can combine with water according to the following reactions to produce acids. These may later appear as what has been described as an acid rain (2).

$$SO_2 + H_2O \rightarrow H_2SO_3$$
$$SO_3 + H_2O \rightarrow H_2SO_4$$

There are also several so-called natural items that can be used.

(4) Leaves. Oak leaves, in particular, liberate a large amount of acid when they decompose.

(5) Pine Needles.

(6) Citrus Fruit Peels. The peels contain citric acid.

(7) Nut Shells. These were used to make inks in colonial times and some of the pages of manuscripts on which these inks were used are being eaten through with time due to the action of tannic acid. Tannic acid is a component of most tree barks and nut shells.

WHEN ACIDITY IS TOO HIGH

Soils are more likely to be too acid rather than too alkaline. The ideal pH for general gardening is 6 to 8. As a general guide, consider the information given in Table 9.5. If nothing is specified, the plant will usually prefer a neutral soil.

Acid soils can be corrected by addition of several things.

(1) Lime. To a chemist, lime (also known as quicklime) is calcium oxide, CaO, which is extremely alkaline and will burn plants. Slaked lime is calcium hydroxide, $Ca(OH)_2$, and is equally harmful. In fact, lime is converted to slaked lime when it comes in contact with water according to the following equation:

$$CaO + H_2O \longrightarrow Ca(OH)_2$$
$$\text{lime} \qquad\qquad\qquad \text{slaked lime}$$

To the farmer, lime is calcium carbonate, $CaCO_3$, which is better known as lime-

TABLE 9.5   pH REQUIREMENT FOR PLANTS AND VEGETABLES (9a)

ACID LOVING PLANTS pH 4–6	NEUTRAL pH 7	ALKALINE SOIL PLANTS pH > 7
azalea	apple	asparagus
blackberry	cornflower	bean
blueberry	gardenia	beet
chrysanthemum	pansy	cabbage
cranberry	pumpkin	cantaloupe
huckleberry	rice	carnation
lily	turnip	cauliflower
marigold		celery
oak		cucumber
peanut		lettuce
radish		nasturtium
raspberry		onion
rhododendron		pea
spruce		rhubarb
strawberry		squash
sweet potato		
watermelon		

stone. Thus, "agricultural lime" is the same as limestone. In addition to correcting an acid pH, limestone also provides calcium. The neutralizing action of $CaCO_3$ is described by the following.

$$CaCO_3 = Ca^{++} + CO_3^{--}$$

$$\underset{\substack{\text{carbonate} \\ \text{ion}}}{CO_3^{--}} + H^+ \longrightarrow \underset{\substack{\text{bicarbonate} \\ \text{ion}}}{HCO_3^-}$$

$$HCO_3^- + H^+ \longrightarrow \underset{\substack{\text{carbonic} \\ \text{acid}}}{H_2CO_3} \text{ (unstable)}$$

$$H_2CO_3 \longrightarrow H_2O + CO_2$$

(2) **Dolomite.**   Dolomite is a mixture of magnesium carbonate, $MgCO_3$, and $CaCO_3$. Limestone usually contains some $MgCO_3$.

Both $CaCO_3$ and $MgCO_3$ dissolve slowly in the soil, thereby providing acid neutralizing materials for several years with one application. In addition to their alkaline properties, calcium and magnesium are secondary plant nutrients.

(3) **Wood Ashes.**   Ashes contain 1–10% potash plus up to 1½% phosphorus. Potash is potassium oxide, K_2O. It occurs in ashes as K_2CO_3, a soluble material that

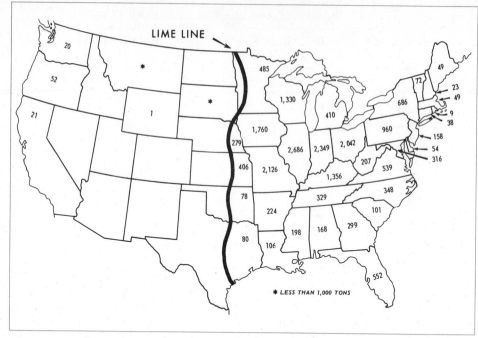

Figure 9.2   Consumption, in 1955, of liming materials, expressed as limestone equivalent, in thousands of tons of limestone. Based on data supplied by the National Agricultural Limestone Institute and the National Lime Association. The "lime line" roughly divides the lime-requiring regions from those where liming is not generally needed. Heavier agricultural use and heavier rainfall in the eastern United States has caused greater depletion of lime from these soils (USDA).

will leach out of soil easily. K_2CO_3 is very alkaline and can neutralize an acid soil. Potassium salts are all quite soluble and are easily washed away, so they must be applied repeatedly. Wood ashes are so basic that they are usually not allowed to come in contact with plants and leaves directly, but are used as a side dressing. Besides potash, wood ashes contribute certain other nutrients discussed in the following.

Nitrogen

Table 9.2 provided a list of plant nutrients. Of these, the most important are the macronutrients. It was already noted that carbon, hydrogen, and oxygen are obtained from water and photosynthesis. The following paragraphs will consider common sources of the other macronutrients.

Nitrogen is the nutrient that is often considered to be the most important, since nitrogen is required for the synthesis of plant proteins. The organic and the nonorganic (chemical) gardeners use different sources of nitrogen as well as other nutrients.

Organic gardeners look to several sources of nitrogen. The most concentrated source is bloodmeal (15% nitrogen). Others are hoofmeal and horndust (12.5% nitrogen), cottonseed meal (7% nitrogen), and manure. Manure is a frequent source but not a very concentrated one as seen in Table 9.6. Raw garbage

TABLE 9.6   ANALYSIS OF MANURE (9b)

MANURE	% N	% P_2O_5	% K_2O
Rabbit	2.4	1.4	0.6
Hen	1.1	0.8	0.5
Sheep	0.7	0.3	0.9
Steer	0.7	0.3	0.4
Horse	0.7	0.3	0.6
Duck	0.6	1.4	0.5
Cow	0.6	0.2	0.5
Pig	0.5	0.3	0.5

normally contains 1–3% nitrogen. Sewage is another source. Processed sewage is called sludge and contains about 2% nitrogen. Sludge contains a higher concentration of nitrogen than does raw sewage. The processing involves bubbling air through raw sewage. This promotes the growth of bacteria that undergo aerobic respiration, thus consuming some of the carbohydrate content. This process leaves the nitrogen in a more concentrated form, which coagulates and settles out. This sludge is dried to form activated, heat-treated sludge with 5–6% nitrogen content and 3–6% phosphorus content. Fresh grass clippings have about 1% nitrogen. Compost (discussed later) has a good nitrogen content, if properly made.

In general, the organic farmer will use unprocessed chemicals such as limestone and dolomite, which are natural minerals and need only be crushed into the proper consistency before use. The organic farmer avoids the use of synthetic chemicals, especially insecticides and herbicides, which is the secret of success of the nonorganic farmer. The common chemical sources of nitrogen are described in the following paragraphs.

(1) **Ammonia, NH_3, 82% Nitrogen.**   Nonorganic farmers find the greatest source of nitrogen in ammonia. NH_3 (boiling point $-33°C$) is used as a water solution or is pumped into the ground as a gas. Ammonia is 82% nitrogen, soluble in water up to 28% NH_3, binds readily to many components of the soil (particularly H^+, which converts it to NH_4^+ so that it is no longer a gas), and is readily and quickly converted to useable plant food. Nitrogen is taken up by plants in the form of nitrate, NO_3^-, which is produced from NH_3 by a complex oxidation process carried out by bacteria in the soil.

Ammonia is produced on a commerical scale by the Haber-Bosch process described by the following equation.

$$3H_2 + N_2 \xrightarrow[\substack{catalyst \\ 500°C}]{} 2NH_3$$

500–1000 atmospheres pressure

Ammonia is a very alkaline substance, but the alkalinity largely disappears as the ammonia is converted to nitrate. Ammonia will also react with many metal ions to render them soluble and accessible to plants. Ammonia is toxic, injurious to living tissues, and difficult to handle. It is, however, very inexpensive and

Figure 9.3   Pumping ammonia into the soil. (Photo courtesy of Grant Heilman.)

extremely high in nitrogen. It adds no organic matter to the soil. In fact, the nitrogen from ammonia is either used very quickly by plants or washed away, but the increased growth of plants due to ammonia is spectacular. Consequently, NH_3 is the single largest source of nitrogen for agricultural use.

(2) Ammonium Nitrate, NH_4NO_3, 35% Nitrogen. Ammonium nitrate is a solid granular material that is quite soluble in water. To maintain an effective nitrogen concentration in the soil, this substance must be applied frequently, especially in areas of heavy rainfall. Ammonium nitrate has some other problems associated with it. It is a powerful explosive reacting according to the equation:

$$NH_4NO_3 \rightarrow N_2O + 2\, H_2O$$

This reaction need only be initiated to cause an explosion to occur. It requires no external chemical or reaction. The danger involved can be understood if we recognize that K^+ and NH_4^+ are very similar in size. Properties associated with size are some of the major properties of these species, and a well-known property of KNO_3 is illustrated by its use as a major component of gunpowder and blasting powder. NH_4NO_3 will burn without exploding and it needs reasonably serious conditions to explode, but the precise nature of these conditions is not known—a fact that makes it difficult to assure safe handling. There have been some serious NH_4NO_3 explosions, including one in which a shipload exploded and destroyed a major part of the city of Texas City, Texas in 1947. More than 400 people were killed in the incident that touched off explosions and fires in nearby petroleum plants (10).

(3) Urea, $(NH_2)_2CO$, 47% Nitrogen. Urea is another good source of fertilizer nitrogen. It is a metabolic waste product of animals. It is easily synthesized on a commercial scale, soluble, relatively nontoxic, and easily handled. Urea is a relative latecomer to the fertilizer scene due to difficulty in manufacturing in the past. Urea hydrolyzes slowly to release ammonia over a period of time, thus providing a longer term supply of NO_3^-.

Biuret and triuret are slight modifications of urea that are also convertible to nitrate by the action of bacteria in the soil. The water solubility decreases from

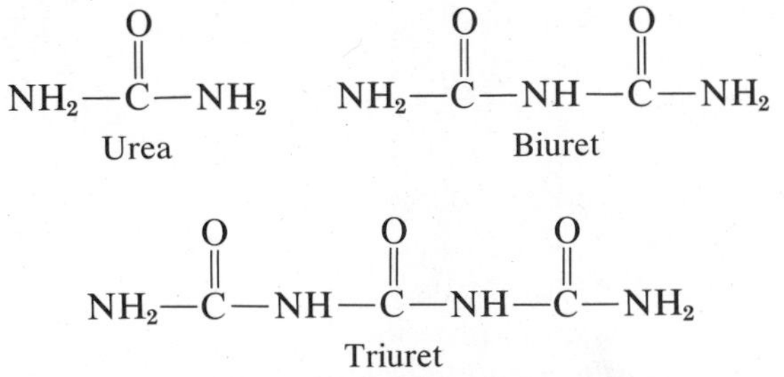

urea to biuret to triuret. This behavior provides a more prolonged release of the nitrogen, more in line with the growth pattern of many plants. More specifically, it has been shown that the triuret nitrogen becomes completely available in 6–12 weeks (11).

Encapsulation is another approach to controlling the rate of release of fertilizers. Some success has been reported on the encapsulation of urea (12). This has been done by developing a coating of sulfur (15% of the final product), wax (3–5%), a microbicide (trace), and a conditioner (1%). The process involves spraying urea granules with atomized sulfur plus the other constituents. The wax acts to seal up the pores in the sulfur coating but also makes the coating a tasty morsel for soil microbes, which may eat holes in the coating so quickly that they negate any attempt at controlled release. Therefore, a microbicide is added to prevent attack. The conditioners used included inert materials such as clay and vermiculite, which counteract the tackiness introduced by the wax. A dissolution rate of 1% per day, which corresponds to a 14–16 week total release, is considered ideal. Dissolution rates of 0.3 to 3% per day were achieved in several formulations. As much as 20% increase in crop yield was achieved over the uncoated urea although it was estimated that the cost of the urea product would increase by about 20% with the coating. Thus, the reduced frequency of application is the major advantage.

(4) Ammonium Sulfate, $(NH_4)_2SO_4$, 21% Nitrogen. This compound is the most important solid nitrogen fertilizer used throughout the world. It contains readily available nitrogen as NH_4^+, is soluble in water, and is not explosive.

(5) Others. There are several other ammonium salts that are used. In fact, almost any ammonium salt with a beneficial or at least innocuous anion can be used as a nitrogen fertilizer. One of the most important ones is ammonium phosphate, the major multinutrient (N plus P) fertilizer in use today.

Many ammonium salts (including ammonium sulfate and nitrate) have an acidifying effect that is usually countered by addition of limestone. To illustrate the magnitude of this problem, it is claimed that for every 100 pounds of ammonium nitrate, 59 pounds of calcium carbonate (limestone) is needed for neutralization of the acid (13).

A major part of the nitrogen fertilization story is the contribution due to *nitrogen fixation*. The process whereby NH_3 is produced commerically can be described as nitrogen fixation but several plants are able to fix nitrogen from the air (80% N_2) directly. Such plants are known as legumes. The fixation process is actually carried out by bacteria that reside in nodules in the roots of leguminous plants. Alfalfa is the most potent nitrogen fixer followed by clovers, soybeans, other beans, peas and peanuts.

There are some bacteria (e.g., *azotobacter*) that are found in soil separate from root nodules but their contribution to the total fixation of nitrogen is minimal.

The nodular nitrogen-fixing bacteria exist in a symbiotic (mutually beneficial) relationship with plants. The plant contains chlorophyll and produces carbohydrates, while the microbes utilize the carbohydrate as fuel for fixing nitrogen, which ultimately ends up as plant protein.

Years ago, it was quite common to grow clover or alfalfa and plow it under to prepare the soil for the following year. In this way, both nitrogen and organic matter was added. Now, there is a tendency to try to utilize the ground year round, e.g., by planting a corn crop followed by winter wheat. This practice puts a much greater demand on the soil, adds no organic matter, and contributes greatly

Figure 9.4 Nitrogen is taken from the air by bacteria that live in the nodules on the roots of alfalfa, clover, and other legume plants. (Courtesy of USDA, Soil Conservation Service.)

to leaching of nutrients. Large-scale leaching ultimately contributes greatly to pollution of lakes and streams (14).

Soybeans are a leguminous crop of tremendous importance. Like other legumes, they do not require nitrogen fertilization. In fact, if nitrogen fertilizers are applied, the bacterial fixation process is often suppressed. This is a source of frustration to plant breeders who have succeeded in making other crops more responsive to fertilizers while the productivity of soybeans has improved very little.

Soybeans have several uses. Soybean oil (discussed in Chapter Fifteen) is in common use. Soybean cake and soybean meal, which are produced when the oil is removed, are very high in protein and are important feeds for livestock. Soy protein is also available as a meat extender for hamburger and other ground meat dishes at about half the cost. It has been estimated that imitation bacon (e.g., Bacos), which is soy protein, can be produced using one-tenth as much cropland as is needed to produce natural bacon (15). Even more dramatic is beef, which requires 15 to 20 times as much land for production of an amount of edible protein equal to that produced by soybeans (40–45% protein). Thus, a greater direct use of soybeans (and other crops) would free more land for growth of more crops that might help control the price of many different food products.

Needless to say, legumes do not contribute to water pollution due to runoff of nitrogen fertilizers, although they do require application of other fertilizers.

Phosphorus

There are several good sources of phosphorus. It is usually found as some form of phosphate, e.g., PO_4^{-3}, HPO_4^{-2}, or $H_2PO_4^-$. The percent is commonly quoted as percent P_2O_5,* which reacts with water to give phosphoric acid, H_3PO_4. The organic sources are bone meal (calcium phosphate, 22-25% P_2O_5), dried shrimp wastes (10%), raw sugar wastes (8%), dried ground fish (7%), activated sludge (3-6%), dried blood (1-5%), wool wastes (2-4%), nut shells (1%), manure (see Table 9.5), wood ashes (1.5%), and tankage from slaughter houses (up to 3%). Up until this century, these organic sources were the primary sources of phosphate. Thereafter, phosphates were obtained from calcium phosphate ores known as *phosphate rock*. These ores are found in extensive deposits in beds that were originally ocean floor. In general, they are a combined phosphate fluoride associated with $CaCO_3$, and are symbolized as $Ca_{10}F_2(PO_4)_6$, although the composition varies with the source. The principal commercial deposits are in North Africa, the United States, and the U.S.S.R. These mineral ores are the primary source of all phosphate fertilizers used today. Organic gardeners use phosphate rock directly without any chemical modification. It is regarded by them as acceptable, since it is a naturally-occurring material. However, the need for increased efficiency of fertilizer for transport and application has led to the production of materials from phosphate rock but with a higher phosphate content. These are described in the following paragraphs.

(1) Phosphoric Acid, H_3PO_4, 73% P_2O_5. Phosphate rock can be converted to H_3PO_4 by the action of sulfuric acid according to the following equation.

$$Ca_{10}F_2(PO_4)_6 + 10\ H_2SO_4 + 20\ H_2O \rightarrow$$
$$6\ H_3PO_4 + 10\ CaSO_4 \cdot 2\ H_2O^\dagger + 2\ HF$$

(2) Superphosphate, 16–22% P_2O_5. Superphosphate is also produced from phosphate rock by the action of sulfuric acid but under different conditions, as illustrated in the following reaction.

$$Ca_{10}F_2(PO_4)_6 + 7\ H_2SO_4 + 3\ H_2O \rightarrow$$
$$3\ Ca(H_2PO_4)_2 \cdot H_2O + 7\ CaSO_4 + 2\ HF$$

---

* %P $\times$ 2.29 = % P_2O_5.
  %P_2O_5 $\times$ 0.437 = % P.

$\dagger$The formula $CaSO_4 \cdot 2\ H_2O$ indicates that there are two molecules of water present for every one calcium and one sulfate in a crystal of calcium sulfate.

The $CaSO_4$ is not removed and the total product has 16–22% P_2O_5. This is normal superphosphate.

(3) Triple Superphosphate, 44–47% P_2O_5. If the original phosphate rock is treated with phosphoric acid rather than sulfuric acid, a product known as triple superphosphate is formed.

$$Ca_{10}F_2(PO_4)_6 + 14\ H_3PO_4 + 10\ H_2O \rightarrow 10\ Ca(H_2PO_4)_2 \cdot H_2O + 2\ HF$$

It has a P_2O_5 content of about three times that of normal superphosphate, since no $CaSO_4$ is present in the final product.

(4) Ammonium Phosphate. Phosphoric acid can also be combined with ammonia to give ammonium phosphate, which is a mixed fertilizer, mentioned previously under *nitrogen fertilizers*. Both the nitrogen and phosphorus contents are variable. They are dependent on the production process. As the following reactions show, the relative amounts of NH_3 and H_3PO_4 determine the final product.

$$H_3PO_4 + NH_3 \rightarrow NH_4^+\ H_2PO_4^-$$
$$H_3PO_4 + 2\ NH_3 \rightarrow (NH_4^+)_2\ HPO_4^{-2}$$
$$H_3PO_4 + 3\ NH_3 \rightarrow (NH_4^+)_3\ PO_4^{-3}$$

The second product, known as diammonium phosphate, is a very common fertilizer.

(5) Basic Slag. Slag is a by-product of steel-making when the steel has a high phosphorus content. Phosphorus is removed by use of dolomite or limestone producing a slag (magnesium and calcium phosphates) that runs about 16% P_2O_5.

   The solubility of phosphate fertilizers is an important characteristic. Phosphate rock is relatively insoluble in neutral solutions, although the exact nature of the ore depends on the source. Ores from North Africa are somewhat different and more soluble than ore from Florida. Liming is often important for efficient utilization of phosphates, since they are most soluble near pH 7. At pH 6 or above, the iron and aluminum oxides are largely precipitated so that the iron and aluminum cannot precipitate the phosphates. Phosphates also tend to be inaccessible due to binding to clay. There is a phenomenon known as *phosphate fixation,* but, unlike nitrogen fixation, this phrase merely describes the precipitation of phosphates (by iron and aluminum), which makes them inaccessible.

Potassium

The percentage of potassium (K) is commonly reported as *percent potash** (K_2O), just as percent phosphorus is reported as percent P_2O_5. This unusual practice is followed even though neither type of fertilizer actually contains either P_2O_5 or K_2O.

---

* % K $\times$ 1.2 = % K_2O.
 % K_2O $\times$ 0.83 = % K.

The United States.

To all to whom these Presents shall come. Greeting.

Whereas Samuel Hopkins of the city of Philadelphia and State of Pensylvania hath discovered an Improvement, not known or used before, such Discovery, in the making of Pot ash and Pearl ash by a new Apparatus and Process; that is to say, in the making of Pearl ash 1st by burning the raw Ashes in a Furnace, 2d by dissolving and boiling them when so burnt in Water, 3d by drawing off and settling the Ley, and 4th by boiling the Ley into Salts which then are the true Pearl ash; and also in the making of Pot ash by fluxing the Pearl ash so made as aforesaid; which Operation of burning the raw Ashes in a Furnace, preparatory to their Dissolution and boiling in Water, is new, leaves little Residuum; and produces a much greater Quantity of Salt: These are therefore in pursuance of the Act, entitled "An Act to promote the Progress of useful Arts", to grant to the said Samuel Hopkins, his Heirs, Administrators and Assigns, for the Term of fourteen Years, the sole and exclusive Right and Liberty of using, and vending to others the said Discovery, of burning the raw Ashes previous to their being dissolved and boiled in Water, according to the true Intent and Meaning of the Act aforesaid. In Testimony whereof I have caused these Letters to be made patent, and the Seal of the United States to be hereunto affixed. Given under my Hand at the City of New York this thirty first Day of July in the Year of our Lord one thousand seven hundred & Ninety.

G Washington

City of New York July 31st 1790. —
I do hereby certify that the foregoing Letters patent were delivered to me in pursuance of the Act, entitled "An Act to promote the Progress of useful Arts"; that I have examined the same, and find them conformable to the said Act.

Edm: Randolph   Attorney General for the United States. —

(Endorsement on back of grant)
Delivered to the within named Samuel Hopkins this fourth day of August 1790.

Th Jefferson

Figure 9.5   The first United States Patent Grant, July 31, 1790. Note the signatures of Washington and Jefferson. The patent was issued to Samuel Hopkins for his process for making potash and pearl ash (purified potash).

Potash, K_2O, is a strongly alkaline substance found in wood ashes in the form of K_2CO_3. The high alkalinity makes it somewhat less useful than most of the other potassium fertilizers.

All of the common potassium fertilizers are acceptable to both organic and nonorganic farmers since they are all naturally-occurring materials.

Granite dust is an excellent source of slow-working potash. The potash content varies from 3–5%, although a Massachusetts quarry yields a granite dust with 11% potash. Granite also supplies various other nutrients (phosphorus, calcium, magnesium, iron, and manganese).

Greensand (also known as greensand marl) is another source of potassium. It is a naturally-occurring marine deposit made up of iron, potassium, silicon, and oxygen. It contains 6–7% potash.

Manures are also an excellent source of potassium and since the potassium cannot decompose, a livestock farm on which manure is handled properly requires little or no potassium fertilization except to make up for losses due to leaching.

The most concentrated natural sources of potassium are deposits of KCl, with major quantities in central Europe, the western United States, and Canada.

Other common ores are potassium sulfate, K_2SO_4, and potassium nitrate, KNO_3. The sulfate is also produced commercially by combining KCl with magnesium sulfate, $MgSO_4$. This is an expensive process, which limits the demand for K_2SO_4. It is primarily used for fertilization of tobacco.

Mixed Fertilizers

Mixtures of the three macronutrients are commonly used. It is generally less expensive and safer to combine these chemicals at the production level than on the farm. Obviously, a mixture precludes the necessity of several applications to assure the full complement of nutrients. A familiar formulation is 5-10-5 and it must be remembered that this signifies 5% N, 10% P_2O_5, and 5% K_2O, which is actually 5% N, 4.4% P, and 4.2% K.

Other Nutrients

Of the other nutrients (secondary and micro) listed in Table 9.2, several were mentioned in the coverage of the major nutrients, since they often accompany the major nutrients or are introduced when modifying pH. For example, liming with dolomite introduces the secondary nutrients calcium and magnesium, whereas potassium or ammonium sulfate, as well as superphosphate, supply sulfur (also a secondary nutrient).

Micronutrients are normally found in sufficient amounts in mineral deposits that are used for fertilization or pH adjustment, either directly or after chemical modification. Phosphate rock (contains molybdenum) and granite dust (contains iron and manganese) are examples.

Many of these micronutrients are an integral part of plant enzymes, e.g., those involved in nitrogen fixation (Mo), photosynthesis (Zn), and respiration (Fe).

THE ORGANIC WAY

Up to this point, there has been an occasional mention of how the organic farmer satisfies specific requirements of fertilization and pH control. Table 9.7 gives a fairly complete compilation of the acceptable and unacceptable components of the organic farmer's arsenal.

Consider some of the items listed in Table 9.7. First and foremost is **compost.** Compost is essentially new top soil. It is the product of a wide variety of organic matter, which contains many minerals and has been subjected to decomposition by the action of aerobic and anaerobic bacteria. These bacteria live on this organic matter and, in doing so, they reduce the carbon content, thus increasing the percentage of nitrogen and other nutrients. Compost is acclaimed as the perfect soil conditioner. The most often cited research on compost was conducted by an Englishman, Sir Albert Howard, who proposed the *Indore Method* of com-

TABLE 9.7 THE ORGANIC APPROACH

ACCEPTABLE	UNACCEPTABLE
compost	insecticides
phosphate rock	herbicides
dolomite	chemical fertilizers
granite dusts	
greensand marl	
pulverized limestone	
ashes	
manures	
green manures	
crop rotation	
earthworms	
oil sprays (for insect control)	
companion planting	
mulching	
plant extractives	

post making. It is one of five methods listed by the editors of *Organic Gardening and Farming Magazine* (16). This method and the others that follow differ primarily with regard to technique. The starting materials for compost vary only slightly. They include weeds, grass clippings, kitchen garbage, garden residues, manure, sewage sludge, bone meal, blood meal, dried fish, leaves (oak leaves for acidity), and lime (for alkalinity).

The five methods of composting are:

1. The Indore Method
2. The 14-Day Method
3. The Anaerobic Method
4. The Sheet Composting Method
5. The Earthworm Method

For the Indore Method, Sir Albert Howard found that layering different organic materials causes rapid decomposition of this material to useable soil material. He laid down a five- to six-inch layer of green material, plant residues, grass clippings, etc., then a two-inch layer of manure, blood meal, sewage sludge, or other high protein material. This was followed by a layer of topsoil, limestone, and phosphate rock. The layers were repeated until the pile reached a height of five feet and a width of ten feet. Stakes that were positioned during the pile formation were removed to provide ventilation to the interior of the pile. The pile was watered plentifully, but not so it became soggy. In three months, the pile had converted to a sweet smelling, dark, crumbly soil.

Within 24 hours after this kind of compost pile is constructed, the interior begins to build up heat. This heat buildup is due to the bacterial action. It is a sign of the decomposition occurring within the pile. The heat builds up to temperatures approaching 82°C within a few days. The temperature will stay in this range for several weeks, then begin to decrease. Ventilation is important at this stage be-

A compost heap.

cause the bacteria are aerobic. If the bacteria cannot get sufficient air, an anaerobic putrefaction may set in. As the pile cools, it should be turned. This allows the material on the surface of the pile to come into the interior. The temperature of the pile will then begin to rise again. The higher the temperature, the shorter the time required to finish the decomposition, although the temperature cannot be raised indefinitely, since the bacterial activity will be adversely affected.

During the decomposition, many changes occur in the pile. A compost heap constructed by the Indore Method goes through five distinct changes to convert the ingredients to finished compost (17a).

> *First Change:* Chemical oxidation produces heat in the same manner as burning. In order for plants to participate, their protective outer layering must be removed. The best way to accomplish this is to shred the components. During the oxidation, the pile decreases in height from five feet to three and one-half feet.

Second Change: Aerobic fungi penetrate the plant material and cover it with white growth, which are new generations of fungi. The growth of fungi is facilitated by heat, moisture, and air. The heat is provided by the first change, the moisture is added, and the air is provided by the ventilation. This phase is complete in about three weeks and is very similar to the rotting of fruit once the skin is broken. If the skin is intact, the fruit keeps much longer.

Third Change: Aerobic bacteria take over from the fungi by eating into the plant material, following the path of the dead fungi. This further breaks down the cells of the components. The bacteria work for about three weeks, at the end of which time most of the heat is lost and the heap is a brown color.

Fourth Change: By this phase the heap is compost and excludes most atmospheric oxygen. Now mostly anaerobic bacteria take over the decomposition, although there is a need for some oxygen, which is provided by slow diffusion through the pile. The heap will permit this diffusion as long as it has not been tamped down or allowed to become sodden. If oxygen is excluded, some bacteria will begin to break down nitrate to obtain oxygen. This will result in the formation of ammonia, which may escape, thus cutting down on the amount of nitrogen present in the compost. This phase is complete in 4 to 8 weeks, depending on the initial components. Cold weather will retard this process.

Fifth Change: If enough lime has been added to prevent the heap from becoming too acid, nitrogen-fixing bacteria will begin to fix appreciable quantities of nitrogen into the heap if air is allowed to penetrate into the pile.*

The overall changes that occur during composting are illustrated by data cited in Table 9.8. The major change is the oxidation of carbon to CO_2, which is lost from the pile. This loss is a favorable one, since it causes an increase in the concentration of several other nutrients that must be obtained from the soil. In contrast, atmospheric CO_2 is a ready source of carbon.

Such compost can be used as a fertilizer, side-dressing, and general soil conditioner. There can be no analysis for compost because of the wide variety of materials that can be used. In general, it is a rich and versatile fertilizer. In fact, compost alone is usually too rich for most plants. Stories are told about vegetables grown in compost alone, such as the single tomato plant that grew five feet by five feet and yielded 100 pounds of tomatoes, or the 129 pound squash grown in a compost bin or the 63 foot squash vine.

The Indore Method can be altered somewhat to speed up the decomposition process. The essential change is the grinding or pulverizing of all components and is the basic modification that constitutes the 14-Day Method. The pulverizing has several effects on the compost. The major change is an increase of the surface area so that the microorganisms can multiply at a greatly increased rate. Aeration is also improved because shredded material has less tendency to pack down. Moisture control is improved due to better absorption by the increased surface area and the heap is easier to turn. No layering is required in the formation of the heap. The pile is usually turned every 2 to 3 days. The graph in Figure 9.6 describes the progress of a compost pile constructed via the 14-Day Method. As the graph shows, this particular case was one in which the temperature reached peaks of 81°C and 74°C and cooled in 12 to 14 days. The compost is sufficiently decayed at this stage for garden use. As in the Indore Method, manure and adequate moisture are essential.

* Reprinted with permission of Rodale Press, Inc., Emmaus, Pa., from *The Complete Book of Composting,* by J.I. Rodale. Copyright 1950 by J.I. Rodale.

TABLE 9.8   TYPICAL CHANGES DURING COMPOSTING (17b)

	BEGIN	END
% Ash	7.6	17.4
% C	41.7	34.5
% N	1.66	2.45
% P	0.13	0.27
% K	0.80	1.65
C/N	24.9	14.1

Compost can also be made entirely by the Anaerobic Method. The chief difficulty in anaerobic composting is effectively eliminating air from the pile. The simplest approach is to cover the heap with plastic but this is difficult for the home gardener. Some European cities compost trash and garbage anaerobically. The method often includes innoculation of the pile with anaerobic bacteria. By this method, it is possible to avoid any wasteful loss of oxidation products and liquids. These liquids, sometimes known as compost tea, are very rich in nutrients. They need not be lost in other composting methods if suitable provisions are made for their recovery.

The Sheet Composting Method is one in which manure and other organic matter are introduced directly into the soil during nongrowth periods. Decay then proceeds in the soil. This is a slower process but one which avoids loss of nutrients (mostly nitrogen), due to the high temperatures of some compost piles.

The Earthworm Method is an interesting one. Worms eat all kinds of organic matter, digesting it and leaving their waste matter, called castings. These

Figure 9.6   The temperature changes in a compost pile made by the 14-day method (17c).

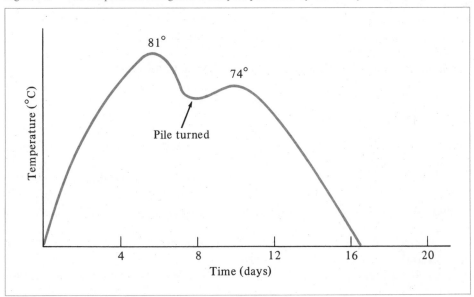

castings are very rich in nitrogen and phosphorus. Worms also work on compost piles. They can be used as a follow-up to the Indore Method after the heat has dissipated, or they can be used on shallow piles (two feet maximum), which prevent the heat buildup. The worms can be purchased by the thousands from seed companies and other garden suppliers. They multiply very rapidly so that only a small fraction of the total needed have to be introduced into the pile. Their castings replace the manure normally added in other methods.

Mulching is another technique used by the organic farmer. It provides several advantages. A layer of material is laid down on the soil. The material may be straw, plastic, rocks, leaves, compost, wood chips, gravel, bark chips, or a variety of other possibilities. Mulching discourages weeds, preserves moisture, keeps low-growing plants clean and dry, acts as insulation by keeping the soil warmer in winter and cooler in summer, and may also fertilize. Of these, the most notable may be the effect on weeds, so that chemical weed killers can be avoided.

Oil sprays and certain plant extractives (e.g., rotenone or pyrethrum dust) are used to a limited extent for insect control. More often companion planting of things like garlic cloves or marigolds is done. Other approaches to insect control, some of which are acceptable to the organic farmer, are discussed in Chapter Ten.

By avoiding the use of chemicals, the organic farmer obviously assures that such materials will not be found in foods. The organic farmer also argues that chemicals not only destroy harmful insects, microorganisms, and weeds, but also may destroy beneficial species. For example, various microorganisms known to produce antibiotics, which may serve to defend plants from disease, may be destroyed by insecticides. A second example is the destruction of worms, which provide both fertilization and better aeration of the soil. The strongest argument is that healthier, organically-grown plants are more resistant to both pests and disease.

Of the items listed in Table 9.7, manure is the final one to be discussed. Typical percent composition of various manures were given in Table 9.6. Although manure is not a concentrated source of nutrients, it is a versatile one, which provides many nutrients as well as organic matter.

Surprisingly, manure requires some careful attention to prevent heavy losses of nitrogen and minerals. The nitrogen may be lost as ammonia due to bacterial attack, whereas minerals may be leached out and washed away.

RIPENING

Another aspect of plant development is the ripening of fruit. Most fruits do ripen on the vine although some (e.g., bananas and pears) are normally picked green and ripened under controlled conditions. The avocado is an example of a fruit that will never ripen if left on the tree. There is evidence that as long as it remains attached, there is a continual supply of some substance (possibly a hormone) that prevents it from ripening.

In many fruits, the start of ripening is accompanied by a distinct increase in respiration. Recalling that dark respiration is useful in providing energy for the plant, it is reasonable that there should be a high level of respiration as the plant matures. This increase is known as the *climacteric rise*. Before the climacteric, the plant is resistant to attack by diseases and microorganisms. It is partly for this reason that it is often desirable to pick the fruit green (e.g., bananas), perhaps even inhibit the ripening in some way, and then induce ripening under controlled conditions.

Inhibition of the ripening can be achieved by keeping the fruit in an atmosphere of only 5–10% oxygen, compared to the normal 20% found in air. This apparently prevents the climacteric rise in respiration and delays the ripening. Refrigeration is an alternate approach to slowing the climacteric phase.

The climacteric rise is also accompanied by an increased production of ethylene in some plants. Moreover, ethylene is known to speed the climacteric rise even at concentrations as low as 1 ppm.* In fact, it is a common practice to ripen green bananas in ripening rooms for three to five days in an ethylene atmosphere.

Since ethylene is produced naturally by some plants and induces ripening, it is often described as a plant hormone.

This form of ripening has been criticized for use on tomatoes in which case the process has been termed "gas reddening" rather than ripening. Although a tomato may reach full size while still green, if picked green, it can be made to change color but not to mature. An immature tomato is often tough and tasteless much like a "rubbery gob of cellulose (18)."

HERBICIDES

Chemical weed killers are in widespread use. They serve a useful purpose and, sometimes, are regarded as culprits.

Widespread herbicide use began in the forties following the discovery of the potential of certain chemicals and has led to a multimillion dollar business.

Weeds are very costly. In addition to lowering the quality and quantity of crops, they may poison livestock, induce off-flavors in milk, reduce the flow of irrigation waters and hinder mechanization of crop production.

The selectivity of weed killers is of major importance. If applied in high dosage, herbicides kill all plants. At low dosage, no plants die. In between, some plants are killed and some are not affected. A waxy layer, known as a cuticle and found on leaves, serves to prevent plants from becoming dehydrated when exposed to low humidity conditions. This layer becomes very important in the action of certain herbicides, since the herbicides may simply roll off the leaf. A **surfactant** (discussed in Chapter Sixteen) is a substance that allows oil and water to mix (i.e., a soap). If a surfactant is added to a herbicide formulation, penetration through the cuticle is possible so that the herbicidal action can take place. In the absence of surfactant, some herbicides act by having their spray droplets bounce off of the leaves of certain grassy cereal crops but wet the broad leaves of certain types of weeds. With surfactant present, the same weed killer may be lethal to the grasses.

The best known weed killers are the chlorophenoxys, 2,4-D and 2,4,5-T and others. The **phenoxys** are growth regulators with hormone-like activity. 2,4-D selectively attacks certain broad-leaf weeds but is harmless to cereals and other grassy crops. The resistance of these grasses may be traceable to the well-known high levels of certain detoxifying enzymes in these plants.

* ppm (parts per million) = milligrams (mg) of substance dissolved per 1000 grams of solution.
  ≅ mg/liter of solution for water solutions.

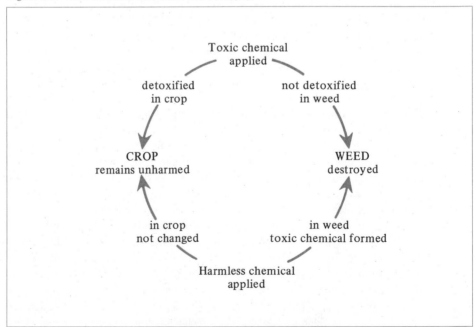

2,4-D
2,4-dichloro
phenoxyacetic acid

2,4,5-T
2,4,5-trichloro-
phenoxyacetic acid

Diuron

The kind of chemical selectivity described in the previous paragraph has been generalized according to the scheme in Figure 9.7 in which a harmless chemical may become lethal, or a toxic chemical may become harmless, due to chemical transformations carried out by either weed or crop plants (19).

Herbicides may also act on weeds by contact in the soil. For this reason, the solubility in soil water can be a major factor in determining the selectivity and success of a herbicide. Compounds of low solubility (e.g., diuron) tend to be strongly fixed in soils and are most effective in killing shallow rooted weeds, while leaving deeply rooted plants unaffected. The opposite is true for more soluble herbicides. Figures 9.8 and 9.9 illustrate the two situations.

Relatively little is known about the mode of action of herbicides, although several possibilities have been suggested. Proliferative growth, due to chromosome aberrations, of certain parts of a weed may cause the weed to literally choke

Figure 9.7.   How selective herbicides can work (19).

Toxic chemical
applied

detoxified
in crop

not detoxified
in weed

CROP
remains unharmed

WEED
destroyed

in crop
not changed

in weed
toxic chemical formed

Harmless chemical
applied

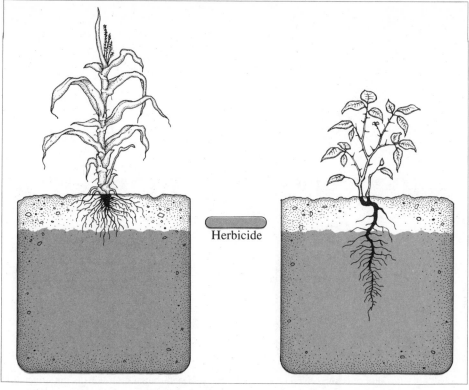

Figure 9.8.   Position of a soluble herbicide in soil (20). Shallow-rooted crop (left) remains alive if herbicide moves beyond its rooting zone. Deep-rooted weed (right) is killed when herbicide is leached into the deeper zones of the soil.

itself to death by crushing or blocking the flow of nutrients and water. This may be the result of uncontrolled synthesis of protein. In some cases, there may be an inhibition of enzymes that are involved in photosynthesis, respiration, or other metabolic processes.

A compound known as TCDD (2,3,7,8-tetrachloro-dibenzo-p-dioxin) has stirred a controversy in the use of certain herbicides. It is a teratogen (causes birth defects) and causes a severe skin disease known as chloracne.

TCDD   DCDD

In 1969, 2,4,5-T was also found to cause some very dramatic teratogenic effects in laboratory animals. This has since been attributed to the presence of TCDD as an impurity in the herbicide. It can be formed as a by-product in the synthesis of 2,4,5-T, if temperature and alkalinity are not carefully controlled. DCDD is the corresponding dioxin that could potentially form in the synthesis of 2,4-D. This formation is less likely, but even if it does occur, the dioxin (DCDD) is not teratogenic. In spite of this and the success in eliminating TCDD from 2,4,5-T, both

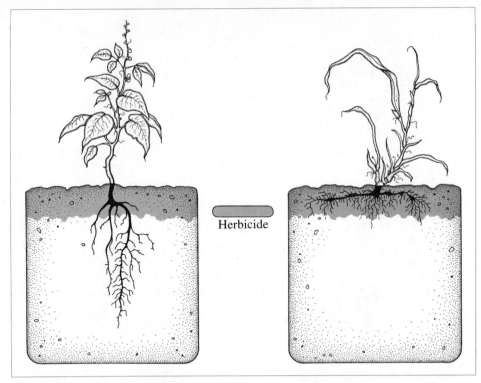

Figure 9.9   Position of a low-solubility herbicide in soil. Deep-rooted crop (left) is not affected by herbicide that remains near soil surface. Shallow-rooted weed (right) is killed by herbicide that stays near surface.

herbicides have been partially banned from use. It has also been concluded that the environmental impact of these two dioxins is not very great. They are so insoluble that they do not leach into groundwater, they decompose in soils, and DCDD may be lost by volatilization (21a). There also appeared to be no buildup of dioxins in a particular food chain (see Chapter Ten for coverage of DDT) that was studied (21b).

SUMMARY

Research on the chemical pathways of photosynthesis and respiration suggests that the rate of net photosynthesis may be affected by the use of chemicals, i.e., growth regulators.

Soils are primarily composed of oxides of silicon, aluminum, and iron, and are commonly designated as clay, silt, or sand in order of increasing particle size. Clay has the greatest affinity for nutrients and water. Sand has the least. Soils that are predominantly clay or sand are the most challenging to work with, whereas loam is the easiest. Organic matter also contributes to soil fertility. The live organic matter is largely microorganisms, which improve soil fertility via mineralization.

The pH of the soil is a major concern when planting a crop. Some crops ·grow very well in alkaline soils, whereas others grow well in an acidic environment. In addition, some nutrients are more available within certain pH ranges. The pH of the soil can be altered in a variety of ways, some of which are even acceptable to organic farmers, who avoid the use of synthetic chemicals.

Fertilizers may be introduced into the soil to act as nutrients for plants that utilize mineral nutrients to produce protein and other organic nutrients that are food sources for animals. Plant nutrients are commonly categorized as macro-, secondary, and micronutrients, in line with their relative importance to plants. In most instances, it is the macronutrients that are of concern when planning fertilization. These are carbon, hydrogen, oxygen, nitrogen, phosphorus, and potassium. The first three are provided by moisture and photosynthesis. The last three must be provided externally and are available in a wide variety of forms.

Encapsulation is a technique for increasing the efficiency of fertilization while, at the same time, cutting down the number of applications that are necessary during a growing season.

Mixed fertilizers are commonly used by the home gardener and contain varying amounts of nitrogen, phosphorus, and potassium described in terms of the percent N, percent P_2O_5, and percent K_2O (potash), respectively. The secondary and micronutrients are less of a problem but are often present in the substances used as fertilizers or for adjustment of the soil pH.

Organic farmers have quite different methods for conditioning soils. For the home gardener, compost is invariably a major part of his arsenal, although substances such as dolomite and phosphate rock are unprocessed minerals and are considered acceptable for organic gardening. Insecticides and herbicides are avoided by organic farmers unless the substance has been extracted directly from a natural source. For the most part, other methods are used to fill these needs.

The Indore Method is the best-known method for making compost. Variations are the 14-day method, the anaerobic method, the sheet composting method, and the earthworm method. Composting is a complex series of aerobic and anaerobic processes, carried out by microorganisms that thrive on the organic matter and convert it into a very fertile topsoil by reducing the carbon content with the resultant increase in the percentage of the useful nutrients.

Some of the complex chemistry of ripening is also used to advantage in initially delaying and later promoting ripening when the time is right.

Herbicides are also very important in influencing productivity. Damage by weeds can be prevented by proper choice of type and quantity of weed killer, although the decisions require a knowledge of the chemistry of the herbicides, in order to minimize damage to the crop while causing maximum damage to weeds.

PROBLEMS

1. Give a definition or example of each of the following:
   - a) photosynthesis
   - b) respiration
   - c) aerobic
   - d) mineralization
   - e) lateritic soil
   - f) dolomite
   - g) urea
   - h) encapsulation

i) legumes
j) phosphate rock
k) carbon fixation
l) nitrogen fixation
m) phosphate fixation
n) potash
o) Greensand

p) compost
q) compost tea
r) climateric
s) herbicides
t) cuticle
u) teratogen

2. Write the reaction for photosynthesis. Define net photosynthesis.
3. What are the major components of soil?
4. What are the major differences between clay, silt, sand, and loam?
5. What are the advantages and disadvantages of clay soil?
6. What does the organic matter found in soils consist of?
7. What are the two major functions of a fertilizer?
8. What are the advantages and disadvantages of ammonia (NH_3) as a fertilizer?
9. What does "5-10-5" mean in regard to a fertilizer?
10. What is the major difference between the Indore Method and the 14-day Method for making compost?
11. How can the ripening of fruit be inhibited or promoted?
12. Why has the average tomato produced today lost its flavor?
13. What effects do weeds have on the farming industry?
14. Explain how solubility in soil water can be a major factor in determining the selectivity and success of a herbicide?
15. Which of the following happen during composting, and why are these changes beneficial?
    a) the carbon content increases
    b) the nitrogen content increases
    c) the phosphorus content increases
    d) the potash content increases

REFERENCES

1. Zelitch, I. *Photosynthesis, Photorespiration and Plant Productivity.* N.Y.: Academic Press, 1971.
2. Marx, J.L. "Photorespiration: Key To Increasing Plant Productivity?" *Science 179* (1973): 365.
3. VanOverbeek, J. "The Control Of Plant Growth." *Scientific American 219* (1) (1968): 75.
4. Salisbury, F.B. "Plant Growth Substances." *Scientific American 196* (4) (1957): 125.
5. "Without Leaves, A Tree Cannot Grow." *Chemistry 47* (1) (1974): 5.
6. Grime, J.P. "Shade Tolerance In Flowering Plants." *Nature 208* (1965): 161–163.
7. Slack, A.V. *Defense Against Famine.* Garden City, N.Y.: Doubleday, 1970, pp. 40–42.
8. Likens, G.E., and Bormann, F.H. "Acid Rain: A Serious Regional Environmental Problem." *Science 184* (1974): 1176.
9. (a) Rodale, J.I. *Encyclopedia Of Organic Gardening.* Emmaus, Pa.: Rodale Press, 1959, p. 8.
   (b) *ibid.,* p. 689.
10. "If You Heat NH_4NO_3, Will You Laugh Or Cry." *Chemistry 38* (4) (1965): 33.

11. Hays, J.T., and Hewson, W.B. "Controlled Release By Chemical Modification Of Urea: Triuret." *Journal of Agricultural and Food Chemistry 21* (1973): 498.

12. Rindt, D.W., et al. "Sulfur Coating On Nitrogen Fertilizer To Reduce Dissolution Rate." *Journal of Agricultural and Food Chemistry 16* (5) (1968): 773.

13. Bear, F.E. *Soils and Fertilizers.* New York: Wiley, 1953, p. 222.

14. Likens, G.E. "The Nutrient Cycle Of An Ecosystem." *Scientific American 223* (1970): 92.

15. Dovring, F. "Soybeans." *Scientific American 230* (2) (1974): 14.

16. Gerras, C. *300 Of The Most Asked Questions About Organic Gardening.* New York: Bantam Books, 1974, Chap. 4.

17. (a) Rodale, J.I. *The Complete Book Of Composting.* Emmaus, Pa.: Rodale Press, 1960, pp. 54–6.
    (b) *ibid.,* p. 87.
    (c) *ibid.,* p. 115.

18. "Why Tomatoes You Buy This Winter May Be Tough, Tasteless, Costly." *Consumer Reports 38* (1) (1973): 68.

19. Wain, R.L. "Selective Herbicidal Activity." *Chemtech,* June 1975, pp. 356–360.

20. Ashton, F.M., and Crafts, A.S. "Mode Of Action Of Herbicides." Wiley-Interscience, 1973, p. 21.

21. (a) Kearney, P.C., et al. "Environmental Significance Of Chlorodioxins." In E.H. Blair, Chlorodioxins–Origins and Fate, Amer. Chem. Soc., Washington, D.C., Advances In Chemistry Series, No. 120, 1973, Chap. 11.
    (b) Woolson, E.A., et al. "Dioxin Residues In Lakeland Sand And Bald Eagle Samples." *ibid.,* Chap. 12.

In this chapter, the agricultural story continues with the coverage of one of the farmer's greatest problems. Losses due to insect damage have been characterized in several ways. It is often stated that one-third of everything grown or stored is consumed by insects. In addition, it is estimated that one-half of all human deaths and deformities due to disease is traceable to insects. The boll weevil reportedly causes losses of $200–300 million annually (1,2). It is the most significant farm pest, but many others cause other dramatic losses.

Malaria is the biggest killer of man, even though only a small percentage of those who contract the disease die from it. The *anopheles* mosquito that transmits malaria is regarded as the most harmful disease-carrying pest (3). In parts of the world where malaria is common, it is so widespread that it cripples a large percentage of the work force, with serious consequences to the economics of many countries.

Consideration will be given to some of the behavioral characteristics of the boll weevil, the *Anopheles* mosquito, and several other insects. The emphasis throughout will be on the role of chemicals as they affect the behavior of insects and as they can be used to control the population of insects.

Insect control is a multidisciplinary subject. Scientists from several fields within the general areas of biology and chemistry have been actively engaged in research, often combining efforts to take advantage of their respective areas of expertise. New techniques and instrumentation developed by and for the chemist have been of major importance in aiding progress in this field since they have allowed separation, identification, and synthesis of components of complex mixtures obtained from insects and other natural sources.

Control of pests can take many forms. The most obvious and most common approach is the use of insecticides. This purely chemical approach, some purely biological approaches, and several intermediate ones will be considered in the following paragraphs.

INSECT CONTROL: CHEMICAL COMMUNICATION

INSECTICIDES: THE BULLDOZER APPROACH

In addition to the *Anopheles* mosquito and its role in the spread of malaria, the louse is the carrier of typhus, the rat flea transmits the plague bacillus, and the tsetse fly has caused major epidemics of sleeping sickness (4).

DDT (dichlorodiphenyltrichloroethane) is the best known, most effective insecticide, but is only one of many components of a very versatile arsenal of insecticides that have been used very effectively. In many instances, minimizing crop losses due to insects is of great importance, but the most dramatic examples of insect control have been in combating epidemics of disease.

In Greece in the 1940s, there were two million malaria cases each year resulting in about 10,000 deaths. Application of DDT has reduced the malaria to about 50,000 cases per year in recent years.

In India, mosquito eradication programs were put into effect in 1953 at which time 50 to 100 million cases of malaria were reported each year. By 1968, the figure had dropped to only 300,000 per year. Life expectancy rose from 32 in 1948 to 52 in 1970. This form of death control is a major factor in the population explosion (5).

In Ceylon, malaria was a major problem, which was reduced from over two million cases in 1946 to only 17 cases in 1963 through the use of DDT. Unfortunately, total eradication of the mosquito was not achieved before the use of DDT was discontinued and malaria returned to the tune of over two million cases in 1968 and 1969.

The effectiveness of DDT in combating malaria is partly due to the behavior of *Anopheles* mosquitos and their means of transmitting malaria. Prior to

Figure 10.1   A mosquito drawing blood. (Courtesy of Edward S. Ross.)

laying eggs these mosquitos must have a blood meal. Ten to twelve days later the mosquito is capable of transmitting the malaria parasite picked up from the first bite. During this time, the mosquito routinely rests on interior walls, a fact that provides the means for control. It has been a routine practice to spray interior wall surfaces with DDT, so that when the mosquito or other insect does set down and come into contact with the DDT, it is killed. The exact mode of action of DDT on insects is not known, but it appears to be some kind of attack on the central nervous system.

DDT was first synthesized in 1873, but its insecticidal properties were not discovered until 1939 by J.R. Geigy. Geigy received the Nobel Prize in 1948 for the discovery. DDT is very inexpensive and easy to apply and its effectiveness in combating diseases has been dramatic, but it has also been termed "the world's most expensive chemical at 17 cents per pound (6)." Its use and the use of other insecticides have also been described as "dropping nuclear bombs on pick-pockets (7)."

There are two major problems associated with insecticides. The first is the development of a resistance to the insecticides. The second is environmental and takes many forms.

Resistance

Resistance to a very potent insecticide is often considered particularly serious, since it seems to imply that the resistant insect is a very sturdy species that is selectively bred into prominence when it no longer has to compete with the nonresistant species. It is often feared that sturdier breeds may also be able to withstand all insecticides. Fortunately, this is not the case. In fact, *many DDT resistant insects have been effectively killed by DDT*. This became possible once the chemical basis of resistance was understood. The resistance was traced to an enzyme that is

Copyright © 1965 Ed Fisher. Reprinted with permission.

"Have you ever noticed? DDT
strikes down only the wicked."

INSECT
CONTROL:
CHEMICAL
COMMUN-
ICATION

185

able to change the DDT to DDE. The enzyme is known as DDT-dehydrochlorinase or simply DDTase. DDE is the major breakdown product of DDT.

$$\text{DDT} \xrightarrow[-HCl]{\text{DDTase}} \text{DDE}$$

DDT
dichlorodiphenyltrichloroethane*

DDE
dichlorodiphenyldichloroethylene*

It is not an insecticide. When an insect has the DDTase in its arsenal, it can render the DDT harmless. However, such an insect can be made vulnerable to DDT by using a DDT synergist. Two synergists are shown below. Their structures are very

Chlorfenthol

p-Chlorobenzene-N,N-dibutyl-sulfonamide

similar to DDT and apparently either one can become bound to the same enzyme (DDTase). This binding serves to inhibit the enzyme and prevent it from attacking DDT. This inhibition of the enzyme causes the DDT to remain intact and effective, since the resistant insect remains susceptible to DDT (5).

In the case of the malaria-carrying mosquito, the problem of resistance is not so great as with certain other insects because the gene that is responsible for synthesis of the DDTase protein is recessive, which means the gene is often inactive even when it is present.† Unfortunately, in other insects and even in some *Anopheles* mosquitos, the DDT resistance is not recessive.

Environmental Problems

In spite of the success in adding synergists to DDT to restore its full insecticidal activity, DDT has been almost totally banned for environmental and health-related reasons.

One major problem with DDT is its lack of specificity in killing insects and other organisms. Only a small fraction of the insect population is harmful. Most insects are either helpful or have no apparent value, although in any given situation they all contribute to the natural biological balance of nature. If, in the course of eliminating an undesirable insect, all of its predators are killed off, the harmful insect could find itself in a sort of *biological vacuum* that might allow it to multiply

---

* A phenyl radical forms when hydrogen on a benzene ring is replaced by another atom or group. The names given for DDT and DDE are common rather than IUPAC names.
† The situation is like having a person with one gene for brown eyes and another for blue. The gene for brown eyes is dominant and the gene for blue eyes is recessive, so that, if an individual has both genes he will have brown eyes.

TABLE 10.1   A TYPICAL FOOD CHAIN WITH RESULT-
ANT MAGNIFICATION OF DDT (10)

LOCATION OF DDT	CONCENTRATION (ppm)
Lake Michigan	0.000002
amphipods	0.410
fish	3–6
gulls	99

far beyond its original population when the DDT use slackens or if resistance develops. Debach cites many examples of insects that changed from minor to major problems when insecticides were used (8). He refers to insecticides as "the ecological narcotics," since the common sequence of increasingly higher doses and frequency of use, plus the variety needed for control, as resistance develops, is much like that experienced by drug addicts.

DDT and several other chlorinated hydrocarbon pesticides came into prominence in the late 1940s. Of these, DDT was the first to be banned in this country, followed by aldrin, dieldrin, and several others.

Creation of a biological vacuum is one problem but pollution is the major reason for these bans. DDT is a major pollutant because of its tendency to undergo "biomagnification (9)." Biomagnification, illustrated in Table 10.1, is a result of the solubility properties of DDT in water and fatty solvents or tissues. DDT has a solubility of about 100,000 ppm* in fat but only 0.0012 ppm in water at 20°C (68°F) (5). In other words, if 1 gm of DDT were placed in a container with equal amounts of a liquid fat and water and mixed thoroughly, 10^9 molecules of DDT would dissolve in the fat for every one that dissolved in the water.

In a living organism, this means that DDT will tend to build up in fatty deposits rather than in the blood (water) from which it could be eliminated. As shown in Table 10.1, this inability to eliminate DDT causes a tremendous magnification of the concentration proceeding from the lake water up the predatory food chain to gulls at the top of the chain. In this example, the gulls have a DDT concentration *in their fatty tissue* about 10 million times that in the lake water, which is the beginning of this particular chain.

Now that we have seen the subtle way in which low levels of DDT can be magnified into very high concentrations, the next logical question is whether DDT and other insecticides are harmful. The answer depends on whom you ask. If you were to ask a representative of a company that makes a sizeable income from the sale of DDT, the answer might be very different from that obtained from an independent scientist (11). The latter would likely refer you to a multitude of literature on the effect of DDT on the reproduction of many species of birds. There are apparently several species of predatory birds, which are high on the food chain, that have vanished from areas where they were once prominent. The most common explanation comes from the observation of very thin egg shells in some of these species. One study found that peregrine falcon eggs collected since the

* ppm (parts per million).

INSECT
CONTROL:
CHEMICAL
COMMUN-
ICATION

187

late 1940s show a sharp drop in shell thickness, averaging 19% (12). Eggshells are largely crystalline calcium carbonate ($CaCO_3$) and an enzyme called *carbonic anhydrase* catalyzes the conversion $CO_2 \rightarrow HCO_3^-$ (bicarbonate ion), which makes carbonate available for eggshell formation. It is currently thought that this enzyme is subject to coprecipitation by DDT, which interferes with the shell formation (13).

DDT has been shown to be a carcinogen in rats, mice, and trout. As for man, there is no conclusive evidence other than the statistical finding that cancer victims have more than two times more DDT in their fatty tissues than the normal population (12). In the United States, the average inhabitant carries a load of 2.3 to 4.0 ppm of DDT and 4.3 to 8.0 ppm of DDE (the breakdown product of DDT) in his body fat. In India, where the use of DDT has been very heavy, 16 ppm DDT and 10 ppm DDE are average. Does this make DDT a serious health hazard? The answer cannot be determined. The kind of testing needed cannot be carried out on humans. Certain manufacturers of pesticides argue that studies done on rats or mice cannot be extrapolated to humans. Unfortunately, when one recalls that pesticide sales were around $500 million in 1966 alone, it seems doubtful that such a source of information can be entirely objective. The Environmental Protection Agency (EPA) upheld the validity of using rats and mice in such a study when they banned certain pesticides, although a ban typically means a ban on further production and not necessarily on the sale of existing inventories (14, 15, 16). In any case, such animal studies are certainly the most practical alternative to studies on humans. Considering the vast amount of information that biochemists have found in studying rats and mice, there is good reason to fear that DDT and several other insecticides could be a serious health hazard, particularly in view of the biomagnification phenomenon.

The use of insecticides has been termed the *bulldozer approach* (17), for the reasons cited. The next question is, what are the alternatives to this purely chemical approach to insect control? There are several possibilities to be considered ranging from purely biological control methods to biochemical approaches. It is the latter type that this coverage will concentrate on, but before addressing these methods directly, it is necessary to consider some of the fascinating ways in which chemicals play a role in the behavior of certain types of insects. A knowledge of this chemistry suggests some very logical ways to turn insects against themselves.

CHEMICAL COMMUNICATION

A wide variety of chemical compounds play a very dramatic role in controlling the behavior of insects, as well as other living systems. Compounds that are sources of energy (carbohydrates, fats, etc.), structural components (proteins), catalysts (enzymes), DNA, hormones, and many others, are common to all living organisms from bacteria to man.

Although man would seem to be the most sophisticated and most complex living organism, intelligence is his most unique characteristic, and intelligence actually eliminates the need for certain elaborate chemical mechanisms that are essential for the survival of lower animals. One rather obvious example is defense.

The following discussion will cite several examples of chemicals (venoms,

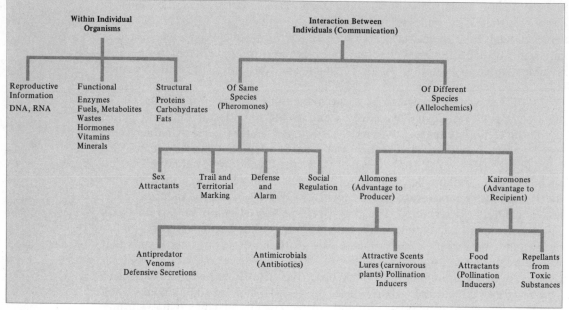

Figure 10.2   A flow sheet description of the role of chemicals in insects and other organisms.

repellents) that are used by insects to defend themselves from would-be predators. Chemical sex attractants are often essential to assure mating and propagation of the species in many insects.

Variations of the scheme in Figure 10.2 have been proposed by several workers in the field as a way of categorizing known or suspected chemical compounds used by insects and other living organisms for communication, defense, etc.

In the following paragraphs, various insects will be considered to determine how they fit into parts of the scheme outlined in Figure 10.2. Some of the insects are very undesirable ones (e.g., the boll weevil and the gypsy moth) and some of these will also be considered in the section titled Modern Methods of Pest Control.

The Honeybee (*Apis mellifera*)

The honeybee is one of the best understood and most dramatic examples of complex chemical communication in insects. As with other insects, vision, hearing, and touch each play a role, but this coverage will concentrate on the chemistry related to some of the categories in Figure 10.2.

The social structure of the honeybee colony is an unusual one consisting of only one developed female, the *queen*. The other inhabitants of the hive include the males, called *drones,* whose sole purpose is to mate with the queen. The drone is absent from the nest a large part of the time. Many thousands of *workers* make up the bulk of the population. The workers, which are underdeveloped females, do not serve a reproductive role unless a queen becomes lost. In view of the relative darkness of the beehive, it is not surprising that several chemicals are used to

Figure 10.3 Queen honeybee on comb with workers. (Courtesy of Edward S. Ross.)

regulate the social order of the nest and to assist in many functions that work for the good of the nest (18–20). These substances are called **pheromones** (see Figures 10.2 and 10.4).

The most thoroughly studied pheromones are the first two listed in Figure 10.4. The 9-oxo compound is commonly known as *queen substance*. It is a multi-purpose substance that is produced by the mandibular gland. There is evidence that queen substance may work together with other substances that have the same properties or are synergists since the isolated pure substance does not exhibit its full biological effect. One such effect is the inhibition of ovary development in the workers. This substance appears to inform the workers of the presence of the queen. If she is removed, the workers will build queen cells to rear new queens from eggs previously laid by the queen. In addition, the ovaries of some of the

Figure 10.4   Pheromones of the honeybee (21, 22).

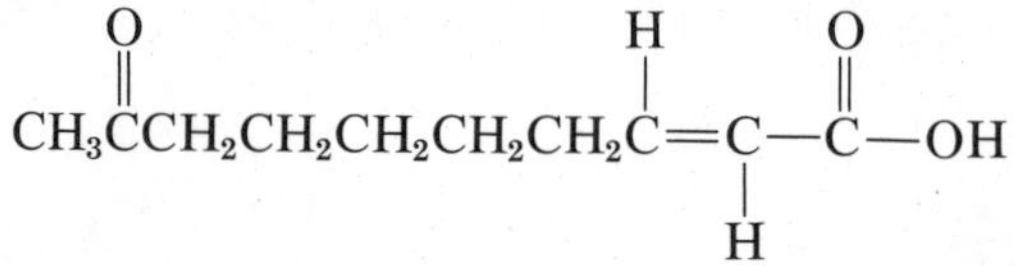

9-oxo-*trans*-2-decenoic acid
Shorthand notation: 9-oxo

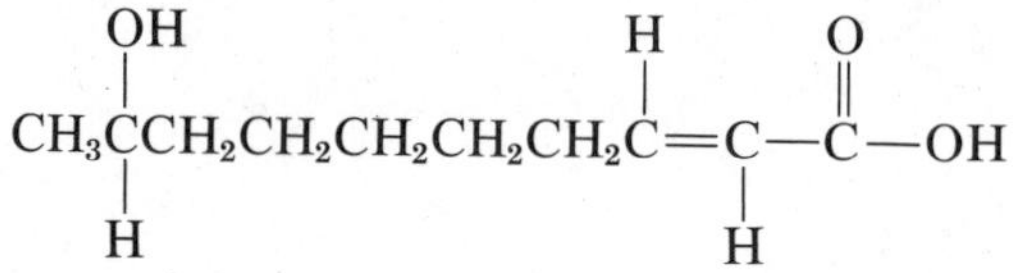

9-hydroxy-*trans*-2-decenoic acid
Shorthand notation: 9-hydroxy

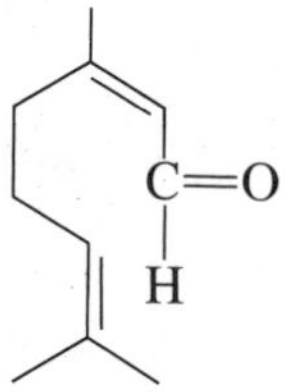

citral

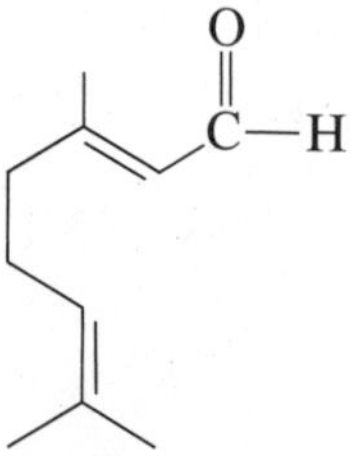

geranial

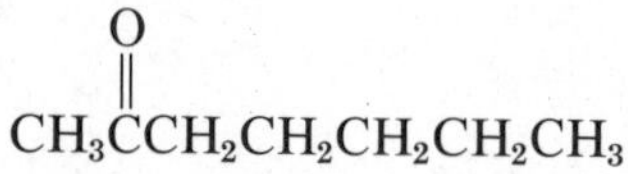

2-heptanone

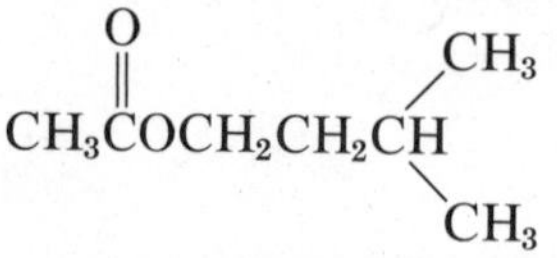

isoamyl acetate

INSECT
CONTROL:
CHEMICAL
COMMUN-
ICATION

191

workers ripen and they begin to lay eggs that develop into drones (23). If a queen is removed from a hive and queen substance is supplied, ovary development is significantly inhibited (24, 25).

Queen substance also serves as a sex attractant used by the queen to attract drones when she leaves the nest to go on her nuptial flight. It is known that the drone honeybee can only see a queen from about 1 meter or less but can be lured from as much as 60 meters downwind.

The 9-hydroxy compound has been called *queen scent* and appears to function as a cohesion pheromone. It, too, is released from the mandibular gland and helps to keep the workers clustered around the queen. If queen substance alone is introduced into the nest, the workers are inhibited from developing, but the 9-hydroxy compound has a quieting effect on the workers.

Citral and geranial are the most important substances released by workers from the Nassanoff gland when they locate a food source. These substances are very attractive to other foraging bees and thereby increase the efficiency of locating food for the entire colony. The pheromones released from the Nassanoff gland are efficiently broadcast by fanning currents of air with the wings (26).

Defense is also of great importance for the survival of the honeybee colony. It has long been known that a single sting increases the likelihood of attack by more workers. It is now known that this is due to the release of isoamyl acetate, a sweet-smelling component of banana oil, which elicits aggressive behavior in other workers and seems to direct them to sting in the same location. Isoamyl acetate is only released during stinging, whereas 2-heptanone, which is another pheromone, is released whenever workers are excited and is apparently of major importance in organizing defense against robber bees or other attack. Certain robber bees in search of food exude citral, which serves to disorganize host workers when they

Figure 10.5   Honeybee broadcasting pheromone. (Courtesy of Edward S. Ross.)

are under attack. If a particular colony is not affected by citral, it is not vulnerable to attack by these robbers (27).

Male bumblebees mark their territory with a pheromone that attracts both males and females so that mating probability is increased (27).

Worker honeybees are also able to distinguish between a mated laying queen and a strange virgin. Some unidentified pheromone reportedly will incite the workers to kill the stranger or even a native worker contaminated with the substance (26).

Finally, the venom released by the sting contains certain proteolytic (protein cleaving) enzymes that attack proteins involved in nerve impulses. This may cause paralysis and death in some recipients. It has also been reported that hemorrhaging and hemolysis (cleavage) of red blood cells are other toxic effects of bee venom (28, 29).

The Gypsy Moth (*Porthetria dispar*)

The gypsy moth is an insect that has caused heavy losses of forest timber. Up to now the destruction has occurred primarily in the northeastern United States, but there is evidence of infestation as far south as Alabama and as far west as Wisconsin. It first became established in the northeast because it was imported in 1869 into Massachusetts by a scientist who was hoping to breed a sturdier silk moth

Figure 10.6 (a) Male gypsy moth; (b) female gypsy moth; (c) egg mass.

Figure 10.6 (*Continued*) (d) Larvae. (Photos courtesy of United States Department of Agriculture.)

(30). Unfortunately some moths escaped and the species flourished in the absence of its natural enemies. The larvae and eggs can be carried long distances on cars and other vehicles, which accounts for its designation as a gypsy.

In the spring, the larvae emerge from eggs laid the previous year. The larvae pass through many stages of development and finally change into the pupal stage from which the moth emerges. It is the larvae or caterpillars which cause the damage. It is reported that a single 5 centimeter caterpillar eats about 0.1 square meter (1 square foot) of leaf surface a day (31, 32). Some trees are killed by even a single defoliation and many more are killed by two successive defoliations (33).

Adult moths do not feed, and live only a short time but it is the adult moths that mate and propagate the species. Only the male gypsy moth flies. The female is loaded down with eggs and is unable to fly. To assure successful mating, the female moth produces a potent attractant pheromone, which the male detects. The male then flies upwind to find the female.

The sex attractant is heavier than air but the male moth tends to fly close to the ground. This improves his chances of picking up the scent of the female, which tends to locate on tree trunks.

Three compounds have been associated with the attraction by the female. The compounds are known as gyptol, gyplure, and disparlure. Gyptol was isolated from the female gypsy moth and appeared in most cases to be attractive to males. It was desirable to try to synthesize the compound for further studies but gyptol is relatively difficult to make. A very similar compound known as gyplure was more easily synthesized from a component of castor oil, but it was found to be unattractive to males.

$$CH_3(CH_2)_5\overset{\underset{|}{OCCH_3}}{CH}CH_2\overset{H}{C}=\overset{H}{C}(CH_2)_6OH$$

d-10-acetoxy-cis-7-hexadecen-1-ol
gyptol

$$CH_3(CH_2)_5\overset{\underset{|}{OCCH_3}}{CH}CH_2\overset{H}{C}=\overset{H}{C}(CH_2)_8OH$$

d-12-acetoxy-cis-9-octadecen-1-ol
gyplure

$$CH_3(CH_2)_{10}\overset{H}{C}-\overset{H}{C}(CH_2)_4CH(CH_3)_2$$

cis-7,8-epoxy-2-methyloctadecane
disparlure

The bioassays needed to determine the attractiveness of this and other pheromones are frequently the most difficult part of the research and results are often conflicting or confusing (30). In some cases, the problems have been traced to the daily behavioral patterns of certain insects, e.g., the finding that some insects only mate at certain times of the day or night so that a sex attractant may only be effective a few hours a day. In any case, both the natural gyptol and the synthetic gyplure are now known to be unattractive. The confusion over the attractiveness of gyptol is attributed to trace contamination of the isolated gyptol, with the real attractant known as *disparlure*. It is an extremely potent attractant (34).

The Boll Weevil (*Anthonomus grandis*)

It was previously noted that the boll weevil is the most destructive agricultural pest. In this insect, the males produce an attractant pheromone that is best described as an aggregation pheromone, since it is equally attractive to both males and females. The males are only attracted to females from a few inches away, presumably by vision alone, whereas the male pheromone is detectable up to 30 feet.

Figure 10.7   The boll weevil. (Photo courtesy of United States Department of Agriculture.)

The male pheromone is known as *grandlure* and is a combination of the following four compounds.

196

A cotton diet leads to greater attractiveness of the males than an artificial diet. It is thought that some constituent of the cotton is converted to one or more components of the pheromone mixture.

Other Insects

The successful mating of the silkworm moth, *Bombyx mori,* is assisted by the use of a female sex attractant pheromone known as bombykol. The pheromone is not attractive to other females. The level of the bombykol is at a maximum when the moth emerges and the production is suppressed after copulation (24).

$$CH_3CH_2CH_2\overset{\overset{\displaystyle H}{|}}{C}=\overset{\overset{\displaystyle H}{|}}{C}-\overset{\overset{\displaystyle H}{|}}{C}=\underset{\underset{\displaystyle H}{|}}{C}(CH_2)_9OH$$

Bombykol *trans*-10-*cis*-12-hexadecadien-1-ol

The common housefly, *Musca domestica,* mates with the help of a female sex attractant known as muscalure, which is weakly attractive to sexually mature males (35).

$$CH_3(CH_2)_{12}\overset{\overset{\displaystyle H}{|}}{C}=\overset{\overset{\displaystyle H}{|}}{C}(CH_2)_7CH_3 \qquad \begin{array}{c} cis\text{-9-tricosene} \\ \text{(muscalure)} \end{array}$$

The American cockroach, *Periplaneta americana,* has both a female sex attractant and another very potent substance, called *Seducin,* which is put out by courting males and acts as an arrestant (36).

There have been suggestions that mosquitos are attracted to humans by a substance(s) that has kairomone activity (see Figure 10.2), but none has been proven to exist. Mosquitos are stimulated to fly by an increase in the CO_2 level. When it encounters a warm moist air current, the mosquito is attracted to fly upwind. The common mosquito repellents have been found to interfere with the

INSECT
CONTROL:
CHEMICAL
COMMUN-
ICATION

197

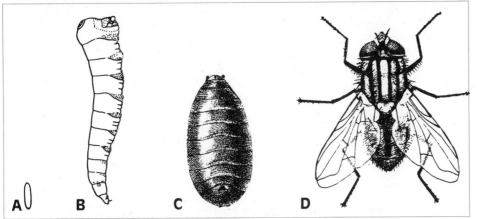

Figure 10.8   Life stages of the house fly, *Musca domestica*. A, egg; B, larva; C, puparium; D, adult. (Courtesy of United States Department of Agriculture.)

mosquito's search for the source of the warm, moist air. The term *repellent* is regarded as a misnomer since mosquitos are not really repelled. Instead the repellents seem to block the sensors that the mosquito uses for direction so the mosquito is unable to locate his target (37).

DEFENSE

Chemical defense has been studied in many species of insects and certain animals. The best known defensive secretion is the one given off by an irritated skunk. This secretion has only recently been found to consist of a ratio of 4:3:3 respectively of the following three compounds (38).

In insects, the pheromone system is more subtle but just as effective in pre-

Figure 10.9   Insect repellents. (Photo by James J. Kane, Jr.)

$$\begin{array}{c} H_3C \\ \\ H \end{array} C = C \begin{array}{c} H \\ \\ CH_2SH \end{array}$$ *trans*-2-butene-1-thiol

$$\begin{array}{c} CH_3CHCH_2CH_2SH \\ | \\ CH_3 \end{array}$$ 3-methyl-1-butanethiol

$$\begin{array}{c} H \\ \\ CH_3S-SCH_2 \end{array} C = C \begin{array}{c} CH_3 \\ \\ H \end{array}$$ methyl-1-(*trans*-2-butenyl) disulfide

venting or defending against the attack by certain would-be predators. Long-or short-range sprays or surface secretions are well known in several species of insects (39, 40). Alarm pheromones and sting venom in the honeybee have already been discussed.

Certain species of millipedes store a compound known as *mandelonitrile,* which is a source of *hydrogen cyanide,* according to the following reaction.

$$\begin{array}{ccc} OH & & O \\ | & & \| \\ C_6H_5-C-H & \longrightarrow & C_6H_5-C-H & + & HCN \\ | & & \\ CN & & \end{array}$$

mandelonitrile   benzaldehyde   hydrogen cyanide

When attacked by a predator, the millipede carries out this reaction and dis-

Figure 10.10   Grasshopper with defensive froth. (Courtesy of Edward S. Ross.)

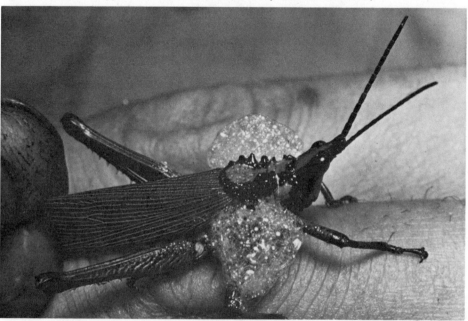

charges the toxic repellent (41). Other millipedes ooze a mixture of compounds when irritated. The major repulsive components are known as *quinones* (42).

The whip scorpion, which is only a few centimeters long, is able to accurately direct a spray many times that distance. A would-be predator receives a unique mixture of acetic acid, caprylic acid (also known as octanoic acid), and water. The acetic acid is the irritant but its effectiveness is enhanced by the presence of the fat-soluble caprylic acid, which permits penetration of the waxy coating found on many recipients. The acetic acid alone would have relatively little effect on some enemies.

$$H-\overset{\overset{\displaystyle O}{\|}}{C}-OH \qquad CH_3\overset{\overset{\displaystyle O}{\|}}{C}-OH \qquad CH_3(CH_2)_6\overset{\overset{\displaystyle O}{\|}}{C}-OH$$

formic acid   acetic acid   caprylic acid

Formicinae ants defend themselves by discharging a spray of formic acid as far as 20 centimeters. Other ants use acetic and other acids, as well as some more complicated chemicals (42).

The monarch butterfly has an effective defensive chemical substance which its larvae acquire by feeding on milkweed plants. Predatory birds vomit violently

Figure 10.11 Monarch butterfly. (Courtesy of Edward S. Ross.)

Figure 10.12   Toad being repelled by a bombardier beetle. The toad (A) eyes the beetle, (B) strikes at it with its sticky tongue, and (C) rejects it. (Courtesy of Thomas Eisner.)

after consuming this species and quickly learn not to repeat their mistake. The monarch is brightly colored, which helps potential predators to clearly identify and avoid this species.* If these same butterflies are raised on cabbage they are safely consumed by predators. In fact, if predatory birds are conditioned to consume monarchs that are raised on cabbage, the predators will also attack the unsafe monarchs and suffer the consequences (43–46).

The bombardier beetle produces a quinone secretion when attacked. In

* Apparently other similarly colored butterflies that can be mistaken for the Monarch may also be avoided by potential but conditioned predators that fear a repeat of a previously unpleasant experience.

Figure 10.13   A bombardier beetle is shown in the moment that it is "firing" a defensive spray forward in response to having a front leg pinched with forceps. For purposes of photography, the beetle has been attached to a hook with wax. (Courtesy of Thomas Eisner and Daniel Aneshansky.)

Figure 10.12 (A), a predatory toad is shown about to consume a tasty morsel. Unfortunately for the toad, the defensive secretion is detected when the toad strikes the prey (B), and the beetle is quickly rejected (C) (47). The defensive chemical spray of a bombardier beetle is also shown in Figure 10.13.

Whirligig or gyrinidid beetles are protected from attack by fish by secretion of a milky, odorous fluid, which has been identified as gyrinidal (48).

gyrinidal

In some instances, insects use chemicals for offensive purposes. Robber bees, previously mentioned, are able to disrupt host worker bees, thus permitting the theft of food supplies (27).

Certain species of ants have been observed using chemicals as offensive weapons. Slavery is practiced by several species of ants that raid ant colonies, discharge certain chemicals at defending workers, and capture the worker pupae. When these pupae emerge from the pupal stage, they accept the slavemakers and assist in the work of the foreign colony (49).

PHEROMONE DUPLICATION IN DIFFERENT INSECTS

One intriguing aspect of the use of pheromones is the finding that several different insects may use the same pheromone for the same purpose (50–52). This would seem to be perfectly reasonable except when sex attractant pheromones are involved, in which case confusion could have serious consequences. There are several trivial reasons why this would not normally be a problem. For one thing, duplication of the sex attractant pheromone would not be a problem if the insects have different daily or seasonal cycles such that mating occurs at different times of the day or year. Different geographical location is another obvious way of avoiding confusion. This kind of separation could even be the result of their attraction for different plants. Morphological incompatability for copulation is another obvious solution to the problem. Genetic incompatability is a more chemical way in which confusion may be avoided.

The cabbage looper and the alfalfa looper utilize the same sex attractant, called *looplure,* but these two insects seem to have a very subtle mechanism to

Figure 10.14   The cabbage looper. (Courtesy of Edward S. Ross.)

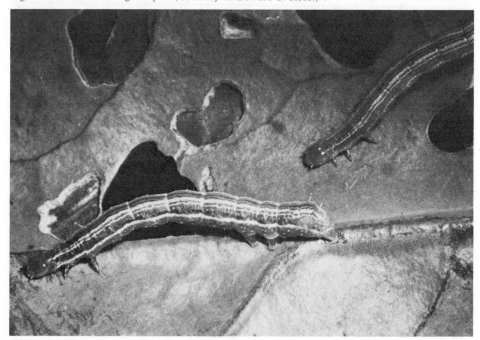

INSECT
CONTROL:
CHEMICAL
COMMUN-
ICATION

203

provide separation. In this instance, the amount of pheromone released by the females and the sensitivity of the corresponding males seem to be attuned to one another. It has been shown that the male cabbage looper is less sensitive, i.e., requires more of the pheromone for attraction, than the alfalfa looper (51, 52). This observation nicely accounts for the low attraction of the male cabbage looper for the female alfalfa looper (which puts out very little sex attractant). It does not, however, account for the failure of the female cabbage looper (which puts out a higher level of pheromone) to attract the male alfalfa looper. It has been shown that there is not a total avoidance of one species by the other but a very strong preference between the males and females of each individual species. But even the strong preference leads to the conclusion that the detection system of the very sensitive male alfalfa looper may become saturated by a high concentration of female cabbage looper pheromone at a point when the male is far downwind from the female. In such a situation, further travel upwind would certainly increase the amount of pheromone reaching the male, but if the detection apparatus is already saturated, the male would not be able to detect any progress.

Another subtle means of separation is used by the Indian meal moth and the almond moth, which also use the same sex attractant:

cis-9-trans-12-tetradecadien-1-ol acetate

However, the female almond moth also puts out cis-9-tetradecen-1-ol acetate, which stimulates almond male moths and inhibits Indian meal moths. In the case of the Indian meal moth, the female produces cis-9-trans-12-tetradecadien-1-ol, which inhibits the male almond moths. Thus, both a synergistic and a masking effect help in preventing confusion (51).

cis-9-tetradecen-1-ol acetate

cis-9-trans-12-tetradecadien-1-ol

CHEMICAL DEFENSE MECHANISMS OF PLANTS

A subtle balance of nature also controls the interaction of many plants and insects or other animals that feed on these plants. Pyrethrin is a naturally-occurring insecticide present in chrysanthemums. In fact, it is even considered an acceptable insecticide by some organic gardeners because of its natural origin. Pyrethrin is a potent toxin for many insects although some are unaffected by it. This latter group is known to have certain detoxification enzymes capable of metabolizing this and perhaps other plant toxins. This situation is analogous to DDT resistance due to DDTase and it may be overcome in some cases by the plants themselves, which may produce a natural synergist (53). *Sesamin* is a known synergist of pyrethrin in chrysanthemums.

An insect, usually in the larval stage, that has the necessary detoxification enzymes to overcome the natural plant defenses will often be a natural feeder on such a plant. This would seem to account for the natural preference that certain insects have for certain plants (45, 53).

Catnip is an intriguing substance produced by certain plants in the mint family. Its effect on cats is interesting but its purpose is that of an insect repellent (54).

It has been suggested that some plants may even practice chemopsychological warfare as a defense mechanism. For example, "does an insect that has fed on a fungus containing LSD mistake a spider for its mate, or does a zebra that has eaten a plant rich in alkaloids become so intoxicated that it loses its fear of lions (43)?"

MODERN METHODS OF INSECT CONTROL

An expert in entomology (a branch of zoology that deals with insects) will not approve of the title of this section since several of the methods to be described are by no means modern. Perhaps it might more accurately be titled "post-insecticide era methods of insect control" because many of these traditional methods were largely shelved when the easier and less expensive DDT approach came into prominence in the 1940s.

The following list of methods is a fairly complete one that indicates the tremendously versatile arsenal that is available. These methods are discussed in the paragraphs that follow.

Pesticides
Pheromones (56, 57)
Sterilization of males (58–61)
Antifeeding compounds (43, 62)
Hormones (26, 63)
Biological control
  (a) predators (60, 64–66)
  (b) parasites (59, 60)
  (c) pathogens (60, 66)

Pesticides

Chemical pesticides will undoubtedly continue to play a role in insect control. Some of the available methods are relatively ineffective in dealing with large infestations of insects. Therefore, in some cases, it may be appropriate to use a

INSECT
CONTROL:
CHEMICAL
COMMUN-
ICATION

205

two-phase attack in which insecticides are used to kill off a large percentage of an insect population, after which another more ecologically oriented approach can be used to maintain control.

Even heavy use of insecticides can be kept under greater control. Biodegradable insecticides are preferred as a means of minimizing pollution. Controlled release or encapsulated insecticides should permit a lower level and frequency of application (67–69).

Pheromones

Perhaps the most intriguing approach to insect control lies in the use of pheromones. By now it is clear that these too are chemicals and that their use could also have an environmental impact, but the problems are really minimal when compared with the advantages. The most obvious advantage of pheromones is their low toxicity. The kind of biological vacuum previously mentioned as a side effect of pesticides is a relatively unimportant phenomenon in the use of pheromones for insect control. The low toxicity, coupled with the minute amounts normally required, results in a very slight environmental impact. The sensitivity that insects have to

Br'er Weevil and the Pheromone Baby.

certain pheromones, which allows the use of such small quantities, is almost unbelievable (70).

Biodegradability is a common characteristic of the pheromones. This is a normal property of most naturally-occurring substances.

The wide variety of approaches using pheromones have one thing in common. Each approach tries to take advantage of natural communication mechanisms, some of which are essential to survival and propagation, and either interfere with them or turn them against the insect.

The attractant pheromones, such as looplure, grandlure, and disparlure, almost seem to imply the obvious approach of using these substances to lure unwanted insects into traps. Such traps could also contain insecticides in an arrangement that would totally prevent pollution problems. A sticky substance that catches the insect once it is lured inside a trap could be used. A chemosterilant could be present in which case the insects would later be released (see later discussion).

Development of resistance, which has so greatly hampered the use of insecticides, seems almost incomprehensible for pheromones. If an insect developed the ability to metabolize its own sex attractant so as to resist its effect, the propagation of the species would be greatly hampered. In other words, development of resistance to a sex attractant would adversely affect an insect. It is interesting to speculate as to whether such a resistance could be induced. It might be possible to breed such a characteristic into a small population of confined insects that were kept in such close contact that an attractant was unnecessary. However, once such a breed was released, they would simply die off.

Even in the insect's normal environment, heavy infestations may not be controllable using pheromone baited traps. When the population density is high, vision plays a more significant role in mating. This is why insecticides may sometimes be necessary before employing other techniques. However, even when pheromones cannot be directly used for control, they may be useful in detecting or even predicting future infestations, which might be dealt with in some other way, once the danger is recognized.

Confusion and *masking* are two specific ways of employing pheromones. The confusion method is illustrated by a study (31) in which small pieces of paper treated with disparlure were air-dropped over a wide area at the rate of 5000 pieces per acre; 20 mg of disparlure was used per acre. Baited traps containing either unmated female gypsy moths or disparlure attracted only a small fraction of the number of males trapped in control areas that were not air-dropped. Apparently, the male gypsy moths had become confused by the abundance of sex attractant and were unable to locate either the females or the traps baited for analytical purposes.

Masking is another way of taking advantage of pheromones for purposes of insect population control (57). The idea of this method is to introduce another substance that will eliminate the attractiveness of the normal sex attractant. It was previously noted that the almond and Indian meal moths utilize the same sex attractant but avoid confusion due to the presence of a masking agent that inhibits the attractiveness of the pheromones.

The pink bollworm moth provides another illustration of the potential for masking. The male of the species is attracted to a substance known as propylure, which is the *trans* isomer of the following structure.

INSECT
CONTROL:
CHEMICAL
COMMUN-
ICATION

207

$$(CH_3CH_2CH_2)_2C{=}CH(CH_2)_2\overset{\overset{\displaystyle H}{|}}{C}{=}\overset{}{C}\overset{\overset{\displaystyle O}{\|}}{}(CH_2)_4O\overset{}{C}CH_3 \qquad \text{Propylure}$$

trans

It has been shown that the *cis* isomer inhibits the attractiveness of the *trans* and could, therefore, provide a means of masking the normal attractant (27, 71).

Sterilization

This technique has been termed "autocidal control." It consists of trapping and sterilizing the male by the use of chemosterilants or ionizing radiation. In the latter approach gamma radiation from a cobalt-60 source has usually been used. Sterilization of females would have little more effect than simply killing them, whereas, a sterile male, which may mate many times, could prevent conception by many females that may mate only once.

The use of such sterile males has been an important method used in the southwestern United States to prevent infestation by the Mexican fruit fly (60). Screwworm control by release of sterile males has also been used in Texas and California, particularly along the Mexican border (61).

The main drawback to this approach is the difficulty of rearing large numbers of sterile males. Obviously, they cannot reproduce, so large-scale rearing followed by sterilization is necessary. Another problem arises from the low selectivity of ionizing radiation and even chemosterilants. In some cases the sterilization process may even make the male unable to mate or at least unable to compete with other males, in which case he has little or no usefulness for purposes of control (59).

Antifeeding Compounds

Certain compounds have been found useful in preventing insects from feeding on plants that they normally consume. In this way, the insect becomes susceptible to starvation in the presence of an abundance of food. Some compounds chosen for this purpose are potential pollutants unless they are biodegradable, but since they do not even directly kill the insect they are not likely to be harmful to predators or parasites. It is intended that such compounds will make the plants unpalatable to the insect and may possibly even make the insect turn to weeds as a source of food, thus providing a double-barreled advantage for the plants (62).

Hormones

In addition to exerting influence over the behavior of insects by subjecting them to external factors, it is possible to alter the normal development of some insects by exposing them to substances that are normally present within the insect.

A substance known as *juvenile hormone* (JH) or certain other chemicals, which mimic JH or other insect hormones, seem to hold a key to success in controlling some insects (72, 73). The juvenile hormone controls the various phases of development of insects as they pass through several larval stages, the pupa stage, and finally mature into an adult. The larval stages require JH, but in order for a mature larva to change into a sexually mature adult, the hormone supply must be cut off. The insect does this automatically but man can interfere at this stage. This stage has been termed "the Achilles heel" because, although man cannot prevent the flow of hormones when the insect needs them, he can supply them when the insect cannot tolerate them. Such is the case with JH. In the presence of an external source of JH, a susceptible insect continues to develop as a larva or it may change into an immature adult, which retains many larval characteristics and is incapable of reproduction.

The end result is comparable to that of an insecticide but with some very significant advantages. The primary advantage is specificity. Unlike many pesticides, JH is not likely to kill off beneficial soil microorganisms, although it could eliminate certain beneficial insects.

Development of resistance is unlikely, since JH is essential to the insect in some stages of development. Finally, since JH is a natural substance, it and other hormones should be biodegradable and therefore relatively nonpolluting.

There is one potential problem with the use of JH. As previously noted for the gypsy moth, the larvae are the big eaters that do the greatest damage to crops, so, if the larval stages are prolonged, crop loss could temporarily increase until the total insect population decreases.

Biological Control

Predators, parasites, and pathogens make up the arsenal commonly known as *biological control*. Some of the other methods already described are supposed to interfere with, or take advantage of, biological development or activity of certain insects, but they are not routinely included under the heading of biological control.

Predators, in this context, consist of toads, frogs, lizards, birds, fish, and certain mammals. A predator normally consumes many prey during its life. In contrast, a *parasite* is generally a smaller organism that establishes itself on or within prey and depends on the prey to supply nourishment. In many cases the prey suffers a slow, agonizing death.

Control by natural enemies like predators and parasites can be very effective, even though these species require some maintenance of the insect population to permit survival and propagation of the enemy. Of particular interest is the unusually great effectiveness of this approach in combating heavy infestations of insects. This is a weakness of several other methods.

Pathogenic (disease-causing) microorganisms and viruses are also very effective in controlling insect populations. Just as blight and other diseases can devastate large areas of cropland, insects are also susceptible. This form of control is also effective in combating heavy infestations.

Other Methods

Other methods and combinations of methods will likely be developed in the future. Perhaps the most intriguing idea to come along recently has been the use of dogs, aptly described as *mothhounds* (74), which are to be trained to pick up the scent of gypsy moths. The idea is to detect the moths in areas which are not infested and, thus, prevent them from becoming established in those areas.

INSECT
CONTROL:
CHEMICAL
COMMUN-
ICATION

209

SUMMARY

Insects are capable of causing tremendous damage by spreading disease and by consuming crops. Insecticides, and particularly DDT, have been the major defense against insects but have fallen into disfavor due to adverse environmental effects. In some cases, insecticides have actually promoted the establishment of harmful insects by killing off harmless predators and because of the development of resistance. Resistance can sometimes be traced to enzymes that convert the insecticide to a harmless substance, e.g., DDT to DDE, and this may make it necessary to use a synergist with the insecticide in order to overwhelm the resistance mechanism. Biomagnification may cause insecticides to become serious hazards.

The honeybee, the gypsy moth, the boll weevil, and most other insects are attracted to mate, to feed, to defend themselves, and for other purposes by chemicals termed *pheromones, allomones, kairomones,* etc. It is possible to use this information to turn insects against themselves. An example is the use of sex attractants for trapping or confusing insects. Sterilization of trapped males may follow in some cases and be used to interfere with normal species propagation. Pheromones can also be used in combination with insecticides in methods that avoid environmental pollution by the latter. Masking can be used to interfere with normal attraction.

Juvenile hormone (JH) is another natural substance that is required for normal development of insects. It can also be turned against insects. Biological control can be carried out by using predators, parasites, or pathogenic organisms.

PROBLEMS

1. Give a definition or example of each of the following:
   a) biomagnification
   b) biological vacuum
   c) pheromone
   d) allomone
   e) kairomone
   f) synergist
   g) hemolysis
   h) entomology
   i) biodegradability
   j) disparlure
   k) grandlure

2. Name some diseases that are transmitted by insects.

3. How did DDT become a factor in the population explosion in India?

4. How does an insect develop a resistance to DDT?

5. How can DDT kill an insect that is resistant to DDT?

6. Why was DDT banned by the United States government?

7. Why, if a person receives one bee sting, is he likely to get several more from the same hive?

8. In what stage of development does the gypsy moth do its greatest damage?

9. What is the most significant agricultural pest? Why?

10. What is the function of *queen substance* in a honeybee hive?

11. How do mosquito repellants work?

12. What is an example of an insect that derives an important allomone from its normal food source?

13. How do cabbage loopers and alfalfa loopers differentiate between the species for mating purposes?

14. What is the major advantage of using pheromones rather than insecticides to control insect populations?

15. Confusion and masking are two approaches to the use of pheromones for insect control. What is meant by each?

16. How does biological control of insect populations differ from other types of control?

17. What is one major technique for sterilization of insects? What is the purpose? Why is it done to males rather than females?

REFERENCES

1. Carter, L.J. "Eradicating the Boll Weevil: Would It Be a No-Win War?" *Science 183* (1974): 494.

2. Cross, W.H. "Biological Control and Eradication of the Boll Weevil." *Annual Review of Entomology 18* (1973): 17.

3. Gillett, J.D. "The Mosquito: Still Man's Worst Enemy." *American Scientist 61* (1973): 430.

4. Rothschild, M. "Fleas." *Scientific American 213* (1965): 44.

5. Metcalf, R.L. "A Century of DDT." *Journal of Agricultural and Food Chemistry 21* (4) (1973): 511.

6. Wright, J.W., and Wurster, C.F. "DDT: It is Needed Against Malaria, But For The Whole Environment?" *Smithsonian 1* (7) (1970): 40–49.

7. Shapley, D. "Mirex and the Fire Ant: Decline of Fortunes of a 'Perfect' Pesticide." *Science 172* (1971): 358.

8. Debach, P. *Biological Control by Natural Enemies.* London: Cambridge Univ. Press, 1974, pp. 11–19.

9. Carter, L.J. "Pesticides: Environmentalists Seek New Victory in a Frustrating War." *Science 181* (1973): 143.

10. Metcalf, R.L. "Pests and Pollution." In *Wednesday Night At The Lab,* edited by K.L. Rinehart, Jr., W.O. McClure, and T.L. Brown, New York: Harper & Row, 1973, pp. 102–3.

11. "DDT Proponents Challenged," *Science 171* (1971): 522.

12. Peakall, D.B. "Pesticides and the Reproduction of Birds." *Scientific American 222* (1970): 72.

13. "Enzyme Inhibition by DDT Questioned." *Chemical and Engineering News,* 18 January 1971, p. 20.

14. Carter, L.J. "Controversy Over New Pesticide Regulations." *Science 186* (1974): 904.

15. "EPA Plans To Ban Two More Pesticides." *Chemical and Engineering News,* 25 November 1974, p. 5.

16. "EPA Orders Suspension of Two Pesticides." *Chemical and Engineering News,* 12 August 1974, p. 4.

INSECT
CONTROL:
CHEMICAL
COMMUN-
ICATION

211

17. Jacobson, M. "Chemical Insect Attractants and Repellents." *Annual Review of Entomology* (11) (1966): 403.

18. Gray, N.E. "Pheromones of the Honey Bee, *Apis mellifera* L." In *Control of Insect Behavior by Natural Products,* D.L. Wood, R.M. Silverstein and M. Nakajima. New York: Academic Press, 1970, pp. 29–53.

19. Wenner, A.W. "Sound Communication in Honeybees." *Scientific American 210* (1964): 116.

20. Butler, C.G., and Callow, R.K. "Honeybee Pheromones." *Chemistry and Industry* (London) (1965): p. 883.

21. Regnier, F.E., and Law, J.H. "Insect Pheromones." *Journal of Lipid Research* 9 (1968): 541.

22. Wilson, E.O. "Animal Communication." *Scientific American 227* (3) (1972): 52–60.

23. Vollmer, J.J., and Gordon, S.A. "Chemical Communication." Part II, *Chemistry 48* (4) (1975): 6–11.

24. Wigglesworth, V.B. *Insect Hormones*. San Francisco: W.H. Freeman, 1970, p. 137.

25. Karlson, P., et al. "Pheromones in Insects." *Annual Review of Entomology 4* (1959): 39.

26. Butler, C.G. "Insect Pheromones." *Biological Reviews 42* (1967): 42–87.

27. Beroza, M. *Chemicals Controlling Insect Behavior*. New York: Academic Press, 1970, p. 87.

28. Beard, R.L. "Insect Toxins and Venoms." *Annual Review of Entomology 8* (1963): 1–18.

29. Hodgson, N.B. "Bee Venom: Its Components and Properties." *Bee World 36* (1955): 217–222.

30. Jacobson, M., and Beroza, N. "Insect Attractants." *Scientific American 211* (1964): 20.

31. Beroza, M., and Knipling, E.F. "Gypsy Moth Control With the Sex Attractant Pheromone." *Science 177* (1972): 19.

32. Reese, K.M. "Gypsy Moth Doing Well." *Chemical and Engineering News,* 22 October 1973, p. 32.

33. Kulman, H.M. "Effects of Insect Defoliation Growth and Mortality of Trees." *Annual Review of Entomology 16* (1971): 289–324.

34. Leonard, D.E. "Recent Developments in Ecology and Control of the Gypsy Moth." *Annual Review of Entomology 19* (1974): 197.

35. Carlson, D.A., et al. "Sex Attractant Pheromone of the House Fly: Isolation, Identification and Synthesis." *Science 174* (1971): 76.

36. Roth, L. "Male Pheromones." *McGraw-Hill Yearbook of Science and Technology* New York: McGraw-Hill, 1967, pp. 293–5.

37. Wright, R.H. "Why Mosquito Repellents Repel." *Scientific American 233* (1) (1975): 104.

38. "Noxious Chemicals In Skunk Scent Detailed." *Chemical and Engineering News,* 16 September 1974, p. 16.

39. Remold, H. "Scent-Glands of Land-Bugs, Their Physiology and Biological Function." *Nature 198* (1963): 764.

40. Eisner, T., and Blumberg, D. "Quinone Secretion: A Widespread Defensive Mechanism of Arthropods." *The Anatomical Record 134* (1959): 558–9.

41. Eisner, T., et al. "Cyanogenic Glandular Apparatus of a Millipede." *Science 139* (1963): 1218–20.

42. Schildkrecht, H. "The Defensive Chemistry of Land and Water Beetles." *Angewandte Chemie. International Edition in English 9* (1) (1970): 1–9.

43. Ehrlich, P.R., and Rowen, P.H. "Butterflies and Plants." *Scientific American 216* (1967): 104.

44. Brower, L.P. "Ecological Chemistry." *Scientific American 220* (1969): 22.

45. Leopold, A.C., and Ardrey, R. "Toxic Substances in Plants and the Food Habits of Early Man." *Science 176* (1972): 512–3.

46. Brower, L.P. et al. "Cardiac Glycosides of the Monarch Butterfly." Proceedings of the National Academy of Sciences of the United States of America 57 (4) (1967): 893.

47. Eisner, T., and Meinwald, J. "Defensive Secretions of Arthropods." *Science 153* (1966): 1341–50.

48. Meinwald, J., and Ophem, K. "Chemical Defense Mechanisms of Arthropods XXXVI Stereospecific Synthesis of Gyrinidal, a Norsesquiterpenoid Aldehyde." *Tetrahedron Letters* (4) (1973): 281–4.

49. Wilson, E.O., and Regnier, Jr., F.E. "Chemical Communication and 'Propaganda' in Slave-maker Ants." *Science 172* (1971): 267–9.

50. Schneider, D. "The Sex Attractant Receptor of Moths." *Scientific American 231* (1974): 28.

51. Tumlinson, J.H., et al. "Sex Pheromones and Reproductive Isolation of the Lesser Peachtree Borer and Peachtree Borer." *Science 185* (1974): 614.

52. Kaal, R.S., et al. "Pheromone Concentration as a Mechanism for Reproductive Isolation Between Two Lepidopterous Species." *Science 179* (1973): 487.

53. Krieger, R.I., et al. "Detoxification Enzymes in the Guts of Caterpillars: An Evolutionary Answer to Plant Defenses?" *Science 172* (1971): 579.

54. Eisner, T. "Catnip: It's Raison d'Etre." *Science 146* (1964): 1318–20.

55. Jones, J.C. "The Sexual Life of a Mosquito." *Scientific American 218* (1968): 108.

56. Marx, J.L. "Insect Control (I): Use Of Pheromones." *Science 181* (1973): 736.

57. Seltzer, R.J. "Role of Insect Sex Lure Isomers Probed." *Chemical and Engineering News* 20 August 1973, p. 19.

58. Ling, L. "Plant Pest Control on the International Front." *Annual Review of Entomology, 19* (1974): 177.

59. Proverbs, M.D. "Induced Sterilization and Control of Insects." *Annual Review of Entomology 14* (1969): 81.

60. Bosch, R. vanden, and Messenger, P.S. *Biological Control.* New York: Intext Educational Publishers, 1973.

61. Calman, J., and Smith, R.H. "Screwworm Control." *Science 182* (1973): 775.

62. Geier, P.W. "Management of Insect Pests." *Annual Review of Entomology 11* (1966): 471.

63. Truman, J.W. "How Moths 'Turn On': A Study of the Action of Hormones on the Nervous System." *American Scientist 61* (1973): 700.

64. Djerassi, C., Shih-Coleman C., and Diekman, J. "Insect Control of the Future: Operational and Policy Aspects." *Science 186* (1974): 596.

65. Berg, C.O. "The Fly That Eats The Snail That Spreads Disease." *Smithsonian 2* (6) (1971): 8.

66. Kilgore, W.W., and Doutt, R.L. *Pest Control: Biological Physical and Selected Chemical Methods.* New York: Academic Press, 1967.

67. "A Microencapsulated Pesticide." *Chemical and Engineering News,* 22 July 1974, p. 12.

68. Dvorchik, B.H., Istin, M., and Maren, T.H. "Does DDT Inhibit Carbonic Anhydrase." *Science 172* (1971): 728.

69. "Controlled-release Pesticides Attract Interest." *Chemical and Engineering News,* 30 September 1974, pp. 20–22.

70. Beroza, M. "Insect Sex Attractants." *American Scientist 59* (1971): 320–5.

71. Jacobson, M. "Sex Pheromone of the Pink Bollworm Moth: Biological Masking by Its Geometrical Isomer." *Science 163* (1969): 190.

72. Williams, C.M. "Third-Generation Pesticides." *Scientific American 217* (1967): 13.
73. "Ideal Mosquito Larvicide." *Chemical and Engineering News,* 29 November 1971, p. 9.
74. "Mothhounds On The Way." *Chemical and Engineering News,* 19 August 1974, p. 44.

INSECT
CONTROL:
CHEMICAL
COMMUN-
ICATION

213

FOOD

PART FOUR

N OW that we have traced food through much of the chemistry of growing it and protecting it from insects, consideration will be given to modifications of certain types of food. In some cases, the modification may irreversibly alter the food from a nutritionally useless or undesirable form (e.g., wheat) to an edible form (e.g., bread). In other cases, foods may be reversibly modified in order to preserve them and make safe storage possible.

The chemistry underlying the production of alcoholic beverages (beer, wine, and distilled beverages) will be considered first. Much of the chemistry involved in the production of alcoholic beverages is similar to that involved in the production of some dairy products, baked goods, and fermented foods, in the spoilage of many foods, and in the production of antibiotics as discussed in Chapter Eighteen.

BREWING

The production of beer is a major industry throughout the world. It is a very complex chemical process. Although some of the minute details of the chemistry are unknown, many of the phenomena are well understood.

If one is to trace brewing all the way to the origin of the starting materials, it is necessary to start by considering barley, which is the major grain used. Barley is first malted. The malt is then subjected to mashing, addition of hops, fermentation, lagering, and packaging.

Malting

Barley is the grain of choice for beer production, in part because of the relative ease of malting. For one thing, barley provides the enzymes, called α- and β-amylases,* which are required for efficient conversion of the plant starches into fermentable sugars. It is also significant that barley hulls remain solidly attached to the kernel even during threshing so that a good sturdy compartment is available for some of the stages of malting.

Malting is a three-step process following the cleaning and sizing of the kernels. The steps are known as steeping, germination, and kilning. The purpose of malting is to develop enzymes that will convert the starch into sugars, which can be fermented to produce alcohol.

* The formal name for α-amylase is α-1, 4-glucan 4-glucanohydrolase. The formal name for β-amylase is α-1, 4-glucan maltohydrolase.

ALCOHOLIC BEVERAGES

Figure 11.1 Barley, a field of beer in the making. (Photo courtesy of Anheuser-Busch.)

Steeping consists of soaking the barley kernels in water at about 10°–15°C for 48–72 hours. During the soaking, a water content of 45–50% is achieved in 24–48 hours. A uniform water content provides conditions suitable for germination.

Germination is encouraged by placing the barley in a location with provisions to maintain both humidity and temperature (typically 15°–25°C). The germination is allowed to proceed for about a week, during which much of the rigid structural starches and proteins are broken down and enzyme systems are developed. The product is known as *green malt*. Its formation is accompanied by respiration and root development. The former is necessary to provide energy for synthesis of the enzymatic proteins and other changes that occur. Root development is a wasteful phenomenon although the roots are a high-protein material that is used as a livestock feed.

Kilning (drying) reduces the water content of the barley to about 10%. Temperatures of 65°–80°C for 1–3 days yield light-colored malt. If temperatures up to 100°C are used, dark colored malts are produced. They are used for porters, stouts, ales, and other dark beers. The lower temperatures favor retention of a maximum of the enzyme activity developed during germination. High temperatures destroy more of the enzyme activity but cause development of colors and flavors that are characteristic of some beers.

Following kilning, the malt is cleaned (roots removed) and either stored or used directly. Malt contains less moisture than barley grain. This makes it more suitable for storage and grinding.

The malt provides several things required or at least desirable for beer production. These include amylase activity to convert the malt starches and any other added starches to fermentable sugars (mostly maltose) and limit dextrins. The latter will be discussed along with fermentation. The malt also provides other proteins that contribute to the body of the beer and the foam. Amino acids, essential for good yeast nutrition, and flavor ingredients are also contributed by the malt.

Mashing

The actual brewing process begins with preparation of the mash, which is done by crushing the malt to expose all of its starches and proteins to the enzymes and water. The mash is cooked for one to two hours, during which time the enzymatic attack by the amylase enzymes on the starches is completed. Much of the barley starch has already been converted to maltose, glucose, and limit dextrins (discussed later) during the malting process. During the mashing, this is completed and adjunct starches are also broken down by attack of the amylase enzymes.

The structure of starches, maltose, glucose, and other carbohydrates was described in Chapter Eight. The reader is encouraged to review that material before proceeding.

Recall that starch consists of both amylose and amylopectin. An amylose molecule is a linear chain of glucose units with only 1,4 linkages between the units. The more complex amylopectin has some branching of the main chains. On the average, the branches are about 12 glucose units long and are attached to the main chain by 1,6 linkages at about every twelfth unit of the chain.

The cleavage of amylose by the amylase enzymes forms maltose (a disaccharide, Glu_2*) plus glucose. The β-amylase is a very specific enzyme that attacks the amylose by successively cleaving off maltose units from one end of the chain to the other.

$$\text{Glu—Glu}\ \text{Glu—Glu}\ \text{Glu—Glu}\ \text{Glu—Glu}\ \text{Glu—Glu}\quad \text{Amylose}$$

$$\downarrow \beta\text{-amylase}$$

$$n\ \text{Glu—Glu}$$

Maltose

The α-amylase is a less specific enzyme that randomly cleaves the amylose chains. This causes formation of many fragments with an odd number of glucose units, which eventually are cleaved into both maltose and glucose.

Neither the α- nor the β-amylase are able to completely break down amylopectin, since these enzymes are unable to cleave the 1,6 linkages found at each of the branches of this polysaccharide. The result is the formation of species known as *limit dextrins*. They are produced when the amylases have acted to near the limit of their capability. The limit dextrins commonly range from di- to about hexasaccharides. They are not fermentable and, therefore, remain even in the finished beer. Along with other components of the malt, they contribute to the stability of the foam.

The product of the mashing step is known as a **wort.** It contains some solids, which are allowed to settle and are removed.

The wort is then boiled with the hops for two to three hours, filtered, and placed in a fermentation tank to which yeast is added. The addition is commonly described as *pitching* or *injection*. The fermentation is allowed to proceed for several days.

* Glu (glucose).

Figure 11.2   A brewing kettle. (Courtesy of the Canadian Brewers Association.)

Fermentation

Two categories of yeast are in common usage. They are known as *top-fermenting* and *bottom-fermenting*. During the vigorous stages of the fermentation, when large amounts of CO_2 are being produced, the yeasts tend to remain suspended, but as the fermentation subsides, the top-fermenting yeast rises to the top and the bottom-fermenting yeast settles.

Typical conditions for fermentation are 4°–10°C for 7-11 days when using a bottom-fermenting yeast, and 10°–16°C for 6 days or less for a top-fermenting yeast. The bottom-fermenting yeast is the type used in the production of the common lager beer that predominates in the United States.

The most important chemical change that occurs during fermentation may be described by the following equation.

$$C_6H_{12}O_6 \longrightarrow 2\ CH_3CH_2OH + 2\ CO_2$$

glucose   ethyl   carbon
                          alcohol   dioxide

Figure 11.3   A fermentation tank being filled with wort. (Courtesy of Paul Mueller Company.)

Unfortunately, this is a tremendous oversimplification. The details of the fermentation process (also called *glycolysis*) are given in Figure 11.4 and, although all the details are not essential, it is important to consider part of the process since we are going to encounter much the same sequence in subsequent coverage of baking, cheese making, and curing of other foods. The important features of the sequence are highlighted.

In the first place, it should be emphasized that each of the 12 steps leading to ethanol (ethyl alcohol) requires a separate enzyme, which is provided by the yeast. The word enzyme originates from the Greek *en* (in), *zume* (yeast). Since the yeast is a living microorganism, the reaction pathways represent chemical reactions, which are necessary for the survival of the yeast cells.

Of greatest importance to the yeast is the production of *adenosine triphosphate, ATP,* which is the most important source of chemical energy for all living

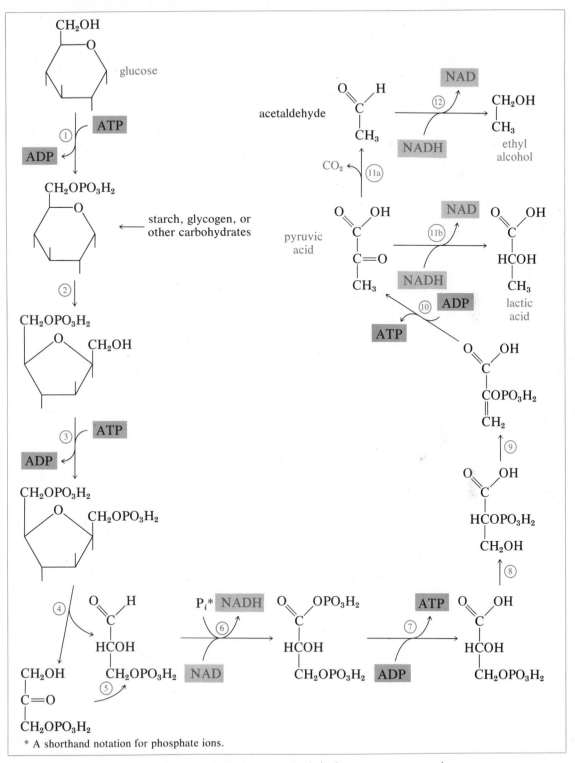

Figure 11.4 Fermentation (glycolysis). Only the key compounds in the sequence are named.

organisms. ATP is the common currency of energy in biological systems. In fact, your brain is utilizing the energy of ATP as you read this page. When ATP is consumed, it is converted to *ADP, adenosine diphosphate,* by removal of the terminal phosphate group (shown in color).

adenosine triphosphate (ATP)

It is convenient to think of ATP as the charged form of the battery used to power living systems. The ADP is the discharged form which must be charged by conversion back to ATP (1).

In the fermentation sequence, during the conversion of glucose to ethyl alcohol, there is a net production of 2 ATP, which explains why yeast and other organisms (including humans) carry out all or part of this sequence. One ATP molecule is consumed at steps 1 and 3, whereas one ATP is produced for each 3-carbon fragment that passes through steps 7 and 10, so that a net 2 ATP is produced per molecule of glucose that is metabolized.

One special quality of yeast is its ability to survive either aerobically or anaerobically. The difference in the chemistry under the two conditions begins with *pyruvic acid,* which is the product of step 10 of the breakdown of glucose. The presence or absence of oxygen has no effect on steps 1-10, but does determine the fate of the pyruvic acid. If no oxygen is available, the pyruvic acid is converted in two steps to ethanol plus CO_2. If oxygen is available, the pyruvic acid is completely metabolized to 3 CO_2 by the steps in the series of reactions known as the Krebs Cycle, shown in Figure 11.5. The deciding factor is the need to regenerate a substance known as *nicotinamide adenine dinucleotide* or simply NAD. It can be seen that NAD is consumed (converted to NADH) at step 6. If the pyruvic acid is converted to ethyl alcohol, the NAD is regenerated in step 12.

An alternative method of regenerating NAD is by conversion of pyruvic acid to lactic acid. This result is called *lactic acid fermentation.* This is the most common pathway used by anaerobic microorganisms, except yeasts, which carry out *alcoholic fermentation* under anaerobic conditions.

The third method of regeneration of NAD is by metabolizing the pyruvic acid via the Krebs Cycle. NAD is later regenerated in a process that requires oxygen (discussed later). Humans utilize this more efficient aerobic pathway rather than carrying out alcoholic fermentation, although the latter might be more exciting. Under conditions of extreme physical exercise (e.g., running), the demand for

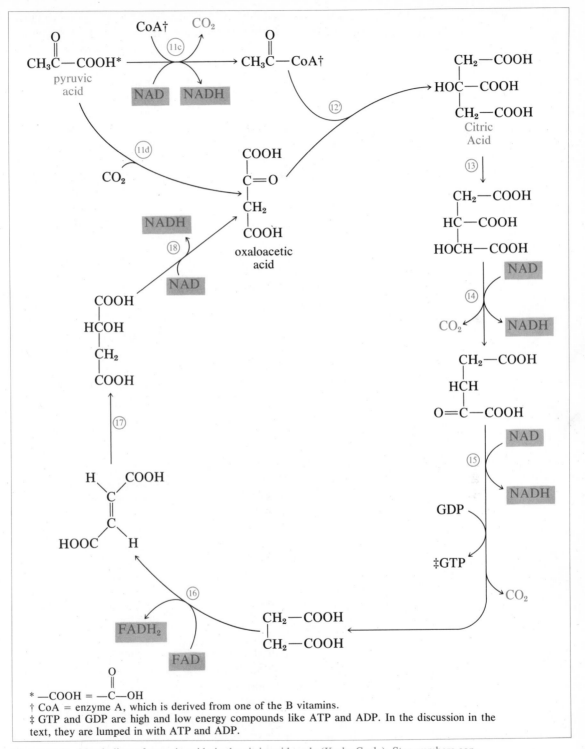

* —COOH = —C—OH
  O
  ‖

† CoA = enzyme A, which is derived from one of the B vitamins.

‡ GTP and GDP are high and low energy compounds like ATP and ADP. In the discussion in the text, they are lumped in with ATP and ADP.

Figure 11.5.   Metabolism of pyruvic acid via the citric acid cycle (Krebs Cycle). Step numbers continue from the glycolysis scheme (Figure 11.4).

oxygen by the muscles often exceeds the ability of the body to supply the oxygen. The oxygen is carried by the blood from the lungs to the muscles. In response to exercise, the heart beats faster and the blood flow is increased in order to maximize the efficiency of the oxygen delivery to the muscles, but often the efficiency is not great enough. Under these conditions, the muscles may become partly anaerobic and begin to carry out lactic acid fermentation. The buildup of lactic acid in the muscles is the cause of muscle cramps (e.g., leg cramps) under heavy loads.

NAD and NADH are two forms of the B vitamin known as niacin, which is found in virtually every commercially available vitamin supplement. For the sake of discussion, NAD will be referred to as **active niacin** and NADH as **inactive niacin.** Active niacin is required for survival of all living things. As noted above, yeast can regenerate active niacin under both aerobic and anaerobic conditions, but it should be obvious that the brewer prefers the anaerobic pathway, since he receives two ethyl alcohol molecules for every glucose that enters the scheme. The yeast grows more vigorously under aerobic conditions but at the expense of ethyl alcohol production. Consequently, the brewer or wine maker routinely limits the supply of oxygen during the fermentation.

Under aerobic conditions, the regeneration of the active niacin occurs by the following reaction, which is a shorthand description of some seven steps.

$$NADH + O_2 \xrightarrow{\text{7 steps}} NAD + H_2O$$
$$3\ ADP \quad 3\ ATP$$

In this way, all of the NAD consumed in step 6 (of glycolysis) and in the Krebs Cycle is regenerated. For each NAD formed, 3 ATP are also formed. This latter fact accounts for the rapid growth of yeast cells under aerobic conditions when ATP is abundant and provides the energy required for growth.

FAD and $FADH_2$ also appear in the Krebs Cycle. They are two forms of the B vitamin known as riboflavin, which is also commonly found in vitamin supplements. Here again, FAD will be described as *active riboflavin* and $FADH_2$ as *inactive riboflavin.* FAD may be regenerated only under aerobic conditions according to the following process in six steps.

$$FADH_2 + O_2 \xrightarrow{\text{6 steps}} FAD + H_2O$$
$$2\ ADP \quad 2\ ATP$$

REVIEW OF FERMENTATION

In summary, the key questions that can be answered regarding the metabolic pathways given in Figures 11.4 and 11.5 follow.

    1. How many steps are involved?
      Answer:
      (a) Anaerobically:
          12 steps from glucose to ethyl alcohol.*
          11 steps from glucose to lactic acid.*

* All reactions beginning with step 6 are utilized twice per glucose. That is, the 6-carbon compounds

(b) Aerobically:
10 steps from glucose to pyruvic acid plus 1 step to the starting material of the Krebs Cycle plus 7 steps in the Krebs Cycle plus 7 steps to regenerate each active niacin (NAD) plus 6 steps to regenerate each active riboflavin (FAD) plus 36 steps for production of 36 ATP equals *149 steps per glucose metabolized.* *

2. What is the purpose of these pathways?
Answer:
The process generates useful energy (ATP), which the yeast or any other living organism needs to drive a wide variety of energy-consuming processes.
(a) Anaerobically:
2 ATP are produced per glucose metabolized.
(b) Aerobically:
Via the Krebs Cycle and the subsequent steps, an additional 36 ATP are produced. The aerobic process is much more efficient in fulfilling the purpose, but if a supply of O_2 is not available, or if an organism is lacking the necessary enzymes, one of the anaerobic pathways must be utilized.

3. Why does each pathway end where it does?
Answer:
(a) Anaerobically:
Both anaerobic pathways proceed until active niacin (NAD) is regenerated. Yeasts carry out alcoholic fermentation, whereas other organisms and animals carry out lactic acid fermentation under anaerobic conditions. The choice of pathway depends on what enzymes are available for catalysis.
(b) Aerobically:
The aerobic pathway ends when the fuel (glucose) is completely oxidized to CO_2 and water, all the active niacin (NAD) and riboflavin (FAD) are regenerated, and all the available energy (ATP) is synthesized.

The ethyl alcohol and lactic acid produced in the anaerobic pathways represent a waste of energy. Each compound has some energy content that cannot be utilized by anaerobic organisms. This is indicated by the production of only 2 ATP per glucose anaerobically, as compared to 38 ATP per glucose aerobically.

THE COURSE OF FERMENTATION

Figure 11.6 provides some additional information on the changes that occur during the fermentation. Overall, fermentation is the step in which the wort is changed into beer. The alcohol content is determined by the amount of fermentable sugar contributed by the malt and adjuncts. Since alcohol content is a greater concern in making wine, the relationship between fermentable carbohydrate and alcohol produced is discussed under wine making.

The progress of the fermentation is routinely monitored by use of a hydrometer, sometimes called a saccharometer, which measures the specific gravity of the wort. **Specific gravity** is the weight of a given volume of liquid. The reader may be more familiar with the term *density,* which is approximately the same as specific gravity. During the fermentation, high specific gravity glucose is being replaced by low specific gravity ethanol. The result is illustrated by Figure

that are involved in steps 1-3 are split into two 3-carbon fragments at step 4, and each 3-carbon fragment is carried through from step 5 to the end. Step 11d is not counted, since it only occurs occasionally. This is true because the product of step 11d is continually regenerated at step 18 of the cycle.

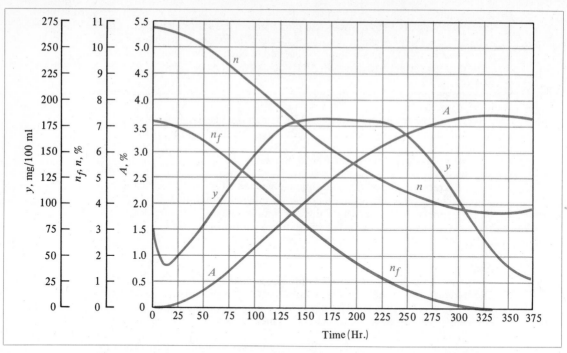

Figure 11.6   Changes that occur during fermentation. The curves are identified as follows: A, alcohol; n, extract; n_f, fermentable extract; y, suspended yeast solids. From Kirk-Othmer, *Encyclopedia of Chemical Technology*, John Wiley & Sons, Inc., 1963. (Reprinted by permission of John Wiley & Sons, Inc.)

11.7, which shows the hydrometer sinking farther down into the liquid when most of the sugar has been replaced by alcohol. The effect is comparable to a swimmer in fresh versus salt water. He is buoyed up much more by the high specific gravity salt water.

Recall that the primary sugar contributed by the malt is the disaccharide maltose, which is cleaved into two units of glucose by the enzyme maltase.

Lagering And Packaging

Lagering (storage) follows the fermentation. It involves storing the beer, typically two to six months, during which time some additional fermentation may occur. The beer also clarifies and some of the flavor characteristics are developed. The last of these changes is a subtle and complex one, although a major suspect is the formation of esters by combination of certain acids and alcohols that are formed during the fermentation.

Following lagering, proteolytic (protein cleaving) enzymes are often added to help digest and solubilize any proteins that might tend to precipitate out of the beer when it is cooled. This step is called "chillproofing."

Finally the beer is packaged. This includes adding carbon dioxide under pressure and pasteurization. Later, when the beer is opened and poured, the CO_2

TABLE 11.1 THE CHARACTERISTICS OF VARIOUS TYPES OF BEER

TYPE OF BEER	CHARACTERISTICS
Lager	Light, bottom yeast, 3–4% alcohol, hops content variable
Munich	Dark brown*, aromatic, sweet, high hops, up to 5% alcohol
Bock	Heavy, very dark, variable alcohol
Ale	Heavy, slightly dark, top yeast, up to 6% alcohol, high hops
Porter	Dark, low hops, 4–6% alcohol, heavy rich malty flavor
Stout	Very dark, heavy, strong malt taste, 5–7% alcohol, high hops

* Darker colors are derived from darker malt produced by high temperature kilning of the malt.

pressure is released, thus causing a foam to build. Unlike carbonated soft drinks, the head is retained with the help of various proteins and limit dextrins, already mentioned.

There are many types of beer in production throughout the world. Table 11.1 provides a fairly complete list including some of the characteristics of each.

Figure 11.7   The use of a hydrometer in monitoring the course of fermentation. *Left*: Example of hydrometer reading of 1.042. This is the average specific gravity of beer before yeast is added. *Right*: Hydrometer reading is 1.010, the typical level near the end of the fermentation.

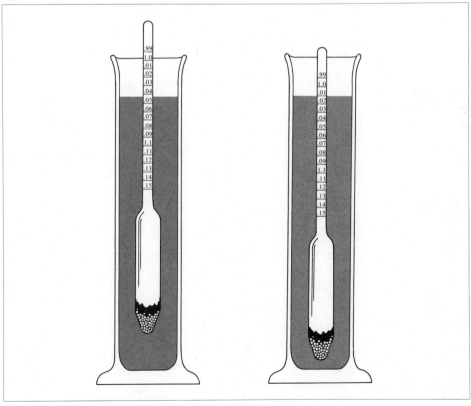

Wine is fermented grape juice. The juice of other fruits, e.g., peaches or apples, may also be fermented but it is common practice to identify the source, i.e., peach wine or apple wine. The grapes are crushed to produce what is called a *must* in which the fermentative enzymes are able to attack the available sugar.

Climate is the major factor in determining the development of grapes. Certain areas of the world are well-known for their wines. Even within the United States, California and eastern wines are quite different. The more temperate climate makes the California wines the better natural (unfortified) wines.

A "good year" is one in which the climate is warm enough to permit full ripening of the grapes, which means that the grapes develop a high level of fermentable sugar. At the same time, if the climate is too warm, low acidity of the wine may be the result. Acidity is considered to be a major factor in determining the quality of a wine. The major acids are tartaric and malic (Latin *malum*, apple) acids and even though they are present in small amounts, 0.2–0.8% and 0.1–0.5% respectively, they have a great influence on the character of a wine. In too-warm climates, the sugar content may be sufficiently high but the acidity may be too low for a good dry (not sweet) wine. On the other hand, high-sugar, low-acid grapes may be used for production of sweet dessert wines.

Grapes grown in the eastern United States generally tend to be high in acid and low in sugar. This low-sugar defect is routinely corrected by addition of

Figure 11.8   Vineyard, finger lakes region of New York State. (Courtesy of Taylor Wines.)

Figure 11.9   (a) The Wilmes Press. (b) Wilmes press with covers (foreground) removed, showing the inflatable rubber sleeve (bladder) inside the cylinder. In operation, the cylinder is filled with grape pulp, covers are fastened in place and the cylinder is revolved. A trough beneath the press collects the juice. When the free-run juice ceases to flow, the bladder is inflated with compressed air which presses the pulp against the inside of the cylinder. The pressing action expresses the remainder of the juice. Solids (pomace) are discharged by removing the covers and rotating the cylinder. A second trough containing a screw conveyor is moved into place beneath the press and conveys the pomace away. These presses hold about 4 tons of pulp and the complete pressing cycle is about 30 minutes. (Courtesy of Taylor Wines.)

TABLE 11.2   TYPES OF WINES AND TAXATION

TYPE OF WINE	PERCENT ALCOHOL	TAX (per gallon) (2)
Still*	up to 14	$0.17
Still	14–21	0.67
Still	21–24	2.25
Champagne or sparkling wine	——	3.40
Artificially carbonated wine	——	2.40

* Nonsparkling.

sugar, a technique known as *fortification,* but this does not introduce any flavor components, which might develop naturally if the climate were better. The high acidity may be corrected by encouraging a secondary fermentation known as the malo-lactic fermentation, which is discussed later.

Alcohol Content

The alcohol content is a major concern in wine making. As shown in Table 11.2, wines are categorized and taxed accordingly.

As might be suspected, most wines have an alcohol content somewhat less than 14% because of the tax schedule. These wines are commonly referred to as "table wines." So-called "dessert wines" have an alcohol content above 14%. In most cases such wines are fortified by addition of alcohol, usually brandy (discussed later).

In some dessert wines, fermentation is stopped with some sugar remaining. Examples of such sweet wines are muscatel (10–15% sugar) and port (9–14% sugar).

Figure 11.10   Fermentation tanks. (Courtesy of Taylor Wines.)

Figure 11.11   Surface of fermenting wine. (Courtesy of Taylor Wines.)

Because of the taxation, the table wine producer does not want to exceed 14% but the producer routinely tries to approach this level. Approximately 9% is considered necessary for stability and this will be produced during fermentation, if the sugar content of the must is 16.4% sugar. In more general terms, each 1% sugar ferments to 0.55% alcohol.* Therefore, 25% sugar, which may be obtained in good years, will yield a wine of 13.75% alcohol. In some parts of Europe, high humidity encourages the growth of a grape dehydrating fungus, *Botrytis cinerea*. The loss of water from these botrysised grapes may increase the sugar content to 30–40%.

Sulfur Dioxide

The fermentation process differs only slightly from that discussed under brewing. The chemical changes are the same but a few technical details vary. For one thing, there are "wild yeasts" that routinely grow on the skins of the grapes. Since these wild varieties may provide a different and perhaps less desirable group of enzymes, it is common practice to introduce sulfur dioxide,† SO_2, to the *must* to kill off the undesirable microbes. Fortunately, the common wine yeasts are not sensitive to the SO_2. The SO_2 acts as a general antiseptic by inhibiting the growth of any bacteria that might cause spoilage even after fermentation. Wine yeasts also have a good tolerance of the high concentrations of alcohol found in wines. They also

---

* It is common practice to express the percent sugar as a percent by weight (i.e., grams of sugar per 100 grams of must). The percent alcohol is normally expressed in percent by volume (i.e., liters of alcohol per 100 liters of wine).
† Sulfur dioxide is a gas that can be pumped directly into the must. The SO_2 can also be provided by the use of Campden tablets, which are commonly used by the home wine maker. The tablets are sodium metabisulfite, $Na_2S_2O_5$, which reacts in water according to the following reaction.

$$Na_2^+ S_2O_5^{-2} + H_2O \longrightarrow 2\ Na^+HSO_3^-$$

sodium
metabisulfite

sodium
bisulfite

Sodium bisulfite is the half-neutralized form of sulfurous acid, H_2SO_3, which is formed when SO_2 is dissolved in water.

tend to clump together, which aids in settling. This helps assure that no undesirable flavors due to dead yeast cells will be present in the wine.

Sulfur dioxide also combines with any unconverted acetaldehyde that would give the wine an undesirable character. Acetaldehyde is the immediate precursor to ethyl alcohol in the fermentation scheme.

Sulfur dioxide is also an antioxidant, i.e., prevents oxidation. Oxidation is a process that may take many forms. In connection with alcoholic beverages, the oxidation that is of major concern is the conversion of ethyl alcohol in two steps to acetic acid, as shown.

Acetic acid is the ingredient of vinegar that gives it a sour taste, which may be fine for vinegar but not for wine.

Oxidation of organic compounds is often a complex process. One or both of two phenomena generally occur. These are incorporation of oxygen and/or dehydrogenation (removal of hydrogen). The conversion of ethyl alcohol to acetaldehyde is easily seen to be a dehydrogenation, whereas, the second reaction in the sequence above seems to be a simple oxygen insertion. The opposite of oxidation is reduction.

Niacin is the major vitamin required for oxidation-reduction reactions in living systems. It is called a *coenzyme,* because it acts in cooperation with several enzymes. As noted previously, active niacin (NAD) is required for the breakdown of glucose. It is commonly regenerated in the presence of oxygen or by the reverse of step 1 above. Therefore, if step 1 is to proceed in the forward direction (i.e., from left to right), NAD can only be regenerated when the oxygen supply is adequate. Since both steps 1 and 2 are undesirable in wine, it is necessary to either totally exclude oxygen, which is difficult, or to add an antioxidant (SO_2), which functions by intercepting oxygen and being oxidized itself according to the following equation.

$$2 \, SO_2 + O_2 \longrightarrow 2 \, SO_3$$

Other Characteristics of Wine

White wines are fermented free of the skins, whereas red wines derive their color from pigments extracted from the skins. Sparkling wines are produced when excess CO_2 is formed by fermentation of sugar added after the primary fermentation. The home wine maker is apt to provide too much sugar, causing too much CO_2 to form, and the bottles may explode. Carbon dioxide may also be introduced under pressure from an external source. This is known as *artificial carbonation*.

Some carbonation may also be obtained during the *malo-lactic fermentation*. The chemical change which occurs is given in the following equation.

malic	lactic	malo-lactic
acid	acid	fermentation

According to this conversion, some carbonation is generated, but more important, malic acid with 2 acid groups is converted to lactic acid, which has only one acid group. This means that a decrease in acidity of the wine accompanies the change and the wine mellows.

The malo-lactic fermentation may either be encouraged or discouraged depending on the method of handling the wine following the primary fermentation. High-acid eastern wines (e.g., Concord and Catawba) are invariably made to undergo this secondary fermentation. California wines and other wines usually only require malo-lactic fermentation in cool years when acidity is relatively high. When ripening occurs, it is thought that malic acid is translocated from the grapes into the leaves. High SO_2 concentrations inhibit growth of microbes that carry out malo-lactic fermentation.

DISTILLED ALCOHOLIC BEVERAGES

As the name suggests, distillation is the distinguishing feature of these beverages. At some point following fermentation, the mash is heated to boiling and distilled. During distillation, all solids and other high-boiling materials are left behind as a residue, while the volatile components, including alcohol, are boiled off and collected.

Various grain starches may be used as a source of alcohol and may be handled in different ways. In making Scotch whisky, malted barley is dried over peat

TABLE 11.3 CONGENERIC CONTENT OF MAJOR TYPES OF DISTILLED ALCOHOLIC BEVERAGES (3)*

COMPONENT†	AMERICAN BLENDED WHISKY	CANADIAN BLENDED WHISKY	SCOTCH BLENDED WHISKY	STRAIGHT BOURBON WHISKY	BONDED BOURBON WHISKY	COGNAC BRANDY
fusel oil	83	58	143	203	195	193
total acids (as acetic acid)	30	20	15	69	63	36
esters (as ethyl acetate)	17	14	17	56	43	41
aldehydes (as acetaldehyde)	2.7	2.9	4.5	6.8	5.4	7.6
furfural	0.33	0.11	0.11	0.45	0.90	0.67
total solids	112	97	127	180	159	698
tannins	21	18	8	52	48	25
total congeners, wt/vol %	0.116	0.085	0.160	0.292	0.309	0.239

† Grams per 100 liters at 50% alcohol.

From Kirk-Othmer, *Encyclopedia of Chemical Technology,* John Wiley & Sons, Inc., 1963. (Reprinted by permission of John Wiley & Sons, Inc.)

fires. This gives the malt a smokey flavor that carries through into the final product. Other grains are commonly blended into the mash for both Scotch and other whiskies. Canadian whisky is a blend of corn, rye, and barley malt. Rye whisky is made from a mash that is predominantly rye grain. Gin is usually a blend of various grains plus certain berries, most notably juniper, which provides flavor constituents. Bourbon is a blend in which corn predominates, although real corn whisky must be made from a minimum of 80% corn.

Brandy is distilled from grapes. Other brandies normally specify the source, e.g., apricot brandy. It was previously noted that brandy is often added in making high alcohol dessert wines.

Rum is a distillate derived from fermented juice of sugar cane, sugar syrup, and molasses.

Vodka is also distilled from fermented grain mash and the distillate is charcoal filtered. This treatment removes all pigments and certain other constituents.

Congeners are significant in distilled as well as other alcoholic beverages. They are side products of fermentation or other incompletely converted species along the main route to ethyl alcohol. They may react with one another or remain unchanged. Either way some are carried over during the distillation. Table 11.3 gives some typical figures.

Aging contributes to full development of congeners, which may even include things extracted from storage containers, e.g., oak barrels.

SUMMARY

Brewing is a major worldwide industry. Prior to the actual brewing process, barley is malted in a three-step process, which includes steeping, germination and kilning. It is during the germination step that structural starches and proteins are broken down and complex enzyme systems are developed. These enzymes are required for the breakdown of barley starches and adjuncts into fermentable sugars. The green malt, which results from germination, is kiln dried. High temperature kilning yields dark-colored malts that are used in dark beers, ales, and other products.

Hops are added to give beer a bitter character that counteracts the sweetness of the carbohydrates. Complete breakdown of starches is accomplished in the mashing step by the action of α- and β-amylases. The product is called a *wort,* which is then subjected to fermentation.

Fermentation is catalyzed by enzymes provided by yeast, which use fermentable sugars as food, and convert them to ethyl alcohol and carbon dioxide under anaerobic conditions. Under aerobic conditions, the sugars are largely oxidized to carbon dioxide and water. The progress of fermentation can be monitored with an hydrometer, which measures the specific gravity of the wort as it becomes a beer. The beer is then lagered, chillproofed, and packaged with carbonation.

Wines are made by fermentation of sugars found in fruits, usually grapes. Although some wines are fortified to increase the alcohol content, better wines are usually produced when the climate during the growing season is good and grapes mature naturally to produce a satisfactory balance of sugar and acidity. Low sugar and high acidity are common characteristics of eastern United States wines, which can be modified by addition of sugar and by encouraging malo-lactic fermentation, which lowers the acidity. When the alcohol content exceeds 14%, taxation increases dramatically, as in the case of dessert wines.

Sulfur dioxide is often added to wine. It acts as an antioxidant, it suppresses wild yeasts and other microorganisms, and it combines with and removes acetaldehyde.

Distilled alcoholic beverages are usually made from grains, except for brandies, which are distilled from fermented fruits. Congeners are side products of fermentation. They greatly influence the flavor and aroma of all alcoholic beverages.

PROBLEMS

1. Give a definition or example of each of the following:
   - a) anaerobic
   - b) hydrometer
   - c) congeners
   - d) malting
   - e) steeping
   - f) kilning
   - g) hops
   - h) limit dextrins
   - i) wort
   - j) enzyme
   - k) ATP
   - l) NAD
   - m) specific gravity
   - n) lagering
   - o) proteolytic
   - p) dehydrogenation
   - q) must

2. What is the major grain used in brewing? Why?

3. Why does yeast ferment glucose? Why does the fermentation proceed all the way to ethyl alcohol rather than stopping at pyruvic acid?

4. Give one reaction that occurs during the fermentation of glucose. How is it catalyzed?

5. What are adjuncts? What is their purpose?

6. What is the alcohol content of the typical American beer?

7. What step in the production of beer determines whether the final product will be light or dark?

8. What is the cause of leg cramps at times of heavy exercise?

9. What is the function of limit dextrins in beer?

10. What is the source of fermentable sugar for wines?

11. Why is sulfur dioxide added to wine?

12. What are Campden tablets? Why are they used?

13. What is the major difference between the procedures used to make white wines and red wines?

14. What does the wine maker mean by "a good year?" What can he do in the event of a bad year in order to make a good wine?

15. Why is the 14% alcohol level important in wine making?

16. What is the primary source of fermentable carbohydrate used in making Scotch whisky? Rye whisky? Gin? Bourbon? Brandy? Rum?

17. What is the origin of the "smokey" flavor of scotch?

18. Confirm the figure 149 for the number of steps in the metabolism of glucose under aerobic conditions.

REFERENCES

1. Carson, R. *Silent Spring*. Boston: Houghton-Mifflin, 1962, p. 202.
2. Amerine, M.A., Berg, H.W., and Cruess, W.V. *The Technology of Wine Making*. Westport, Conn.: AVI, 1972, p. 750.
3. *Kirk-Othmer Encyclopedia Of Chemical Technology,* 2nd edition. Vol. 1. New York: John Wiley & Sons, 1963, p. 510.

I n many respects, the chemistry involved in baking is very similar to that involved in the production of alcoholic beverages. Yeast enzymes were seen to
catalyze the fermentation of sugars to produce ethyl alcohol plus carbon
dioxide and, except for certain carbonated beverages, the alcohol is the major
product. In baking, it is the CO_2 that is the important product because of its role in
the leavening (raising) of certain baked goods.

FLOUR

Once again, it is necessary to first consider the characteristics of the grains that
are milled to produce flour. Wheat flour is by far the most important for baking.
Rye flour is used, although it is generally used in combination with wheat flour.
Triticale, a hybrid cross between wheat and rye, seems to hold some promise for
the future although there seems to be some controversy over its suitability.

So-called *common wheat,* which accounts for more than 90% of the wheat
grown in the United States, is divided into several major classifications, some of
which are listed in Table 12.1. They are listed approximately in order of decreasing importance.

The winter wheats are planted in the fall and harvested in the spring. They
are generally grown in areas where the winters are not too harsh. The spring
wheats are planted in the spring and harvested in the fall. They are generally
grown in the cooler northern areas.

The distinction between *hard* and *soft* (see Table 12.1) relates to the texture
of the endosperm, which is the chief location of the starchy material that is used
for making flour. More important, hard and soft wheats have very different
amounts of protein, which is a major concern in deciding their suitability for different baked goods. Although there are many different kinds of proteins found in
wheat, two of the most important proteins are called *gliadin* and *glutenin.* When
the gliadin and glutenin are combined in the presence of water these proteins take
on water, i.e., become hydrated, and form a complex mass called *gluten.* Gluten,
with other components of the dough embedded in it, provides the unique structural framework for trapping gases in leavened baked goods.

The gliadin appears to provide elasticity. The glutenin provides strength to
the dough mixture. The hard wheats with their high protein content yield flour that
produces a high gluten dough. The high gluten content makes the dough suitable
for use in making breads and rolls which are highly leavened. Low gluten doughs
do not have sufficient strength to allow them to trap the quantity of gas needed to
produce what we consider normal bread, although some flat breads made from

BAKING

TABLE 12.1   SOME TYPES AND CHARACTERISTICS OF COMMON WHEAT

CLASSIFICATION	MAJOR GROWING AREAS IN THE U.S.	PERCENT PROTEIN	USED IN MAKING
Hard Red Winter	Central and Southern Great Plains	12–16	bread, rolls
Soft Red Winter	Ohio, Indiana	7–9	cakes, cookies, pastries
Hard Red Spring	Northern Great Plains	12–16	bread, rolls
White	Pacific Coast, New York, Michigan	<7	cakes, cookies, pastries
Durum	North Dakota, South Dakota		macaroni products

soft wheat flours are common in certain parts of the world. More frequently, the soft flours are used in making cakes, pies, and other pastries, as well as crackers, in which relatively little leavening is required.

Hard and soft wheats are said to yield *strong* and *weak* flour, respectively,

Figure 12.1   Wheat acreage by regions. (Courtesy of United States Department of Agriculture.)

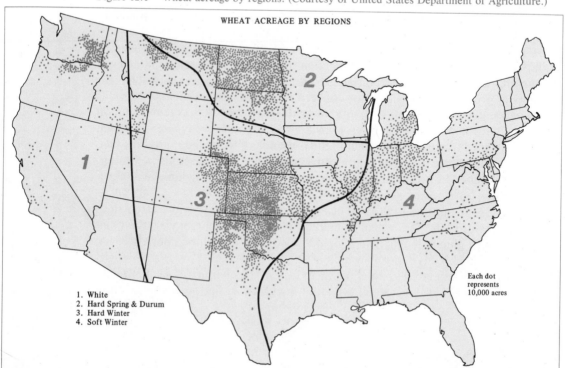

WHEAT ACREAGE BY REGIONS

Each dot represents 10,000 acres

1. White
2. Hard Spring & Durum
3. Hard Winter
4. Soft Winter

Figure 12.2   Types of wheat. (a) Soft red winter wheat. (b) Hard red spring wheat. (c) Durum wheat. (Photos Courtesy of United States Department of Agriculture.)

in which the strength is dependent on the amount of gluten in the final product. Some of the strong wheats are often too good (strong) for normal use. They are commonly blended with the weaker varieties. Common white flour is a relatively weak blend.

Full development of dough, achieved by mechanical or hand kneading, involves a series of complex chemical changes, during which the arrangement of the fibers in the dough proteins is modified. There is thorough interaction with other dough components, e.g., shortening, and the yeast is mixed thoroughly. In cases when flours are enriched, these components are also mixed in thoroughly. The major physical changes, which are a consequence of many complex chemical changes, are an improvement of the gas-retaining properties and formation of many minute air cells that can later expand when the dough rises during fermentation or chemical leavening.

After the bread dough is given an opportunity to rise, the dough is generally punched down to eliminate large pockets of gas that might appear as large holes in the finished product. At the same time, a fresh supply of oxygen becomes available to the yeast cells. Home baking generally involves milder handling so that softer flours are generally suitable.

Flour may be modified in several ways before use. Besides blending strong and weak flours, the baking properties of flour are improved by aging. During the aging, a bleaching process occurs due to the oxidation of highly-colored pigments, called *xanthophylls*. This change and other changes that improve the flour during aging can be mimicked by chemical treatment, which is a common practice due to

Figure 12.3. A wheat combine. The reel (1) holds the crop against the guards until the knife (2) cuts the material and lays it onto the header. The auger (3) delivers the cut material to the feeder (4). Fingers (5), extending out of the auger at the opening to the feeder, help feed material to the feeder (4), which carries it to the cylinder (6) and concave area (7). The function of the cylinder and concave is to create a rubbing action that removes the head of the grain from the straw. After this threshing action, the grain and chaff fall through the concave grates to the grain pan (11). The straw that passes over the concave is then stripped from the cylinder by the cylinder beater (8), and transferred across the cylinder beater grate fingers (9) which separate some of the threshed grain trapped in the straw, allowing it to fall through the fingers to the grain pan (11). As the straw continues across the beater grate fingers, it is transferred to the straw racks (10), which continue to separate grain from straw as they move the straw to the rear of the combine. As the straw moves to the rear, the cylinder beater check flap (not shown) is used to regulate and control the depth of material in the racks. The rear straw rack check flap (not shown) also provides retarding action for thorough separation. At the rear of the rack, the straw falls to the ground. The separated grain that falls through the open bottom straw racks is returned to the front of the chaffer sieve (12) and onto the cleaning system by the straw rack return pan (13).

The cleaning system is made up of the following components: Cleaning fan (14); chaffer sieve (12) and shoe sieve (15). The function of the cleaning fan is to supply an adequate air blast to provide separation of chaff and grain. The air velocity must be sufficient to lift the chaff, but must not be so great that it lifts the grain and blows it out over the sieves. The plenum box (16) is the distribution system for the air blast supplied by the cleaning fan, and divides the air into the proper proportions to the chaffer and shoe sieves.

As the grain and chaff travel to the rear over the chaffer sieve, grain falls through the chaffer, and the chaff is blown out through the rear of the combine. The grain and unthreshed pieces, or heads, fall onto the shoe sieve (15), which does the final separation. The size of the particle that falls through this sieve is determined by the opening between the sieve louvers. Unthreshed heads are carried to the rear of the shoe sieve and fall into the tailings auger (17), and are carried to the front of the cylinder for reprocessing (18). The clean grain that falls through the shoe sieve is transferred by the clean grain auger (19) to the clean grain elevator (20). The grain elevator carries the grain to the grain tank (21). (Courtesy of International Harvester.)

the expense of storage and the demand for wheat flour. The chemical changes that improve the dough are complex but apparently involve the S—H functional groups found on certain amino acid components of the flour protein.

Chemical aging has been a frequent subject of criticism. The treatment destroys many of the natural nutrients found in wheat flour and, although a few nutrients may be added to enriched flour, the nutritional value is decreased.

In making whole wheat flour, the bran and germ of the wheat are included. Whole wheat flour is generally considered to be the most nutritious, although it has been argued that the bran content of whole wheat flour, which has a high cellulose content, causes a lowering of the nutritional value. The cellulose is not digestible but, more important, it provides bulk, called roughage, which stimulates contraction of the intestines and speeds the passage of food through the system. If this is carried too far some digestible food may also be wasted.

OTHER DOUGH INGREDIENTS

Sugar is added to the dough to provide some sweetness. It also acts as yeast food. During baking, caramelization of sugar on the surface of the bread is partly responsible for browning of the crust. Caramelization occurs above 135°C. This process was noted previously in production of dark beers as partial caramelization of barley malt. An interaction between the starch and protein also contributes to the browning.

Shortening or cooking oil, discussed in Chapter Fifteen, is another important ingredient in baking dough. Shortness or richness in baked products is characterized by a crumbly texture. The oil also contributes flavor and assists in the leavening by helping to trap expanding gas bubbles. The germ constitutes only about 3% of the wheat kernels and remains in whole wheat flour. It has a high oil content and, consequently, decreases the shelf life of the flour since the oil is prone to spoilage.

Wheat flour is sometimes enriched. Wheat protein is frequently lacking in certain essential amino acids, e.g., lysine. This is a minor problem when other dietary components make up for the deficiency, but in some areas of the world other protein may be scarce. In these areas, hybrids are a partial solution, but enrichment with either amino acids or soybean protein, i.e., soy flour, may be more practical. The latter has been studied and blends of wheat and up to 8% soy flour were found to give acceptable breads. Certain vitamins are commonly used to enrich flour even in the United States. They are thiamine (vitamin B_1), riboflavin (vitamin B_2), and niacin (vitamin B_3).

Sourdough is used in making rye and pumpernickel breads. A *sour* is a culture of old dough containing wild yeasts and other bacteria that provide flavor, aroma, and texture to the finished product. Other products, like the congeners in alcoholic beverages (Table 11.3), are formed even with regular yeasts, but different microbes can cause formation of differing amounts of these substances and, thus, control the characteristics of the bread.

Blends of rye and wheat flour are commonly used in making rye and pumpernickel breads. Rye flour alone has too little gluten to permit the normally desired amount of leavening. A typical formula consists of one-third rye flour and two-thirds hard spring wheat flour. Caramel color is often added, especially to commercial pumpernickel breads. Westphalian pumpernickel bread is made from 100% rye flour. Leavening is poor and the bread is very dense, dark, and chewy (1).

LEAVENING

Leavening action is caused by one or a combination of the following: yeast, chemical leavening, air expansion during baking, and steam production during baking.

In Chapter Eleven, we considered many of the details of the utilization of glucose by yeast cells. Among other things, it was noted that yeast can function either aerobically or anaerobically. For maximum production of ethyl alcohol, the latter is encouraged, since each glucose molecule yields two molecules of ethyl alcohol and two molecules of CO_2 under purely anaerobic conditions.

Under aerobic conditions, each glucose molecule may be broken down to six molecules of CO_2. Quite obviously, for leavening this would be ideal but not attainable since there cannot be an adequate supply of oxygen in the interior of a bread dough in order to maintain aerobic conditions. Nevertheless, some air is trapped in the interior of bread dough when it is prepared so that the metabolism of available sugar is partially aerobic and partially anaerobic. Either way, much CO_2 is liberated, and even the alcohol liberated during anaerobic fermentation may be vaporized during baking and thus contributes to leavening.

Chemical Leavening

Chemical leavening is used in many baked goods including cakes, cookies, pies and other pastries, pancakes and pizza. Baking soda, $NaHCO_3$, is the most important chemical leavening agent. It releases carbon dioxide according to the following reaction.

$$Na^+HCO_3^- + H^+ \longrightarrow Na^+ + H_2CO_3 \quad \text{(unstable)}$$
$$\downarrow$$
$$CO_2 + H_2O$$

Baking powder is a mixture of several components. Baking soda is the active ingredient, which acts according to the reaction shown above. The other major ingredients are an acid (the source of H^+) and some inert ingredients to prevent premature reaction between the acid and the baking soda. Calcium sulfate, calcium lactate, and starch (mostly corn starch) are commonly used as inert ingredients.

The formulation must provide an amount of CO_2 equal to at least 12% of the weight of the baking powder, which requires that at least 23% $NaHCO_3$ be present. Most baking powders contain 26–30% $NaHCO_3$.

Several acids are used in baking powder formulations. Some are very fast-acting, some are slow. The difference is usually due to solubility, since some of the acids are quite insoluble, and therefore inactive, at low temperatures, but become soluble, and therefore active, as the temperature rises. Combinations of fast- and slow-acting acids can provide leavening action throughout the baking period.

One of the acids often used in baking powder formulations is sodium aluminum sulfate, which may be written as $Na_2SO_4\text{-}Al_2(SO_4)_3$ or $NaAl(SO_4)_2$. It is acidic by virtue of the chemical reactions indicated for $Al(H_2O)_6^{+3}$ (see Chapter Nine). It is relatively slow-acting and is found in many baking powder formulations to provide slow but continuous leavening action.

Figure 12.4   Baking soda (courtesy of Church and Dwight) and baking powder (courtesy of General Foods). Photo by James J. Kane, Jr.

Several phosphate salts and their analogs are used as acids. The situation is like that described in Chapter Nine, in which ammonia was shown acting as an acceptor of H^+ to produce either $H_2PO_4^-$, HPO_4^{-2}, or PO_4^{-3}. Both $H_2PO_4^-$ and HPO_4^{-2} salts are weakly acidic and, thus, applicable for use as acids in baking powders. Salts of pyrophosphoric acid and tartaric acid (see the following) are also frequently used to provide acidity in baking powder formulations.

$$Na_2H_2P_2O_7$$
sodium acid pyrophosphate

$$KHC_4H_4O_6$$
potassium acid tartrate
(cream of tartar)

Finally, ammonium bicarbonate is a good chemical leavening agent by virtue of the reaction that it undergoes when heated.

$$NH_4^+ HCO_3^- \longrightarrow NH_3 + H_2CO_3 \quad (\text{unstable})$$

$$\downarrow$$

$$H_2O + CO_2$$

Ammonia gas and CO_2 both provide leavening action but since the ammonia cannot be tolerated in the final product, ammonium bicarbonate is used only in cookies, eclair and cream puff shells, and crackers, in which the end product is baked to a very dry consistency so that the water soluble ammonia is completely expelled.

Air Expansion

Air trapped in the bread dough also provides significant leavening as it expands in response to heating. The inside of the bread dough can reach a temperature up to 100°C. It does not exceed 100° because of the formation of steam at that temperature from trapped water.

The leavening contribution of the air can be estimated by calculating the volume change predicted by the "ideal gas law," which describes the influence of pressure (P), temperature (T), and the amount of gas (described as n, the number of moles*) on the volume of the gas.

$$PV = nRT$$

In this equation, R is a constant with the value of 0.082 liters atmospheres/moles degrees. The complex units are explained by cross-multiplying the equation to give

$$\frac{PV}{nT} = R$$

in which P has the units atmospheres, V is expressed in liters, and temperature is expressed in degrees absolute (also known as degrees Kelvin). In order to express the temperature in degrees absolute for use in this equation, simply add 273 to the temperature given in degrees Celsius.

$$°A \text{ (or } °K) = °C + 273$$

For example 20°C (68°F)† would be equal to 293°A.

It is often stated that 1 mole of any gas occupies 22.4 liters at standard temperature and pressure (STP). This may be confirmed using the ideal gas equation.

---

* n equals the number of moles. One mole equals the weight of a substance equal to its molecular weight. The molecular weight equals the sum of the atomic weights. For Example, O_2 has a molecular weight equal to twice the atomic weight of oxygen. Consequently, 16 grams of O_2 would be 0.5 mole.
† Interconversion between Fahrenheit and Celsius can be carried out using the following relationship

$$F = \frac{9}{5}C + 32$$

where C is the temperature expressed in degrees Celsius and F is the temperature in degrees Fahrenheit.

at STP:

$$P = 1 \text{ atm}$$
$$T = 0°C = 273°A$$

$$PV = nRT$$

so

$$V = \frac{nRT}{P} = \frac{(1)(0.082)(273)}{1} = 22.4 \text{ liters}$$

Problem. Calculate the percent change in the volume of air trapped in bread dough when heated from 20°C (68°F) to 100°C (212°F).

$$R = 0.082 \text{ 1 atm/mol deg.}$$

Since P, *n,* and R will remain constant, they may be left in the equation. Writing the ideal gas equation in the appropriate form

$$V = \frac{nRT}{P}$$

$$V_{20°} = \frac{nR}{P} 293$$

whereas

$$V_{100°} = \frac{nR}{P} 373$$

therefore

$$\frac{V_{100}}{V_{20}} = \frac{373}{293} = 1.27 \text{ or } 27\% \text{ increase}$$

Notice that it makes no difference what gas is trapped in the dough. Air, composed mostly of N_2 and O_2, as well as any gases from leavening agents, are affected in the same way.

Steam Production

The generation of steam in bread dough and in other baked goods is important in leavening.

Problem. Calculate the change in the volume when 1 mole of liquid water at 20°C is heated to 100°C to produce steam during baking. The density of liquid water at 20°C is about 1g/ml. Express the change as a ratio.

For liquid water at 20°C:

$$
\begin{aligned}
\text{molecular weight} &= \text{sum of atomic weights for } H_2O = 18 \\
\text{volume in ml} &= \text{weight} \div \text{density} \\
&= 18g \div 1g/ml \\
&= 18 \text{ ml}
\end{aligned}
$$

For water vapor at 100°C, assuming P = 1 atm:

$$
V = \frac{nRT}{P}
$$

$$
= \frac{(1)\,(.082)\,(373)}{(1)} = 30.6 \text{ liters}
$$

or

$$
\frac{V_{100}\ (\text{vap.})}{V_{20}\ (\text{liq.})} = \frac{30.6 \text{ liters}}{18 \text{ milliliters}} = \frac{30{,}600 \text{ ml}}{18 \text{ ml}} = 1700
$$

This is just one of many illustrations of the tremendous expansion encountered when a liquid (or solid) is vaporized. The result is somewhat of an exaggeration in trying to evaluate the importance of steam formation in leavening since all of the water may not vaporize. However, even if a small portion of the water trapped in the dough, cake batter, etc. is converted to steam, it can have a dramatic effect.

CHANGES DURING BAKING OF BREAD

Following the final fermentation done at about room temperature or slightly above, the raised dough is placed in an oven. Different recipes specify different oven temperatures, mostly in the range from 375–450°F (190–230°C).

The first event is a phenomenon known as "oven spring," in which the dough rises sharply due to the expansion of gases (CO_2 and air) already present in the dough. The previous discussion of leavening explains this. As the temperature of the dough increases, the yeast activity increases and then finally stops as the yeast are killed at higher temperatures.

As the temperature inside the dough reaches 175°F (80°C), the ethyl alcohol produced during fermentation vaporizes and aids in the leavening. The inside of the dough finally reaches 212°F (100°C), which converts some of the moisture into water vapor and aids in leavening (see previous discussion).

When the yeast is killed, its enzymes are inactivated and fermentation stops. This is followed by coagulation of all proteins and gelatinization of starch, which give the expanded air cells a rigid structure and cause the bread to remain leavened even when cooled.

The high temperatures at the surface (crust) of the bread cause browning to occur. The browning is attributed to two processes. One is caramelization of the sugar as described in the drying of malt used in dark beers. The other process is known as Maillard browning and is a complex interaction between the protein and starch (2).

Throughout the baking, a wide variety of complex chemical reactions occur and cause development of many flavor and aroma ingredients, very much like the

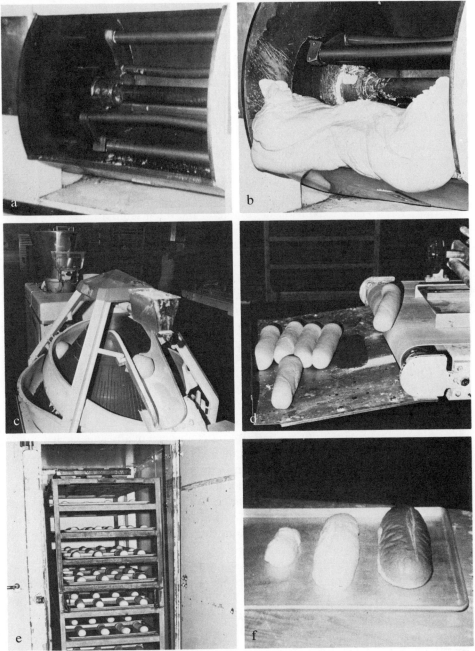

Figure 12.5   Steps in making bread. (a) Heavy-duty mixer stands open and ready to be used for thorough mixing of dough ingredients. Note the heavy iron bars that tumble the dough to assure complete mixing and proper development of the gluten. (b) The conditioned dough emerges from mixer. (c) Dough is then fed through funnel in background into machinery that divides and shapes dough into balls of specific size. (d) The balls are then shaped to be made into bread. (e) The dough is then "proofed" (allowed to rise) in a steam room at 98°F for about 45 minutes and then transferred to an oven for baking for 15–20 minutes at 400–425°F. (f) A tray holds the product at its three stages: after shaping, after proofing, and after baking. The proofed dough in the middle is extremely delicate and cannot be handled. (Courtesy of C. Nardone & Sons, Inc.)

congeners formed in ethyl alcohol fermentation and subsequent aging of alcoholic beverages.

OTHER BAKED GOODS

Yeast sees little use outside of breadmaking. *Chemical leavening, air expansion,* and *steam formation* are the major types of leavening. Pie crusts and matzos are examples of unleavened products. Eclair shells, cream puff shells, and popovers are examples of products that are highly leavened. A mixture of ammonium bicarbonate and baking powder is generally used as the leavening agent in these products. The ammonium bicarbonate is particularly useful because it is fast-acting. In addition, much of the leavening action is due to steam formation. Very hot ovens (450–475°F) are used to cause rapid formation of steam to assure expansion before the crust is set. After expansion is complete, the oven temperature may be reduced to prevent burning.

Cakes

Cakes contain a number of ingredients. Flour, water, shortening, whole eggs, egg whites, flavorings, sugar, and a leavening agent (e.g., baking powder) are the major ones. Different types of cakes have widely differing amounts of these ingredients. In comparison to bread, soft, low-gluten flours are used and in relatively small amounts. This causes the soft, crumbly texture associated with cakes.

Pound, sponge, chiffon, and angel food cakes illustrate the diversity available by suitable modification of the recipe.

The pound cake is traditionally based on a formulation consisting of one pound (454 grams) each of flour, butter, whole eggs, and sugar. Pound cake has a very crumbly texture resulting, in part, from the absence of any specific leavening agent. The leavening that does occur is due to expansion of gases (steam and air) that are trapped by the shortening and eggs, particularly the egg whites.

In making a sponge cake, very little shortening is used, but a chemical leavening agent is—egg whites are a major ingredient. They are normally beaten in order to incorporate large amounts of air, which contributes to leavening.

Even greater amounts of egg whites are normally used in chiffon cakes. Shortening is also present along with some baking powder for oven spring. Angel food cakes also have a high content of egg whites but contain no shortening or chemical leavening agent. Here again, steam and air expansion are responsible for leavening.

SUMMARY

Wheat flour is the major ingredient used in baking. The most important types of wheat are designated hard and soft, for which the major difference is the protein content. When the proteins gliadin and glutenin are combined in the presence of water, a unique structural network, called gluten, is formed. It gives dough a strong, elastic quality required for making breads and rolls, which are highly leavened. Hard wheats yield high gluten, strong doughs. Soft wheats are used for making baked goods with little leavening and a more crumbly texture.

During the baking process, many complex changes occur. The most important is leavening, which may be attributed to production of gas (yeast, baking soda, steam) and expansion of gases (air, CO_2) on heating. Baking powder is a complex mixture containing baking soda, an acid, and other ingredients. Proper selection of the acidic ingredient permits a controlled production of carbon dioxide.

Steam is often a major cause of leavening. The conversion of liquid water into water vapor has the potential for very dramatic leavening. This process is a major part of the leavening action in making eclair and cream puff shells, as well as certain types of cakes that include egg whites as a major ingredient for the purpose of trapping expanded gases (air and steam).

PROBLEMS

1. Give a definition or example of each of the following:
   a) endosperm
   b) gluten
   c) leavening
   d) "oven spring"
   e) Maillard browning
2. What grain is the most important source of flour for baking?
3. What are the differences between hard and soft wheats?
4. What function does gluten perform in baking?
5. Give three common agents used for leavening baked goods.
6. What are the major ingredients in baking powder? How does baking powder differ from baking soda?
7. When can ammonium bicarbonate be used as a leavening agent?
8. Why is sugar added to bread dough?
9. Why does bread brown on the outside?
10. What is the function of shortening in bread dough?
11. What is sourdough bread?

REFERENCES

1. Matz, S.A. *Bakery Technology and Engineering.* Westport, Conn.: AVI, 1972, p. 215.
2. Shallenberger, R.S. "Browning Reactions, Nonenzymic." In *Encyclopedia of Food Technology,* Vol. 2, A. H. Johnson and M. S. Peterson. Westport, Conn.: AVI, 1974, pp. 136–9.

CHAPTER THIRTEEN

M ILK and several by-products of milk are important foods. In this chapter, the composition of milk and the chemistry of the processes used to make some of the by-products will be considered. Butter is a major by-product that will only be discussed briefly in this section but will be considered in detail in Chapter Fifteen.

MILK

Whole milk contains approximately 87% water, 5% carbohydrate, 4% fat (called butterfat), 3% protein, and 1% minerals.

Carbohydrate

The carbohydrate found in the milk of humans, cows, and all other animals is **lactose.** It is about one-sixth as sweet as common sugar (sucrose). Lactose was described in Chapter Eight. It is a dissaccharide made up of the monosaccharides glucose and galactose with a β linkage between the units.

Many people in the world are unable to digest milk properly because of the lactose. It has been noted previously that humans are unable to metabolize cellulose due to the lack of the enzyme called *cellulase,* which cleaves the β linkages between the glucose units in the polysaccharide.

In the discussion of alcoholic beverages, it was noted that cereal grain starches, which have α linkages between the glucose units, have to be broken down to both maltose and glucose in order to be metabolized by the enzymes contributed by yeast. The same is true for bread yeasts, which require that sugar be added to dough or that malt (containing amylase enzymes to break down the starch) be present in the flour to provide food for the yeast.

The enzyme β-*galactosidase,* or simply *lactase,* is necessary for the digestion of milk sugar. A person lacking lactase is suffering from a mutation like that discussed in Chapter Two. When the enzyme is present, it is found in the small intestine. When the enzyme is missing, animals frequently show an intolerance to lactose, or more specifically milk, which is the main source of lactose.

This is rather surprising in view of the fact that powdered milk has seen widespread use in many areas of the world in dealing with problems of malnutrition.

In addition to the intolerance problem, the practice of reconstituting milk by mixing with water has been subject to the problem of unsanitary water supplies

DAIRY PRODUCTS

that are often encountered in underdeveloped countries where malnutrition problems are usually most severe.

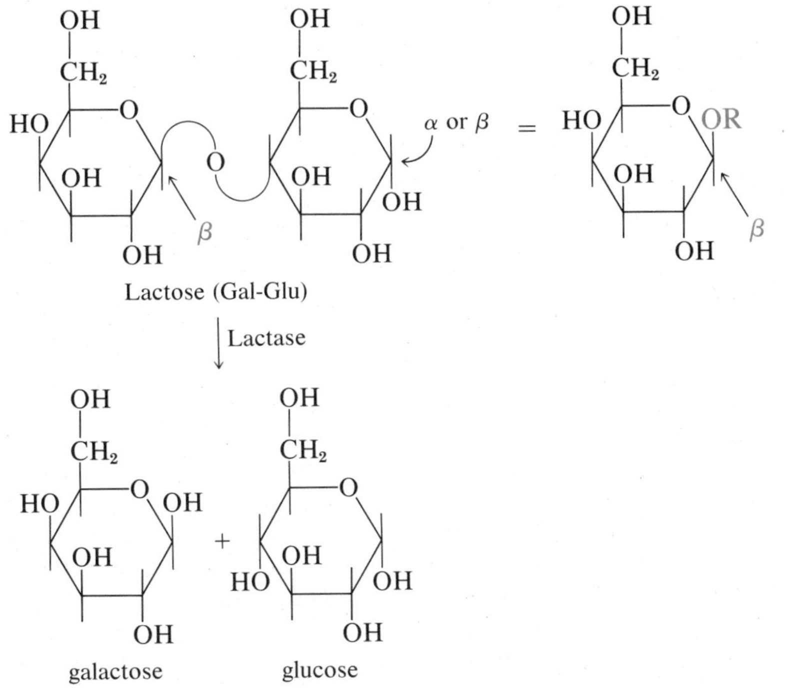

Lactose (Gal-Glu)

galactose   glucose

The common symptom of intolerance is diarrhea, which may possibly be attributed to a net flow of fluid into the intestine in response to a buildup of lactose that cannot be metabolized. The effect is analogous to the experimentally useful techniques of dialysis or osmosis previously considered. An alternative explanation, which may accompany the above, is that some lactose might be broken down by the enzymes of certain bacteria found in the digestive tract by a fermentation process, accompanied by the production of CO_2 and organic acids, which might contribute to a bloated feeling, cramps, and diarrhea. It is very likely that these symptoms have often been attributed to unsanitary water, when in actuality the water was not the problem at all.

Clinically, the intolerance is readily monitored by administering an oral dose of lactose and testing the blood glucose level after about 15 minutes. The blood glucose level will rise in response to lactase-catalyzed breakdown of the lactose to produce the glucose.

Adults of all animal species, except man, lack lactase. Even humans usually lack the enzyme by about ages 2–4. In fact, tolerance seems to be the exception rather than the rule for all but northern Europeans and their descendants, including many white Americans. Certain African tribes can also be characterized as tolerant.

In rats, the lactase attains a maximum level at birth and drops to a very low level immediately after weaning.

Human milk contains about 7.5 grams lactose per 100 milliliters, whereas cow's milk, which is not as sweet, contains 4.5 grams per 100 milliliters. Certain mammals totally lack carbohydrate in their milk. These are classified as Pinni-pedia (i.e., seals, sea lions, and walruses) and have milk with an unusually high fat content, 35 grams per 100 milliliters or higher. Whales and polar bears also have a low-lactose, high-fat content in their milk. Such milk is very much like heavy cream.

The origin of intolerance is an interesting subject. Many areas of the world have been ravaged by epidemics of sleeping sickness, transmitted by the tsetse fly, which is responsible for destroying cattle populations and making milk unavailable. Thus, a long term evolutionary pattern can be imagined in which peo-ple had no milk available and lost the ability to synthesize the lactase protein. In fact, these people would have a slight advantage over those persons who synthe-sized a protein that had no function, since protein synthesis consumes energy. On the other hand, in areas where cow's milk was routinely available, an intolerant person would be at a disadvantage, since he would be unable to utilize the food value of the lactose.

It is interesting that many bacteria have been found to exhibit a very so-phisticated response to either the presence or absence of lactose. If these bacteria are provided with a nutrient medium containing lactose, they are found to contain lactase, whereas in the absence of lactose, no lactase is present. Some very so-phisticated studies of a set of genes called the "lactose operon" have explained how genetic information found in DNA provides these bacteria with the ability to adapt and tolerate lactose on demand or to be economically efficient by not syn-thesizing the lactase when it is not needed.

Surprisingly, humans do not seem to have a similarly economical system for handling lactose. In fact, one interesting study was conducted on six Nigerian medical students who had no tolerance for lactose. They were fed steadily increasing amounts of lactose over a period of six months during which time they developed a tolerance as witnessed by the absence of the usual symptom of diar-rhea, but they showed no increase in blood glucose following a dose of lactose. This contradiction suggests that intolerance may be equated to the absence of the enzyme but that tolerance cannot necessarily be equated to the presence of the en-zyme. This latter condition has been attributed to the possible development of bacteria in the gastrointestinal tract, which are thriving, in part, on a diet of lac-tose and which seem to be causing a false tolerance (1).

In response to lactose intolerance, active research is underway to make milk digestible by treatment with lactase (probably isolated from yeast) to break down the lactose (2).

Fat

The fat or lipid content of milk is commonly referred to as butterfat. It exists as a suspension of minute droplets or globules that will rise to the top of the liquid to form a cream layer, under certain conditions. The cream rises due to the low den-sity of the fat. Commercially, centrifuges are used as cream separators. **Homoge-nization** is the process whereby these fat globules are reduced in size so that they remain suspended. The homogenization process consists of forcing the milk through a nozzle at high pressure. Churning has the opposite effect. If the cream is skimmed off the milk and churned, the globules aggregate into larger particles that coagulate to form butter, which is about 80% fat.

Figure 13.1   A milk homogenizer. This particular model can process up to 3500 gallons per hour; other models have a capacity in excess of 10,000 gallons per hour. (Courtesy of Gaulin Corporation.)

The amount of fat in milk and milk by-products is a major factor in determining the physical and flavor characteristics. Heavy whipping cream is at one extreme for fluids with a fat content around 35%. Light cream usually contains at least 18%. A product known as "half-and-half" contains 10–12% fat. Low-fat milk may contain up to 2% butterfat, whereas skim milk contains less than 0.5%.

It will be seen shortly that the fat content of cheese is quite variable but often even exceeds that found in heavy cream.

The chemical structure of butterfat and other fats and oils will be considered in detail in Chapter Fifteen.

Minerals

A wide variety of metal salts are found in trace amounts in milk, but calcium is of major importance, not only because of its nutritional value, but also due to its role in the production of other dairy products, e.g., cheese. Of the calcium present in milk, most is tightly bound to protein.

Milk protein is often referred to as *casein*. This is a slight oversimplification since only about 80% of the protein is casein. The rest is a mixture of soluble proteins that include the albumins and the globulins. These are commonly known as the *whey proteins,* as described in the following.

Casein is actually four different proteins, commonly referred to as α-, β-, γ-, and κ-caseins, of which the α-, β-, and κ-are the major types comprising roughly 50%, 33%, and 15% of the casein, respectively.

The α- and β- caseins have similar and very unusual properties. Each is insoluble in the presence of calcium ions (Ca^{+2}) at the concentration normally found in milk. The κ-casein is both soluble and solubilizing with respect to the α- and β-caseins. In other words, when the κ-casein is present and intact, it keeps the α- and β-caseins in solution.

CHEESE

We now come to the section that ''Little Miss Muffet'' has been waiting for. Her snack of ''curds and whey'' is well-known. Curdled milk is also well-known to anyone who has handled any badly-spoiled milk or cream. The heavy unappetizing lumpy material is called the *curd*. The curd forms by *coagulation* (often called *clotting*) of the α- and β-caseins. In spoiled milk, clotting occurs as a result of the lactic acid produced by bacterial fermentation of the lactose by the pathway described in Figure 11.4. Recall that lactose can be broken down into glucose plus galactose and the latter can be converted to glucose so that the entire disaccharide is subject to the lactic acid fermentation. Lactic acid or any other acid causes the κ-casein to dissociate from the other caseins. Consequently, the α- and β-caseins become insoluble and separate out as a curd.

During this clotting, some of the lactose and most of the butterfat is entrapped in the curd. The liquid that remains is the *whey*. The whey contains much of the lactose, plus minerals, plus the soluble noncasein proteins simply known as *whey proteins*. Most of the calcium is retained in the curd.

The clotting process is the basic chemistry involved in making cheese, since cheese is merely curd that is formed and treated under controlled conditions. Lactic acid formation could occur if the milk were simply aged to give bacterial enzymes time to develop and function, and the lactic acid would cause the milk to curdle. This approach is time-consuming and would require facilities to store the milk, but more important, certain undesirable changes may occur. Instead of this procedure, it is common practice to add *rennet* to the milk. Rennet is a crude extract of gastric juices generally obtained from calves and contains an enzyme called *rennin*. The rennin is a proteolytic (protein-cleaving) enzyme that attacks the κ-casein by cleaving the protein chain. This cleavage of the κ-casein causes it to dissociate from the complex with α- and β-casein, and clotting occurs.

There is a legend regarding the origin of cheese making that suggests that an Arabian merchant made a trip across the desert on a hot day. He carried a supply of fresh milk in a pouch that was made from a sheep's stomach lining. When he stopped to drink some of the milk, he found it contained a thin watery

Eating her curds and whey . . .

liquid and a soft curd with a pleasing taste. It is supposed that some rennin present in the sheep stomach had caused the curd formation.

Addition of lactic or other acids to promote curd formation also causes a loss of certain salts. The situation is similar to that described in the coverage of agricultural chemistry in which certain mineral nutrients may be so insoluble, due to the pH in the soil, that they are virtually unavailable. Addition of an acid or base can often increase the solubility of these nutrients and make them more readily available. Similarly, the calcium in milk is mostly fixed in a calcium caseinate-calcium phosphate complex and becomes trapped in the curd. When the pH is more acidic, much of this calcium becomes soluble and ends up in the whey. This result not only makes the cheese less nutritious but the lower pH also restricts some microbial growth that may be essential for proper *ripening* of the cheese in order to achieve the desired flavor and texture.

Cheese takes many forms. There are hard cheeses, soft cheeses, cottage cheese, cream cheese, roquefort cheeses, sharp cheeses, cheddar cheeses, American cheeses, processed cheeses, and many more. Quite obviously, rennin alone cannot provide such a wide variety of end products. The characteristic properties and production methods of some of the above are described in the following paragraphs.

Each cheese has some unique characteristics due to the many variations in technique used in producing cheese. Texture, fat content, flavor, and aroma are some of the gross variables that are determined by a multitude of chemical differences.

The steps in cheese making may include some or all of the following with a variety of alternatives available at every step. Some typical variables are noted.

1. Coagulation (with acid and/or rennet)
   Variables:
   (a) Type of milk
       (1) cow, sheep, other
       (2) whole or skim
       (3) raw or pasteurized
   (b) Temperature
   (c) Rennet addition (often follows formation of some acid via lactic acid fermentation)

2. Cut and press curd
   Variables:
   (a) Higher pressure for harder varieties
   (b) Amount of whey removed
   (c) May follow microbial inoculation

3. Addition of bacteria or mold culture and ripening
   Variables:
   (a) Type of microorganism
   (b) Culture may be added before coagulation
   (c) Time ranges from zero to many months
   (d) Addition of salt (inhibits growth of certain undesirable microbes)
   (e) Texture (long-term ripening often softens the cheese due to extensive cleavage of proteins)

The method of *ripening* is of prime importance in determining the type of cheese that forms. A wide variety of bacteria and molds are either applied to the surface of the curd, injected into the curd, or incorporated into the milk prior to curd formation.

One of the first chemical changes to occur during and even before ripening is fermentation of any trapped lactose, catalyzed by the enzymes of most microorganisms. This is particularly significant in view of the lactose intolerance noted earlier. In certain areas where preservation of whole milk was virtually impossible, it was, nevertheless, a simple matter to preserve much of the nutritional value by converting the milk to cheese, which is more easily preserved. Due to the lactic acid fermentation, the lactose is totally lost. This is considered a very good feature by people on low-carbohydrate diets, for whom cheese is often a major component of the diet.

Following and accompanying the lactic acid fermentation, many other complex chemical changes occur and it is these changes that make each cheese unique, since each microorganism provides its own complement of enzymes, which catalyze a multitude of chemical reactions. It is virtually impossible to describe, or even know, all the chemical reactions carried out by even a single microbe. Some interesting studies have been done on the chemical changes induced when *Penicillium roquefortii* is used to inoculate the curd. In Roquefort and most other cheeses some of the fat is broken down. Fatty acids (see Chapter Fifteen) such as octanoic acid are an integral part of fats. Many of these acids have distinctive flavors and aromas and are also convertible into other substances that contribute flavor and aroma. The following sequence is apparently operative since

$$CH_3CH_2CH_2CH_2CH_2CH_2CH_2\overset{\displaystyle O}{\overset{\|}{C}}\!-\!OH \qquad \text{octanoic acid}$$

$$\Big\downarrow \text{3 steps}$$

$$CH_3CH_2CH_2CH_2CH_2\overset{\displaystyle O}{\overset{\|}{C}}CH_2\boxed{\overset{\displaystyle O}{\overset{\|}{C}}\!-\!O}H \qquad \text{3-oxo-octanoic acid}$$

$$\Big\downarrow CO_2$$

$$CH_3CH_2CH_2CH_2CH_2\overset{\displaystyle O}{\overset{\|}{C}}CH_3 \qquad \text{2-heptanone}$$

Figure 13.2   Swiss or Emmenthaler cheese. (Courtesy of the Switzerland Cheese Association.)

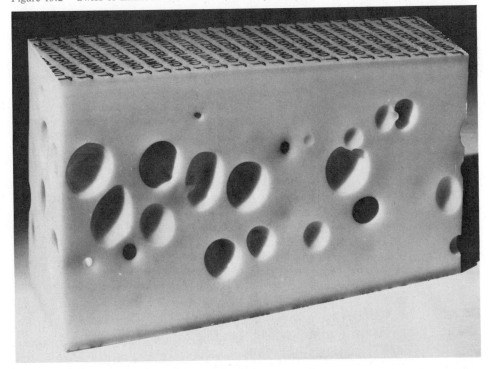

2-heptanone has been shown to be a major ingredient in blue and Roquefort cheeses and is easily recognizable to anyone who is familiar with these cheeses. It is interesting that 2-heptanone was previously described as an alarm substance of the honeybee (see Chapter Ten).

Cottage Cheese

Cottage cheese is the unripened curd from skim milk produced by acid or by the combined action of acid and rennet. Since most of the fat is removed before clotting, cottage cheese is high in protein but low (0.5%) in fat content. A somewhat tastier product, called creamed cottage cheese, is made by adding cream to bring the fat content up to 2–6%.

Figure 13.3 Making processed cheese. (1) The process begins with fresh milk to which a starter (microorganism) and rennet are added. The milk is allowed to curdle in a vat until it is the consistency of custard (2) and is ready to cut with special curd knives. The curd is then stirred and heated and the whey is allowed to drain (3), while simultaneously packing the curd against the sides of the vat. The curd is later shredded and placed in 500-pound barrels for curing and aging, after which the cheese is

Cream Cheese

Cream cheese is also a soft, unripened cheese. Unlike cottage cheese it is made from a mixture of milk and cream. The final product generally contains more than 30% fat.

Cheddar Cheese

Cheddar cheese is a hard cheese ripened by bacteria. As with all hard cheeses, the cheese is curdled at higher temperature and the whey is pressed out of the curd under high pressure. This treatment lowers the moisture content to 30–40%, which is typical of all hard cheeses and improves the keeping properties. In contrast, soft cheeses may contain as much as 75% water. Coagulation temperatures are gener-

cut into workable pieces (4). Cheeses of various stages of flavor development are selected and blended in a grinder (5) to produce a desired flavor and texture in the finished product. The cheese then moves into a steam cooker where it is heated to 165°F (75°C) or higher and other ingredients are added (6). The hot cheese then moves over a quick-chill roll (7). Narrow ribbons of cheese flow from the chill roll onto a belt to be cut into slices (8). (Courtesy of Pauly Cheese Company.)

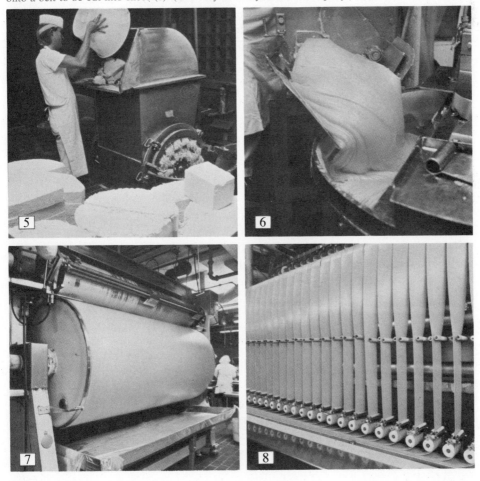

ally lower and the product is more perishable. Cottage cheese is an example of a soft cheese whose shelf life is particularly low because it is also unripened. The length of the ripening is quite variable for cheddar cheeses. Mild cheeses are ripened for shorter times. Sharp cheeses often require more than a year to develop the full flavor.

Roquefort and Blue Cheeses

In addition to the use of the special mold *P. roquefortii,* Roquefort cheese is made from sheep's milk. Blue cheese is made from cow's milk.

Swiss Cheese

Swiss or Emmenthaler cheese is a hard cheese ripened by bacteria, which generate CO_2 and cause the formation of eyes (holes) in the cheese.

Processed Cheese

This type of cheese is actually a blend of cheeses (mostly cheddar). Blending permits the cheese maker to produce a more consistent product. The components of the mixture are finely ground, melted together, and reformed (See Figure 13.3.). Coloring, seasoning, and other additives may be introduced in the processing.

OTHER MILK PRODUCTS

A number of other dairy products are consumed throughout the world. Yogurt, koumiss, kefir, sour cream, and cultured buttermilk, like cheeses, are examples of products made by fermentation.

Yogurt

Yogurt is also a product of the fermentation of milk. Skim milk is generally used and the fermentation is carried out by two microorganisms, *Lactobacillus bulgaris* and *Streptococcus thermophilus.* Prior to inoculation, the milk is heated to boiling to kill all other microorganisms so that only the unique set of enzymes provided by the desired microbes are active. Yogurt itself is often used for inoculation.

Koumiss and Kefir

These fluid milk products are actually alcoholic beverages. Both are made by a double fermentation, which consists of the combined action of a lactic acid bacterium and an alcohol fermenting yeast. Consequently, these products are both somewhat sour and slightly carbonated. Consumption of both drinks is heavy in the USSR. Very little is consumed elsewhere.

Koumiss is a fermented beverage made from mare's milk. In contrast to cow's milk, mare's milk has a very low casein content and does not tend to clot. Cow's milk is not suitable due to the ease of clotting. Acidity (due to lactic acid) may run as high as 1.2% and alcohol as high as 2.5% in Koumiss.

Figure 13.4   Yogurt before and after processing. (a) Each tank holds 60,000 lbs. of milk ready for processing into yogurt. (b) Containers with cultured milk are wheeled into incubators to gel into yogurt. (Courtesy of Dannon Milk Products.)

Kefir is a slightly effervescent beverage made by fermentation of milk from goats, sheep, or cows. It contains less fat due to formation of a soft curd which is removed before bottling.

Sour Cream and Cultured Buttermilk

Sour cream is produced by the action of a bacterial culture on cream. The final product normally contains about 18% butterfat.

Buttermilk is the fluid that is left after churning milk or cream to produce butter. It has a variable, but normally low, fat content. Commercial buttermilk is frequently cultured with a lactic acid bacteria.

Evaporated and Condensed Milk

There are a variety of unfermented milk products sold commercially. Evaporated milk and condensed milk are both produced by evaporation of some of the water in whole milk.

Evaporated milk must contain at least 7.9% butterfat and 25.9% total solids to meet federal standards. It is sterilized by heating for preservation.

Condensed milk is supposed to contain not less than 8.5% fat and 28% total milk solids but it is unsterilized and is therefore quite perishable. It is commonly marketed as "sweetened condensed milk," which is made by adding sucrose to give about 18% sugar content prior to evaporation and about 42% sugar after evaporation. The high sugar content serves as a preservative. This phenomenon is explained in the following chapter.

SUMMARY

Milk is a very versatile food which may be consumed directly or converted to a wide variety of products. The major components of milk are water, lactose, butterfat, protein, and minerals (mostly calcium).

The lactose is an item of great concern because the majority of the world's adult population cannot tolerate it, due to the absence of the enzyme, lactase, which is required for metabolism.

The butterfat content may be handled in a number of ways. Milk may be homogenized by breaking up the fat droplets to keep them suspended or the fat may be allowed to rise and form a cream layer, which may be skimmed off. The cream may be marketed in a variety of forms ranging from heavy cream (35% fat) to half-and-half (10–12% fat). Skim milk contains less than 0.5% fat. The cream layer may also be churned into butter.

The proteins in milk consist of 80% casein and 20% whey proteins. The caseins separate out in the curd and trap most of the calcium and fat. Rennin can be used to initiate clotting by attack on the κ-casein with elimination of the solubilizing effect on the α- and β-caseins. Lactic acid, from fermentation of the lactose, may also cause curd formation.

Curd formation is the initial step in making cheese. Many variations in technique during and following curd formation are responsible for the wide variety of cheeses that are available. Among the most important variables are the choice of milk (whole, skim), the pressing of the curd (higher pressure yields harder cheese), and the ripening (bacteria, molds, none). The method and time of ripening is usually the most important in determining the flavor that develops.

Other important by-products of milk are yogurt, koumiss, kefir, sour cream, and buttermilk, all of which are products formed by fermentation with microorganisms. Evaporated and condensed milk are unfermented products made by evaporation of milk. Condensed milk is usually preserved by addition of sugar.

PROBLEMS

1. Give a definition or example of each of the following:

a) lactose	i) processed cheese
b) lactase	j) cream cheese
c) butterfat	k) yogurt
d) curd	l) sour cream
e) whey	m) koumiss
f) rennet	n) kefir
g) cottage cheese	o) buttermilk
h) cheddar cheese	p) homogenization

2. What is the composition of whole milk?

3. Why are a great number of people unable to digest milk?

4. What is the role of rennin in cheese formation?

5. Why is lactic acid fermentation in cheese making important to people with an intolerance to milk?

6. Why is sugar added to condensed milk?

7. Cheese has been described as a "cured curd." What is wrong with that description?

REFERENCES

1. Kretchmer, N. "Lactose and Lactase." *Scientific American 227* (4) (1972): 70.
2. Reese, K.M. "Agricultural Research Service Seeks Low-lactose Dairy Products." *Chemical and Engineering News,* 23 July 1973, p. 74.

U P to this point in Part Four, the discussion has centered on food processing. In some cases, processing means converting an already appetizing food into another form, e.g., converting milk to cheese, which increases the versatility of the original food. In other cases, an unpalatable food can be processed into an appetizing one, e.g., the conversion of wheat to bread. While processing milk into cheese, there is also the important aspect of preservation involved; several cheeses have a very long shelf-life, particularly when compared to milk. In this chapter, greater emphasis will be placed on food processing techniques, which are carried out primarily for the sake of preservation.

Control or modification of biochemical processes that could result in spoilage of foods is a major concern. In several areas of prior coverage, it has been shown that microorganisms can be very beneficial. Still more examples will follow, but it is now time to look at some harmful effects that microbes can cause and to consider how such problems can be prevented.

FOOD MICROBIOLOGY

Many volumes of information have been written concerning this subject. Biology has been described as "a kind of superchemistry," which suggests the complexity of the chemistry involved and the difficulty of trying to comprehend many of the effects, favorable and unfavorable, which microorganisms have on food (1).

Some microorganisms make their presence known in very obvious ways. Examples are *Neurospora crassa* (bread mold) and the lactic acid bacteria found in milk that cause the milk to sour and eventually curdle.

Other microbes may be more subtle but may also be much more dangerous. One example of this type of microbe is *Clostridium botulinum,* which produces a dangerous toxin called *botulism*. It is an anaerobic microbe which can thrive inside sealed containers. The bacteria itself is not harmful but the toxin, which is a protein, causes paralysis, often resulting in death, by interfering with a reaction that is of great importance in transmission of nerve impulses. This effect is quite similar to that known to be caused by certain snake venoms. It has been said that 14 ounces of botulism toxin would be sufficient to kill the entire world population (2).

The danger of this organism is increased by its tendency to form spores. A spore is a dormant state of a microorganism, which is formed in response to condi-

FOOD PRESERVATION

tions that are unsuitable for normal growth. In the case of *Cl. botulinum*, the spores are particularly resistant to conditions that are lethal to most other microorganisms. When suitable growth conditions are restored, these spores are able to germinate and multiply and continue to produce their deadly toxin. Being a protein, the toxin is quite sensitive to heat. On the other hand, the microbe itself may be able to withstand prolonged heating even at boiling temperatures. It is, therefore, common practice to cook certain foods at temperatures above boiling by using a pressurized cooker prior to canning. This process is routinely used in commercial canning, but not in home canning. *Cl. botulinum* spores are completely destroyed in ten minutes at 250°F (121°C).

The acidity of food is of major importance in deciding what conditions are suitable for heat treatment. Low pH (high acidity) reduces the survival time of the spores. Many fruits are quite acidic so that *Cl. botulinum* spores can be completely destroyed in about $1\frac{1}{2}$ hours at normal boiling temperatures. Unfortunately, this is a much longer time than that normally called for in most home canning recipes (3).

Tomatoes are of particular concern since they are frequently processed by home canners. Tomatoes are normally thought to be very acidic, thus allowing the use of very mild treatment in order to kill harmful bacteria and their spores. Unfortunately, many of the newer hybrid varieties have very low acidities, which makes them particularly susceptible to spore formation and stability. Consumers' Union recommends heating tomatoes at 250°F at 15 pounds per square inch (psi) for 15 minutes to insure safety (3). As noted previously, the toxin is destroyed by

Figure 14.1   A Pressure Canner (Courtesy of Presto Industries)

TABLE 14.1   TEMPERATURE AND PRESSURE IN A PRESSURE COOKER

TEMPERA-TURE		ACTUAL PRESSURE†		PRESSURE IN EXCESS OF NORMAL PRESSURE‡	
°C	°F	POUNDS PER SQ. IN	ATM*	POUNDS PER SQ. IN	ATM*
100	212	15	1.0	0	0
109	227	20	1.35	5‡	0.35
116	240	25	1.70	10‡	0.70
121	250	30	2.05	15‡	1.05

*A common unit of pressure is atmospheres (atm).
†Normal atmospheric pressure is approximately 15 pounds per square inch.
‡Typical settings used in pressure cookers.

even a few minutes treatment at normal boiling temperatures. The microorganism does not multiply if it is consumed. Its danger lies solely in its ability to produce a toxin prior to being eaten. *Cl. botulinum* is often abundant in soils, which accounts for its relative significance as a cause of food poisoning.

Salmonella is another type of bacteria that has often been associated with food poisoning. There are many types of *Salmonella*. The *Salmonella* bacteria do not form spores and they are readily destroyed by heating at 140°F (60°C) for 15-20 minutes. Their presence is usually not obvious in foods and they may multiply after being eaten. The bacteria also produce a toxin, which is not as hazardous as botulism but can cause unpleasant symptoms. The *Salmonella* toxin is only produced after the microorganism has infected the body.

Salmonellae are aerobic organisms, which cannot multiply in a closed container (e.g., a can). Mortality is very low (less than 1%) among diagnosed cases of *Salmonella*. It is believed that there are many mild cases diagnosed as a 24-hour virus or something comparable, since *Salmonella* causes symptoms known as gastroenteritis, or intestinal upset. The infection commonly lasts about one week (4).

If food is kept cold, the *Salmonella* organism cannot multiply. The danger of infection and multiplication arises when cooked food, e.g., meat, is allowed to remain at room temperature for several hours. These conditions are ideal for multiplication of *Salmonella*.

As of June 1, 1975, the FDA put a ban on the sale of pet turtles because they have been found to carry the *Salmonella* organism.

Staphylococcus organisms are another frequent cause of food poisoning much like *Salmonella*, in that the primary danger is due to production of a toxin once the organism is established in the intestinal tract. Cream-filled pastries are probably the best-known food subject to *Staphylococcus* infection, since high temperatures are not used in their production. This microorganism does not form spores so it is readily destroyed at boiling temperatures, but cream pastries and several other foods are never heated to such high temperatures. The toxin of the strain most often associated with food poisoning is unusually heat stable, so even thorough cooking may not solve the problem for a food that is highly infected. That is, the toxin may also cause food poisoning if ingested in large quantities. As with *Salmonella,* the primary symptom is intestinal upset, which can be very unpleasant but is rarely fatal. Proper sanitation is one of the best means of avoiding both *Salmonella* and *Staphylococcus* infections.

A wide variety of disease-carrying microorganisms are less frequent, but, nevertheless, serious causes of infection that can be transmitted in food or water.

Trichinosis is a food-borne disease, which is not categorized as food poisoning. *Trichinella spiralis* is a parasite often found in raw pork. It is killed by thorough cooking but, if ingested intact, it can establish itself in the intestinal tract and cause very severe symptoms.

Ptomaine poisoning is an old-fashioned and incorrect term for food poisoning. Ptomaines are diamino compounds formed during the decay of dead, protein-containing substances. They are not toxic and they are not associated with any common form of food poisoning, although they are very foul-smelling substances.

The best known ptomaines are called *cadaverine* and *putrescine.* They are formed during putrefactive decay in which certain amino acids suffer a loss of CO_2 (decarboxylation).

$$
\begin{array}{cc}
\begin{array}{l}
CH_2NH_2 \\
| \\
CH_2 \\
| \\
CH_2 \\
| \\
CH_2 \\
| \\
CHNH_2 \\
| \\
C-OH \\
\| \\
O
\end{array}
&
\xrightarrow{-CO_2}
\begin{array}{l}
CH_2NH_2 \\
| \\
CH_2 \\
| \\
CH_2 \\
| \\
CH_2 \\
| \\
CH_2NH_2
\end{array}
\\
\text{Lysine} & \text{Cadaverine}
\end{array}
$$

$$
\begin{array}{cc}
\begin{array}{l}
CH_2NH_2 \\
| \\
CH_2 \\
| \\
CH_2 \\
| \\
CHNH_2 \\
| \\
C-OH \\
\| \\
O
\end{array}
&
\xrightarrow{-CO_2}
\begin{array}{l}
CH_2NH_2 \\
| \\
CH_2 \\
| \\
CH_2 \\
| \\
CH_2NH_2
\end{array}
\\
\text{Ornithine} & \text{Putrescine}
\end{array}
$$

As in the case of ptomaine formation, an almost endless list of microorganisms comes into play in causing food spoilage. Most do not produce a toxin but may cause changes that merely make food unpalatable. These organisms, the three described earlier and a few others that cause some degree of food poisoning, must be either controlled or killed, if certain foods are to remain edible.

TECHNIQUES OF FOOD PRESERVATION

Although the preservation techniques are many and varied, the following list includes most of the major methods used, some of which are often combined in handling specific foods.

> Cold Storage and freezing
> Dehydration: including freeze drying
> Heat processing: canning
> Irradiation
> Chemical alteration: fermentation, pickling
> Use of additives

Cold Storage and Freezing

Preservation of foods by storing at low temperatures is a common practice. The purpose is to suppress the growth of decay microorganisms. The low temperature slows down chemical processes and, when frozen, the water present in food is unavailable as a growth medium for these microorganisms. It is often desirable to inactivate some of the enzymes present in microbes and the food itself before freezing. This is done by heating briefly.

The freezing process itself is important. If the freezing occurs slowly, large ice crystals form and many of the cells in the food are broken. On thawing, the tissue collapses and the texture of the food is altered. If the food is quick frozen, the ice crystals are smaller and tissue destruction is less. Fruits may be frozen in order to preserve them, but quick freezing is essential to retain the natural texture of the fruit. More often, fruits are canned because of this problem.

The texture of ice cream is also greatly affected by the rate of freezing. When originally frozen, ice cream has a smooth creamy texture which is gradually lost after repeated opening and closing of a container. Each exposure to room temperatures causes ice cream to suffer some melting and, when refrozen, the ice crystals grow larger. Frequent repetition causes a very dramatic change in the texture of the product.

Dehydration and Freeze Drying

The basic idea behind these techniques is the same as that described for freezing. If water is unavailable, microorganisms cannot function normally.

Heat Processing

The use of heat in food preservation is primarily important in canning, although a brief heating step often precedes drying or freezing of some foods.

Commercial and home canning techniques differ somewhat, but the principle involved, i.e., control of the growth of spoilage organisms, is the same. Commercial canning dates back into the early part of the nineteenth century and has been a process subject to continual change as new techniques evolved and a better understanding of spoilage and other processes was gained. Major technical improvements have been in the direction of shorter cooking times at higher temperatures to retain fresher flavor, firmer textures, and higher nutrition. The latter point is important since cooking can destroy certain vitamins naturally present in foods.

Major advances in processing food have been the use of pressurized cookers and solutions containing high concentrations of salt. It is common knowledge that salt lowers the freezing point of water. Ice-salt mixtures have often been used to reach temperatures well below freezing. Homemade ice cream is routinely made using an ice-salt mixture to provide sufficient cooling. Salt also raises the boiling point of water, thus allowing one to achieve higher boiling temperatures in processing food. Shorter cooking times also increase the rate of commercial production.

Home canning may involve pressure packing, hot packing, or cold packing. In general, the less acid the food, the longer the heating time or the higher the temperature required for safe canning. Acid is a natural preservative found in food. In general, fruits are high in acid and can often be safely cold packed. Vegetables contain relatively little acid and are hot packed or even pressure packed. In pressure packing, a pressure cooker is employed to achieve very high temperatures (see Table 14.1). Meats are preserved by pressure packing, freezing, or in an entirely different way described in a later section.

Jams and jellies are cold packed. Their stability is due to natural acidity in the fruit or fruit juice used. High levels of sugar also act as a preservative of these and other foods, such as canned fruit, which is routinely packed in a heavy sugar syrup.

It is an interesting contradiction that foods can be protected from microbial spoilage by using such an excellent food source (sugar) for these microbes. Sweetened condensed milk, described earlier, is also preserved by a high sugar content. This phenomenon could almost be equated to drying the food, since at very high sugar concentrations the water is all effectively tied up by the sugar and is, therefore, unavailable as a growth medium for decay-causing microorganisms.

In commercial hot packing, a microorganism known as PA (putrefactive anaerobe) 3679 is used as a test of the effectiveness of heat processing. PA 3679 is an extremely resistant, spore-forming microbe. Even very resistant *Cl. botulinum* is more susceptible to heat than PA 3679, so that safety is assured if conditions are adequate to destroy the latter.

Irradiation

This technique has been termed "cold sterilization" (5). There are some significant factors that limit its usefulness, although it has been successfully applied to a few foods.

On the negative side, it is well-known that spores tend to be much more resistant to radiation than rapidly-dividing (vegetative) cells. This kind of behavior may be put to good use in treating cancer, but is a major drawback in using ionizing radiation as a method of food preservation. The two applications are not exact opposites. In radiation therapy, normal cells must survive. In food preservation, one is not concerned with allowing any microbial cells to live. The problem is that the kind of chemical alterations induced by irradiation, which is lethal to microbes, may cause undesirable changes in a food. In fact, when the radiation dose is high enough to destroy all spores, the odor and flavor of the food is often altered.

High-protein foods are particularly susceptible to change. The texture of meats and vegetables may be altered. Irradiation of milk reportedly causes changes similar to those encountered when milk is heated (5). An "off-flavor" develops in both cases.

On the positive side, radiation can be used to kill certain pathogenic microorganisms, such as *Salmonella* and *Staphylococcus*. Organisms that do not form spores are very susceptible to radiation. Control of trichinosis and tapeworms is also possible using low doses of radiation.

One unusual application is the control of sprouting by certain root crops. Potatoes, carrots, and onions have been treated successfully.

Concern is often expressed over the danger of inducing radioactivity in foods. The problem is a real one if neutrons are used, since they do induce radioactivity, but since neutron emission is not a form of spontaneous radioactive decay by any natural isotopes, it is easily avoided. Gamma-, x-, and some β-radiation are sufficiently penetrating and do not induce radioactivity in foods.

Chemical Alteration

In the following paragraphs, vinegar, meats, and the general process of pickling will be considered.

VINEGAR

The production of vinegar is a subject largely covered in the earlier discussion of alcoholic beverages. In wine making, the following set of reactions must be avoided or the acetic acid that forms will cause the wine to taste sour. On the

$$CH_3CH_2OH \longrightarrow CH_3\overset{\displaystyle O}{\overset{\|}{C}}-H \longrightarrow CH_3\overset{\displaystyle O}{\overset{\|}{C}}-OH$$

ethyl alcohol   NAD   NADH   NAD   NADH   acetic acid

acetaldehyde

other hand, vinegar is required by law to contain at least 4% acetic acid. Consequently, it is allowed to form, even from wine in the production of wine vinegar, by encouraging the action of *Acetobacter*, which is an aerobic microorganism. This means that making vinegar is a two-step process. The first step is the anaerobic, alcoholic fermentation (see Chapter Eleven) carried out by yeast. The second step is then carried out under aerobic conditions to assure an adequate supply of NAD.

The acetic acid in vinegar is used as a preservative against the growth of other microbes in foods that are normally sour (e.g., sour pickles and sauerkraut) or foods that are normally sweetened to counter the sour taste (e.g., sweet pickles and watermelon rind).

Some very interesting chemistry comes into play in processing and preserving meats for both short- term consumption and long-range storage. Cold storage and freezing are obvious ways of preventing spoilage. Pickling, curing, and corning are all essentially the same and are also effective ways to preserve meats and other foods by controlling the changes that occur in the food.

Even before spoilage becomes a problem, fresh meat undergoes some chemical changes that, if properly controlled, may greatly affect the characteristics and keeping quality of the meat.

The first event following slaughter is the onset of lactic acid fermentation. Prior to the kill, the circulatory system of an animal continually carries oxygen from the lungs to the tissues by using a pickup and delivery system in the form of a protein known as hemoglobin. Once killed, the oxygen supply is immediately cut off and the tissues become anaerobic. For a time, all of the natural enzymes necessary for either aerobic or anaerobic breakdown of glucose remain active. Recall that the aerobic and anaerobic pathways are identical through pyruvic acid formation, and they continue to operate until the fuel supply is depleted or the enzymes are inactivated. In effect, the enzymes do not know the animal is dead, so they continue to carry out their normal function, which is to metabolize glucose to provide energy (ATP). In fact, the demand for energy to carry out many chemical changes is even greater after slaughter since the efficient aerobic pathway for ATP production shuts down in the absence of oxygen.

In animal cells, the glucose is stored as the polysaccharide, glycogen. The amount of glycogen present determines how much lactic acid is formed. This is particularly significant since the lactic acid, like the acetic acid in vinegar, acts as a preservative to suppress the growth of microorganisms, thus slowing spoilage and increasing the keeping quality of the meat. So, the amount of glycogen present in the tissues is very important and is determined by the condition of the animal. Unlike fat, an animal can only store a limited amount of glycogen. Consequently, a well-fed, rested animal will have a higher glycogen level than one that is hungry or has been exercised before slaughter. Meat from the former will keep much better due to the higher lactic acid content that develops in the meat. Under the best conditions, the pH will drop from about 7 to 5.4–5.7, at which point the process will stop even if some glycogen remains (6).

A bull killed in a bullfight, an animal killed in a hunt, or a fish caught after a long fight have low glycogen levels and produce very little lactic acid. Feeding prior to slaughter very effectively restores the glycogen supply.

The second major change in the meat following slaughter occurs over a period of several days and causes the beef to become more tender. Immediately after death, the meat is very soft and hard to work with. Rigor mortis soon sets in. At this point the meat is quite tough and remains so for a day or two. If the meat is allowed to hang somewhat longer, natural proteolytic enzymes break down some of the protein and cause the meat to become quite tender. The efficiency of the process may be increased and the required time may be decreased by injection of enzymes prior to slaughter. This procedure provides a uniform distribution of the enzymes throughout the bloodstream into all of the tissues with a resultant effect of tenderization.

Figure 14.2　A worker is shown cutting steaks from a naturally-aged loin of beef. Aging of beef has a tenderizing effect due to the action of natural proteolytic enzymes. Ribs (for prime rib) and loins (for steaks) are routinely aged in coolers at 34°F for 21–28 days. Sale of this well-aged beef is generally confined to restaurants. (Courtesy of Esmark, Inc.)

Figure 14.3　Products containing papain. (Courtesy of Adolph's Ltd.)

In the home, this process can be mimicked by using a meat marinade that contains *papain,* a proteolytic enzyme obtained from the unripe fruit of papaya. The papain will cleave some of the protein fibers that can cause toughness. Most recipes call for a brief treatment, usually about 15 minutes, to allow the papain to penetrate the meat. Much of the tenderization occurs in the early stages of cooking, when the warmer temperatures increase the activity of the papain.

Color is an important characteristic of meat. The familiar red color of meat is often taken as an indication of freshness. Some darkening of color may be attributed to water loss at the surface of the meat, which causes the pigments to become more concentrated and thus appear darker. More complex chemical changes are responsible for browning of meat. The site of the changes is the protein myoglobin, which acts as a storage depot for oxygen in muscle tissue. It is a dark-colored protein because of the presence of an iron atom, as Fe^{+2}, which is tightly bound to the protein and is also bound to a highly-colored compound called *heme.* The myoglobin can exist as simple myoglobin, oxymyoglobin, or metmyoglobin, as shown in the following sequence. When a large piece of beef is sliced, the exposed surface is initially purple due to Mb. This quickly changes to bright red due to oxygenation to form MbO_2.

$$Mb(Fe^{+2}) \xrightarrow{O_2} MbO_2(Fe^{+2})$$

myoglobin (purple)   oxymyoglobin (bright red)

$$\downarrow \text{oxidation}$$

$$Mb^+(Fe^{+3})$$

metmyoglobin (brown)

Preservation of the red color is a major goal in handling meats. Prior to cutting, there is a small surface area exposed, but once the meat is cut, the bright color that develops must be retained if the meat is to be prized by the consumer. Success seems to depend on keeping the myoglobin in the form of oxymyoglobin, which prevents oxidation (conversion of Fe^{+2} to Fe^{+3}) to form the brown-colored metmyoglobin. This problem presents a rather strange situation in which it is necessary to provide an adequate supply of oxygen to prevent oxidation.

One major obstacle arises when meat is cut and packaged. Before the meat is sealed in the package the myoglobin becomes oxygenated but, once sealed up, aerobic bacteria present on the meat consume some of the oxygen causing the surface oxymyoglobin to become deoxygenated and thus susceptible to oxidation to the brown metmyoglobin. The solution to this problem is to use packaging which allows oxygen to enter (and CO_2 to leave). This type of packaging is known as a breathing film.

Meat is sometimes cured to preserve it and to produce some desirable new forms, e.g., corned beef. Here again, color is important but the chemistry involved is a bit more complicated. A major ingredient used in curing meats is sodium nitrite, $NaNO_2$. In the presence of acid (usually either citric or acetic), the nitrite (NO_2^-) is converted to nitrous acid (HNO_2), which reacts with myoglobin to produce nitrosomyoglobin (MbNO), which contributes the pink color normally associated with cured meat.

The major ingredient used in curing meats is salt (NaCl) and the primary function of the salt is to control the type of fermentations that occur. In effect, the salt determines which microorganisms are permitted to grow, with the salt concentration being a major deciding factor. The action is primarily dehydration by tying up water that could serve as a growth medium for undesirable microbes. In general, lactic acid bacteria are more tolerant of salt so that, here again, lactic acid fermentation is a major event and the lactic and other acids that might form act as preservatives.

Quick curing is commonly carried out commercially by pumping the salt solution into the meat through 100 or more hollow needles. Hams are commonly cured after pumping salt solution into the meat by way of a main artery so that the salt is distributed evenly throughout the ham. (See Figure 14.4.)

The final product of the cure is relatively dry and dense and has a high salt concentration. To make it appetizing, it must be reconstituted or freshened before eating by repeated soaking in warm water, which removes much of the salt and makes the meat juicier. Sugar and other spices are also present to control the flavor of the final product. Pork shoulders, hams, and briskets of beef are often cured in this way. In curing beef, this corning process normally takes 4–5 weeks.

Figure 14.4 Artery pumping of ham. A worker is shown pumping a ham with a solution of brine (NaCl) containing other ingredients (including sodium nitrite for color) that are necessary to assure that the ham will cure properly. The worker pumps the solution into the main artery by way of a needle so that the solution is distributed throughout the entire ham to assure a uniform cure. The ham is pumped until it increases in weight by a specific amount. It is possible to see the salt solution spraying out of several places at the surface of the ham. (Courtesy of Percy A. Brown & Company.)

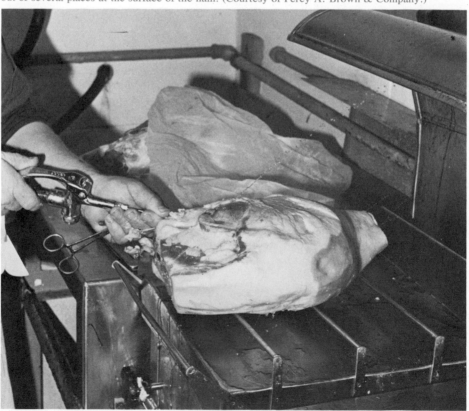

It is also a very common practice to smoke meats after curing, both for flavor and as a means of preservation. It is believed that the smoke dehydrates the surface of the meat and also deposits a surface coating, which consists of formaldehyde and other chemical compounds that prevent the growth of microorganisms.

PICKLES

A major use of fermentation is the production of pickles from cucumbers. Like many other foods, cucumbers have a very high water content. For this reason, it is a common practice to put a wax coating on the surface to prevent loss of water and help preserve the cucumber. For pickling, such a coating is undesirable, since it is necessary to have free passage of salts, spices, sugar, and water in and out of the cucumber.

As in meat curing, salt is the major ingredient used in making pickles. The concentration of salt used is commonly expressed in *degrees salometer*. These units are based on the concentration of saturated NaCl in water at room temperature, which is 25% NaCl, defined as 100° salometer, so that 1% NaCl equals 4° salometer.

In making pickles, a solution of salt that is 8–10% NaCl (32°–40° salometer) is added initially and fermentation is allowed to proceed for about a week. The usual dialysis occurs, in which water flows out and salt enters. Once again, the salt controls the types of fermentation that may occur, with lactic acid fermentation being a major pathway. At the end of the first week, the salt concentration is increased about 1% per week up to 16% salt (64° salometer). After 4–6 weeks, the fermentation is complete. The salted, fermented cucumber is called a salt stock and the pickles are well preserved in this state. Before packaging or eating, the salt stock is freshened by repeated washing with water to leach (dialyze) out most of the salt. If the product is to be sour pickles, vinegar is added to the freshened salt stock to give at least 2.5% acid. For sweet pickles, a sweet, spiced vinegar solution is added. Dill pickles are handled like sour pickles except that dill herb is introduced during the fermentation or to the freshened salt stock.

SAUERKRAUT

Saukerkraut is fermented cabbage. Salt (2–2.5 pounds per 100 pounds of cabbage) is added to shredded cabbage and fermentation is allowed to proceed for several weeks. The pickling process for sauerkraut has been thoroughly studied and found to proceed through three distinct stages of fermentation carried out by three microorganisms that work in succession. As acidity develops, one microbe is suppressed and another takes over until only one microbe is active toward the end of the fermentation. The microorganisms are naturally found in cabbage.

Use of Additives

In this section, consideration will be given to a wide range of additives used in foods. Things like sugar, salt, and smoke belong in this category and have been discussed previously.

Food additives have been the subject of much controversy, pro and con. Among those favoring the use of additives are chemical companies, which make huge profits from the sale of chemical additives, and many sectors of the food industry, which employ these additives in their products. These companies repeatedly claim that the consumer benefits greatly from the use of additives and, in many cases, this is true. The unfortunate thing is that there is a clear tendency to overuse additives to cover up defects or cut costs by replacing natural foods with chemical substitutes. The industry seems to justify this by noting that all foods are chemicals.

The other major problem with some food additives is inadequate or at least questionable testing. In part, the problem is the same as that discussed in Chapter Ten. Testing done on laboratory animals is the best alternative to the use of human guinea pigs, but the validity of many experimental findings is often attacked by industries that want to market an additive or a food that contains an additive. The classic example is the use of cyclamate as an artificial sweetener. In 1962, Metrecal (with cyclamate) was the largest selling food substitute and cyclamate also found its way into many other foods. After more than 15 years of heavy use, cyclamate was removed from the GRAS (generally recognized as safe) list by the Food and Drug Administration due to some tests that revealed an abnormally high incidence of bladder cancer in laboratory rats that had been fed large amounts of cyclamate.

Certain segments of the food industry and individuals who repeatedly caution against the heavy use of sugar discount the findings that led to the ban by claiming that the amount of cyclamate fed to these animals was much too high to reflect a realistic danger. This is a convincing argument that is effectively countered by those who argue that typical laboratory animals may be less susceptible to chemicals than humans. Test animals are bred and kept under carefully controlled conditions (including diet) that prevent exposure to infectious diseases, other chemicals, etc., which might very well enhance the dangers of cyclamate and certain other food additives.

Fortunately, most food additives are safe and do help preserve foods, make them more appealing to the general public or to certain parts of the population (e.g., artificial sweeteners for diabetics), or simply increase the convenience of handling many foods. It is the positive side of the story that will be emphasized, for there are a multitude of sources available, some of which are listed at the end of the chapter, which discuss the negative side of the story and often emphasize a political as well as a scientific aspect of the controversy over the use of some food additives (18–25).

The following list demonstrates the wide variety of applications of additives. Some of the items have already been mentioned. They are included below along with several new ones that are discussed in varying detail.

Anticaking agents
Clarifying agents
Coloring or bleaching agents
Flavorings
Foam regulators

Incidental additives
Leavening agents (see Baking)
Nutrients
pH adjusters
Preservatives
Release agents
Sequestering agents
Texture modifiers: emulsifiers, gellers, thickeners, stabilizers, firming agents, enzymes, humectants

Figure 14.5 Silica. From Advanced Inorganic Chemistry, Third Edition, by F. A. Cotton and G. Wilkinson, Copyright © 1972 (John Wiley & Sons, Inc.). Reprinted by permission of John Wiley & Sons, Inc.

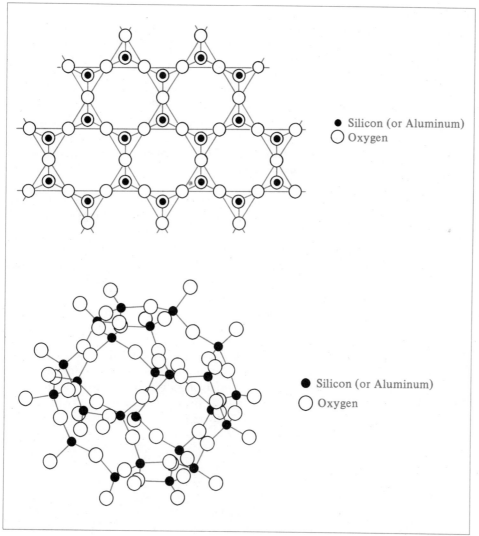

ANTICAKING AGENTS

Anticaking agents may be added in small amounts to finely powdered or crystalline food. Calcium silicate, calcium stearate, sodium aluminosilicate, and sodium ferrocyanide are some examples. These compounds can act by either or both of two mechanisms. The first, illustrated by sodium aluminosilicate, finds the anticaking agent acting as a dehydrating agent by virtue of its ability to incorporate water molecules into its crystal structure so that the water cannot adversely affect the food product. The mode of action is illustrated by Figure 14.5. The diagrams on the left and right are representations of a crystal of silica $(SiO_2)_n$. Silica has a complex, three-dimensional, repeating structure in which each silicon atom is surrounded by four oxygen atoms and each oxygen atom acts as a bridge between two silicon atoms. A two-dimensional cross section of this structure is shown on the left and a portion of the full three-dimensional structure is shown on the right. The symbol $(SiO_2)_n$ indicates that silica contains two oxygen atoms for every silicon atom, with the subscript n indicating a repeating structure. SiO_2 is the usual formula given for sand, which is silica (plus oxides of iron and aluminum) with a particular range of particle size.

Sodium aluminosilicate is often used as an anticaking agent. Silica and sodium aluminosilicate differ in that some of the silicon atoms may be replaced by aluminum atoms. In this structure, each aluminum will carry a negative charge (anion) and require some cation (e.g., Na^+) to balance the charge. These charges give the crystal a great affinity for water and the large holes in the crystal enable water molecules to enter and become trapped, thus minimizing any caking of the salt to which the sodium aluminosilicate is added.

Sodium ferrocyanide acts on common table salt (NaCl) by altering the shape of the salt crystals. NaCl normally forms cubical-shaped crystals that can pack tightly together, thus making them very susceptible to caking. In the presence of certain concentrations of sodium ferrocyanide, the salt plus the additive forms star-shaped crystals that do not cake as readily.

The old, familiar technique of placing a few grains of rice in a salt shaker accomplishes the same purpose. Anticaking agents are supposed to make this practice unnecessary.

CLARIFYING AGENTS

The acceptability of certain liquid foods is decreased by turbidity (cloudiness). In brewing, the final product is chillproofed by addition of proteolytic enzymes that cleave proteins found in beer. The beer might become cloudy if these proteins come out of solution when the beer is chilled. Fruit juices, wine, and vinegar are other examples of food products that may form sediments or suspensions if not clarified.

COLORING AND BLEACHING AGENTS

There are many examples of coloring agents and bleaching agents that are used to assure that a food product has whatever color is considered appetizing. Previous discussion centered on the color of meat, which can be controlled by proper packaging of uncured meat and the use of sodium nitrite as an additive in cured meats.

The orange pigment in carrots is called β-carotene or provitamin A. It is an example of a naturally-occurring substance that is often used as a food color additive, e.g., in margarine to provide a yellow color. Several other natural, as well as synthetic, pigments are in use as food colorings.

The use of bleached flour for baking has already been discussed.

FLAVORINGS

It has been estimated that two thirds of the food additives used in the United States are natural or synthetic flavorings (8). The list of artificial food flavorings is endless. For this reason, the discussion will be restricted to two items, flavor enhancers and artificial sweeteners.

Flavor Enhancers. Flavor enhancers are substances that have little or no taste of their own but accentuate the flavor of other foods. The best known flavor enhancer is monosodium glutamate (MSG). The parent compound, glutamic acid, is one of the 20 common amino acids found in proteins.

MSG is particularly effective in enhancing the flavor of meat-containing foods, which accounts for its use in frozen meat, fish, and chicken dinners, dry soup mixes, canned stews, sauces, as well as ham- and chicken-salad spreads.

It is theorized that MSG works by increasing the sensitivity of the taste buds or, perhaps, by stimulating greater saliva flow, which begins the breakdown of food and releases the flavor.

$$\underset{\substack{| \\ NH_2}}{HO-\overset{\overset{O}{\|}}{C}CH_2CH_2CH\overset{\overset{O}{\|}}{C}-O^-Na^+} \qquad \text{monosodium glutamate (MSG)}$$

MSG has been the subject of attack by many critics, who claim it is often used to enhance the flavor of food products that contain less of an expensive ingredient, thus cutting production costs. There has also been much concern over the use of MSG in baby foods, apparently to make them more appealing to mothers, even though the safety of MSG for babies may not have been clearly established.

MSG has also been associated with symptoms characterized as "Chinese restaurant syndrome," so named because of the relatively heavy use of MSG in Chinese foods. Heavy use of MSG may cause a kind of upset characterized by pain and a burning sensation in the neck, forearms, chest, and head.

The sodium salts of two 5'-nucleotides, 5'-IMP and 5'-GMP, known as sodium inosinate and sodium guanylate, which have been marketed under such names as Mertaste (Merck), Ribotide (Takeda), and Corral (Pfizer), are also used as flavor enhancers of meat. They are more expensive than MSG and more effective (20 times as great) than MSG, but their effect on taste is somewhat different, although they are also used in enhancing the flavor of meats and poultry. IMP and GMP are nucleotides, which are the building blocks of DNA.

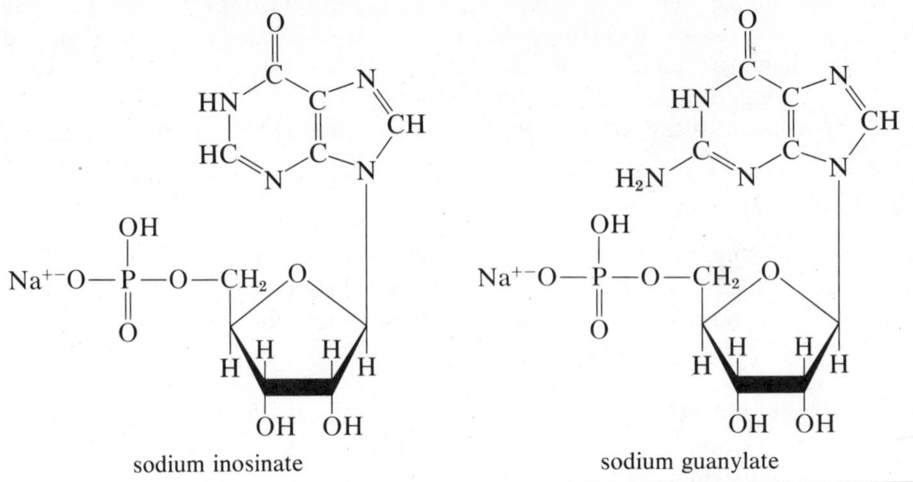

sodium inosinate   sodium guanylate

Maltol and ethyl maltol are flavor enhancers that are used in fruit-, vanilla-, and chocolate-flavored foods.

Artificial Sweeteners. Artificial sweeteners, often described as non-nutritive sweeteners, are the most studied, discussed, and treasured food additives. The overweight individual, the diabetic, or any other individual who might wish to limit sugar intake is a prospective consumer of artificial sweeteners.

Calcium cyclamate is one of many artificial sweeteners that have been marketed over the years and is probably the most appealing. It has a sweetness about

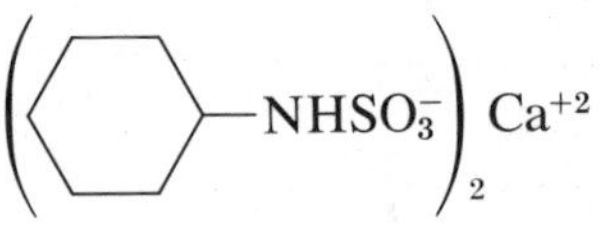

calcium cyclamate

30 times as great as sucrose. In 1969, the Food and Drug Administration removed cyclamate from the GRAS (generally recognized as safe) list and banned its use because of the finding that large doses caused a high incidence of bladder cancer in rats (9). It was also reported that injection of cyclamate or cyclohexylamine (CHA), a frequent metabolic breakdown product of cyclamate, into chicken eggs caused deformities of embryos taken from the eggs. It was also found that CHA may cause breakage of chromosomes in laboratory animals (10).

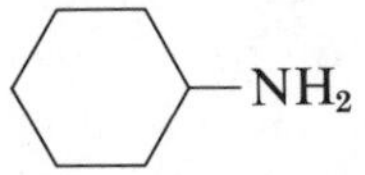

cyclohexylamine

Bladder cancer was the only finding cited in removing cyclamate from the GRAS list because it was already covered under the terms of the Delaney Amendment of the Food, Drug and Cosmetic Act, which became law in 1958, and stated that anything found to cause cancer in man or in animals should not be added to food. However, the laboratory studies and the conclusions that were drawn from them have been severely criticized and questioned ever since the ban became effective, as the amount of cyclamate used in the study was extremely large (11).

The FDA has since commented that while two thorough studies using rats were negative (no tumors), three other studies of bladder tumors caused by cyclamate "had to be considered at least equivocal" and justified continuation of the ban on cyclamate (12). Since even the FDA has not termed the positive findings "unequivocal," the validity of the results remains open to attack by several scientists and certain other groups, some of whom might profit from reinstatement of cyclamate. This is clearly an economic and even political aspect of the story, but on the opposite side of the issue is the claim that the sugar industry is exerting pressure against the use of artificial sweeteners. About the only thing that is really clear on this issue is that there is more to be heard, but for the moment, cyclamate is restricted from the GRAS list.

Saccharin has been the major alternative since the ban on cyclamate. It has a much longer history than cyclamate. The sweetening effect of saccharin was first reported in 1879, and it has been produced commercially in the United States since 1902. The Sherwin-Williams Company is now the sole producer of saccharin.

Saccharin

Saccharin has a sweetness about 300 times that of sucrose. Before cyclamate was banned, saccharin and cyclamate were often combined in artificially sweetened products. Saccharin is used in food and beverages, cosmetics, toothpastes, drugs, and animal feed. The soft drink industry accounts for about 74 percent of the saccharin used in food and beverages. Other food uses include powdered juices and drinks, canned fruits, dessert toppings, cookies, gum, jams, candies, ice cream, and puddings. It is also available as a sugar substitute for direct use in cooking and in coffee, tea, cereal, and other foods.

Saccharin is characterized as having a bitter aftertaste. It was because of this that saccharin was often used in combination with cyclamate, since cyclamate does not present this problem.

The more recent history of saccharin has been a major controversy over its safety, or more specifically its danger as a carcinogen. In early 1977, the controversy became a very public issue when the Food and Drug Administration (FDA) announced its intention to ban saccharin as a food additive based on the finding of bladder cancer in rats in a Canadian study (13).

This was not the first time that concern had been expressed over the safety of saccharin. In 1972, the FDA removed saccharin from the GRAS list, but at that time they did not ban the use of saccharin. Instead, the FDA issued an interim food additive food regulation that froze uses at then-current levels and recommended certain limitations on the use of saccharin.

In 1972, there were no definitive studies that linked saccharin and cancer, but some studies had raised some uncertainty and the FDA issued the interim regulation pending completion of further studies.

The Canadian study reported in early 1977 was severely criticized as being invalid because of the general approach used, which was similar to the approach used on cyclamate. In fact, the approach was the very standard procedure in which laboratory animals (in this study rats were used) are fed very large doses of the suspected carcinogen. To be more specific, a group of 100 rats was fed saccharin at the level of 5 percent of the total diet in a study that was conducted over a period of two years. In the study, three of the first 100 rats developed bladder tumors. One hundred of the offspring of these rats were also fed the standard dose of saccharin so that these rats were exposed to saccharin both before and after birth. Of the 100 offspring studied, 14 developed bladder tumors. Thus, the concern over the hazard to future generations is particularly great.

On the other hand, there are critics who believe that the entire study is absurd. The five percent dietary level does not sound excessive until this is projected to a similar human intake of diet soda, which would consist of 800 twelve-ounce bottles per day for a lifetime. Some critics have suggested that drowning would be far more likely than the development of cancer if such an experiment were conducted on humans. Other critics have suggested that saccharin-containing foods might be labeled in ways such as: *Warning: The Canadians have determined that saccharin is dangerous to your rat's health!*

With a typical intake of saccharin by a person using saccharin in coffee, drinking diet soft drinks, and eating food sweetened with saccharin, the diet contains about 0.1 percent saccharin, i.e., about fifty times less than that used in the Canadian study.

The reasoning behind the use of large doses of a chemical for studies with test animals relates to their relatively short life span. It is presumed that if large doses cause tumors in a significant number of animals during their short lifetime, a smaller dose may be expected to cause tumors over the much longer human lifetime. In addition, it is only practical to study a small number of test animals, whereas a large percentage of humans could be adversely affected.

There has also been a report of an epidemiological study that links saccharin and cancer. This study was conducted by the Canadian National Cancer Institute. It was a retrospective study that examined 480 men and 152 women who developed bladder cancer in the period 1974–76 and comparable groups without cancer. Men using artificial sweeteners, cyclamate and saccharin, were found to have a 60 percent higher incidence of bladder cancer than nonusers. Female users were not found to have any increased risk. Most of the men had consumed only saccharin and the level of use seemed to influence the incidence of cancer (14).

On the other hand, there have been numerous studies, both experimental and epidemiological, that have revealed no significant risk from saccharin, so that the matter of hazard to animals and the relationship that this may have to a possible danger to humans is very unclear.

Among the questions that cannot yet be answered is the matter of whether there may be a threshold or no-effect level for carcinogenic substances. Many substances are toxic, but can be consumed safely in low quantities. Whether the same is true for carcinogenic substances is not known and probably will not be known until the details of the chemistry involved in the development of cancer are known. In any case, the final fate of saccharin may not be decided for many years to come.

A relative newcomer to the list of artificial sweeteners is called aspartame, which is a compound derived from aspartic acid and phenylalanine—two

$$NH_2-CH-\overset{\overset{\displaystyle O}{\|}}{C}-NH-CH-\overset{\overset{\displaystyle O}{\|}}{C}-OCH_3$$

(Aspartic Acid | Phenylalanine | Methyl Ester)

naturally-occurring amino acids found in proteins. It is 200 times as sweet as sucrose, with good flavor and no unpleasant aftertaste and, unlike most other artificial sweeteners, it has some nutritional value because it is metabolized to the amino acids. It is interesting that the two separate components, aspartic acid and phenylalanine are flat- and bitter-tasting, respectively. This emphasizes the difficulty in predicting sweetness or any other taste.

Aspartame would seem to be safe, since it will break down into the amino acids, and many studies support this idea (14). In July of 1974, the FDA approved the use of aspartame as a table sweetener and in dry dessert wines. In December of 1976, the FDA removed aspartame from the market, pending further studies.

Synsepalum dulcificum is a plant native to West Africa that yields a small red berry known as miracle fruit, which contains a glycoprotein (combination sugar and protein) known as miraculin. It has no significant taste of its own but causes sour foods to taste sweet.

The Miralin Corporation has succeeded with some breeding work aimed toward production of the miracle fruit on a large enough scale to become commercially important. Laboratory synthesis is impractical on a large scale for this type of compound. It is reported that the miraculin effect can persist for several hours (15). As an example of its effect, it is reported that a lemon tastes like an orange after a person has chewed a miracle fruit berry (16).

Two other proteins known as monellin and thaumatin are described as intensely sweet substances, 3000 and 1000 times that of sucrose, respectively. The first has a persistent sweet aftertaste, whereas the second has a licorice aftertaste.

Somewhat like the action of miraculin, artichokes contain substances that cause water to taste sweet, apparently by affecting the tongue.

An enzyme called *glucose isomerase* catalyzes the interconversion of glucose and fructose. Consequently, the glucose obtained from starch can be converted into a mixture of glucose and fructose, similar to that of *invert sugar*. Because of the greater sweetness of fructose, about 1.5 times as sweet as sucrose, the mixture can be used as a sweetener where sucrose is normally used. Glucose, alone, is not competitive with sucrose because it is not as sweet as sucrose. Corn syrup (from corn starch) is now treated in this way and used in several food products (17). Such a sweetener is not considered artificial, even though it does differ from normal sugar, because it has the same calorie content as sugar.

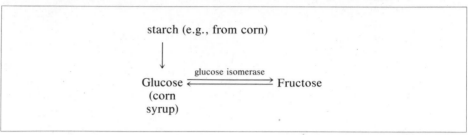

FOAM REGULATORS

Foaming agents are added to certain foods such as canned whipped cream to make it foam as it passes out of an aerosol can. Dextrins and cellulose gums are among the substances used for this purpose.

Antifoaming agents are used in various liquid food products such as pineapple juice, which would otherwise tend to foam vigorously. Certain silicones are very effective antifoaming agents.

INCIDENTAL ADDITIVES

Substances that often find their way into food include pesticide residues and drugs, such as antibiotics and growth hormones, that are used to improve the general health and development of animals used as sources of food. Another class of incidental food additives are substances that can be leached out of packaging materials, such as antioxidants, driers, drying oils, plasticizers, and release agents.

NUTRIENTS

The primary nutrients added to foods are amino acids and vitamins. The importance of amino acids as the building blocks for proteins was first discussed in Chapter Two while considering the chemical basis of radiation damage.

PRESERVATIVES

Most of the major preservatives used in foods are discussed in the coverage of specific types of food throughout Part Four. Table 14.2 provides a list of preservatives in the order in which they are covered and some of the foods in which each is used.

TABLE 14.2   MAJOR FOOD PRESERVATIVES

PRESERVATIVE	USED IN
SO$_2$	alcohol beverages
sugar	condensed milk, canned fruits, jams and jellies
salt	fermented foods (meats, cucumbers, sauerkraut)
smoke	meats
acids	fruits, cured meats, pickles
antioxidants*	fats

*SO$_2$ also acts as an antioxidant.

SUMMARY

Food processing often means conversion into a stable form that can be stored safely for extended periods. Invasion by microorganisms can cause spoilage of food and, in some cases, make the food dangerous. The most dangerous example is the microbe that produces a toxin known as botulism; but *Salmonella, Staphylococcus,* and trichinosis are also serious problems. Ptomaine poisoning is a myth.

Cold storage, freezing, and dehydration are all convenient methods for preventing or slowing the development of microorganisms. Heat processing, such as that used in canning, can be used to kill microorganisms in food, which is then sealed up to prevent further contamination. The microbe that releases botulism forms spores, which makes it much more resistant to heat, although acidity makes it more susceptible. Consequently, high acid foods may often be safely cold packed. Normally-present or added sugar is also an effective preservative when present in large amounts.

Ionizing radiation can sometimes be used for food preservation but undesirable side effects are not uncommon. Prevention of sprouting has been one successful approach using radiation.

In many cases, some degree of fermentation can be used for preservation. Cured meats (e.g., corned beef), pickles, sauerkraut, and many other foods are preserved by methods analogous to those used to ripen cheese, which employ the use of salts to control the activity of microorganisms and, thus, prevent the growth of microbes that might cause spoilage. In most cases, acids are added and/or produced by the fermentation and act as preservatives. Meat from animals that are well-fed and rested prior to slaughter is most easily preserved. The various colors observed in meats may be attributed to myoglobin (purple), oxymyoglobin (red), metmyoglobin (brown), and nitrosomyoglobin (pink). In handling uncured meats, it is necessary to encourage oxygenation in order to avoid oxidation.

Food additives greatly increase the safety, utility, and quality of many foods, although they are sometimes overused. Controversy still rages over the ban on cyclamate and has contributed to the push to develop other artificial sweeteners such as aspartame, miraculin, monellin, and thaumatin. In the meantime, saccharin dominates the market, although not without its problems.

Glucose isomerase seems destined to have a great impact on the sweetener market because of its ability to produce a mixture very much like invert sugar from glucose. This makes corn syrup more competitive with sucrose as a natural sweetener.

Flavor enhancers are another member of the long list of food additives, which includes anticaking agents, foam regulators, texture modifiers, and preservatives.

PROBLEMS

1. Give a definition or example of each of the following:
   a) salometer
   b) flavor enhancers
   c) miraculin
   d) ptomaines
   e) MSG
   f) botulism
   g) vinegar
   h) trichinosis
   i) putrefaction

2. Why it is necessary to use a pressure cooker when canning some foods?

3. What effect does acidity have on spore survival?

4. What food is known to cause a Staphylococcus infection?

5. What is the value of quick-freezing?

6. What is the idea behind dehydration and freeze-drying?

7. How does a high sugar content act to preserve food?

8. Why does meat from a well-fed, rested animal keep much better than meat from a hungry, exercised animal?

9. How does a "meat marinade" work?

10. What is the cause of the bright red color of fresh meat? What is the cause of the pink color of cured meats, e.g., ham or corned beef?

11. What is sauerkraut?

12. What is the preservative in each of the following: wines, jams and jellies, meats, pickles?

13. How is a pickle produced from a cucumber?

14. How does an anticaking agent work?

REFERENCES

1. Lehninger, A.L. *Biochemistry*. New York: Worth, 1970, p. 4.
2. "Botulism Research." *Chemistry 37* (1) (1964): 27.
3. "Canning Tomatoes? Here's Something To Worry About." *Consumer Reports*, August 1974, p. 569.
4. Weier, H.H., Mountney, G.J., and Gould, W.A. *Practical Food Microbiology and Technology*. Westport, Conn.: AVI, 1971, p. 313.

5. Desrosier, M.W. *The Technology of Food Preservation*. Westport, Conn.: AVI, 1970, Chapter 10, pp. 313–364.

6. Borgstrom, G. *Principles of Food Science,* Vol. 1. New York: MacMillan, 1968, pp. 86–9.

7. Cotton, F.A., and Wilkinson, G. *Advanced Inorganic Chemistry*. New York: Interscience, 1972, pp. 323–4.

8. Jacobson, M.F. *Eater's Digest*. Garden City, N.Y.: Doubleday, 1972, p. 71.

9. Turner, J.S. *The Chemical Feast*. New York: Grossman, 1970, Chapter 1, pp. 5–29.

10. Price, J.M., et al., "Bladder Tumors In Rats Fed Cyclohexylamine Or High Doses Of A Mixture of Cyclamate and Saccharin." *Science 167* (*1970*): 1131.

11. Inhorn, S.L., and Meister, L.F. "Cyclamate Ban." *Science 166* (1969): 685.

12. "Abbott, FDA Argue Cyclamates Safety." *Chemical and Engineering News,* 18 November 1974, p. 7.

13. "Saccharin and Its Salts," *Federal Register* 42(73) 19996-20010 (1977).

14. "First Human Cancer Link To Saccharin Found," *Chemical and Engineering News,* June 27, 1977, p. 7.

15. "African Berry Yields Sweetest Substance." *The Journal of the American Medical Association 220* (2) (1972): 180.

16. "Chemical Technology Key To Better Wines." *Chemical and Engineering News,* 2 July 1973, p. 4.

17. Skinner, K.J. "Enzymes Technology." *Chemical and Engineering News,* 18 August 1975, p. 22.

18. Sanders, H.J. "Food Additives, Part I," *Chemical and Engineering News,* 10 October 1966, pp. 100–120.

19. Sanders, H.J. "Food Additives, Part II," *Chemical and Engineering News,* 17 October 1966, pp. 108–128.

20. Majtenyi, J.Z. "Food Additives—Food For Thought." *Chemistry 47* (5) (1974): 6.

21. Verrett, J., and Carper, J. *Eating May Be Hazardous To Your Health*. New York: Simon and Schuster, 1974.

22. Jacobson, N.F. *Eater's Digest*. Garden City, New York: Doubleday, 1972.

23. Turner, J.S. *The Chemical Feast*. New York: Grossman, 1970.

24. Inglett, G.E. *Symposium: Sweeteners*. Westport, Conn.: AVI, 1974.

25. Hunter, B.T. *Consumer Beware! Your Food And What's Being Done To It*. New York: Simon and Schuster, 1971.

I N earlier chapters, consideration has been given to three major categories of biological molecules: proteins, polynucleotides (DNA), and carbohydrates. In this chapter, the structure and properties of fats are described.

CHEMICAL STRUCTURE

In Chapter Six, it was noted that petroleum-based oils are complex mixtures of linear and branched hydrocarbons. As such, they may be burned to produce heat but they have no fuel value for living organisms. This characteristic is the major distinction between petroleum-based oils and food oils (e.g., corn oil) since the latter, which are also predominantly hydrocarbon, contain certain functional groups of atoms that make them biodegradable, i.e., susceptible to attack so that living organisms are able to derive food value from them.

Both food oils and fats have the same general chemical structure. They differ only in their melting points, with an oil having a melting point below room temperature, so that it is a fluid, whereas a fat is a solid under normal conditions. Oils can be solidified if cooled sufficiently and solid fats (e.g., butter) will change to oils if heated.

The more general term, *lipids,* includes both fats and other fat-like substances, e.g., steroids (see Chapter Twenty), all of which are predominantly hydrocarbon.

A fat molecule is formed by the reaction of glycerin (glycerol) with three fatty acids. The product is commonly called a *triglyceride*. As described in

fatty acids   glycerin   a triglyceride

FATS AND OILS

Chapter Six, the symbol R describes the hydrocarbon portion of the fatty acid. In fatty acids, the groups represented by R are linear chains of carbon atoms (plus hydrogens) of variable length. The length of these chains and the amount of unsaturation, if any, determine the physical properties of the fat, i.e., whether it will be a solid or a liquid. The discussion of margarines considers this point in more detail.

The fatty acids that are incorporated in butterfat and coconut oil triglycerides tend to be relatively short-chain, saturated fatty acids, whereas the predominant fatty acids found in soybeans, peanuts, and other oilseeds are predominantly long-chain with considerable unsaturation. It can be seen in Table 15.1 that the fatty acid composition of butterfat is quite varied, ranging from C_4 to C_{18} fatty acids. It should be noted that fatty acids almost always contain an even number of carbon atoms, i.e., the hydrocarbon groups (represented by R) contain an odd number of carbons. This characteristic is due to the series of reactions used for synthesis of fatty acids in which the long chains are built up two carbons at a time.

A slightly saturated fatty acid.

TABLE 15.1   SATURATED FATTY ACIDS (1)[*]

COMMON NAME	SYSTEMATIC NAME	NOTATION*	TYPICAL FAT SOURCE
Acetic	Ethanoic	2:0	——
Butyric	Butanoic	4:0	Butterfat
Caproic	Hexanoic	6:0	Butterfat
Caprylic	Octanoic	8:0	Butterfat, coconut oil
Capric	Decanoic	10:0	Butterfat, coconut oil
Lauric	Dodecanoic	12:0	Coconut oil
Myristic	Tetradecanoic	14:0	Butterfat, coconut oil
Palmitic	Hexadecanoic	16:0	Most fats and oils
Stearic	Octadecanoic	18:0	Most fats and oils
Arachidic	Eicosanoic	20:0	Peanut oil
Behenic	Docosanoic	22:0	Rapeseed oil

*A notation such as 12:0 indicates a fatty acid with 12 carbons and no double bonds.

The wide variation of fatty acid content makes it impossible to draw a molecular structure for a fat molecule from any particular source. One can draw a molecule such as trilaurin, which is the triglyceride made from glycerin and three lauric acid (C_{12}) residues, but even coconut oil, which has a particularly high percentage of lauric acid residues (ca. 50%), contains many triglyceride molecules in which one or more of the lauric acid groups is replaced by another fatty acid. In

Trilaurin

fact, it is because fats, as well as petroleum oils, are mixtures of different compounds, that they tend to form amorphous rather than crystalline solids when cooled below their melting points.

UNSATURATION

The presence of carbon-carbon double (or triple) bonds dramatically changes the physical properties of organic compounds such as fats. The popularly advertised *polyunsaturated* salad or cooking oil consists of triglycerides with a high content of unsaturated fatty acid residues. An important example is oleic acid (9-octadecenoic acid), which can be designated by the shorthand notation 18:1 to indicate it is an 18-carbon fatty acid containing 1 double bond. Stearic acid (18:0) is

**TABLE 15.2 SOME UNSATURATED FATTY ACIDS (1) FOUND IN
NATURAL FATS**

COMMON NAME	SYSTEMATIC NAME	NOTATION	TYPICAL FAT SOURCE
Caproleic	9-Decenoic	10:1	Butterfat
Lauroleic	9-Dodecenoic	12:1	Butterfat, coconut oil
Myristoleic	9-Tetradecenoic	14:1	Butterfat
Palmitoleic	9-Hexadecenoic	16:1	Animal fats, seed oils
Petroselinic	6-Octadecenoic	18:1	Parsley seed oil
Oleic	9-Octadecenoic	18:1	Most fats and oils
Vaccenic	11-Octadecenoic	18:1	Butterfat, beef fat
Linoleic	9,12-Octadecadienoic	18:2	Most seed oils
Linolenic	9,12,15-Octadecatrienoic	18:3	Soybean oil, linseed oil
Elaeostearic	9,11,13-Octadecatrienoic	18:3	Tung oil
Gadoleic	9-Eicosenoic	20:1	Fish oils
Arachidonic	5,8,11,14-Eicosatetraenoic	20:4	Fish oils
Erucic	13-Docosenoic	22:1	Rapeseed oil

the fully saturated 18-carbon fatty acid, whereas linolenic acid (18:3) is at the other extreme—an 18-carbon triunsaturated fatty acid. The major unsaturated fatty acids are listed in Table 15.2.

Two representations of oleic acid are shown in the following structure in the *cis* form, which is the normal arrangement of all double bonds of fatty acids isolated from natural fats and oils.

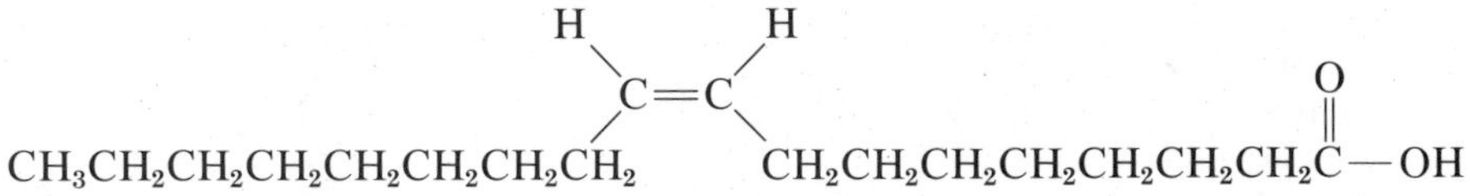

cis

Oleic acid (18:1)

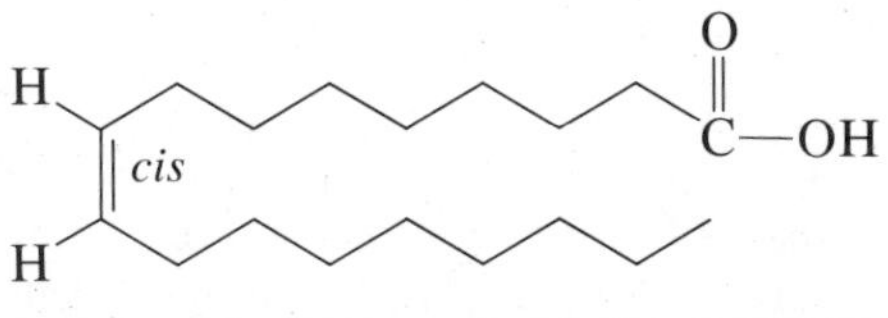

Since unsaturation is so important in determining the properties of the fat and since unsaturated fat is considered desirable in the diet, it is useful to have a quantitative measure of the amount of unsaturation. A convenient parameter used for this purpose is the **iodine value,** which is based on the reaction between iodine

$$RCH{=}CH(CH_2)_n\overset{\displaystyle O}{\overset{\|}{C}}{-}OH + ICl \longrightarrow RCH{-}CH(CH_2)_n\overset{\displaystyle O}{\overset{\|}{C}}{-}OH$$
$$\quad\quad\quad\quad\quad\quad\quad\quad\quad\quad\quad\quad\quad | \quad\quad |$$
$$\quad\quad\quad\quad\quad\quad\quad\quad\quad\quad\quad\quad\quad I \quad\quad Cl$$

TABLE 15.3 MAJOR FATS AND OILS

FAT OR OIL SOURCE	IODINE VALUE	MAJOR FATTY ACIDS* (GREATER THAN 25%)	ORDER OF DOMESTIC PRODUCTION†	ORDER OF DOMESTIC USE†
Soybean	120–141	(18:2) (18:1)	1	1
Cottonseed	97–112	(18:2) (16:0)	2	4
Lard (hogs)	58–68	(18:1) (16:0)	3	2
Lard	>85‡			
Butterfat	25–42	(18:1) (16:0)	4	3
Tallow (beef)	35–48	(18:1) (16:0)	5§	7§
Tallow (lamb)	48–61	(18:1) (16:0)		
Corn	103–128	(18:2) (18:1)	6	8
Peanut	84–100	(18:1) (18:2)	7	9
Safflower	140–150	(18:2)	8	10
Coconut	7.5–10.5	(12:0) (14:0)	¶	5
Palm	44–58	(16:0) (18:1)	¶	6
Olive	80–88	(18:1)	¶	
Chicken fat	64–76	(18:1) (16:0)	—	—
Human fat	57–73	(18:1) (16:0)	—	—
Linseed (flaxseed)	175–202	(18:3)	#	#

FATTY ACIDS	IODINE VALUE	SHORTHAND NOTATION
Oleic	89	(18:1)
Linoleic	180	(18:2)
Linolenic	272	(18:3)

*Listed in order of decreasing content.
†For 1973, reference 3, order may vary.
‡From hogs finished on peanuts.
§Edible tallow only.
¶No domestic production.
#A nonfood oil.

chloride (ICl) and an unsaturated fatty acid. In this reaction, the ICl can be seen breaking one of the bonds of the carbon-carbon double bond and adding across the two carbon atoms. It is an instantaneous reaction that occurs quantitatively (100% efficiency). It is generally conducted by using an excess of ICl. The leftover ICl is then measured to determine the amount of ICl that reacted with the fat. The iodine value is generally given in terms of the number of grams of iodine* that reacts with 100 grams of fat. *The higher the iodine value, the greater the amount of unsaturation.* A so-called *polyunsaturated oil* is one with a relatively high iodine value. Table 15.3 contains a list of most of the important fats, their iodine values, and the

* Iodine (I_2) does not react with carbon–carbon double bonds. It is, therefore, common practice to employ the more reactive ICl and then calculate the amount of I_2 that would have reacted by the same pathway.

fatty acids that make up 25% or more of the total. The iodine values of the 18:1, 18:2, and 18:3 fatty acids are included for comparison. The other information presented in Table 15.3 will be discussed in the following paragraphs.

A close examination of Table 15.3 reveals a number of interesting facts, which are discussed throughout the chapter. Consider a few at this point.

Soybeans are the most important commercial source of oil. Soybean oil is one of several so-called polyunsaturated vegetable oils that are processed for direct use or converted into another form, e.g., margarine or shortening. The other major oils that are often converted into other forms are cottonseed, corn, palm, and safflower, with safflower oil being the most unsaturated of this group. As we will see in the discussion of margarines and shortenings, olive and peanut oils are not suitable for conversion to these forms, but are used directly as salad and cooking oils. In a later discussion, consideration will be given to the major uses of many of the fats and oils listed in Table 15.3.

Linseed oil (also listed in Table 15.3) is the most highly unsaturated oil listed. It is not used as a food, either directly or in a converted form, but it is a major component of most oil-based paint formulations.

It is also clear from the iodine values given in Table 15.3 that vegetable oils are generally more unsaturated than animal fats, although coconut oil is a dramatic exception.

Diet is a major factor in determining the characteristics of fat stored by an animal, as indicated by the iodine values for lard from hogs on a standard diet as compared to hogs fed on peanuts for a period of time prior to slaughter. The latter store fat which becomes increasingly like that found in peanuts, i.e., more unsaturated.

In addition to the edible tallow tabulated in Table 15.3, inedible tallow is also a major raw material used in the United States fats and oils industry. It finds major markets for such uses as animal feeds, soaps, lubricants and as a source of fatty acids, which, along with paraffin (from petroleum), are a major component in making candles. Soaps and lubricants are discussed later.

ISOLATION

Isolation of vegetable oils is carried out by pressing or extraction of the oilseeds. In some instances, pressing, which removes most of the oil, is followed by extraction of the residual "cake" to remove most of the remaining oil. Hydrocarbon solvents, such as hexane, are normally used for extraction of oils since the oils, which are predominantly hydrocarbon, readily dissolve in these solvents.

Isolation of animal fats is called **rendering** and involves *grinding,* followed by *heating,* which coagulates the protein, and *centrifugation,* during which the lower density fat rises and separates.

TRENDS IN PRODUCTION AND
USE OF FATS AND OILS

In Table 15.3, the last two columns show the relative importance of the major fats in terms of both production and consumption. Production trends are described by the data in Table 15.4, which indicate a drop in the production of animal fats and a

TABLE 15.4 DOMESTIC PRODUCTION TRENDS OF COMPETING
FOOD FATS AND OILS (3, 4)

FAT OR OIL	1949	1959	1969	1973	CHANGE SINCE 1949	
		(MIL. LB.)			(MIL. LB.)	(PCT.)
Butter	1701	1435	1125	917	− 784	− 46
Lard	2626	2726	1825	1339	−1287	− 49
Edible beef fat	168	347	575	650	+ 482	+286
Animal fats	4495	4508	3525	2906	−1589	− 35
Corn oil	242	332	480	528	+ 286	+118
Cottonseed oil	1799	1832	1350	1552	− 247	− 14
Peanut oil	217	92	185	195	− 22	− 10
Soybean oil	1937	4338	7825	8995	+7058	+364
Edible oils	4195	6594	9840	11270	+7075	+168
Percent distribution						
Animal fats	52	41	26	20		
Edible oils	48	59	74	80		

very sharp increase in the production of vegetable oils. The data also emphasize the tremendous increase in production and virtual dominance of vegetable oil production by soybeans. This fact, coupled with the tremendous importance of soybeans as a source of protein, has made them one of the most important agricultural commodities.

As for the domestic use, the picture is somewhat similar to production except that coconut, palm, and olive oils, which are not produced domestically, are, nevertheless, used in various edible and inedible forms in the United States. The trends of usage are summarized in Table 15.5 with projections to 1985, according to types of food products, with the data given on a per capita basis so as to cancel out the effect of increased population during the period covered. Somewhat the same information is depicted in Figure 15.1. Both forms of the data indicate several significant trends. The most striking is the increase in total fat consumption

TABLE 15.5 FOOD FAT USES, 1950–1985 (3, 5)

FOOD FAT	POUNDS PER CAPITA			
	1950	1960	1972	1985 PROJECTION
Butter (fat content)	8.6	6.2	4.1	3
Margarine (fat content)	4.9	7.5	8.9	10
Shortening	11.0	12.6	16.8	19
Lard (direct use)	12.6	7.5	3.8	3
Salad and cooking oils	6.9	9.2	17.1	22
Other edible uses	1.9	2.3	2.4	3
Total (fat content)	45.9	45.3	53.1	60

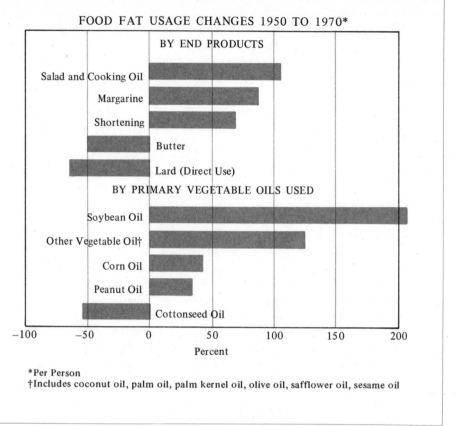

Figure 15.1 Food fat usage changes (1950 to 1970). Courtesy of USDA, Economic Research Service.

per capita, which is largely attributable to growth in the fast-food industry featuring hamburgers, french fries, chicken, and fish products, all of which are cooked in oil and add to the consumption of fat. Other data in Table 15.5 and Figure 15.1 are considered in the following sections.

MAJOR FAT PRODUCTS

Margarine and Butter

These two table spreads are the most familiar food fat products. In the 1930s, butter cornered about 90% of this market, whereas today, margarine consumption is about double that of butter.

Both products contain about 80% fat. Butter is obtained by churning cream to cause the fat globules to clump together. The remainder of the butter is about 16% water plus a few percent of milk solids (lactose, casein, and minerals) and salt, which acts as both a flavoring and a preservative.

The fat content of margarines, derived almost totally from vegetable oils, is regulated by law. A conventional margarine must contain a minimum of 80% fat.

Like butter, the remainder is about 16% water, plus small amounts of salt and nonfat milk solids. Various additives (discussed later) are also included to improve appearance and stability. Diet margarines, also called imitation margarine, may be marketed with as little as 40% fat.

There are a number of reasons cited for the increasing popularity of margarines. Near the top of the list is cost, which until recently has been much lower than the price of butter. As recently as 1973, the retail cost of butter was well over twice that of margarine. Since then, the gap in cost has closed dramatically. The implication of saturated fats as a factor in heart disease has also increased the popularity of the more unsaturated margarines. Various improvements in the techniques for processing vegetable oils into margarine have also increased their popularity. These include some important chemical changes and are discussed in this chapter. Both butter and margarine are described as *plastic fats* to signify that they retain their shape at normal or refrigerator temperatures, but they may be deformed when one cuts into them with a knife or tries to spread them. The *plastic range* is an important property of any table spread. It is the temperature range over which the spread retains its shape but may be readily deformed, and with respect to the plastic range, many margarines are more convenient than butter because they may be easily spread even when cooled to refrigerator temperatures. Equally important, both margarine and butter must have the upper limit of the plastic range just below body temperature so that they will melt in the mouth.

There are several steps in processing an oil from isolation to packaging, but a significant chemical alteration is carried out in converting liquid oils into a solid margarine. The process is known as **hydrogenation** and is the addition of hydrogen (H_2) to double-bonded carbons, as illustrated for ethylene—the simplest unsaturated compound.

$$H_2C=CH_2 + H_2 \longrightarrow H-\underset{\underset{H}{|}}{\overset{\overset{H}{|}}{C}}-\underset{\underset{H}{|}}{\overset{\overset{H}{|}}{C}}-H$$

The process requires a catalyst (usually nickel) and when applied to an unsaturated fatty acid, it causes one or more of the double bonds to become saturated. This means that the iodine value is lowered, but more important, it increases the melting point of the fat. If carried far enough, the melting point of the oil may be increased to above room temperature, in which case, the product will become a solid, i.e., a margarine or a shortening.

In other words, more saturated fats (e.g., lard and tallow) have higher melting points than highly unsaturated fats (i.e., vegetable oils).

Another factor that influences the melting point of fats is the length of the fatty acid chains. Short-chain fatty acids tend to lower the melting point, whereas long-chain fatty acids raise the melting point. This partly explains why butter and margarine have similar melting points even though butter is a more saturated fat. Butterfat has an iodine value in the range from 25–42, whereas conventional margarines have iodine values of 78–90. One important difference is the presence of a relatively high percentage of short-chain fatty acids in butterfat (see Table 15.1), which counteract the greater saturation by lowering the melting point.

Another factor that helps make margarine so similar to butter is a second change that accompanies hydrogenation. In addition to simply saturating many of the double bonds, partial hydrogenation of diunsaturated or triunsaturated fatty acids often causes a remaining double bond to change (isomerize) into the *trans* arrangement. Recall that all natural fatty acids have the *cis* arrangement about the double bonds and the *cis* arrangement causes the fat to have a lower melting point. Consequently, any conversion of double bonds, which remain in the final product, into the *trans* form will raise the melting point and contribute to hardening the oil into a margarine. Margarines are often labeled "lightly hydrogenated," which means some increase of saturation, accompanied by the *cis* to *trans* isomerization. An example would be the hydrogenation of soybean oil (iodine value 120–141) to bring it into the range of iodine values normal for conventional margarines (78–90), at which point it becomes plastic. In contrast, unhydrogenated peanut oil has about the same range of iodine values (84–100), but it is a liquid at normal temperatures. The major difference between the natural peanut oil and the partially hydrogenated (hardened) soybean oil is that the latter undergoes some *cis* to *trans* isomerization during the hydrogenation, whereas the untreated peanut oil has only the naturally-occurring *cis* arrangement of the double bonds. This picture, in part, accounts for the absence of peanut oil as a component of margarines, since hardening the oil into a margarine would lower the iodine value and make it uncompetitive in a market in which there is a premium on polyunsaturated fats.

To summarize the hardening process, the following factors all tend to give fats a higher melting point:

1. long-chain fatty acids
2. less unsaturation
3. *trans* arrangements of carbon-carbon double bonds

In converting an oil into a margarine, a processor cannot alter (1). The fatty acid content is determined by his choice of oil but (2) and (3) are altered by hydrogenation, during which both factors contribute to the hardening.

Looking at this picture yet another way, the effects of (1), (2), and (3) are illustrated by the melting points of the following fatty acids:

oleic acid (18:1,*cis*) has a melting point like (8:0)
isooleic acid (18:1,*trans*) has a melting point like (12:0)

In other words, the presence of one double bond lowers the melting point of a C_{18} fatty acid so it melts at about the same temperature as the C_8 saturated fatty acid, but the change to the *trans* arrangement counteracts the effect of the unsaturation by raising the melting point (6).

One normally does not think of exerting control over the iodine value of butterfat and, indeed, butterfat is never hydrogenated, but the diet of a cow can influence the physical characteristics of butter since dietary fats can conceivably be incorporated directly into the butterfat. It can be seen from Table 15.3 that butterfat may have an iodine value of from 25–42. Normal butter has a range from 29.4 to 35, whereas a very soft butter forms when the iodine value exceeds 35 and a hard, brittle butter appears when the value is below 29.4 (7). Consequently, a

TABLE 15.6 COMMERCIALLY AVAILABLE MARGARINES (9)

CONVENTIONAL	OILS (A)		
	LIQUID	HARDENED	PARTIALLY HARDENED
A&P Corn Oil (A&P Stores)	——	——	Corn
A&P Nutley (A&P Stores)	——	——	Soybean Cottonseed
Allsweet (Swift & Co., Chicago)	Soybean Cottonseed	——	Soybean
American Beauty (B) (L. Frank & Co., New Orleans)	——	Soybean Cottonseed	——
Blue Bonnet (Standard Brands, Inc.)	——	——	Soybean Cottonseed
Blue Plate (B) (Blue Plate Foods, New Orleans)	——	——	Soybean
Borden Deluxe Process (Borden, Inc., NYC)	——	——	Soybean
Dillon's Top Test (B) (Dillon Co., Hutchinson, Kansas)	——	Soybean Cottonseed	——
Eatmore (B) (Kroger Stores)	——	Soybean Cottonseed	——
Mrs. Filbert's Golden Quarters (J. H. Filbert, Inc., Baltimore)	——	——	Soybean Cottonseed
Fleischmann's (Standard Brands, Inc.)	Corn	——	Corn
Fleischmann's Sweet-Unsalted (Standard Brands, Inc., NYC)	Corn	——	Corn
Food Club Corn Oil (B) (Topco Assoc., Skokie, Illinois)	Corn	——	Corn
Food Club Top Spread (B) (Topco Assoc.)	——	Soybean Cottonseed	——
Gold'n Korn Oil (B) (Shedd-Bartush Foods, Detroit)	Corn	——	Corn
Good Value (B) (Fleming Co., Topeka, Kansas)	——	——	Soybean Cottonseed
Kroger 100% Corn Oil (B) (Kroger Stores)	Corn	——	Corn
Mazola (Best Foods, Englewood Cliffs, N.J.)	Corn	——	Soybean Cottonseed
Mazola Sweet-Unsalted (Best Foods)	Corn	——	Soybean Cottonseed

| | OILS (A) | | |
CONVENTIONAL	LIQUID	HARDENED	PARTIALLY HARDENED
Meadowlake (B) (Anderson Clayton Foods, Dallas)	——	Soybean Cottonseed	——
Oak Farms (B) (Southland Corp., Dallas)	——	Soybean Cottonseed	——
Parkay (Kraft Foods Co., Chicago)	——	——	Soybean Cottonseed
Piggly-Wiggly (B) (Piggly-Wiggly Stores)	——	Soybean Cottonseed	——
Promise (C) (Lever Brothers Co., NYC)	Safflower	——	Soybean Cottonseed
Rice (B) (Rice Food Markets Houston)	——	——	Soybean Cottonseed
Soft			
Blue Bonnet Soft (Standard Brands, Inc.)	——	——	Soybean Cottonseed
Chiffon Sweet-Unsalted (Anderson Clayton Foods)	——	——	Soybean
Food Club Soft (B) (Topco Assoc.)	Soybean	——	Soybean Cottonseed
Good Value Soft (Fleming Co.)	Soybean	——	Soybean Cottonseed
Parkay Soft (Kraft Foods Co.)	Cottonseed	——	Soybean Cottonseed
Promise Soft (C) (Lever Brothers Co.)	Safflower	——	Cottonseed Peanut
Whipped			
Whipped Blue Bonnet (Standard Brands, Inc.)	——	——	Soybean Cottonseed
Soft Whipped Blue Bonnet (Standard Brands, Inc.)	——	——	Soybean Cottonseed
Whipped Miracle (Kraft Foods Co.)	——	——	Soybean Cottonseed
Whipped Miracle Corn Oil (Kraft Foods Co.)	Corn	——	Corn
Whipped Parkay (Kraft Foods Co.)	——	——	Soybean Cottonseed
Diet/Imitation			
Diet Blue Bonnet Soft (Standard Brands, Inc.)	——	——	Soybean Cottonseed

TABLE 15.6 COMMERCIALLY AVAILABLE MARGARINES (9) *(continued)*

CONVENTIONAL	OILS (A)		
	LIQUID	HARDENED	PARTIALLY HARDENED
Diet Chiffon (Anderson Clayton Foods)	———	———	Soybean
Diet Fleischmann's (Standard Brands, Inc.)	Corn	———	Corn
Diet Imperial (Lever Brothers Co.)	Cottonseed	———	Soybean
Diet Mazola (Best Foods)	Corn	———	Corn
Diet Parkay Soft (Kraft Foods Co.)	Cottonseed	———	Soybean Cottonseed
Liquid			
Squeeze Parkay Liquid (Kraft Foods Co.)	Soybean Cottonseed	———	Soybean Cottonseed

(A) According to label information.

(B) Regional brand.

(C) A newer formulation substitutes sunflower oil for safflower.

Copyright © 1975 by Consumers Union of United States, Inc., Mount Vernon, N.Y. 10550. Excerpted by permission from *Consumer Reports,* January 1975.

diet high in polyunsaturates could be expected to cause formation of softer butter. This effect of diet is offset somewhat by the presence of microorganisms found in the rumen of cattle and lambs, which can carry out certain chemical alterations such as the breakdown of cellulose (see Chapter Eight) and hydrogenation of dietary unsaturated fats.* Because of this, butterfat and tallow reportedly contain some *trans* double bonds (1).

The normal range of iodine values of 78–90 for margarines can also be achieved by blending fats from different sources. Table 15.6 is a list of margarines compiled by *Consumer Reports,* which illustrates the major oils used for this purpose and the blends that are often employed to achieve the desired characteristics.

Salad or Cooking Oils and Shortening

Salad oils, cooking oils, and shortening are fat products that are packaged as almost pure triglycerides with small amounts of additives usually present to improve appearance, ease of handling, and stability.

---

* One report (8) has appeared in which an attempt was made to produce polyunsaturated lamb chops. Like most meats, lamb contains a fair amount of fat in addition to that which is clearly visible and can be trimmed away, so it was desired to try to make the fat more highly unsaturated. This was achieved by bottle feeding lambs on a milk high in unsaturated fat since it is known that when lambs suckle, the milk by-passes the rumen and thus avoids hydrogenation.

Shortening is a moderately-hardened pure fat, which, like margarine, can be made by hydrogenation or blending or a combination of both in order to achieve a suitable plastic range. The plastic range of a shortening may be both higher and narrower than a margarine. Shortening is generally not refrigerated so it may be a rigid, nonplastic solid when cold. Since shortening is not used as a table spread, it would seem unnecessary to make shortening soft at room temperature, except that shortening must be easily creamed when used for baking. Unlike margarine, complete loss of shape may occur above body temperature, since shortening is not consumed directly and, therefore, need not melt in the mouth.

Mayonnaise and Salad Dressings

A major brand of mayonnaise is pictured in Figure 15.2, along with its counterpart—a spoonable salad dressing. Mayonnaise and salad dressing, along with the familiar pourable dressings for salads, are major items contributing to food oil consumption in the United States. The amount of oil consumed in these products has been estimated at 5.2 pounds (oil content only) per capita in 1972, compared to 2.6 pounds per capita in 1950 (3).

The pourable dressings for salads bottled as French, Italian, Roquefort, etc. must contain at least 35% oil. They may be in the form in which the water and oil phases separate or as the emulsified viscous type.

Spoonable salad dressings and mayonnaise are in an emulsified and very viscous (semisolid) form, which prevents them from pouring but allows for convenient spreading. Either product is readily blended with other ingredients to form a wide variety of food toppings and sauces. Typical formulations of these two products are shown in Table 15.7. The minimum oil content allowed is 65% in mayonnaise and 30% in salad dressings, although both products generally contain more

Figure 15.2 Typical examples of mayonnaise, spoonable salad dressing, and imitation (diet) mayonnaise. (Courtesy of Kraft Foods.)

TABLE 15.7 TYPICAL COMPOSITION OF MAYONNAISE AND
SPOONABLE SALAD DRESSING (10)

INGREDIENTS	WEIGHT PERCENT IN MAYONNAISE	WEIGHT PERCENT IN SALAD DRESSINGS
salad oil	77–82	35–50
egg yolk	5.3–5.8	4.0–4.5
vinegar	2.8–4.5	9.0–14.5
salt	1.2–1.8	1.5–2.5
sugar	1.0–2.5	9.0–12.5
starch		5.0–6.5
water to make 100%		

(Courtesy of AVI Publishing Co., P.O. Box 831, Westport, Conn. 06880.)

in order to give the final products the viscosity to which the consumer is normally accustomed. The lower oil content of salad dressings is counteracted by addition of starch, which acts as a thickener to provide the same consistency as mayonnaise.

Egg yolk contains certain substances (see discussion of soaps) that act as emulsifiers to stabilize the suspension of oil and water. The emulsion is formed by mixing with high-speed beaters (see Figure 15.3). Salt concentrates in the water phase and acts as a preservative. Sugar, vinegar, and sometimes lemon juice are added to provide both flavor and preservative action. The last two contain acetic and citric acids. Several other spices and flavorings are also added. Salad dressings are generally much spicier and more acidic than mayonnaise. Products marketed as imitation or diet mayonnaise are formulated much like a spoonable salad dressing but with a milder flavor. An example is pictured in Figure 15.2.

APPLICATIONS OF MAJOR FATS AND OILS

In the preceding discussion, attention has been directed to the chemical structure of fats and the ways of altering or utilizing the physical properties of these substances in producing some common fat products. In this section a brief consideration will be given to the fats listed in Table 15.3 to determine how they are applied to the food products already discussed and to certain other specialty products for which some fats or oils are particularly well suited. The major vegetable oils are considered first, followed by the animal fats. Major reviews may be consulted for greater detail (6, 7, 11).

Soybean Oil

In 1973, soybean oil accounted for over 60% of the oil processed into both margarine and shortening. The information in Table 15.6 illustrates the frequency of use. It is also the major oil used in making mayonnaise and salad dressings. It makes an excellent salad oil but is not as good for cooking unless partially hydrogenated. It has a relatively high content of the 18:2 fatty acid (51%), which is subject to oxida-

Figure 15.3   Producing the mayonnaise emulsion. (Courtesy of Gaulin Corporation.)

tive spoilage, called *reversion,* particularly when heated repeatedly to high temperatures (see discussion of oxidation).

Cottonseed Oil

At one time, the production of cottonseed oil was greater than any other vegetable
oil, but its production has become more limited due to the decreased production of
cotton in response to an increased demand for synthetic fibers. It is suitable as
both a salad and a cooking oil and reportedly gets heavy use in frying potato chips
(12). Table 15.6 illustrates the frequency of use in margarines.

Corn Oil

This oil was second in quantity of oil processed into margarine in 1973, accounting
for about 10% of the oil used for this purpose. Corn oil margarines are often
prized in advertising campaigns but actually differ little from any other oil when
hardened (hydrogenated) into a margarine. Like soybean oil, it has a slight tendency to undergo reversion due to the high content (54%) of the 18:2 fatty acid,
although it is generally satisfactory for use as a cooking oil in the home.

Figure 15.4 Major sources of triglycerides. In addition to the fish, beef, peanuts, and corn, other items pictured are soybeans (left front), flaxseed (center front), which yields linseed oil, rapeseed (right front), palm kernels (second row right), and coconuts (center rear). (Courtesy of Ashland Chemical Company.)

Peanut Oil

As noted earlier, peanut oil is not processed into margarine or shortening but has good qualities for use as either a cooking or salad oil.

Safflower Oil

The consumer is often confronted with advertising that proclaims the special virtue of safflower oil, since it is the most unsaturated of all the common edible oils. It has also been blended into margarines (see Table 15.6), but these margarines have been characterized as having a "strong oxidized flavor" apparently due to the very high content (75%) of the 18:2 fatty acid, which also makes safflower oil less desirable for cooking (9).

Coconut Oil

Although it is not produced domestically, coconut oil is fifth on the list of oils used in the United States. It is a very highly saturated oil, which makes it undesirable for use as a salad and cooking oil, although the resistance to oxidation provided by the saturated fatty acids makes it good for frying certain commercial food products that must have a long shelf-life, e.g., cookies and crackers.

More than seventy percent of the fatty acids in coconut oil are the 12:0, 14:0, and 16:0, with the first one predominating. Since these three fatty acids melt over a range of only 19 degrees, the coconut oil has a very narrow plastic range, which makes it particularly suitable for incorporation in coatings on ice cream bars. When cold, the coconut oil is below the plastic range and, therefore, very hard, but it melts in the mouth very quickly. Somewhat the same behavior is observed for cocoa butter, which also has a narrow plastic range and melts just below body temperature. This property, plus its compatability with chocolate, make it ideally suited for use in coatings on candy and other foods.

Coconut oil also has several important nonfood applications. Soap production (see Chapter Sixteen) is a major use. Coconut oil is also frequently used in cosmetics, lipsticks, creams, lotions, hair oil, rubbing oil, and hydraulic brake fluids.

Palm Oil

This oil is not produced in the United States, but domestic use has steadily increased in recent years to where it represents about 5% of the edible oil in use. Palm oil is generally processed into shortening. In 1973, roughly 10% of the oil used for making shortening was palm oil. It was the second most important oil used for this purpose.

Olive Oil

Olive oil is a relatively expensive oil that is used directly, or blended with other oils, as a salad or cooking oil. It is prized by many for its distinctive flavor.

Lard

At one time, lard was the major cooking fat both commercially and in the home for baking and frying. The data in Table 15.4 show that the use of lard has dropped by about 50% since 1949, although it remains in heavy use in commercial baking.

Tallow

Beef tallow is a hard, plastic fat that is often blended with lard or vegetable oils to make shortening. Tallow is also used in making soap. (See Chapter Sixteen.)

Fish Oils

Oils from herring, sardine and whale are of little significance in the United States, but are used extensively in certain parts of the world where fishing is a major industry. Fish oils are characterized as having an unusual content of fatty acids with 4, 5, or 6 double bonds, which makes the oils very susceptible to oxidative spoilage and development of very strong fishy odors and flavors.

The chain that links *saturated fats to cholesterol, cholesterol to atherosclerosis, and atherosclerosis to heart attacks* is the basis of the controversy over cholesterol and the current preference for polyunsaturated fats in the diet.

First, consider the two terms—*cholesterol* and *atherosclerosis*. Details of the structure and properties of cholesterol and other steroids are presented in Chapter Twenty. It is classified as a lipid, which means it is predominantly a hydrocarbon, like triglycerides, with physical properties (e.g., high solubility in hydrocarbon solvents and low solubility in water) typical of fats and oils.

Atherosclerosis, better known as hardening of the arteries, is a disease characterized by the buildup of deposits of cholesterol and other lipids on the walls of the arteries, causing them to thicken and harden. This may result in a partial blockage of the blood flow through the arteries.

Consider the three links in the chain mentioned in the first paragraph above. The first link connects saturated fats to cholesterol. This connection stems from the finding that diets high in saturated fats lead to an elevation of the level of cholesterol in the blood. This condition is typical of the average American, whose diet contains a relatively large amount of animal fat due to heavy consumption of meat and dairy products. It has been estimated that even carefully trimmed, lean beef contains 8% or more "invisible fat." This invisible fat contributes to the tenderness and quality of the meat. Higher-priced cuts of meat generally contain more invisible fat. Most important, this animal fat is highly saturated. A diet high in unsaturated fats tends to lower blood cholesterol levels.

One simple explanation for the elevation of blood cholesterol (in response to a diet high in saturated fats) is the direct consumption of cholesterol along with the saturated fats. For example, an average of 300–400 mg of the 500–1200 mg of daily intake of cholesterol is consumed in meat by the average American. Unfortunately, all the facts do not support this explanation, since blood cholesterol levels are somewhat insensitive to dietary cholesterol unless the total intake is restricted to below 200 mg per day. This apparent contradiction is the result of several phenomena. Humans (and other animals) can synthesize cholesterol and most of the cholesterol found in human tissues is synthesized rather than absorbed from the diet.* Dietary cholesterol also causes an inhibition (called *feedback inhibition*) of cholesterol synthesis in the liver,† which tends to regulate the blood cholesterol when the dietary supply is adequate.

The efficiency of absorption of cholesterol is also influenced by the amount supplied in the diet. When intake is low, absorption is more efficient. On the average, only about one-half of the daily intake is absorbed, but the amount varies considerably.

So, as far as the first link in the chain is concerned, the evidence is that the saturated fat in the diet, rather than the cholesterol intake, is the cause of the elevation of the blood cholesterol.

---

* This has been shown by experiments using radioactive tracers.
† The major site of cholesterol synthesis is generally thought to be the liver but since liver synthesis may be inhibited, it has been suggested that the intestine may be the major source because synthesis there is not subject to this inhibition.

A convenient explanation would be that cholesterol is synthesized only from saturated fatty acids, but this is not the case. In fact, both types of fatty acids as well as carbohydrates and proteins can provide the necessary starting material for cholesterol synthesis. Consequently, the first link is a weak one which, at this point, is unexplained.

The second link is a very logical one since a higher blood cholesterol level should mean a more abundant supply for the buildup of atherosclerotic deposits. However, even this link is open to question. The controversy here is over the cause-and-effect relationship. The implication is that excess cholesterol supply is the cause of atherosclerosis. However, it is conceivable that the opposite is true, i.e., the atherosclerosis may be the result of some other problem and may be the cause of the elevated blood cholesterol. The excessive buildup of cholesterol in arterial walls may be the source of circulating blood cholesterol.

The third link between atherosclerosis and heart attack is generally accepted since heart attack or stroke is caused by a blockage of an artery supplying blood to the heart or the brain. When the supply is cut off, heart or brain cells may die due to insufficient oxygen or nutrients, and paralysis or even death may occur. It is generally conceded that arteries already partially blocked by atherosclerotic

Figure 15.5   Courtesy of Standard Brands.

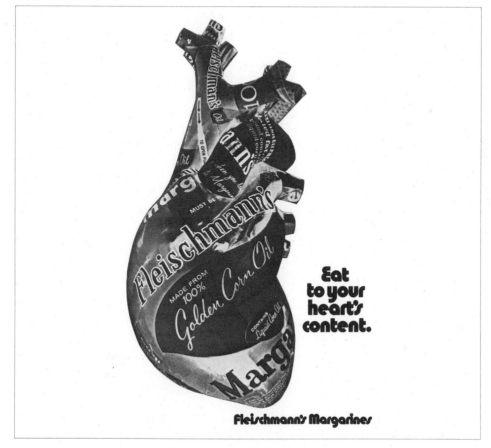

deposits are more prone to total blockage and it may even be that such a deposit could break loose from one site, block another, and cause a heart attack.

In summary, the three-step chain described throughout this section is a logical, although unproven, scheme. The increased incidence of heart disease in the United States is generally blamed on the increased amount of meat and accompanying saturated fats in the average American diet and this explanation is a major argument in support of the chain. Unfortunately, some serious doubts remain over the first two links in the chain and these doubts lead many scientists and physicians to discount the validity of the entire problem.

ADDITIVES IN FAT PRODUCTS

Five additives are in common use in marketing fats and fat products. They are antifoaming agents, colorings, flavorings, antioxidants, and emulsifiers. A discussion of the mode of action of emulsifiers is deferred until the coverage of soaps (Chapter Sixteen). The major use of emulsifiers already noted is in stabilizing the water-oil emulsion in mayonnaise or salad dressings, in which natural emulsifying agents present in egg yolks act as stabilizers.

Antifoaming Agents

Some cooking oils have a tendency to foam when used for frying. To counteract this problem, compounds such as methyl silicone are often added to cooking oils and shortenings.

Colorings

It is generally expected that shortenings should have a clear white color, whereas margarines should resemble butter. The color in butter is due to a substance called β-carotene or provitamin A. It is the same substance that causes the orange color in carrots and it is abundant in nature. The color appears in butter because this substance is a fat-soluble hydrocarbon that is extracted from the milk when butter is prepared. Although not visible, it is a common pigment in most green vegetables (e.g., spinach), where the green color of chlorophyll masks the orange pigment. Even the intensity of the color in butter varies depending on the feed supplied to milk-producing cows. The same substance is often added to margarines to duplicate the color of butter.

Flavorings

Here again, in making margarines, attempts are often made to duplicate butter. A compound known as biacetyl is formed when butter is ripened and is considered the major flavor ingredient. It is sometimes added to margarines.

Antioxidants

Perhaps the most important and most criticized food additives are a group known as *antioxidants,* which are used to prevent the type of oxidative spoilage already mentioned. This coverage will examine the course of the oxidation, the antioxidants used, and the criticisms over their use.

The highly unsaturated fatty acids are the primary culprits in development of oxidative rancidity. Carbon atoms adjacent to carbon-carbon double bonds are prone to oxidation by formation of a hydroperoxide. The reaction is illustrated for a segment of a fatty acid chain.

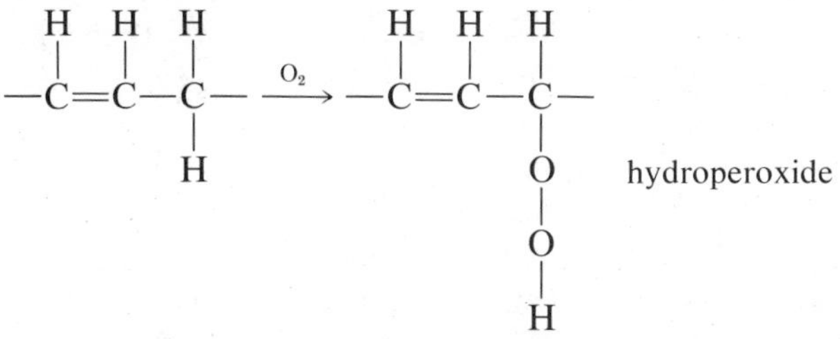

It is a complex process called *autoxidation* and results in insertion of oxygen into a carbon-hydrogen bond with the adjacent double bond activating the C-H bond and making it susceptible to attack by oxygen. The hydroperoxides are then prone to further oxidation involving cleavage of the double bonds and formation of fragments (aldehydes and acids), which have very foul odors and flavors.

The oxidation problem is not too significant for monounsaturated fatty acids, such as oleic acid (18:1), but is particularly acute for diunsaturates, triunsaturates, etc. Linoleic acid (18:2) is a common component of most oils with high iodine values. In fact, the data in Table 15.4 shows that the 18:2 or other even more unsaturated fatty acids must be present whenever the iodine value exceeds 89.

The autoxidation of linoleic acid (18:2) is shown below. The fatty acid is the $\Delta^{9,12}$ isomer, which has the normal arrangement of double bonds. This symbolism indicates that the two double bonds are between carbons 9 and 10 and carbons 12

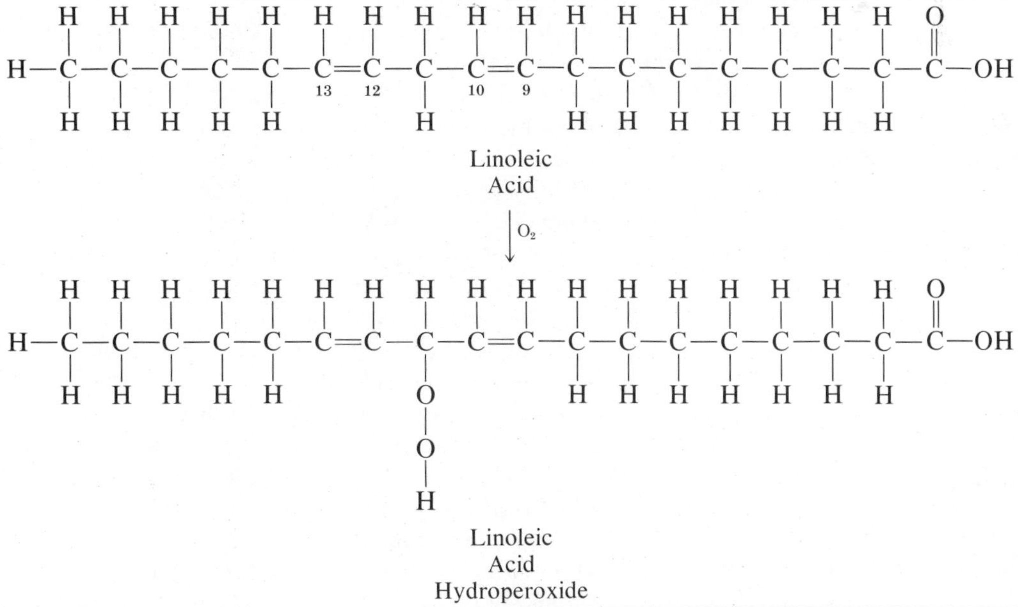

Linoleic
Acid

Linoleic
Acid
Hydroperoxide

and 13. The autoxidation reaction occurs primarily at carbon 11, which is located between the double bonds, because both double bonds tend to activate the C-H bond at that site making it especially susceptible to autoxidation and causing the subsequent collapse of the hydroperoxide to produce foul-smelling products.

In the discussion of the uses of soybean and particularly safflower oil, it was noted that these two oils are not as well suited for cooking as some oils because of the high content of linoleic acid. Soybean oil contains 51% and safflower has 75% of this acid. The autoxidation process is greatly speeded up when an oil is hot. *Consumer Reports* describes a margarine made with safflower oil as having a "strong oxidized flavor (9)."

Obviously, hydrogenation would decrease the linoleic acid content and, thus, improve the keeping quality of an oil high in linoleic acid. In fact, by carefully selecting the conditions for the hydrogenation, the linoleic (or more unsaturated fatty acid) may be the primary one that is hydrogenated. In this way, the oil may be hardened in such a way as to give it a greatly decreased susceptibility to oxidative rancidity.

The ease of autoxidation of linoleic acid explains why the simple blending of saturated and unsaturated fats is not totally satisfactory for making a margarine. While it might be possible to achieve a satisfactory plastic range, the final unhydrogenated product would not keep as well.

Fish oils are very highly unsaturated, with fatty acids having 4–6 double bonds being quite common. This makes the oils very susceptible to oxidation and development of foul odors, as anyone familiar with the odor of dead fish can readily attest.

The prevalence of food oils in the diet obviously make their preservation a major concern. It is a common practice to package cooking and salad oils in cans or brown bottles since light can serve to activate the autoxidation process. Margarines and other oil products are also refrigerated, at least after opening a container and allowing access to atmospheric oxygen.

The use of antioxidants is a common practice. Antioxidants are additives that interfere with the autoxidation process. There are two types of antioxidants in use and they are sometimes used in combination. The first type is the sequestering agent (citric acid or EDTA), which acts by tying up metal ions that are known to catalyze autoxidation. The second group of antioxidants directly interfere with autoxidation. In this group are butylated hydroxyanisole (BHA), butylated hydroxytoluene (BHT), propyl gallate (PG), and nordehydroguaretic acid (NDGA).

There has been much controversy and criticism over the use of these and other additives (13–15). It has been pointed out repeatedly that not all manufacturers include antioxidants in fat-containing foods. This fact has been taken as an indication that these additives are unnecessary. One critic has attributed their use to "habit" or "outmoded manufacturing techniques (14)."

In the early sixties, a flurry of research activity was directed toward a question of safety raised by Australian researchers who reported in 1959 that BHT caused certain birth defects in rats. The safety of BHT has been repeatedly confirmed without an explanation of the study that stirred up all the interest.

Vitamin E is a fat-soluble vitamin that is found in most oils. It is a natural antioxidant that apparently protects the oils *in vivo*. Vitamin C is also an antioxidant, but it is water soluble and normally not present in oils.

SUMMARY

Food fats and oils are triglycerides in which the acid portion consists of long, linear fatty acids. These fatty acids may be completely saturated or highly unsaturated. In general, animal fats are more saturated than vegetable oils. The iodine value can be determined experimentally using the reaction of iodine monochloride (ICl). This test describes the degree of unsaturation quantitatively.

Soybeans are the major source of food oil in use directly as a salad and cooking oil, or in converted form as margarine, shortening, and as a major ingredient in mayonnaise and salad dressings.

Oils are generally isolated by solvent extraction and/or pressing. Animal fats are isolated by a process called rendering.

Along with the dramatic trend away from the use of animal fats, e.g., butter and lard, to the increased use of vegetable oils, there has been a definite increase in the consumption of triglycerides. This is largely attributable to the rapid development of the fast-food industry in recent years.

Margarine is a major fat product that is produced by blending of saturated and unsaturated fats or, more frequently, by hydrogenation to give a product with an acceptable plastic range. The melting point of a fat is increased by long-chain fatty acids, less unsaturation, and *trans* arrangement of the unsaturation. Naturally-occurring fats contain only *cis* double bonds but hydrogenation can cause some *cis* to *trans* isomerization. Unsaturation also makes a fat more susceptible to autoxidation.

Saturated fats have been linked to atherosclerosis and heart attacks, but the connection is somewhat tenuous and is open to criticism. The more highly unsaturated (polyunsaturated) oils seem to lessen the risk.

A wide variety of additives are commonly used in fat products. The most frequent and most controversial are the synthetic antioxidants. Vitamin E is a natural, fat-soluble antioxidant, which is often present in fats.

PROBLEMS

1. Give a definition or example of each of the following:
   a) biodegradable
   b) lipids
   c) triglycerides
   d) unsaturation
   e) amorphous
   f) iodine value
   g) rendering
   h) plastic range
   i) hydrogenation
   j) shortening
   k) atherosclerosis
   l) autoxidation
   m) antioxidants

2. What is the major difference between petroleum-based oils and food oils?

3. What is the main difference between food oils and fats?

4. Draw the structure for the triglyceride derived from lauric acid (12:0), caproic acid (6:0), and linoleic acid (18:2).

5. What is the most important commercial source of food oil?

6. How do the contents of butter and margarine differ?

7. What factors affect the melting point of a fat?

8. Most margarines have an iodine value of about 85. Peanut oil also has an iodine value of about 85. Explain why margarines are solid and peanut oil is a liquid at room temperature.

9. What are the chemical changes that occur when corn oil is hardened into a margarine? Why does it harden?

10. Butter and margarine have about the same melting temperature even though margarine has a much higher iodine value. How do you account for this?

11. Why is coconut oil especially good for coatings on ice cream bars?

12. What is the major reason for the increased per capita consumption of fat in recent years?

13. What are the three most important sources of oil used for making margarines?

14. What are the sources of lard and tallow?

15. Why is cottonseed oil (iodine value 97-112) more suitable for use as a cooking oil than safflower oil (iodine value 140-150), e.g., for frying potato chips?

16. What is the major difference between a solid shortening and a margarine?

17. Why can fish oils smell so bad?

REFERENCES

1. Institute of Shortening and Edible Oils. *Food Fats and Oils*. Washington, D.C., October 1968.

2. "Fats and Oils." *McGraw-Hill, Encyclopedia of Science and Technology*. Vol. 5. New York: McGraw-Hill, 1971, p. 199.

3. *Fats and Oils Situation*. FOS-277, Economic Research Service, U.S. Department of Agriculture, April 1975, p. 11.

4. *Fats and Oils Situation*. FOS-253, Economic Research Service, U.S. Department of Agriculture, June 1970, p. 29.

5. *Fats and Oils Situation*. FOS-267, Economic Research Service, U.S. Department of Agriculture, April 1973, p. 21.

6. Weiss, T.J. *Food Oils and Their Uses*. Westport, Conn.: AVI, 1970, Chap. 1, pp. 1–25.

7. Mattil, T.F. "Butter and Margarine." In *Bailey's Industrial Oil and Fat Products*, D. Swerh. New York: Interscience, 1974, p. 323.

8. "Polyunsaturated Lamb Chops." *Chemistry 46* (9) (1973): 33.

9. "Choosing A Margarine For Taste and Health." *Consumer Reports*, January 1975, pp. 42–7.

10. Weiss, T.J. *Food Oils and Their Uses*. Westport, Conn.: AVI, 1970, Chap. 1, pp. 146, 163.

11. Kromer G.W. "U.S. Food Fat Consumption Trends." *Fats and Oils Situation*. FOS-272, April 1974.

12. Weiss, T.J. *Food Oils and Their Uses*. Westport, Conn.: AVI, 1970, Chap. 1, p. 28.

13. Verrett, J., and Carper, J. *Eating May Be Hazardous To Your Health*. New York: Simon and Schuster, 1974, pp. 29–32.

14. Jacobson, M.F. *Eaters Digest*. Garden City, New York: Doubleday, 1972, pp. 86, 193.

15. "Food Additives." *Chemical and Engineering News*, 17 October 1966, p. 122.

DRUGS:
FROM DENTAL HEALTH
TO MENTAL HEALTH
(PLUS BIRTH CONTROL)

P A R T   F I V E

I N this section, we will direct attention to the chemistry involved in matters pertaining to good health. Chapters Sixteen and Seventeen deal with subjects that could be categorized under the heading of maintenance of good health. In Chapter Sixteen, the emphasis is on good hygiene. In Chapter Seventeen, the chemistry of disease and immunity is discussed in the hope of providing an understanding of the progress and limitations of modern medicine.

In Chapters Eighteen and Nineteen, the emphasis is on drugs used for therapy and drug abuse. At both extremes, an awareness of the chemistry involved is necessary for the consumer to make decisions in matters of personal health and in matters of widespread social concern.

In Chapter Twenty, the subject of steroids and birth control is singled out for special coverage with emphasis on chemical birth control, i.e., "the pill," although a comparison is made with other methods.

DENTAL CHEMISTRY

This section examines the role of chemistry in dentistry. We will consider the structure of tooth enamel with emphasis on how it is broken down and reformed, and how fluoride can play a role in prevention of decay. We will also consider the role of chemistry when all else fails and the dentist enters the picture.

Tooth Enamel

Tooth enamel extends from the surface of the tooth to a depth of about two millimeters. It consists of hydroxyapatite, $Ca_5(PO_4)_3OH$. As with other salts, this formula is not meant to suggest that tooth enamel is made up of discrete molecules with this formula, but rather, it consists of a complex three-dimensional structure, called a *lattice,* in which a series of calcium ions (Ca^{+2}), phosphate ions (PO_4^{-3}), and hydroxide ions (OH^-) are positioned around one another as shown in Figure 16.1. The net charge on the lattice is zero, since there are 5 calcium ions for every 3 phosphate ions and every 1 hydroxide ion in the lattice.

In the environment of the mouth, hydroxyapatite can be dissolved or reformed according to the following equation.

$$Ca_5(PO_4)_3OH \underset{\text{remineralization}}{\overset{\text{demineralization}}{\rightleftharpoons}} 5\ Ca^{+2} + 3\ PO_4^{-3} + OH^-$$

HOME PRODUCTS

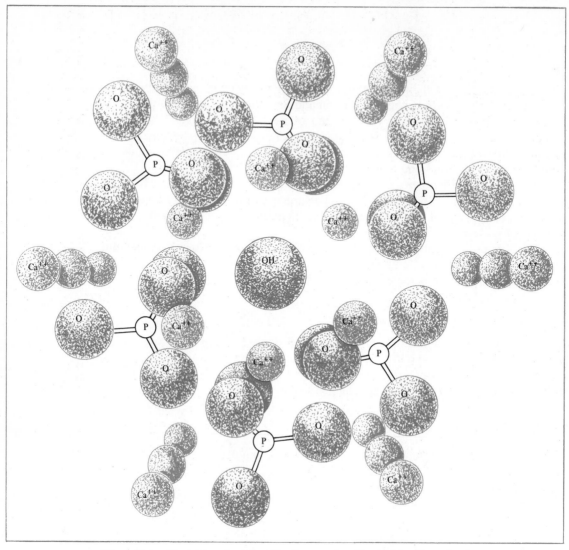

Figure 16.1   The lattice structure of hydroxyapatite. (See text for discussion.)

The process for formation of hydroxyapatite is called *remineralization,* or simply mineralization. The process for breakdown of the enamel is called *demineralization*. The arrows in the previous equation confirm that remineralization is favored over demineralization so that the enamel is maintained.

Localized decay of the teeth, initiated by the demineralization process, is known as *dental caries*. The most common reasons for tooth loss in individuals are peridontitis (inflammation of the gums) and caries. These are the most widespread mouth diseases in man and, despite our scientific accomplishments, their causes are not well understood. In both diseases, a common factor is involved. This is *dental plaque,* which is a gelatinous mass of closely-packed microorganisms attached to the tooth surface. Plaque is just one example of microorganisms that reside in the body. Microbes are normally present in several parts of the body and often exist in a symbiotic (mutually beneficial) relationship with the body. For example, the microorganisms in the large intestine, called the *intestinal flora,* carry out the synthesis of vitamin K and one of the B vitamins and the excess can be absorbed and used by the body. At the same time, these microbes benefit from the supply of nutrients provided in the body.

The microorganisms present in dental plaque are part of the oral flora. In addition, the plaque consists of traces of food that serve as nutrients for the microbes, and a base material that contains complex polysaccharides. The polysaccharides are produced by the microbes and assist in holding the bacteria and substances produced in the plaque right up against the tooth surface, where they can do the maximum damage.

In the development of dental caries, the plaque plays a key role in the demineralization process. Most of the microorganisms in plaque are anaerobic rather than aerobic. Anaerobic microbes are favored by stagnation and decreasing oxygen content that accompanies the buildup (increasing thickness) of plaque. The surfaces below the areas of contact between the teeth are the most stagnant and, therefore, the most susceptible to the onset of caries. As we have seen before, aerobic bacteria break down carbohydrates (and other foods) into carbon dioxide and water, whereas anaerobic bacteria produce simple organic acids, particularly lactic acid, by fermentation of carbohydrate.

The chief culprit in the story is sucrose. Even starchy foods have very little tendency to promote dental caries. It seems that simple sugar is better able to penetrate the plaque where it can undergo fermentation, and each time more sugar reaches the bacteria in the plaque, more acid is produced. Gradually the hydroxyapatite structure is attacked, since demineralization is favored in an acidic environment.

Under normal circumstances, there is an equilibrium between demineralization and remineralization that favors the latter, but as the acidity increases, demineralization is favored more and more. The reason for this is easily seen by looking at the demineralization reaction and focusing attention on the hydroxide ions that are produced in the process. If the pH is unusually low, there is a high concentration of hydrogen ions, which will remove (neutralize) the hydroxide ions and prevent remineralization. In addition, the phosphate ions (PO_4^{-3}), which are released during demineralization, are converted to HPO_4^{-2} and $H_2PO_4^-$ as the pH is lowered, so that the PO_4^{-3} becomes less available for remineralization.

The normal pH of the mouth is 6.8, depending on the foods previously in the mouth, but within the plaque, at the surface of the tooth, the pH may be much lower, in which case remineralization is hindered even more.

The demineralization process also works to the advantage of the microorganisms, which may obtain calcium or phosphorus from the teeth when the saliva contains insufficient amounts to meet their needs. Consequently, we see a never-ending circle of more plaque, which causes more demineralization, which contributes to the development of more plaque.

The tooth enamel can remineralize via the saliva. Calcium and phosphate, which may be removed by bacterial activity, are replenished from the saliva so that repair (remineralization) can occur, but the diet is a major factor in determining whether repair can keep up. If there is a steady stream of sugar supplied to the plaque, both during and between meals, the demineralization process is increasingly favored.

The consistency of carbohydrates in the diet is also important. A sticky type of food, such as candies and confectionary goods, may contribute to caries by adhering to the teeth in the interproximal areas (between the teeth) and along the gingival margins (between the teeth and gums) and also by packing into pits and fissures of the rear teeth where caries are most common.

In effect, cavities occur when the saliva cannot supply calcium and phosphate fast enough to offset demineralization. It should also be obvious that good hygiene plays a role since a buildup of plaque may be minimized. This lowers the amount of acid that is produced and also allows for better access to calcium and phosphate in the saliva, so that remineralization is favored even more.

At the other extreme, as more and more plaque builds up, the saliva may not even be able to supply adequate nutrients for the organisms in the plaque, in which case, the bacteria may resort to attacking gum tissue.

It is interesting that some people develop very few caries, and one or two persons per thousand remain free of caries indefinitely, even though they are exposed to cariogenic bacteria and diets (1). These persons have been designated as caries-immune. The reason for this has not been determined. Familial influences are indicated. Studies show that such caries-free persons are about 40 times as numerous among relatives as in the general population. These same studies show that sex factors are also involved, since caries-free, male adults outnumber females by two to one. Caries-free adults are more numerous in regions where environmental fluoride is naturally high, but are also present in other regions.

FLUORIDATED WATER

One of the most outstanding achievements of dental research during the past four decades has been the finding of the relationship between fluoride in the drinking water and the decreased incidence of dental caries. There seem to be two factors involved in the fluoride effect. It is well known that fluoride (F^-) has the ability to inhibit certain enzymes, such as those that act as catalysts in the fermentation of carbohydrates to produce lactic acid. Studies have indicated that concentrations as low as 0.5 ppm* have a detectable inhibitory influence on the rate of acid production by the oral flora. Although it has been found that the fluoride concentration in saliva is less than 0.25 ppm, recent findings suggest that an accumulation of fluoride occurs within the dental plaque to levels of concentration ranging from 6 to nearly 180 ppm.

The second explanation for the anticariogenic properties of fluoride in drinking water is the one most favored due to a vast amount of accumulated evidence. It has been found that fluoride will substitute into the hydroxyapatite when remineralization occurs. The mechanism appears to be one in which some of the hydroxide ions are replaced by fluoride ions. This is possible, in part, because of the similarity in size of the fluoride and hydroxide ions.

The incorporation of the fluoride may be represented by the following sequence:

* 0.5 ppm of fluoride equals 0.5 mg per kilogram of water.

$$Ca_5(PO_4)_3OH$$

$$\xrightarrow{\text{demineralization}} 5\ Ca^{+2} + 3\ PO_4^{-3} + OH^-$$

$$\xrightarrow[\text{remineralization}]{F^-}$$

$$Ca_5(PO_4)_3(OH)_{1-x}F_x$$

The formula $Ca_5(PO_4)_3(OH)_{1-x}F_x$ represents a structure called fluoridated hydroxyapatite, in which some of the hydroxide ions of the hydroxyapatite structure are replaced, but the sum of the number of hydroxide ions ($1-x$) and the number of fluoride ions (x) equals 1 for every 5 calcium and 3 phosphate ions in the enamel structure.

As more Ca^{+2} and PO_4^{-3} become available, via demineralization and from saliva, the remineralization process is increasingly favored. If free F^- is available, it will become incorporated into the hydroxyapatite structure. Once this occurs, the equilibrium is shifted even farther in favor of remineralization than normal. In other words, the fluoridated hydroxyapatite is less soluble in acid solutions than hydroxyapatite, because fluoride ions in the hydroxyapatite structure help to lock the hydroxide ions more tightly in place. Consequently, we can now understand both how the fluoride is incorporated and why it helps to prevent decay. It has been determined that, at 30% substitution, fluoridated hydroxyapatite is extremely stable to acid attack (2).

TOO MUCH FLUORIDE?

Quite logically, there is much concern over the hazards of fluoride in drinking water. Fluoridated water is a kind of ''compulsory mass medication,'' which many people object to. Environmentalists have also questioned the impact of fluoride on the environment. A very thorough study was conducted by the United Kingdom's Royal College of Physicians, which concluded that there was no hazard to the individual or the environment from fluoride at levels up to 1 ppm, which is greater than the levels used in drinking water (3).

Very high levels, e.g., in excess of 100 ppm, can cause damage to the teeth in the form of a condition known as chronic endemic dental fluorosis, but even here, the damage is only significant when teeth are first developing. Once teeth have fully matured (about 12–14 years), this condition does not develop. In any event, such a condition only arises in areas where natural waters contain fluoride in excess of 100 ppm and bears no relationship to levels below 1 ppm that are used in controlled fluoridation.

OTHER SOURCES OF FLUORIDE

Although fluoridated water is the most common means of providing fluoride, there are alternate methods that have been studied.

Fluoride tablets have been suggested as a means by which individuals might derive the benefits of fluoride when a communal water supply is not available or when the available water supply is not fluoridated. Some studies confirm that fluoride tablets are beneficial, but there is also evidence available that suggests that a large percentage (typically more than 90%) of a large dose of fluoride is rapidly excreted in the urine. Therefore, the fluoride is used much more efficiently by individuals who ingest small quantities of drinking water throughout the day and eat food prepared with fluoridated water.

The use of fluoride tablets also requires that individuals take these pills every day throughout their lives in order to gain the same benefits as those in fluoridated water. It is particularly difficult to assure that youngsters will be conscientious and this is the group that can benefit most from fluoride treatments.

Topical fluoride application is another means of supplying fluoride. This includes both the fluoride treatments given by dentists and the use of fluoride toothpastes. Stannous fluoride (SnF_2), sodium fluoride (NaF), and monofluorophosphate, MFP (PO_3F^{-2}) are the sources of fluoride commonly found in toothpastes. The first two compounds contain free fluoride ions, whereas MFP does not. Phosphate ions near the surface of the tooth, e.g., those released by demineralization, exist primarily as $H_2PO_4^-$ at the pH of the mouth. The similarity of PO_3F^{-2} and $H_2PO_4^-$ makes it possible for the MFP ion to replace some of the $H_2PO_4^-$. MFP can also penetrate and become bound in the plaque.

Simple fluoride ions can be released by the following reaction:

$$PO_3F^{-2} + H_2O \rightleftarrows H_2PO_4^- + F^-$$

Once again the reaction proceeds more efficiently from right to left but, nevertheless, it does provide a steady supply of F^-, particularly at low pH, when the reaction proceeds more efficiently from left to right. In other words, F^- is released most efficiently when it is needed the most, i.e., when the pH is low.

A variety of alternative methods have been suggested for administering fluoride, but there is little information on their effectiveness. Attempts have been made to incorporate fluoride into mouthwashes, chewing gum, salt, milk, and breakfast cereal.

Figure 16.2 Fluoride toothpastes. (Photos by James J. Kane, Jr.)

Dental Amalgams

Cosmetic fillings have been noted in the mummies found in Egypt and, consequently, date back before Christ. There are references made in those same findings to teeth being filled with a cement containing oxides of various metals. However, the majority of dental treatments at that time and even into the 1700s were mainly extractions. The resultant space quite often was replaced by a false tooth which was fashioned from the tooth of an animal or a piece of wood or ivory or in some cases gold. Some restorative dentistry was performed in the Middle Ages by scooping out the decayed material and replacing it with various types of rosins or oxides of metals.

An Italian, Giovanni of Arcoli (1412–84), is credited as the first to use acceptable principles for the filling of teeth with metals. He advised that the cavity be cleared of all decayed matter and then filled with gold leaf. For those who can afford it, this type of filling remains in use even today.

In 1826, combinations of bismuth, lead, and tin were mixed with mercury to form the first dental amalgam. When metals are melted together, the resultant mixture is called an *alloy*. An alloy containing mercury metal is called an *amalgam* and the process of combining the components of the alloy is called *amalgamation*. Mercury is a liquid at room temperature and is alloyed with other metals that then become a solid mass after amalgamation. The original amalgam of bismuth, lead, tin, and mercury had a melting range near that of the boiling point of water and had to be poured into the cavity at that temperature. This procedure would appear to be more painful than the decayed tooth. Later, the formula was

Figure 16.3   A dental amalgam. (Courtesy of Johnson & Johnson Dental Products Company.)

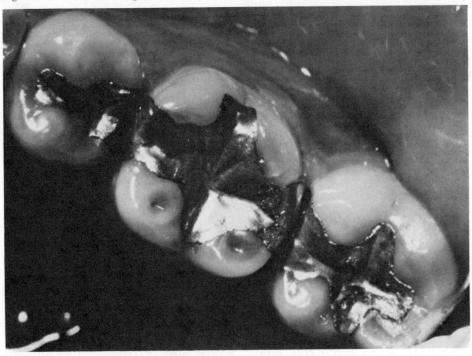

modified to melt at a temperature around 66°C so that it would be put into the cavity cold and then melted and adapted with a hot instrument.

In the late 1800s, G.V. Black (1836–1915) revolutionized dentistry. Although Black had little formal education, even in dentistry, he accumulated a vast store of knowledge of chemistry, dentistry, and related fields, including metallurgy (the science and technology of metals). Black noticed that the amalgam fillings in the larger cavities often became the source of other difficulties, because they became chipped from biting and chewing. Therefore, Black set out to develop a new filling material. He first determined the force required to chew different foods, the force exerted by different people to clench their teeth, and the amount of pressure fillings could withstand without cracking. He found that most jaws exerted 100 to 175 pounds of pressure, but some exerted as much as 325 pounds. Later data indicate this to be approximately 30,000 psi (pounds per square inch). From this information, he resolved to develop an amalgam that would resist such a force.

Figure 16.4   A commercial amalgamator. An amalgamator is routinely used in a dentist's office to prepare an amalgam for filling teeth. A premixed silver-tin-copper-zinc alloy is combined with mercury (liquid) in the capsule in the front right. When activated, the capsule is rapidly vibrated to assure complete mixing of the amalgam components. An extra capsule, funnel, and plastic rod, which is placed in the capsule to assist mixing, appear in the top right. A 30-second timer and switch appear on the left. (Photo courtesy of Toothmaster Company.)

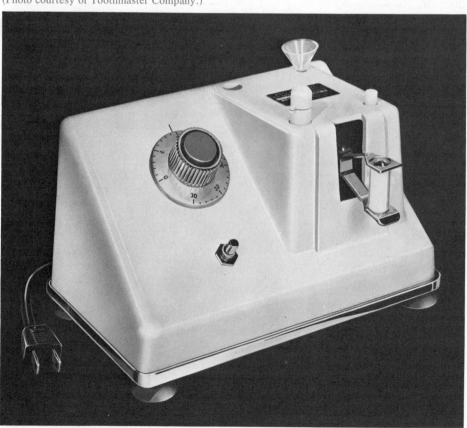

Black developed such an amalgam, consisting of 65% silver, 27% tin, 6% copper, and 2% zinc. Although this basic formula is still used today, the majority of manufacturers are increasing the copper content to approximately 12% with a resultant reduction in tin content. This newer alloy, when mixed with mercury, has superior physical and clinical properties.

The individual components of the amalgam add specific characteristics to the finished product. Silver provides tarnish resistance, strength, and generally slows the set (hardening) of the mixed amalgam. Tin is necessary for the early initiation of amalgamation because it has a high affinity for mercury. This decreases the setting time. Copper is used as a strengthening component and, when increased to the 12% range, it markedly improves clinical performance, specifically in the strength at the margins of the restorations. Finally, zinc is added by some manufacturers to aid in the manufacturing process by reducing the tendency of the other metals to oxidize during melting. It is obvious that the four components should be blended in such a manner that the best physical properties are obtained.

Amalgam alloys are supplied in three forms—as a powder, as a tablet, and as a complete capsule containing the alloy and mercury separated by some type of membrane. The dentist chooses the form that he wishes to use in his practice, and if he uses either the powder or the tablet, he adds the necessary mercury, places it in a capsule, places the capsule in a mechanical device called an amalgamator, and mixes the mercury with the alloy by activating the amalgamator. The amalgamator throws the mercury-alloy mixture back and forth with a sufficient force to properly mix the amalgam. The amalgam is removed from the capsule and placed into the tooth (condensed), carved, and allowed to harden (usually 5-10 minutes). The amalgam continues to harden for a considerable period of time, usually months; however, 96–98% of the hardening has occurred within 48 hours. After 24 to 48 hours, the alloy is usually polished and a restoration of superior quality so obtained can be expected to serve the patient for many years.

SOAPS

Good hygiene is an important contributor to good health. In the previous section, we considered the importance of good hygiene in maintaining good dental health. In this section, we will consider the chemistry of soaps and detergents, including the environmental effects of the latter.

Soaps are metal salts of fatty acids. They can be represented by the formula shown below, where M^+ signifies the metal, which is usually sodium (Na), and the anion is derived from a fatty acid.

$$R-\overset{\overset{\displaystyle O}{\|}}{C}-O^-\ M^+$$

It is reported that the first synthesis of soaps was carried out by the Romans about 2500 years ago (4). They cooked animal fats with wood ashes, which contained potassium carbonate (K_2CO_3), and caused the following set of reactions. The first reaction forms potassium hydroxide (KOH), which causes the breakdown of triglycerides into the component parts, glycerine and fatty acids. In the process, the fatty acid is neutralized by the strong alkali and ends up in the salt

$$K_2^+CO_3^{-2} + H_2O \rightleftharpoons K^+OH^- + K^+HCO_3^-$$

a triglyceride   potassium hydroxide   a soap molecule   glycerin

form. In modern production, sodium hydroxide is normally used, in which case a sodium soap is formed. The soap can be isolated by addition of salt (NaCl)—a procedure known as "salting out," which lowers the solubility of the soap and causes it to separate. A typical soap molecule that is a major component of most hand soaps is sodium myristate, which is the sodium salt of myristic acid (14:0).

$$CH_3CH_2CH_2CH_2CH_2CH_2CH_2CH_2CH_2CH_2CH_2CH_2CH_2\overset{\displaystyle O}{\overset{\|}{C}}-O^-Na^+$$

sodium myristate

While it is correct to describe soap as a type of salt, e.g., common table salt (NaCl), it is the combination of the salt structure, coupled with the properties of the long hydrocarbon chain of the anion, which gives soap properties quite different from that of other salts. Salts in general are very *hydrophilic*. They tend to dissolve in water readily with both cations and anions surrounded (solvated) by water molecules (Figure 7.6). Fats (or oils) are very *hydrophobic*. When oil and water are mixed, they separate into two layers because of the repulsion between water and the hydrocarbon chains of the oil. A soap molecule combines both the hydrophilic nature of the salt and the hydrophobic nature of an oil, and the result is a compound that is capable of interacting with both oil and water and, more important, acting as an intermediary to allow oil and water to mix. This is the most important function of a soap, since dirt invariably has some oil or grease incorporated in it and the oil increases the tendency of the dirt to adhere to surfaces and to prevent penetration by wash water.

Soaps are described as surface-active agents or *surfactants* because they permit water to penetrate the surface of an oil droplet and divide it into many fine droplets, which then become suspended in water. The suspended droplets are prevented from coalescing by interaction with soap in solution.

Milk is an analogous system in which the water-insoluble fat is kept in suspension by the action of milk protein (casein), which acts as a surfactant because of the presence of both hydrophobic and hydrophilic groups on the protein. It is actually more common to describe the casein as an emulsifier and the milk as an emulsion of oil in water. In effect, soaps may be regarded as emulsifiers although this term is generally not used in describing soaps.

The surfactant action of a soap can be explained on a molecular level by describing the arrangement of soap dissolved in water. Unlike simple salts, soap molecules do not exist separate from one another in solution but tend to form micelles (see Figure 16.5). *Micelles* are clusters in which the hydrophobic ends of soap molecules clump together in an arrangement that avoids the unfavorable interaction with water, while the hydrophilic salt-like ends of the soap molecules are on the surface exposed to the water. When cleaning, it is these micelles that interact with oil, with each micelle trapping a minute oil droplet, which is then washed away with the soap. Once the oil is removed, water is able to penetrate into the minute pores of a surface and very thorough cleansing is then possible.

Many advertisements for hand soaps call attention to the use of coconut oil and, indeed, coconut oil is considered the most suitable due to the unusually high content of lauric (12:0), myristic (14:0) and palmitic (16:0) acids. Because of the combination of both hydrophilic and hydrophobic properties in a soap molecule, it is necessary to strike a balance between the two in order to derive the greatest benefit. The best balance seems to come at the C-14 fatty acid. As the hydrocarbon chain is lengthened, the solubility in water decreases. Above C-18, the solubility becomes too low to be of any use. When the chain is shortened, the ability to suspend oil droplets is decreased.

Figure 16.5   The surfactant action of a soap micelle.

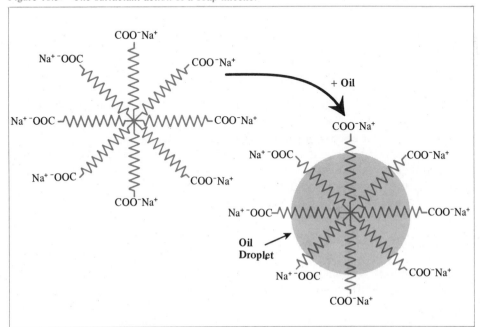

There is no natural oil source that has exclusively myristic acid, but coconut oil has a complement of fatty acids very close to C-14 and is the most natural compromise. The high degree of saturation (low iodine value) is an extra bonus for coconut oil because the susceptibility to oxidation is extremely low.

Tallow from both beef and lamb has often been used in making soap, sometimes in combination with coconut oil. Tallow is predominantly (>90%) palmitic (16:0), stearic (18:0), and oleic (18:1) acids, so it also makes a reasonably good soap. Since higher water temperatures are often used in commercial laundries, the reduced solubility of the long-chain fatty acids is overcome due to the increased solubility in hot water. This has made tallow a major source of soap for this application.

Castile soap is made primarily from olive oil. Since olive oil has almost none of the 10:0 and 12:0 fatty acids, this type of soap is reportedly less irritating to the skin than other soaps and is prized by some people for that reason.

Good lathering is an important characteristic in soaps. Here again the surfactant action is involved, since some soap molecules may tend to align along the surface of the water with the hydrocarbon chains directed toward the surface and the salt-like ends directed into solution exposed to the water. This tends to weaken the surface and promote foaming. Soaps made from coconut oil have the best lathering characteristics.

The density of soap is sometimes lowered by incorporation of air to make it float. Additives such as creams, perfumes, deodorants, abrasives and colorings are also in common use in making hand soaps. Hexachlorophene is an antibacterial agent that has been included in some soaps as well as shave creams, detergents, and even toothpaste. It has also been in common use in hospitals as a 3%

Hexachlorophene

solution that was used for bathing infants. It is very effective in preventing the spread of staphylococcus infections, but in 1972, some concern was generated over some harmful effects observed when hexachlorophene was fed to rats (5). Since then, hexachlorophene was limited to prescription usage by the FDA (6). It has been recommended that its use be restricted to epidemics of staphylococcus, which cannot be dealt with in any other way, although it is suggested for use in hand washing soaps used by hospital personnel (7).

Other Emulsifiers

Now that we have examined the action of soaps and casein on fats and oils, let us consider some additional emulsifying agents that are commercially very important in foods.

Egg yolk was described previously as being a major ingredient in making mayonnaise and salad dressings, because the yolk contains certain substances that act as emulsifiers and stabilize the oil-water emulsion present in these products. The primary compounds involved are the lecithins and some proteins known as lecithoproteins. Lecithins, cephalins, and certain other so-called phospholipids (also referred to as phosphatides) are commonly isolated from soybean oil, which contains about 2% phospholipids.

$$
\begin{array}{cc}
& \quad\quad O & & \quad\quad O \\
& \quad\quad \| & & \quad\quad \| \\
& R_1C-O-CH_2 & & R_1C-O-CH_2 \\
& \quad\quad\quad\quad | & & \quad\quad\quad\quad | \\
& O \quad\quad | & & O \quad\quad | \\
& \| \quad\quad | & & \| \quad\quad | \\
& R_2C-O-CH & & R_2C-O-CH \\
CH_3 \quad\quad & O \quad\quad | & & O \quad\quad | \\
+ | \quad\quad & \| \quad\quad | & & \| \quad\quad | \\
CH_3-N-CH_2CH_2O-P-O-CH_2 & {}^+NH_3CH_2CH_2O-P-O-CH_2 \\
| \quad\quad\quad\quad | & & | \\
CH_3 \quad\quad\quad O_- & & O_- \\
\end{array}
$$

a lecithin   a cephalin

Mono- and diglycerides are also important emulsifiers that are sometimes added to shortenings. They assist in formation of a stable water-oil emulsion in cake batters. This allows for addition of more water to the batter and the greater water content permits incorporation of more sugar.

$$
\begin{array}{cc}
HO-CH_2 & HO-CH_2 \\
| & | \\
& O \\
& \| \\
HO-CH & RC-O-CH \\
| & | \\
O & O \\
\| & \| \\
RC-O-CH_2 & RC-O-CH_2 \\
\end{array}
$$

a monoglyceride   a diglyceride

Other Uses of Soap

Soaps of other metals such as aluminum, cadmium, cobalt, magnesium, nickel, zinc, and lead are used in a variety of applications. They have very low water solubility, which permits use in applications where water might dissolve away a sodium soap. Copper soaps have been used as fungicides in paints in which their green color is not a problem. Zinc soaps are less active but have very little color. Many soaps form gels with mineral oils and make excellent lubricating greases, with a formulation containing 50% soap being typical.

DETERGENTS

The development of the "wash day miracle" has been a series of peaks and valleys for chemists engaged in the pursuit. Several ingenious products have been

Figure 16.6 Detergents. (Photo by James J. Kane, Jr.)

formulated and later scrapped due to environmental or other problems. In fact, it is generally conceded that the consumer is currently in a valley when it comes to choosing a detergent, since none of the currently available options is without some serious drawback. In the following paragraphs, the sequence of developments is traced through some of the modifications that have been made in detergents in the last thirty years in response to problems that have been encountered.

Laundry detergents first came into use due to problems encountered with ordinary soaps, which are unsatisfactory for laundering in hard water. Water "hardness" is attributable to the presence of metal ions, primarily calcium (Ca^{+2}) and magnesium (Mg^{+2}) ions, which form insoluble precipitates when combined with fatty acid anions. A "bathtub ring" is such a precipitate. The precipitate may be tolerable in the bathtub or wash basin, but quite intolerable when washing clothes, since the precipitates often end up as deposits on the clothes. For this reason, the chemical structure of detergents differs slightly from soaps, although they must also be efficient surfactants.

The solution for the precipitation problem was the development of the alkyl benzene sulfonate (ABS) detergent. It is synthesized from propylene, which is available from petroleum. The reactions used for the synthesis follow.

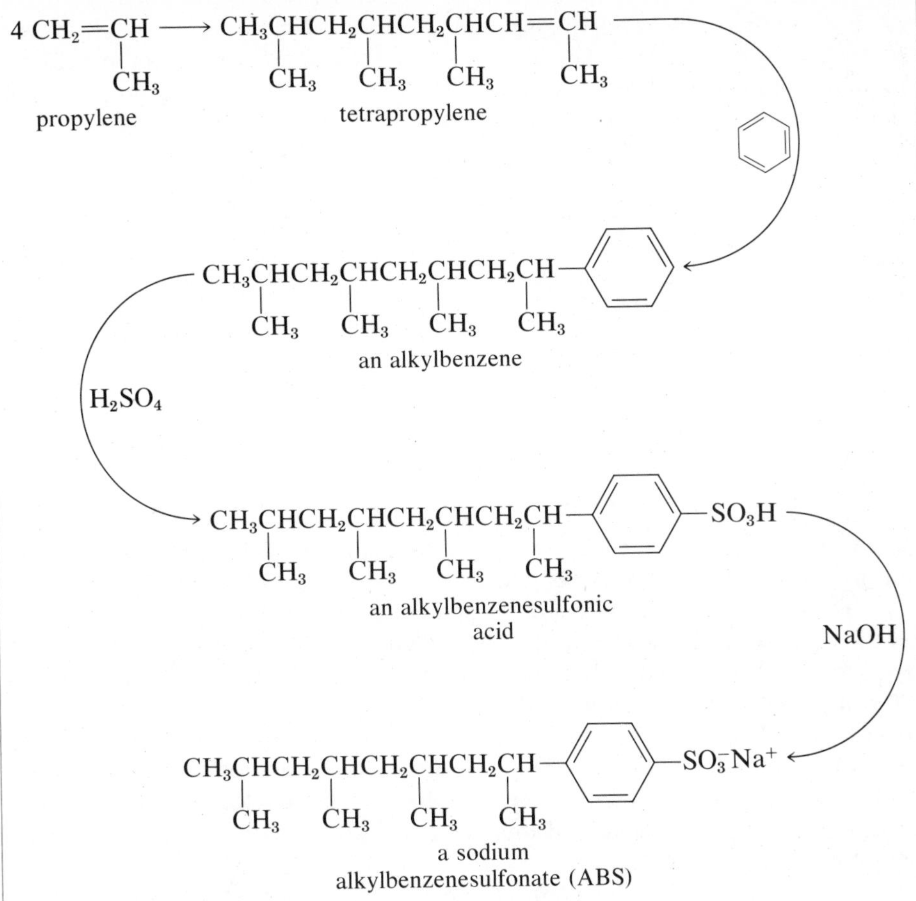

$$4\ CH_2{=}CH \longrightarrow CH_3CHCH_2CHCH_2CHCH{=}CH$$

propylene   tetrapropylene

$$CH_3CHCH_2CHCH_2CHCH_2CH-\!\!\left\langle \text{benzene} \right\rangle$$

an alkylbenzene

H_2SO_4

$$CH_3CHCH_2CHCH_2CHCH_2CH-\!\!\left\langle \text{benzene} \right\rangle-SO_3H$$

an alkylbenzenesulfonic
acid

NaOH

$$CH_3CHCH_2CHCH_2CHCH_2CH-\!\!\left\langle \text{benzene} \right\rangle-SO_3^-Na^+$$

a sodium
alkylbenzenesulfonate (ABS)

Like soap, this synthetic detergent (syndet) is both hydrophilic and hydrophobic so that it can stabilize water-oil emulsions, but unlike fatty acid salts, the sodium can be replaced by calcium or magnesium and the detergent remains in solution.

Unfortunately, merely remaining in solution does not assure that the detergent will exhibit its full surfactant action, since the Mg^{+2} and Ca^{+2} ions are able to bind to the ABS anions and remain in solution but restrict the ability of the anion to suspend oil droplets. In some areas, water is very hard and the problem of detergents is intensified. Water softeners have been installed in some homes, but are usually regarded as a luxury item. After removal of the hardness cations or when water is naturally soft, soap is just as effective in laundering as a detergent.

In any case, the vast majority of homes do not have water softeners but do have some degree of hardness in the water. As a result, in 1947, Proctor and Gamble marketed Tide, which combined the surfactant (15–20%) with a substance(s) called a "builder," which chemically softens water by binding to calcium and magnesium so that the detergent remains free to function as a surfactant. From that point on, the popularity of synthetic detergents (with builders) increased dramatically to where they surpassed soap in sales in 1953 and have since taken over more than 90% of the laundry product market.

The major substances that have been used as builders are sodium pyrophosphate and sodium tripolyphosphate, which are shown as they exist when binding Ca^{+2} or Mg^{+2}. Together or separately these builders are commonly referred to as "phosphates" and, in addition to improving the washing properties of

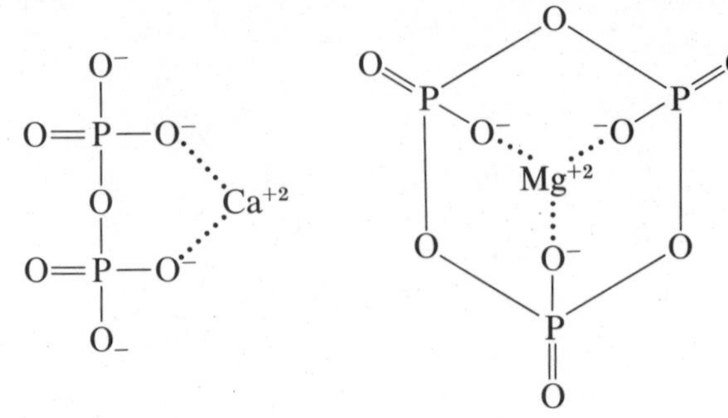

calcium
pyrophosphate

magnesium
tripolyphosphate

detergents, they are also nontoxic. This makes the product particularly convenient for use in the reach of small children.

The combination of ABS and phosphates was the major detergent formulation throughout the 1950s, but in the early 1960s, an environmental problem was traced to these detergents when sudsy foam was found to be building up in the water in many rivers, streams, and even as it came out of the faucet in many homes. It was found that the alkylbenzene sulfonates in detergents do not decompose fast enough during the time they are exposed to microorganisms in sewage treatment plants. In other words, they are not biodegradable, which means they remain in the water and may find their way into streams or back into the water supply of homes in some areas.

Consequently, the ABS-containing detergents disappeared from the market by mid-1965 and were replaced by detergents containing LAS (linear alkylsulfonates) surfactant, which is readily biodegradable and nonpolluting. The structure of LAS is shown below. Like ABS, the LAS surfactant is synthesized from

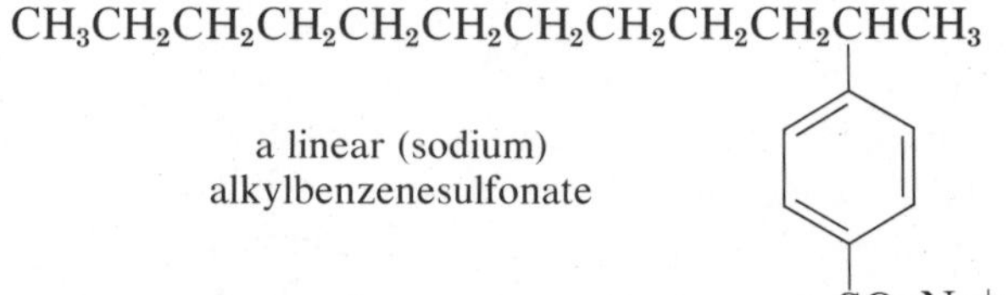

$$CH_3CH_2CH_2CH_2CH_2CH_2CH_2CH_2CH_2CH_2CHCH_3$$

a linear (sodium)
alkylbenzenesulfonate

$SO_3^- Na^+$

starting materials derived from petroleum. The important distinction between the ABS and LAS structures is the branched versus linear hydrocarbon chains, respectively. Like the fatty acid chains, the linear hydrocarbon chain of the LAS makes it susceptible to attack by microorganisms that are unable to metabolize the branched chains found in the ABS detergents. A biodegradable detergent is often called soft, whereas a nonbiodegradable detergent is biochemically hard. The terms *hard* and *soft* used in this context are not to be confused with the same terms as they are used to describe water.

Figure 16.7   The use of the branched-chain alkyl benzene sulfonate (ABS) detergents often caused foaming of fast-moving rivers and streams. This problem was solved by the introduction of biodegradable linear alkyl benzene sulfonate (LAS) detergents. (Photo courtesy of USDA.)

In 1967, attention was focused on another environmental problem that has been attributed to the phosphate builders in detergents. These substances have been blamed for supplying phosphorus and encouraging the growth of aquatic plants (algae) in many lakes. In Chapter Nine, it was seen that phosphorus is one of the major nutrients for plant growth. It is frequently the one limiting growth, since potassium is often abundant and blue-green algae are able to fix nitrogen and carbon (photosynthesis). As nutrients accumulate, a lake may be overrun by "algal blooms"—a condition called *eutrophication*. At first, this may be beneficial to a lake, since the growing algae, while undergoing photosynthesis, may increase the supply of oxygen available to other aquatic plants and animals. As the growth progresses, the oxygen produced by algae at the surface escapes into the atmosphere. Since these and all other algae carry out respiration and cause a net drain on the oxygen supply, the demand, called the *biochemical oxygen demand* (BOD), may exceed the supply and cause the destruction of much aquatic life. The problem is accentuated as algal blooms block out sunlight required by other photosynthesizing aquatic plants. Even dead algae are troublesome, since bacteria that act during the decay of dead algae may also consume oxygen. If carried to extreme, a lake may deteriorate far enough to lead observers to describe it as dead. An article in *Consumer Reports* entitled "Dead Lakes: Another Washday Miracle" summarizes the problem (8). Free-flowing streams and rivers, which were the site of the foam problem, do not suffer from oxygen deficiency, since the movement of the water allows for incorporation of atmospheric oxygen.

The obvious way to minimize the growth of algae is to limit the supply of nutrients, particularly phosphorus. This could conceivably be done in sewage treatment plants by engineering them to remove phosphates from the sewage, but the level of purification required is known as tertiary sewage treatment and, as yet, is only in the experimental stage. Consequently, attention was focused on laundry detergent, since it was estimated that perhaps 40% or more of the phosphorus entering the aquatic environment came from detergents at the peak of their use. A total ban on phosphates was undesirable, since the convenience, efficiency, and safety of the synthetic detergent plus builder is great, but it did seem appropriate to cut down on the amount used, which typically ranged from 20–65% phosphate (as sodium tripolyphosphate) or even higher for some presoaks. Thus, in mid-1970, the major detergent producers announced plans to replace a large percentage of the phosphate with sodium nitriloacetate (NTA), which also effectively binds hardness ions and functions as a builder.

$$\begin{array}{c} \overset{\displaystyle O}{\overset{\|}{}} \\ CH_2C-O^-Na^+ \\ \diagup \\ \overset{\displaystyle O}{\overset{\|}{}} \\ N-CH_2C-O^-Na^+ \qquad \text{sodium} \\ \diagdown \qquad\qquad\qquad \text{nitriloacetate} \\ \overset{\displaystyle O}{\overset{\|}{}} \\ CH_2C-O^-Na^+ \end{array}$$

Unfortunately, by the end of 1970, NTA had already passed out of favor due to the finding that NTA also binds (sequesters) other toxic metal ions such as cadmium and mercury and might release these ions in a location (e.g., across the placental barrier) where the consequences might be very serious.

Since then, some new and some old fashioned ingredients have been used as partial or total replacements for phosphate builders. The simplest is *washing soda,* sodium carbonate (Na_2CO_3), which is water soluble but causes precipitation of the calcium and magnesium ions as calcium carbonate and magnesium

$$Ca^{+2} + Na_2^+CO_3^{-2} \rightarrow CaCO_3 + 2\ Na^+$$
$$Mg^{+2} + Na_2^+CO_3^{-2} \rightarrow MgCO_3 + 2\ Na^+$$

carbonate, respectively. Unfortunately, precipitates formed in a washing machine can deposit on clothes and perhaps even clog the machine. Nevertheless, one major manufacturer of washing soda staged a major "soap and soda" campaign as the answer to the environmental problem. One major drawback to the use of washing soda is the high alkalinity due to the reaction with water. This makes them quite hazardous to use around small children.

$$Na_2CO_3 + H_2O \rightleftharpoons NaOH + NaHCO_3$$

Silicates and soap are other substances that can act as builders in the same way as washing soda, by precipitation, but the precipitate (curd) is undesirable. Nevertheless, in the days prior to the use of detergents, housewives routinely used extra soap that functioned as both builder (by precipitation) and surfactant. Since the amount of hardness in the water varies greatly, one can tell how much soap (or soda) is necessary by observing the amount required to achieve good sudsing. It has been reported that washing with soap also destroys the flame-retardant characteristics of clothing marketed for small children.

Other substances in limited use as builders in detergent formulations are metasilicates, citrates, perborates, and polycarbonates. All have drawbacks such as high alkalinity, poor biodegradability or other environmental uncertainty, adverse effects on flame retardant textiles, or toxicity.

Consequently, a clear, inexpensive choice is not available. Installation of water softeners is an expensive solution which allows the continued use of soap.

Enzyme Detergents

In the late sixties a few pre-soaks and detergents appeared on the market containing enzymes. The enzymes used are very stable ones that can be readily isolated from microorganisms. The enzymes used were both proteolytic (protein-cleaving) and lipolytic (fat-cleaving) and act by breaking down proteins and fats that may bind stains to clothes. The effectiveness of these enzyme-containing products was not significantly better than standard detergent formulations and this, plus some concern over their safety, has caused their disappearance from the market. The safety problem was apparently insignificant for home use, although the enzyme dust in detergent factories caused workers some problems.

DRAIN CLEANERS AND OVEN CLEANERS

Finally, let us consider two products that are related in both composition and mode of action. Clogged drains are generally attributable to fat. In the early discussion of soaps, we saw that strong alkali will react with fat (triglycerides) to produce soap plus glycerine. It is exactly this fact that explains the use of solid sodium hydroxide and concentrated water solutions of sodium hydroxide as drain cleaners, since some of the fat may be converted to soap in the clogged drain. The newly formed soap can be washed away and will also help emulsify any of the remaining fat.

When solid sodium hydroxide is used as a drain cleaner, it offers the additional advantage of liberating a large amount of heat when it comes in contact with water in the drain. The heat alone may be sufficient to break up the clog by melting some of the fat.

Oven cleaners also contain sodium hydroxide. They are usually dispensed in aerosol form with thickeners and propellants. Here again, a greasy residue may be converted to soap on the oven surface and is easily washed away.

It cannot be overemphasized that any consumer product containing sodium hydroxide or related compounds are potentially very hazardous, particularly to the eyes. All too often, users of drain cleaners have become impatient and resorted to using mechanical means for unclogging a drain shortly after adding a drain cleaner and splashed the alkali into the eye, with serious consequences.

SUMMARY

Tooth enamel is hydroxyapatite, $Ca_5(PO_4)_3OH$. It is susceptible to attack by acids due to the release of hydroxide ions by demineralization. Acids are readily available as the products of fermentation by microorganisms found in dental plaque.

Fluoride may be incorporated into tooth enamel by demineralization followed by remineralization, to produce fluoridated hydroxyapatite, which is much more resistant to acid due to lower solubility and decreased tendency to release hydroxide ions. Fluoride may be applied topically by dentists and by individuals who use fluoride toothpastes. Stannous fluoride (SnF_2), sodium fluoride (NaF), and monofluorophosphate, MFP (PO_3F^{-2}), are the common sources of fluoride in toothpastes. The last of these provides a more prolonged release of fluoride ions, particularly when acidity is high.

Alloys of silver, tin, copper, zinc, and mercury are called amalgams. They are used by dentists for filling cavities.

Soaps are metal salts of fatty acids. As such, they are both hydrophobic and hydrophilic and act as surfactants (emulsifiers) by forming micelles to permit oil and water to mix and allow efficient cleansing. Soaps are made by treating triglycerides with alkali (usually NaOH). A soap made from the 14-carbon fatty acid has the best properties. The hydrocarbon chain is short enough to allow good solubility, but long enough to give good surfactant properties. Of all the natural oils and fats, coconut oil is the most preferred, since it has an unusually high content of the 12, 14, and 16-carbon fatty acids.

A number of other compounds are important emulsifiers. Egg yolk contains lecithin and other emulsifiers that are required for stability of various oil-water emulsions, such as that found in mayonnaise.

Detergents are formulated with surfactants, which differ from plain soaps due to the problem of hard water that causes soaps to precipitate. The alkyl benzene sulfonate (ABS) was the first synthetic detergent (syndet). It did not precipitate in hard water, but neither did it clean very well in hard water until it was combined with a substance(s) called a builder, which chemically softens water by binding to the hardness ions (calcium and magnesium). With these ions effectively removed, the detergent is left free to function as a surfactant.

Unfortunately, two environmental problems plagued the detergents formulated with ABS plus builder. The ABS is nonbiodegradable due to branching of the hydrocarbon portion of the molecule. This led to pollution of rivers and streams, which were sometimes found to foam vigorously. Replacement of the ABS detergent with the linear alkyl sulfonates, LAS, solved the problem. The most successful builders used in syndets are phosphates, but the phosphates contribute to eutrophication of waterways. Other builders have been tried as substitutes, but none has been very successful.

PROBLEMS

1. Give a definition or example of each of the following:
   a) demineralization
   b) remineralization
   c) hydroxyapatite
   d) fluoridated hydroxyapatite
   e) dental caries
   f) dental plaque
   g) amalgams
   h) soap
   i) hydrophobic
   j) hydrophilic
   k) surfactant
   l) micelle
   m) hard water
   n) hard detergent
   o) eutrophication

2. How does plaque affect teeth?

3. Why is sugar harmful to teeth?

4. Describe the mechanism for the incorporation of fluoride into tooth enamel.

5. How does fluoride affect tooth decay and why?

6. How safe is fluoride in drinking water? How important is concentration?

7. Write the reaction for the release of fluoride from MFP.

8. What are alternate methods of obtaining fluoride for the teeth? Compare these methods to fluoridated water.

9. How can soap be both hydrophilic and hydrophobic and why is that important?

10. Why is coconut oil especially good for making soaps?

11. What is the advantage of castile soap?

12. How do detergents differ from soaps?

13. Why were ABS-containing detergents removed from the market?

14. Why are phosphates desirable in detergents? Why are they undesirable?

REFERENCES

1. Scherp, H.W. "Dental Caries: Prospects For Prevention." *Science 173* (1971): 4003.

2. Duff, E.J. "Orthophosphates *XIV,* Thermodynamic Factors Influencing The Stability Of Dental Enamel To Decay." *Caries Research 7* (1973): 70.

3. "U.K. Study Vindicates Water Fluoridation." *Chemical and Engineering News,* 26 January 1976, p. 13.

4. *McGraw-Hill Encyclopedia of Science and Technology,* New York, McGraw-Hill, 1977, p. 443.

5. "Hexachlorophene." *Chemistry 45* (9) (1972): 24.

6. "FDA Limits Availability of Hexachlorophene." *The Journal of the American Medical Association 222* (4) (1972): 421.

7. "Study of Hexachlorophene Effects Continues." *The Journal of the American Medical Association 224* (2) (1973): 176.

8. "Dead Lakes: Another Washday Miracle." *Consumer Reports 35* (9) (1970): 528.

O NE of the more controversial issues among scientists in recent years was the matter of mass vaccination against a particular type of influenza commonly known as the swine flu. The question of need for vaccination against any one disease may be settled relatively quickly (although often after the fact) when the seriousness of the threat is known. Unfortunately, the same question must be asked again and again as outbreaks of different diseases occur. Influenza (the flu) outbreaks occur periodically, about every 10–15 years. There was the swine flu pandemic (global epidemic) of 1918–19, which killed millions, and was followed by major outbreaks of other types of influenza in 1933, 1947, 1957 (Asian flu), and 1968 (Hong Kong flu). A very minor outbreak of so-called swine flu occurred in 1976, but its presumed similarity to the 1918–19 disease was the cause for much concern.

The basis for this concern is a major topic to be considered in this chapter along with the chemical basis for the action of the various types of vaccines. The overall theme of this chapter is the chemistry underlying the development of disease and its prevention.

DISEASE

In many instances, disease can be traced to microorganisms that invade the body. This is the basis of the so-called *germ theory* of disease. In earlier and later chapters, coverage is directed to the role of microorganisms in such things as alcoholic fermentation (Chapter Eleven), baking (Chapter Twelve), cheese making (Chapter Thirteen), production of antibiotics (Chapter Eighteen), as well as the less desirable role that microbes play in food spoilage (Chapter Fourteen).

Once again, we have to look at a negative side of the story and consider the role played by bacteria and viruses in causing disease. With that information in hand, we can then proceed to consider how one can go about preventing certain diseases. The importance of good hygiene cannot be overemphasized as a means of simply preventing exposure to harmful microbes. In Chapter Eighteen, we will see how antibiotics may interfere with the development of microorganisms both in the laboratory and in the body.

MICROORGANISMS

Most microorganisms are capable of life by themselves as long as there is an adequate supply of nutrients. Viruses are an exception. They are purely parasitic

IMMUNOCHEMISTRY

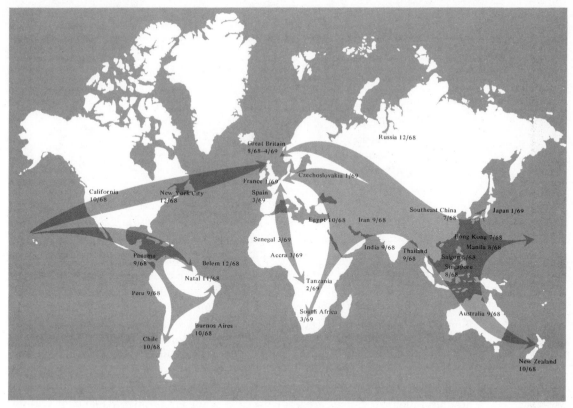

Figure 17.1   The 1968–69 itinerary of the Hong Kong flu. (Adapted with permission from E. D. Kilbourne, *Natural History*, January 1973. Copyright © 1973, The American Museum of Natural History.

particles that develop and reproduce only by invading other organisms—microorganisms, plants, or animals.

Both viruses and other microorganisms may enter the body by way of the respiratory tract (by inhalation), the digestive tract, the genitourinary tract (e.g., venereal diseases), or through the skin via open wounds or insect bites (see Chapter Ten for examples). If the microbe becomes established, it may later enter the bloodstream. This paves the way for invasion of tissues throughout the body. The process by which an organism becomes established is called *infection*. Some of the major infectious diseases are listed in Table 17.1 and are subdivided into bacterial and viral diseases.

Antibiotics are often, although not always, effective against bacterial infections, but they are worthless against infections caused by viruses. The mode of action of some of the major antibiotics will be considered in the following chapter but, as expected, they selectively attack bacterial cells rather than animal cells because of chemical differences between the two. In viral infections, the virus particles attach to normal cells, called host cells, and take over the host cells. In other words, virus particles merely assume control of normal cells and, since antibiotics are ineffective against normal cells, the antibiotics cannot attack virus-infected cells and combat an infection.

TABLE 17.1 MAJOR INFECTIOUS DISEASES

BACTERIAL	VIRAL
Diphtheria	Measles (rubeola)
Tetanus	German measles (rubella)
Whooping cough (pertussis)	Mumps
Tuberculosis	Poliomyelitis
Cholera	Rabies
Typhoid fever	Yellow fever
Plague	Influenza
Syphilis	Common cold
Gonorrhea	

Even some bacterial diseases are not altogether susceptible to control by antibiotics, and so these and many viral diseases are cause for great concern. Fortunately, many such diseases confer a resistance or *immunity* toward repeated infection. For example, once a person has had measles, mumps, polio, or any of a number of other diseases, he develops a lifelong immunity to infection by that same disease. A number of infections do not confer immunity. Examples are syphilis, gonorrhea, and most staphylococcal infections. The common cold may actually result from over one hundred different viruses and, while immunity may result from infection by any one of these, there are always plenty more to go around. Influenza viruses are a special case. They are discussed in a later section.

IMMUNOCHEMISTRY

Immunology is the complex field of investigation of the development of immunity, which is called the *immune response*. Some literature subdivides immunology into immunobiology and immunochemistry, and attention will be directed to the latter.

"Look, baby, I haven't much time—I'm only a 24-hour virus, you know."

Copyright © 1963, Joseph G. Farris. Reprinted by permission.

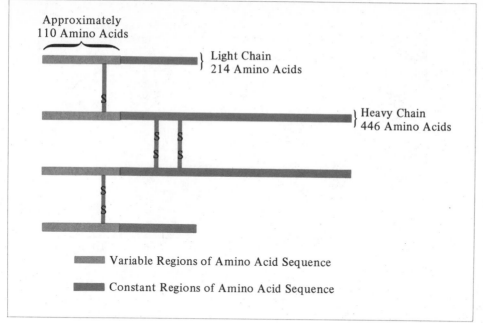

Figure 17.2   An antibody protein (an immunoglobulin). The antibody molecule is a unit consisting of four protein chains (two heavy chains and two light chains) linked by disulfide bonds (—S—S—). All antibody molecules are the same in the portion shown in black, but each type of antibody is unique in the region of the molecule shown in color.

The chemical basis of immunity lies in the body's ability to produce **antibodies,** which are complex proteins that build up in the bloodstream in response to an infection. Substances that stimulate the synthesis of antibodies are called **antigens.**

The antibody proteins are called *globulins*. The globulins may be separated into alpha, beta, and gamma globulins but the antibody molecules are primarily gamma globulins. Because of their importance in the immune response, these gamma globulins are often called *immunoglobulins* and are symbolized Ig. The Ig proteins have the ability to neutralize an invading bacteria or virus particle and, thus, prevent repeated infection. Unfortunately, when the initial infection occurs, it normally takes several days for the immunoglobulins to develop to a level that is effective in combating a disease. During that time, acute symptoms or complications may have serious effects. For example, whooping cough is accompanied by very severe symptoms and is fatal to a large percentage of infants who contract the disease under one year of age. Some of the symptoms result from the release of a toxin by disintegrating bacterial cells. Diseases such as mumps and German measles (rubella) are very mild in infants, but mumps can be quite severe in adults and rubella in a pregnant woman is a serious threat to the fetus.

VACCINES

In order to provide protection against the diseases mentioned above, as well as several others, attempts have been made to mimic the natural immune response.

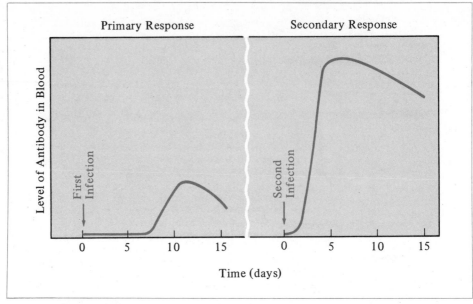

Figure 17.3 The immune response to an infection.

Although antibodies develop slowly upon initial infection, following the disease, a subsequent exposure evokes an immediate and very high level development of antibodies that can neutralize the invading organism and prevent it from causing a full infection.

The cells that are responsible for the synthesis of immunoglobulins are sometimes called *memory cells,* since they seem to remember how to act against substances that had previously caused a particular infection. A typical immune response curve is shown in Figure 17.3. It illustrates how the memory response, at the time of a second infection, is characterized by more efficient antibody production—both faster and in larger amounts. Recall that a substance that evokes an immune response is called an *antigen.*

A vaccine causes the same immune response as the disease. It simulates the normal infecting organism and stimulates the production of antibodies. A subsequent infection by the same organism causes the same rapid and high level immune response—the synthesis of immunoglobulins, which neutralizes the invading microorganism.

Vaccines may take any of three forms.

1. Killed organisms (bacteria or viruses). Examples are vaccines against whooping cough, cholera, rabies, influenza, polio (Salk vaccine).
2. Live attenuated organisms (usually viruses). Examples are vaccines against measles, German measles, mumps, polio (Sabin vaccine).
3. Inactivated toxins called *toxoids* (from bacteria). Examples are vaccines against diphtheria, tetanus.

A killed organism retains many of the chemical features of the live form and, therefore, evokes an immune response resulting in antibody synthesis. As usual, a

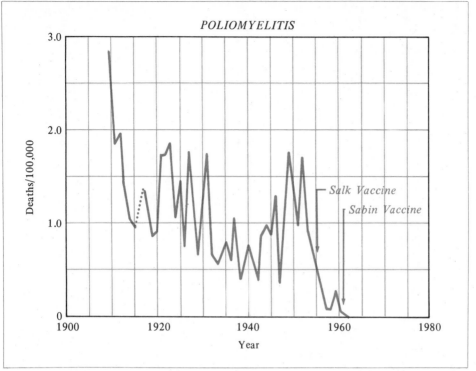

Figure 17.4 Poliomyelitis: Deaths per 100,000 population and impact of Salk and Sabin vaccines. (From *Pills, Profits, and Politics*, Milton Silverman and Philip R. Lee, Copyright © 1974 by The Regents of the University of California. Adapted by permission of the University of California Press.)

later exposure is greeted with a greatly enhanced level of antibodies—the memory response.

A number of viruses have been isolated in hybrid forms. In some cases, these viruses are sufficiently like the normal dangerous type to cause antibody production but lack the ability to cause a serious infection. Such viruses can be given live and are called *live attenuated virus* vaccines. When available, these vaccines are generally superior because they do cause an infection, with either very mild symptoms or no symptoms at all, and a real infection evokes a more efficient immune response resulting in a better and longer-term immunity.

The classic example of a live attenuated virus is that used in the oral polio vaccine. The Salk vaccine (available since 1955) contained killed virus particles and was administered by injection. It has largely given way to the oral, live, Sabin vaccine (available since 1961), which gives a more effective immunity. The normal course of a polio* infection is infection of the digestive tract, followed by invasion of the bloodstream and attack on the brain resulting in paralysis. The attenuated live virus is capable of infecting the digestive tract, where it multiplies and stimulates antibody production, but it cannot invade the bloodstream. The symptoms of the one-stage infection, if any, are very mild.

* There are actually three distinct polioviruses. The vaccine that is in use is a triple vaccine, which provides immunity against all three viruses.

TABLE 17.2   WHEN TO GET A VACCINATION AND WHO NEEDS IT*

DISEASE	AGE WHEN CHILD SHOULD GET THE SHOT	DOSAGE	BOOSTER NEEDED?	SHOULD ADULTS GET IT, TOO?	REMARKS
Diphtheria	About 2 months	Shots, including tetanus and whooping cough, repeated at 4 and 6 months.	At 18 months and again when entering kindergarten or first grade and ages 14 to 16.	Adult toxoid is available for those facing unusual risk.	A booster may be recommended for a child exposed to the disease.
Whooping cough (pertussis)	About 2 months	As above	At 18 months and again when entering kindergarten or first grade.	Not indicated	Protection may be recommended for exposed adults with chronic illnesses.
Tetanus	About 2 months	As above	As above, at ages 14 to 16, then every ten years.	Every ten years; after an injury only if the wound is deep or dirty and last shot was over five years ago.	Consult a doctor for tetanus-prone injuries, deep wounds contaminated by dirt.
Influenza	About 3 months	Currently not recommended except for children with disabling or chronic disease, then two shots, spaced two months apart, for those previously unimmunized.	Once a year.	One dose each fall for older people and those with chronic heart and lung diseases, etc.	Immunization should be completed by mid-November.
German measles (rubella)	12 months to puberty	One shot of live vaccine on a priority basis to pre-schoolers and children in early grades.	See remarks.	Not pregnant women or those who might be pregnant within two months of shot.	Most adults are immune, but a woman of child-bearing age can be tested for immunity.

Disease					
Measles (rubeola)	12 months or older	One dosage	Only if first shot is at age 6 months to 1 year, then at 12 to 16 months.	Rarely, since most adults already immune; a test can check this.	A measles/rubella and a measles/mumps/rubella vaccine can be given at 1 year.
Mumps	Not under 12 months, but particularly recommended before puberty.	One shot of live mumps vaccine	Protection lasts at least several years.	Adults, particularly men who haven't had mumps, though half are probably already immune.	To prevent spread, isolate patient until swelling is gone.
Polio	About 2 months	Triple oral (Sabin) vaccine, three doses, the second two spaced eight weeks apart and a fourth at 18 months.	Triple vaccine preparation on entering school.	Not for adults living in U.S.	Systematic immunization of all exposed people if epidemic occurs.
Typhoid fever	6 months or older	Recommended in U.S. only when there is special risk. Three shots at weekly intervals or two shots four or more weeks apart.	Only in cases of repeated exposure.	Same as foregoing	Sometimes recommended for foreign travel where exposure may occur.
Smallpox	Now recommended only for people in health-care field and travelers to countries that require it.	One vaccination	See next column.	At three-year intervals for travel abroad or before exposure to unusual risk.	Vaccination or revaccination is recommended for people with presumed exposure to smallpox.

* This table is based on data of the U.S. Center for Disease Control and the American Academy of Pediatrics. Physicians may alter the schedules to fit different situations. Adapted with permission from *Changing Times*, copyright © 1974, Kiplinger Washington Editors, Inc., September 1974.

The third type of vaccine is the *toxoid* or inactivated toxin. In this case, one has to grow the microorganism and isolate the toxic chemical substance (usually a protein) that the organism produces. The toxin is then inactivated, normally by treatment with some chemical, and the resultant toxoid may be safely injected. Since the toxoid has much the same chemical structure as the toxin, it will stimulate antibody production and provide protection in the event of a later infection by the microorganism when it releases the toxin.

Table 17.2 is a typical vaccination schedule. A single diphtheria toxoid-tetanus toxoid-pertussis* (killed organisms) vaccine combination, designated DTP, is normally given to combat these three infections.

There have been numerous expressions of concern in recent years over the high percentage of children who have not been vaccinated against the common diseases. This development is a logical consequence of the greatly reduced incidence of many of these diseases. The hazard seems small even for an extremely contagious disease like measles, if most people are immunized and can resist development of the disease. In other words, a person who is immunized against measles not only will not develop that disease but also cannot transmit the disease. Therefore, it would seem that not every single individual needs to be vaccinated against measles. Unfortunately, there are two flaws in this line of reasoning. For one thing, there has been increased concern expressed by the United States Public Health Service (USPHS) and other sources that a large, and in some cases increasing, percentage of the population is not being adequately vaccinated. As of 1974, almost 40% of preschool children had not been vaccinated against polio. The figures also show that approximately 35%, 40%, and 25% of preschool children have not been vaccinated for measles, rubella, and DTP†, respectively. Thus, while a person may expect to be safe from disease because everyone else has been vaccinated, the statistics do not support this kind of thinking.

For another thing, there is the chance that enough people may take the vaccines to retard development of some diseases, and in the process allow a large number of unimmunized persons to reach older age without acquiring immunity. This presents the very serious possibility of an epidemic among these persons at an age when some diseases are considerably more serious than they would have been in childhood.

In any case, there have been reports of outbreaks of polio, diphtheria, and other diseases in various locations in the United States in recent years, particularly in states along the Mexican border.

Antitoxins: Passive Immunity

Antibodies can be regarded as *antitoxins,* since they act against substances that may exhibit toxic effects. Generally one thinks of antibodies that are produced by an animal naturally or by artificial immunization (vaccination) and provide subsequent immunity for the animal. In addition, it is possible to transfer antibodies from one animal to another or from animals to humans. In other words, it is pos-

* whooping cough

† Diphtheria-tetanus-pertussis (whooping cough).

sible to immunize an animal (e.g., a horse), bleed the animal, isolate the gamma globulins, transfer them to a human and provide that person with immunity. Such a procedure is said to confer *passive immunity* to the recipient as opposed to the *natural immunity* that results when an individual synthesizes his own antibodies. Passive immunity is only temporary (typically only four to six weeks) and provides no increased capacity for combating subsequent infections. In other words, there is no memory since the individual does not synthesize his own antibodies in the first place. In short, antitoxin-induced, passive immunity is a poor choice when there is a choice, but there are times when there is no reliable alternative. A major example is the case of an individual with an active tetanus infection or an unimmunized person with a tetanus-prone wound. Administration of tetanus antitoxin may be a life-saving treatment. The treatment of botulism and gas gangrene are also examples of infections for which antitoxins are employed.

Many individuals are staunch supporters of breast-feeding for newborn. Among their arguments is the claim that the child acquires antibodies in the mother's milk, and thus has a greater resistance to any diseases toward which the mother carries antibodies. This would be a special case of passive immunity. *Natural passive immunity* is gained by passage of preformed antibodies through the placenta from mother to fetus. Like all passive immunity, it is temporary (4 to 6 months).

As for passive immunity obtained by breast feeding, it appears to be significant only for a few days following birth, during which time the milk does contain a high level of the antibody proteins. In addition, there is a problem of absorption of the immunoglobulins by the child. Since the digestive system of the newborn infant is poorly developed, breastfed babies may absorb the antibody proteins, but as the digestive system develops there would be a greater likelihood that these proteins would be broken down by the digestive system even if they were still present in the milk.

Rabies

Rabies is an unusual infection. It, too, may be treated with an antitoxin, but rabies has an unusually long development time (incubation period) of 1 to 3 months, which provides an adequate period for active immunization using a killed virus given in daily injections for 2 to 3 weeks. Once symptoms have developed, there is no effective treatment and death invariably occurs. Very often, both antitoxin and vaccine are administered because of the potential severity of the disease.

Influenza

It was noted earlier that the common cold may be attributed to well over a hundred different viruses. Needless to say, this makes immunization impractical. To a far lesser degree, influenza viruses exist in different forms, usually symbolized as types A, B, or C. Of these, the type A virus is the one that is normally associated with influenza epidemics, such as the Asian flu and the Hong Kong flu. This would suggest that it is possible to produce a vaccine against the type A flu virus and, indeed, there are flu vaccines available. Unfortunately, the type A virus is not a single virus but several, designated A_1, A_2, etc. For example, it was the A_2 virus that caused the Asian flu beginning in 1957.

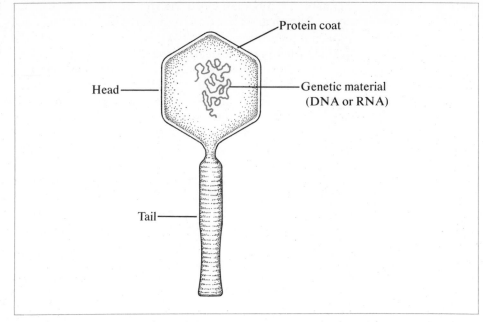

Figure 17.5 A virus particle.

As noted earlier, viruses cannot be treated with antibiotics, although antibiotics are sometimes used to prevent complicating secondary bacterial infections (e.g., pneumonia). The resistance of a virus to antibiotics is attributed to the parasitic nature of viruses. Since a virus is dependent on other cells, a virus is able to exist as a relatively simple particle, which is often described as strands of genetic material, i.e., DNA or a closely related substance symbolized as RNA, surrounded by a protein overcoat. Figure 17.5 is a common representation of the structure of an influenza virus particle, in which the core of genetic material is surrounded by a coating of layers of protein. The outer layer contains proteins that are responsible for much of the difficulty in controlling influenza.

Before pursuing the importance of the outer protein layer, let us consider the process of infection by a virus particle as diagrammed in Figure 17.6. The first step involves binding of the virus particle to the host cell, after which the genetic material of the virus enters the host cell and takes over the chemical functioning of the cell. As usual, the genetic material directs the synthesis of proteins, including enzymes that control metabolism and other chemical processes (see Chapter Two). The situation is analogous to an administrative official taking over an office, getting rid of all the old workers and replacing them with people who will do things to benefit the official. In the case of a virus infection, the cell machinery is turned over to the production of new virus particles that are then released for attack on other cells.

In living organisms, immunity is a major mechanism for combating an invading virus. Antibodies will normally build up following an infection, or a vaccine may be administered. Either way, the antibody proteins appear to act largely by recognizing the surface proteins of a virus and then neutralizing the virus to

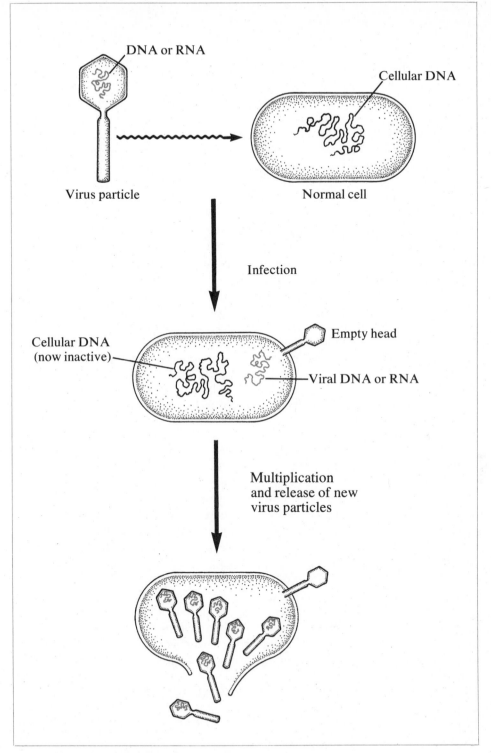

Figure 17.6   The parasitic attack of a virus particle on a normal cell.

prevent infection. Even in a dead virus vaccine, the surface proteins are not greatly altered, so that a later exposure to the live virus evokes the usual memory response resulting in a rapid synthesis of immunoglobulins and neutralization of the virus.

Unfortunately, the recognition step is the weak link in the chain of events for combating influenza infections. The surface proteins of influenza viruses are known to undergo periodic changes or mutations. Minor mutations (a few amino acids) may decrease the efficiency of the recognition so that the immune response is not totally effective and some degree of infection may occur. Minor mutations occur frequently—every year or two. Major mutations may totally obscure recognition of an invading virus and, thus, all existing immunity may be bypassed. These mutations typically occur every 10–12 years and have been blamed for the major flu epidemics in 1918–19, 1933, 1947, 1957, and 1968. It is likely that yet another hybrid is establishing itself while you are reading this section.

It is believed that even the swine flu underwent a minor mutation during the 1918–19 outbreak. It struck in two waves, with the first beginning in Spain in the spring of 1918, after which it spread to many parts of the world. In September of 1918, the second wave of swine flu began. The first wave caused widespread illness but very few deaths, whereas the second wave was devastating, presumably due to a slightly mutated form of the first virus. It was blamed for 20 million deaths throughout the world. Those individuals affected by the first wave acquired immunity that protected them against the mutant virus, since the mutant was similar enough to serve as an antigen and evoke a memory response.

The possibility of mutation of the recent swine flu outbreak, which occurred in January 1976, was also a matter for concern. It was just one factor in the decision to proceed with mass vaccination of the American population in the fall of 1976.

Allergy

Up to this point, the coverage of immunochemistry has viewed antibodies as exerting very positive effects. Unfortunately, there is another side to the story—one in which immunity can cause difficulties ranging from a nuisance like a runny nose for hay fever sufferers, to a severe allergic reaction, called anaphylaxis, in a person given penicillin or other substances to which the person is allergic. The latter may even be fatal if not promptly dealt with.

Recalling that substances that provoke antibody synthesis are called *antigens,* we can see that the problem arises if an individual is confronted with a large amount of an antigen against which the individual has previously produced antibodies. The antibodies will combine with the antigen to remove it. In addition, some immunoglobulins, designated IgE, in combination with antigen molecules, may combine with a certain type of cell and cause the cells to release histamine. Histamine is a potentially potent toxin that can cause contraction of certain types of muscle tissue, primarily in the lungs. In the extreme, it may cause suffocation. In less extreme cases, such as hay fever, an antigen may be inhaled in small quantities causing the release of small amounts of histamine, in which case the histamine causes dilation of blood vessels, swelling, and increased secretion from the eyes and nose. Histamine release is also associated with allergic reactions to

Histamine

Chlor-Trimeton
(an antihistamine)

foods, ragweed, and a variety of other problems, including bronchial asthma.

Some individuals are very susceptible to insect stings and exhibit acute symptoms. Most people are not susceptible because they lack the troublesome antibodies or because they have other antibodies that are more effective in intercepting the allergenic substance but without provoking the release of histamine. The same reasoning applies in the case of penicillin, since many individuals are allergic to this drug.

Although drug treatment is a major topic for consideration in Chapters Eighteen and Nineteen, it should be noted that there are chemicals called *antihistamines* that closely resemble histamine and are capable of blocking the effects of histamine on body tissues. Some very susceptible individuals carry insect sting kits containing an antihistamine or, perhaps, epinephrine (Adrenalin), which can be administered in the event of a sting.

In the rare event that an allergic individual must be given penicillin, a drug is kept on hand to counteract any adverse reaction. Generally, some alternate antibiotic is used.

Finally, a very unfortunate example of immunity shows up in individuals who have received an organ transplant. Since the organ is a foreign substance, it is often rejected by an immune response. In such cases, it is possible to suppress the immune system by administration of drugs, but not without increasing the risk of infection.

SUMMARY

Outbreaks of influenza are just one problem that can be dealt with by stimulating immunity. Diseases caused by microorganisms, such as bacteria and viruses, often provide immunity toward later infection. Immunity can be traced to antibody proteins (immunoglobulins), which can neutralize infecting organisms. When an infection occurs for the first time (the primary infection), the development of antibodies is inefficient, but if another infection (the secondary infection) occurs later on, it evokes a very efficient immune response, called the memory response, which prevents the infection from taking hold.

Vaccines in the form of dead organisms (e.g., whooping cough vaccine), live attenuated viruses (e.g., the Sabin vaccine against polio) and inactivated toxins, called toxoids (e.g., vaccines against diphtheria and tetanus), are commonly used to simulate the primary infection and stimulate immunity. If a real infection should follow, it would then be a secondary infection and the memory response would neutralize the invading organism. In spite of the many vaccines that are available and are recommended, an alarming percentage of the population is not adequately immunized against some very serious but preventable diseases.

Temporary passive immunity may be acquired by receiving antibodies, sometimes called antitoxins, produced in another animal. Natural passive immunity is gained by passage of antibodies from mother to fetus. Passive immunity is also acquired by newborn who are breast fed during the few days immediately following birth.

Since antibiotics and other drugs are ineffective in treating viral diseases, e.g., the common cold and the flu, it is especially important to establish immunity when a viral disease may be life-threatening. Unfortunately, viruses occasionally undergo major mutations. The most significant changes that occur are changes in the protein coat around the exterior of each virus particle. When an individual acquires immunity toward a particular virus, his immune system recognizes the protein coat of the virus and a full-scale immune response occurs and neutralizes the virus. But, when the protein coat undergoes a major mutation, the virus may not be recognized by the immune system and any previously-acquired immunity may be bypassed. In recent years, influenza outbreaks known as the Asian flu (in 1957) and the Hong Kong flu (in 1968) occurred. The Hong Kong flu was a major mutation of the Asian flu, so that individuals who had acquired an immunity to Asian flu were still susceptible to Hong Kong flu.

Hay fever, allergic reactions to insect bites or drugs (particularly penicillin), and rejection of transplanted organs are examples of the immune response working with undesirable consequences.

PROBLEMS

1. Give a definition or example of each of the following:
   a) infection
   b) immunology
   c) antibodies
   d) antigens
   e) memory cells
   f) toxoid
   g) natural passive immunity
   h) anaphylaxis
2. How do microorganisms enter the body?
3. Why are antibiotics worthless against viral infections? Why are they sometimes given anyway when a viral infection occurs?
4. What are some diseases that confer immunity? What diseases do not confer immunity?
5. How does a vaccine work?
6. What are the characteristics of the three types of vaccines?
7. What is the difference between passive immunity and active immunity? Which is better and why?
8. Why do major flu epidemics seem to occur every 10 to 12 years?
9. What are some negative aspects of antibody production?

T HE earlier chapters in this section have been devoted to matters of good health from the point of view of prevention of disease by way of good hygiene, good nutrition and immunity. In this chapter, the emphasis is on the common types of drugs used for curing disease. Chapter Nineteen concerns some of the drugs used for treatment of symptoms and the major drugs of abuse. Chapter Twenty contains a discussion of hormones and hormone-like substances, with emphasis on their role in reproduction and their use for contraception.

Although a number of drugs such as antibiotics, antiviral agents, and anti-cancer agents are used in attempts to cure various diseases, the vast majority of the drugs used by the consumer are for treatment of symptoms such as headache, fever, and nervous tension. Many are available over the counter. Others require a doctor's prescription.

Throughout the following coverage, the complete chemical structures will be given for each of the drugs under discussion. Abbreviations such as those described in Chapter Six will be used, but while the reader should understand what the chemical formulas mean, it will not be necessary to learn these complex structures in order to understand the discussion of each.

DRUG NAMES

The names used to identify drugs are the cause of a major controversy among drug companies, physicians, and consumer organizations. The issue is largely an economic and political one, but an understanding of the chemistry involved is required in order to allow the consumer to make a rational decision.

There are three types of names that can be used to specify a drug. One is its *systematic name,* i.e., the IUPAC name, but these are seldom used outside of the very technical literature. Another is the *generic name,* which is an unsystematic but still somewhat technical name. Generic names are used very frequently. Finally, there is the *brand name* or *trademark name,* which is the name adopted for use by each drug company that markets a drug. In other words, there is only one generic name for each drug but there may be many brand names.

Brand names are capitalized; generic names are not. The brand name frequently appears in the literature with the sign ® at the upper right of the name to

CHEMOTHERAPY
FOR THE TREATMENT
OF DISEASE

indicate that the name is registered and that its use is restricted to the owner of the name. Examples appear below.

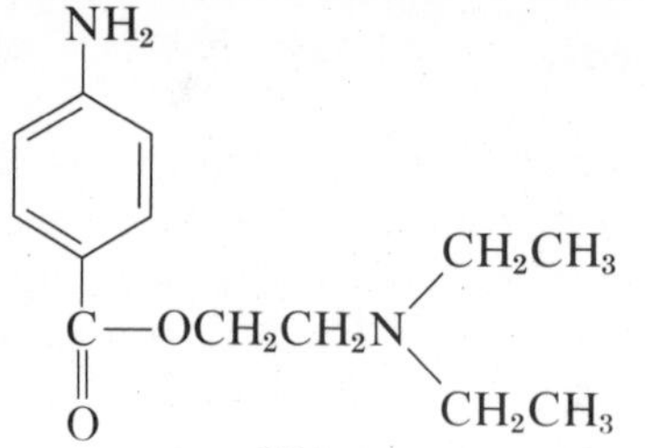

354

procaine
Novacaine®

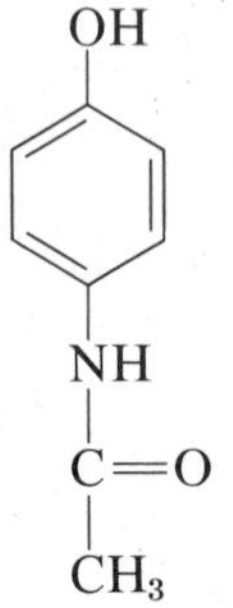

acetaminophen
Tylenol®, Tempra®, Datril®, Liquiprin®, Trilium®

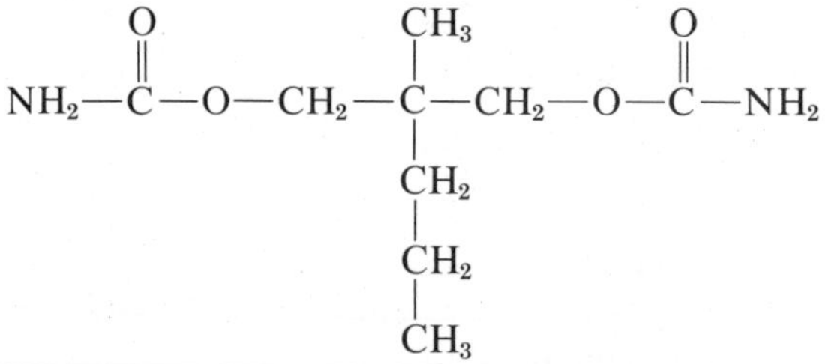

meprobamate
Equanil®, Miltown®

As a company develops a new drug, it may apply for a patent that gives exclusive rights to produce and sell that drug. The patent holder may also sell rights (license) to other companies to produce and market that drug. Even when the original drug company does grant a license to another, the first company maintains a great deal of control over the price of the drug, since each licensee must add the cost of the license to its costs of production and promotion in determining what it must charge for the drug in order to make a profit. In contrast, the company holding the patent not only avoids licensing expense but also derives income by granting licenses. Therefore, the patent holder has a big financial advantage over all its competitors and, in effect, has a sort of monopoly on a drug for the duration of the patent, which is 17 years.

As usual, the patent process is meant to stimulate new discoveries by guaranteeing that the rewards will be substantial when a discovery is an important one. In the case of the drug industry, a very sizeable portion of the income is devoted to research—reportedly five times as much of the sales income as in American industry as a whole (1). Even after a drug has been successfully synthesized, much research must be done to determine its properties. Several of the problems and goals of drug research are discussed in a later section, but none is more costly than the work done to prove the safety of the drug. In most cases, a part of the expense is due to the delay in marketing a drug, even after a patent has been granted, due to the requirements for adequate testing. This may consume several years of the patent period.

CHEMO-
THERAPY
FOR THE
TREATMENT
OF DISEASE

355

Many of the testing regulations have been put into effect since and because of the thalidomide disaster, which was uncovered in the early 1960s. Thalidomide

thalidomide

is a substance that was used in Europe as a tranquilizer from 1957–62. In 1962, serious birth defects were traced to it.

There seems to be little concern over the pricing of drugs during the period when a drug is protected by a patent, but there is a controversy over marketing practices carried out after a patent expires. Some critics argue that the drug industry spends far more for promotion and advertising than it does for research, and part of this promotion is frequently directed to convincing physicians that expensive brand-name drugs are superior to the cheaper, equivalent generics. In other words, the drug industry reportedly emphasizes the old familiar saying that "you get what you pay for" and spends from $3000–$5000 per doctor each year in its promotional efforts (1–3).

In the previous paragraph, those drugs marketed under their generic names were described as equivalent to the same drugs sold under brand names often at several times the price. Is it merely a matter of two or more companies putting the same substance into their own bottles, using their own brand names, and then calling it better, or is there really a difference? Is the drug company with years of experience invested in research on a drug able to market that drug in a form that is superior to the competitor's product? Let us look at the facts.

Following expiration of a patent, the original brand name used by the patent holder is protected by copyright law, but the drug itself may be marketed by any company using either their own brand name or the generic name. The new company may be a small drug company or one of the major pharmaceutical manufacturers. Invariably, a drug sold under its generic name is less expensive and is frequently described as inferior to the more expensive brand-name variety.

How may a given drug differ from one manufacturer to another? When a drug is formulated into a tablet, capsule, or other form, certain inert ingredients are commonly added. The active ingredient is always the same and the amount of the ingredient is usually the same for all brands of any particular drug, but the inert ingredients may vary. In a tablet, fillers, binders, and a coating, along with the active ingredient, may contribute to the size, shape, flavor, and appearance of a drug. These ingredients might logically be called *drug additives*.

Clearly, these additives could make a difference in the performance of a drug (its pharmaceutical action), and the emphasis of the controversy is centered on the question of *bioavailability* or *bioequivalence*. If two brands of a drug are to be equally effective, they must be equivalent once administered, so that the active ingredient becomes available in equal amounts to produce the desired pharmaceutical action. If one drug is better able to reach the target organ following administration, that drug is superior. When a drug is administered orally, it follows the pathway pictured in the following scheme.

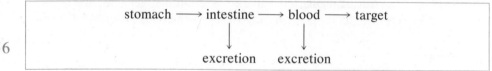

stomach $\longrightarrow$ intestine $\longrightarrow$ blood $\longrightarrow$ target

excretion   excretion

A major factor in determining the efficiency of action of an orally administered drug is the ease with which it enters the bloodstream. This is largely a matter of solubility of both the active ingredient and any inert ingredients. Particle size has a strong influence on solubility. As we have seen earlier (Chapters Four and Nine), smaller particles have a greater surface area per given amount of material, and, in the case of a drug, the greater the surface area, the faster the drug will dissolve due to greater contact with the fluids in the digestive tract. Particle size is primarily important when a drug is given as a powder enclosed in a hard gelatin capsule. In this form, inert ingredients are not required except for the capsule itself. On the other hand, these ingredients can be quite important in controlling the solubility of a tablet, and therefore the rate of uptake of a drug into the blood. As indicated by the absorption scheme, if a drug is not readily absorbed, it may simply be excreted.

Figure 18.1   The marketing of erythromycin (1). Wholesale prices as of April 1977 were $6.50, $10.15, $11.83, and $9.75 for the products in the order shown. (Courtesy of Roy Doty.)

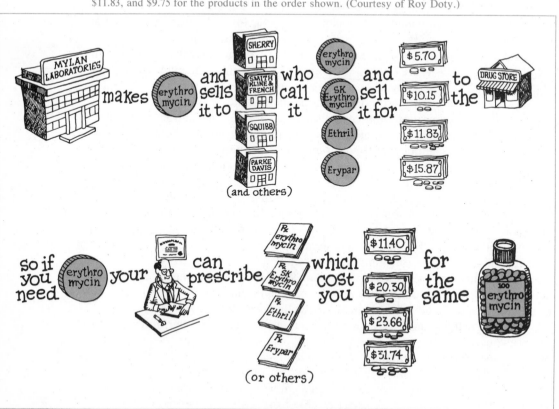

So, it does seem that some chemically-equivalent drugs may not be bioe-quivalent. However, documented examples of differences between brand-name and generic-name drugs are rare (4). For one thing, it is a simple matter to control the particle size of a drug dispensed in capsule form. Secondly, the methods used to compress ingredients into a tablet and the ingredients themselves are quite standard with very little difference from one manufacturer to another. Each company strives to utilize the most economical formulation, and the least expensive method for one manufacturer is usually the least expensive for another. Furthermore, a formulation can easily be copied. Chemical analysis of the ingredients makes it rather easy for one company to duplicate the formulation used by another. The fact that one company has the benefit of years of research on a given drug does not really give them much of an advantage over anyone else once the patent has expired, except that the brand-name drug marketed by that company may be so well known as to give an edge in the promotion, if not in the quality of the drug itself.

Furthermore, there is the more important economic question of how to produce and market a drug in the most economical way. In some instances, one pharmaceutical company may produce a drug in final form and sell it to several other companies, which then market it under their own brand name or under the generic name. A clear example is depicted in Figure 18.1. Erythromycin is an antibiotic frequently used by individuals who are allergic to penicillin. This drug can be purchased under several names, at a wide variety of prices, even though they are produced by the same original laboratory. Invariably, the product sold under the generic name is less expensive. If a physician prescribes an expensive brand name drug, antisubstitution laws prohibit a pharmacist from dispensing any less expensive, but equivalent generic brand.

DRUG ADDITIVES

The use of food additives is a very controversial subject (see Chapter Fourteen). Drug additives are not as controversial but may, nevertheless, be very crucial for increasing the usefulness of a drug. Although many drugs are administered in liquid form, or as sprays, ointments, suppositories, etc., the majority are administered as capsules or tablets. Capsules are more expensive, so pharmaceutical manufacturers prefer to market their products in tablet form. On the other hand, capsules circumvent problems of taste, stability, and disintegration. A capsule is usually made of gelatin and has no flavor. The contents may be quite stable, since they are not in contact with air. At the same time, the drug can be in powdered form so that there is no delay in disintegration after it is consumed. Therefore, the drug can be readily dissolved and absorbed. The gelatin capsule also disintegrates rapidly and the contents are quickly released.

The chief additives used in tablets are *fillers, binders, lubricants, disintegrators,* and *coatings*. Some drugs are so potent that only a small amount of the active ingredient is present in the tablet. In such cases, lactose, sucrose, or some other carbohydrate may be added to increase the size of the tablet up to the desired volume. Tablets are formed by mixing the ingredients and compressing them in a mechanical device under high pressure. Starch paste is often used to provide adhesiveness and thus serves as a binder. Soaps, such as calcium stearate and magnesium stearate are used as lubricants. They are present in small amounts

Figure 18.2   The manufacturing of medication in tablet form. (Photo courtesy of Merck & Company.)

Figure 18.3   Various sizes and numbers of gelatin capsules. Reproduction indicates actual size. (From Bergersen, Betty S., *Pharmacology In Nursing*, 13th edition, St. Louis, The C. V. Mosby Co., 1976.)

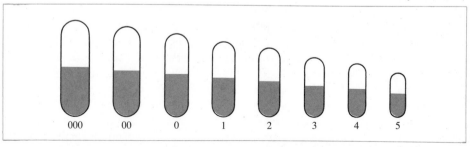

$$(CH_3CH_2CH_2CH_2CH_2CH_2CH_2CH_2CH_2CH_2CH_2CH_2CH_2CH_2CH_2CH_2CH_2\overset{\overset{\displaystyle O}{\displaystyle \|}}{C}-O)_2Ca$$

calcium stearate

and are used to prevent the ingredients from sticking to the machinery used to compress the drug plus additives into a tablet.

Dry starch is often added as a disintegrator. Starch expands when it becomes wet and expansion of a tablet is the reverse of the process used to form (compress) the tablet in the first place. Therefore, starch speeds the breakdown of the tablet and aids in absorption.

Coatings are used for a variety of reasons. They solve problems of taste, appearance, and stability (by avoiding contact with air), and they can be designed to control the location and rate of release of the contents. For example, there are drugs which are enteric* coated. Such coatings may withstand the acidic conditions found in the stomach but disintegrate rapidly and release the ingredients when they encounter the alkaline conditions of the intestine. Thus, if a drug is sensitive to the stomach or if the stomach is sensitive to the drug, the problem can be avoided by preventing the drug from being released in the stomach. Once in the intestine, a drug may be readily absorbed into the bloodstream and delivered to the intended site of action.

In most cases, research is done to find the best combination of ingredients so that a drug may be marketed in the less-expensive tablet form. Until that point is reached, drugs are often marketed in the more-expensive capsule form. Some of the other problems and goals confronted in research on drugs are discussed in the following section.

OTHER GOALS OF DRUG RESEARCH

In the process of developing a new drug, there are many properties that must be studied in order to evaluate the suitability of a drug and to find ways to increase its efficiency. Some of the major properties are listed below and discussed in the following paragraphs.

1. mode of action
2. absorption and excretion characteristics
3. side effects
4. interaction with other drugs
5. toxicity: including possible allergic reactions
6. effective dosage
7. patient appeal

A very important, but often poorly understood, property of a drug is its mode of action. Once one knows how a drug causes its desired effect (analgesic, antibiotic, etc.), there are often many obvious courses of action that may lead to the improvement of a drug or to development of other more effective drugs.

* intestinal

PROFESSIONAL BUILDING

DR. JOHN SMILMAN M.D.
PEDIATRICIAN

DR. WILLIAM BARNS M.D.
INTERNIST

DR. SAMUEL O. MOSS M.D.
EAR NOSE & THROAT

DR. VICTOR B. ROBBINS M.D.
DERMATOLOGIST

DR. HERBERT WINTON M.D.
SIDE EFFECTS

Copyright © 1973, *The New Yorker* Magazine, Inc.

Items (2)–(6) are also factors that describe the action of a drug following administration. The active ingredient must have suitable solubility properties to permit absorption when given orally. Painful injections are both undesirable and inconvenient and are a major impetus for developing drugs in a form suitable for oral administration. Even if a drug is sufficiently water soluble to permit absorption, it may not have adequate fat solubility to permit it to penetrate a target organ even if absorption into and transmission by the blood is very efficient. In addition, the active drug may have adverse effects on the system, e.g., stomach irritation, or the drug itself may be adversely affected. In other words, a drug may be perfectly suited to give the desired pharmacological effect once it penetrates the target organ, but the problems of getting the drug to the target intact and without adverse effects plus the problem of penetration of the target may make a drug unsuitable. In such cases it may become necessary to abandon the use of the drug, synthesize different forms of the drug in hopes of improving its properties while maintaining its efficiency, or synthesize a *prodrug* form. The last of these is the subject of the next section.

PRODRUGS

CHEMO-
THERAPY
FOR THE
TREATMENT
OF DISEASE

361

One of the more intriguing goals of drug research is the development of prodrugs. A prodrug, or drug precursor, is a less active or even inactive form of a drug that is converted to the active form some time after administration. Generally, it is by the action of enzymes in the body that the prodrug is converted to the active drug and, in some cases, the conversion only occurs after the prodrug has located in the target organ. The application of prodrugs is often tried as a way of overcoming solubility problems, either at the site of absorption into the bloodstream or at the site of uptake into the target organ. A prodrug form may also overcome problems such as instability, undesirable metabolism of the active drug before reaching the target, or factors relating to patient acceptance (taste, odor). A bitter-tasting medication may be satisfactory for an adult, even if it has to be given in capsule form, but oral medication for young children must often be chewable and, therefore, pleasant tasting.

Aspirin is a type of prodrug. The compound known as salicylic acid is an effective analgesic (pain killer) and anti-inflammatory agent, but it is too corrosive for general use. This problem was largely overcome by conversion of salicylic acid to acetylsalicylic acid (aspirin), which has become the most commonly used drug.

Prodrugs have also been used to achieve a delayed or sustained-release action. Coated slow-release beads and granules, layered tablets, and prodrugs have been used to control the rate of release, and therefore the rate of absorption. If a drug is absorbed too quickly, its effect may be of very short duration and its level (concentration) in the body may rise and fall very quickly. This is generally described as a "peak and valley" effect. In the case of a pain killer, this behavior would be very undesirable and the same is true of other drugs as well. Therefore, the advantages of sustained-release formulations include (1) reduction of the number and frequency of doses that must be administered, (2) elimination of the peak and valley effect, (3) frequent reduction of the amount needed to achieve the desired results, (4) elimination of the problem of nighttime administration of a drug, in which a patient may have to be awakened, (5) decrease in the number of times a patient must remember to take medication, and (6) reduction of the incidence of side effects on the gastrointestinal tract.

DRUGS IN MEDICINE

In this section, attention will be focused on some of the major types of drugs used in the treatment of disease. In Chapter Seventeen, the emphasis was on the prevention of disease by the use of vaccines, but we have seen that vaccines are often impractical even when they can be produced.

362

The coverage will be divided into two parts—drugs used to effect a cure and drugs used for treatment of symptoms. The latter subject is covered in Chapter Nineteen.

DRUGS USED FOR THE CURE OF DISEASE

As in earlier coverage, the emphasis will be on the chemotherapy of infectious diseases, i.e., those caused by microorganisms. Cancer and heart disease are two major diseases that obviously do not fit this description. Cancer chemotherapy will be considered briefly. The treatment of heart disease is not discussed here, although the reader may want to review the discussion of this subject in Chapter Fifteen (The Cholesterol Controversy).

Antibiotics

The development of antibiotics is one of man's greatest achievements and is a relatively recent success story. In 1907, the German scientist, Paul Ehrlich (1854–1915), discovered a substance called *salvarsan* in his search for a treatment for African sleeping sickness. In spite of this success, it was not until after World War II that the early members of the penicillin group of antibiotics came into general use. In fact, one of the most widely used penicillins, called ampicillin, was not available until 1961. The development of new and better antibiotics is still a major goal of the pharmaceutical industry.

Ehrlich coined the phrase "magic bullet" for salvarsan to denote its lethal effect on disease-causing organisms. This substance was also known as arsphenamine, because it contained arsenic, and salvarsan 606, because it was compound

salvarsan 606
arsphenamine

number 606 that Ehrlich tested (13). Salvarsan also proved effective in the treatment of syphilis and Ehrlich received the Nobel Prize in 1908. He is generally recognized as the founder of chemotherapy.

Ehrlich's work was really quite ingenious. He merely followed up on the earlier (1884) work of the Danish physician, H.C. Gram, who devised a test that is commonly used to classify bacteria. Following treatment with certain dyes and subsequent washing, *Gram-positive* bacteria retain the dyes, whereas the dyes are easily removed from *Gram-negative* bacteria. This suggests that certain chemical groups are present on the cell walls of Gram-positive bacteria, whereas they are lacking in Gram-negative bacteria. Ehrlich reasoned that it might be possible to incorporate a toxic element, such as arsenic, into a dye molecule, which might then become attached to bacterial cells and cause their death. Such was the case with his "magic bullet."

CHEMO-
THERAPY
FOR THE
TREATMENT
OF DISEASE

363

An antibiotic is generally defined as any substance produced by one microorganism, which kills or inhibits the growth of other microorganisms. Those antibiotics that kill microorganisms are described as *bacteriocidal,* whereas those that merely inhibit microbial growth are said to be *bacteriostatic*. The latter type usually acts by inhibiting the synthesis of proteins in the infecting microorganism. Although the bacteriocidal type of antibiotic would seem to be the better one, both types of antibiotic can provide the time necessary for mobilization of normal body defense mechanisms—synthesis of antibodies—while preventing the invading organism from overwhelming the body. Since Ehrlich's magic bullet was a synthetic substance, it is not strictly correct to classify it as an antibiotic.

Antibiotics such as erythromycin, Terramycin®, and Aureomycin® have the suffix -*mycin* to indicate that they were isolated from microorganisms found in the soil. Some of these antibiotics were later synthesized in the laboratory by chemists although, in many instances, it is still more economical to grow the microorganisms and isolate the antibiotics from them.

Several new antibiotics have been obtained by mutation. Just as yeast cells carry out a complex series of reactions, i.e., metabolism or fermentation, to produce ethyl alcohol, other microorganisms carry out reactions leading to formation of antibiotics. By the action of ionizing radiation, microorganisms have been made to undergo mutations which, in some cases, cause the microbes to synthesize other antibiotics or to synthesize the same ones more efficiently.

It has also been possible to alter the production of antibiotics by microorganisms by changing the nutrient media used to sustain the life of the microorganisms. In this method, certain components of the growth media are sometimes incorporated into an antibiotic molecule, which then has a different structure with different properties. Using a slight variation of this technique, it has also been possible to interfere with reactions in these metabolic pathways so that a precursor* to the normal antibiotic could be isolated. Such a precursor could then be converted to other chemical compounds with antibiotic activity. These antibiotics are commonly designated *semisynthetic antibiotics,* to signify that a microorganism does part of the job of synthesis but a chemist carries out the final step(s). In effect, the chemist employs a microorganism to start the job of synthesis, and then completes the rest of the synthesis himself. There have been very dramatic successes using mutants, alteration of growth media, and semisynthetic pathways in the synthesis of members of the penicillin family of antibiotics.

CHOOSING AN ANTIBIOTIC

Despite their failure to combat viral infections, antibiotics have a very impressive record in dealing with microbial infections. This has led many to describe them as the "miracle drugs."

---

* A precursor is a compound that appears early in a metabolic pathway and is normally converted to the end product of the pathway.

Successful application of antibiotics requires the often critical decision as to which antibiotic to use. In some cases, it is possible to obtain a sample of the invading microorganism from an open wound, the throat, the urine, the blood, or other source. This sample may then be *cultured* and examined for sensitivity to antibiotics. The culturing process involves taking the sample of microorganism and placing it in a nutrient medium in order to encourage it to grow and multiply into a larger sample. Testing for antibiotic sensitivity is most easily done by mixing a quantity of melted agar (a complex polysaccharide) with the cultured sample of microorganism. The agar is then allowed to solidify into a gel. Drug companies market circular discs that are impregnated with antibiotics. A number of these discs may be placed on top of the agar gel on a culture plate as shown in Figure 18.4. The antibiotics will then diffuse out of the discs and each antibiotic that is effective against the microorganism will inhibit growth in a ring about that disc.

A typical result is pictured in Figure 18.4, with a large clear zone appearing around some of the antibiotic discs, whereas the microorganism appears to grow almost up to the surface of other discs. Usually, about ten antibiotics are tested. The information is then transmitted to the physician, who chooses the antibiotic to use on his patient.

Unfortunately, the technique does not yield results immediately. Typically, 72 hours or more are required to grow a microorganism in sufficient quantity for testing the antibiotic sensitivity, and that 72-hour period can sometimes be crucial in the development of the microorganism in the body. As noted earlier, the usual goal of antibiotic therapy is to control the development of a microorganism to

Figure 18.4   Testing for antibiotic sensitivity. Paper discs impregnated with different antibiotics are dropped onto the surface of plates inoculated with a microorganism in order to determine the sensitivity of the organism to the various antibiotics. The antibiotic diffuses into the medium, inhibiting the growth of organisms sensitive to the antibiotic. Large clear zones around the antibiotic discs are regions where microbial growth is prevented. From *Basic Microbiology* by W.A. Volk and M.F. Wheeler, Copyright © 1973 by J.B. Lippincott Co., Courtesy of BioQuest, Division of Becton, Dickinson & Co.

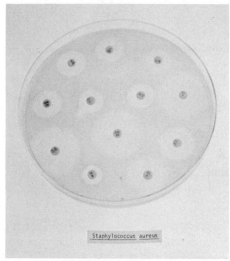

Staphylococcus aureus

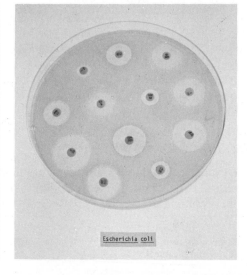

Escherichia coli

allow time for the body to mobilize a defense. Therefore, it usually becomes necessary for the physician to make an "educated guess" and administer an antibiotic before he has the benefit of laboratory results. He makes his choice based on factors such as the symptoms and the site of infection. In many cases, he will choose a *broad-spectrum antibiotic,* i.e., an antibiotic that is known to be effective against a wide variety of microorganisms. A *narrow-spectrum antibiotic* is preferable once the antibiotic sensitivity results are available, but the broad-spectrum drug is like an insurance policy with a wide range of coverage.

CHEMO-
THERAPY
FOR THE
TREATMENT
OF DISEASE

365

It is also possible to administer a combination of antibiotics—called *piggyback antibiotics* when given in a single preparation—in order to provide defense against an even greater number of organisms. But the use of combinations and even single, broad-spectrum antibiotics is not necessarily in the best interest of the patient or the population as a whole. The situation is analogous to the use of insecticides such as DDT. Just as insects may develop a resistance to insecticides (see Chapter Ten), microorganisms may develop resistance to antibiotics.

The development of resistance to antibiotics is actually encouraged by the use of antibiotics. A resistant organism is commonly a mutant or hybrid form of a microorganism that is encouraged to thrive and reproduce when competing organisms are killed off by antibiotics. In other words, the hybrid might be inferior and unable to compete with other microorganisms under normal circumstances but, as a result of antibiotic therapy, this inferior microorganism becomes superior and is able to develop on a large scale, sometimes with very serious consequences.

A *resistant microorganism* is one that has an enzyme available that will chemically change the antibiotic into a harmless substance. A microorganism that develops a resistance to penicillin is said to have the enzyme *penicillinase.* The development of resistance is one of the reasons why there is a continuing research effort geared to the development of new antibiotics. As one source put it, "although the success of antibiotics may warrant their classification as miracle drugs, it is a miracle in constant need of renewal. Resistant variations of pathogenic* bacteria that can withstand effective antibiotic treatment are continually arising, particularly in hospitals (5)."

Broad-spectrum antibiotics or combinations of antibiotics may also have less serious but, nevertheless, annoying consequences. There is normally a heavy concentration of harmless bacteria in the large intestine. This collection of bacteria is called the *intestinal flora.* It is very influential in controlling the passage of wastes through the system. Diarrhea (sometimes persistent) is a frequent side effect of antibiotic therapy, because antibiotics tend to upset the natural population of bacteria in the intestinal tract. This not only argues against the use of antibiotics when they are not really needed, but it also explains why broad-spectrum antibiotics have the greatest effect.

Clearly, the physician is in a difficult situation. He should not administer antibiotics against a wide range of microbes, but he usually does not know what narrow-spectrum antibiotic might be effective, and if a disease is life-threatening or is accompanied by severe symptoms, the choice must be a method that has the greatest chance for success, in spite of the problems that may be created by doing so. Unfortunately, in the long run, the result has been that many previously

* disease-causing

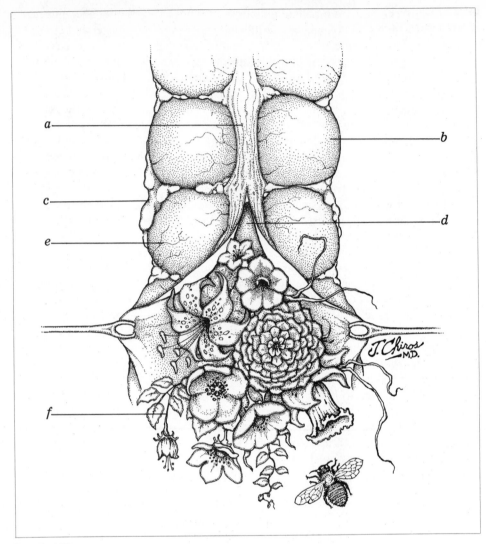

Intestinal Flora

broad-spectrum antibiotics have become narrow-spectrum antibiotics because of the development of resistance.

THE SULFA DRUGS

The mid 1930s marked the real beginning of the wide scale use of the miracle drugs. In spite of the earlier work of Ehrlich and observations on the antibiotic activity of penicillin in 1928, it was not until the antibiotic properties of the early sulfa drugs were discovered that well-organized attempts were directed to development of penicillin and, later on, other antibiotics.

The story of the sulfa drugs began in 1932 with a patent on a new drug called Prontosil® by a German dye manufacturer, I.G. Farbenindustrie. Prontosil is a red substance that had been synthesized for use as a dye, but it was found to

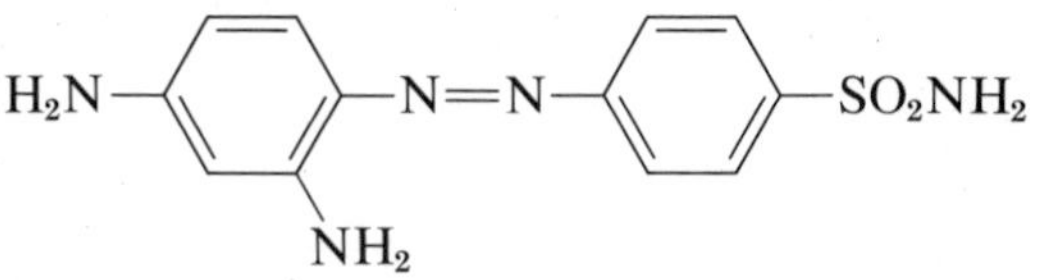

CHEMO-
THERAPY
FOR THE
TREATMENT
OF DISEASE

367

H_2N—⟨ ⟩—$N{=}N$—⟨ ⟩—SO_2NH_2 Prontosil

$\overset{|}{NH_2}$

exhibit antibacterial action when used to dye wool. In 1935, Gerhard Domagk published the results of his work, which showed that Prontosil was effective in combating streptococcal infections in animals. Domagk was awarded the Nobel Prize in 1939 for his contributions to the development of the sulfa drugs.

Soon after the antibacterial properties of Prontosil were reported, it was found that Prontosil was converted to sulfanilamide in the body and that sulfanilamide had the same antibacterial activity as Prontosil. Subsequently, a whole

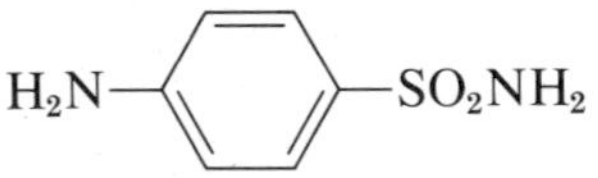

H_2N—⟨ ⟩—SO_2NH_2 sulfanilamide

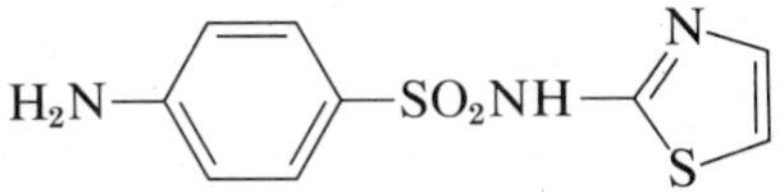

H_2N—⟨ ⟩—SO_2NH—⟨thiazole⟩   sulfathiazole

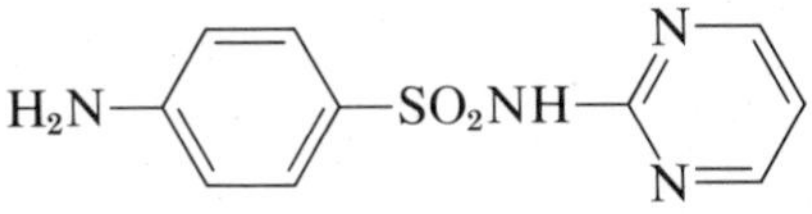

H_2N—⟨ ⟩—SO_2NH—⟨pyrimidine⟩   sulfadiazine

H_2N—⟨ ⟩—SO_2NH—⟨isoxazole⟩   sulfisoxazole

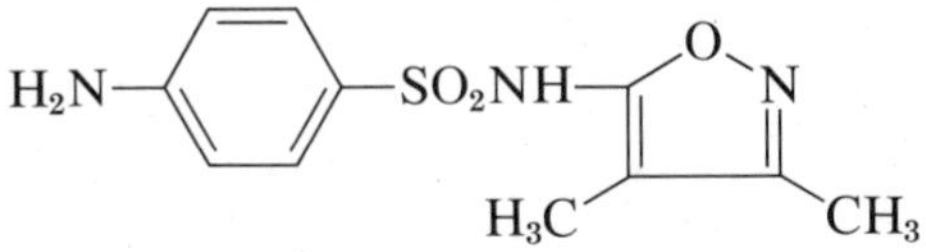

series of derivatives (modifications) of sulfanilamide, called sulfonamides or sulfa drugs, were synthesized.

Strictly speaking, the sulfa drugs are not antibiotics because they are not produced by any living organisms, and yet, their activity is bacteriostatic and they are susceptible to development of resistance by disease-causing microorganisms in the same way as antibiotics.

The sulfa drugs are often said to act as *antimetabolites,* i.e., they interfere with the metabolism of bacteria. Bacteria synthesize a substance called folic acid, which is essential for survival. For animals, including humans, folic acid is a vitamin (a water-soluble B vitamin), which means that animals cannot synthesize the substance, but must consume it in the diet. An important precursor to folic acid is para-aminobenzoic acid (PABA), which becomes incorporated into the folic acid as can be seen in the following.

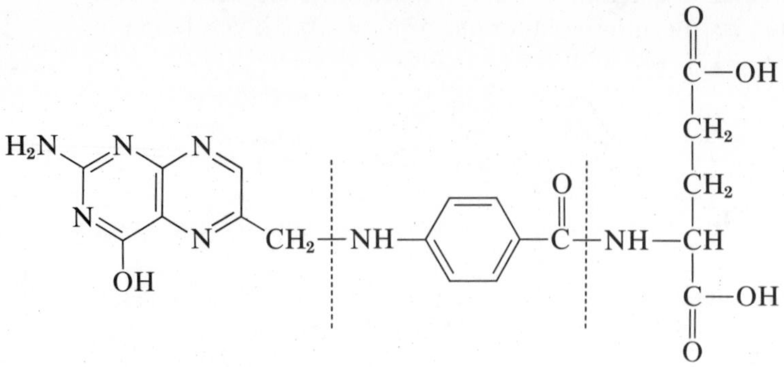

folic acid

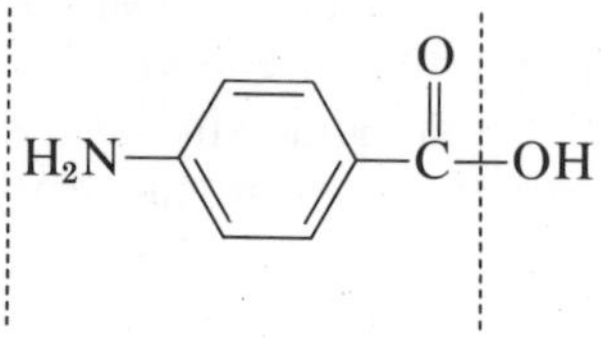

PABA

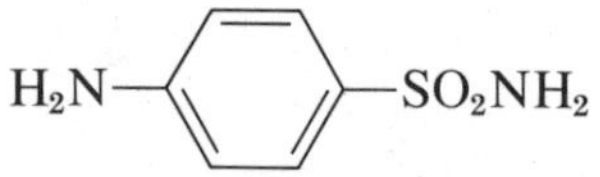

sulfanilamide

Sulfanilamide, and other sulfa drugs, are able to block the incorporation of PABA into the folic acid. As usual, the reaction in which PABA is taken up is catalyzed by an enzyme, and because of the similarity between PABA and the sulfanilamide molecule, this enzyme interacts with the sulfanilamide and becomes

Figure 18.5   The lock and key model for the action of sulfa drugs. An enzyme acts on PABA during a sequence of reactions leading to folic acid, which is an essential compound for bacteria. The enzyme acts like a lock, which may accept keys (PABA or sulfa drugs) that fit. When a sulfa drug enters (structure on right), it blocks the enzyme and prevents further attack on PABA.

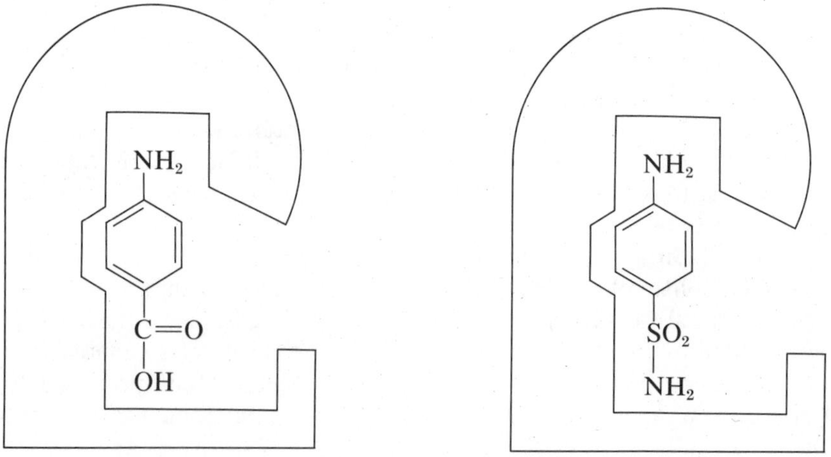

blocked. Molecules interact with enzymes in a sort of *lock and key* arrangement, in which the enzyme acts like a lock and the molecule which is reacting acts like a key (see Figure 18.5). If the key does not fit the lock, no reaction can take place. If the wrong key fits into the lock, it cannot open the lock, i.e., react, but it may jam the lock and prevent the lock from accepting any other keys. In other words, the sulfa drugs are similar enough to fit into the lock and prevent the interaction of PABA to form folic acid. Since folic acid is an essential compound for bacteria, they cannot survive without it. A number of related compounds also act as antimetabolites in microorganisms. Each one of the sulfa drugs has a structure similar to PABA.

CHEMO-
THERAPY
FOR THE
TREATMENT
OF DISEASE

369

The sulfa drugs were once used to combat a wide variety of infections caused by Gram-positive bacteria. In fact, the period from 1935–48 is often known as the sulfonamide era. The sulfa drugs were used heavily in World War II by soldiers who carried packets to sprinkle on open wounds to prevent infection.

Unfortunately, the sulfa drugs cause a number of side effects, particularly kidney damage from prolonged use, so they have largely given way to the penicillins and other antibiotics. Since they are efficiently absorbed and later excreted in the urine, they still see some use in combating urinary tract infections.

THE PENICILLINS

Although virtually everyone has heard of penicillin, few people realize that penicillin may actually be any of a number of different drugs with names like penicillin G, penicillin V, ampicillin, methicillin, oxacillin, cloxacillin, and nafcillin (see Table 18.1). Each one of the penicillins can be distinguished from the others due to differences in the structure of the group represented by the symbol R in the upper left of the formula in Table 18.1.

This is not merely another example of proliferation of names due to the use of both generic and brand names. Brand names do clutter things a bit, but there actually are several similar but distinctly different penicillins. The various penicillins may also be distinguished from one another due to differences in the following properties:

1. susceptibility to attack by stomach acid
2. spectrum of activity
3. effect on resistant (penicillinase-forming) microorganisms

More important, these properties explain why there has been a continued effort directed toward development of new antibiotics, in general, and new penicillins, in particular.

The earliest penicillin to be used on a wide scale was penicillin G (generic). It is still an important antibiotic, but it does suffer by comparison to some of the other penicillins when considered in relation to items (1)–(3), previously listed. Penicillin G is attacked by acid in the stomach, which renders it inactive. This limits its usefulness as an oral antibiotic, although it can be successfully used orally. It is recommended by some physicians as the first choice for oral use, if it is given in large doses and between meals when the stomach is less full. Under these conditions, the level of acid is reduced and there is no delay in passage of the drug

TABLE 18.1 THE PENICILLINS (6, 7)

NAME	YEAR PRODUCED	SIDE CHAIN (R—)	STABILITY IN ACID	SENSITIVITY TO PENICILLINASE	ADVANTAGE
Penicillin G	1940		Poor	Sensitive	Least expensive
Phenoxymethyl penicillin (Penicillin V)	1948*		Good	Sensitive	Better oral absorption than penicillin G
Methicillin	1960		Poor (not given orally)	Resistant	For treatment of infections due to penicillinase-producing organisms
Oxacillin	1961		Good	Resistant	For treatment of infections due to penicillinase-producing organisms

Drug	Year	Structure			
Dicloxacillin	1965		Good	Resistant	For treatment of infections due to penicillinase-producing organisms
Cloxacillin	1962		Good	Resistant	For treatment of infections due to penicillinase-producing organisms
Nafcillin	1961		Poor	Resistant	No advantage over oxacillin or cloxacillins
Ampicillin†	1961		Good	Sensitive	For treatment of infections due to gram-negative, non-resistant organisms
Carbenicillin†	1967		(not given orally)	Sensitive	

* First marketed in 1953.
† Broad spectrum of action.

(Adapted from *Fundamentals of Chemotherapy* by William B. Pratt, Copyright © 1973 by Oxford University Press, Inc. Reprinted by permission.)

out of the stomach and into the intestine where the pH is more suitable (alkaline) for the drug.

The susceptibility of penicillin G to acid has led scientists to synthesize acid-stable penicillins, which can be considered true oral antibiotics. Penicillin V and ampicillin are the major oral penicillins. In fact, ampicillin was the second most prescribed of all drugs of all kinds in 1974 (8). The popularity of ampicillin is also partly attributable to the fact that it has a very broad spectrum of activity. Many of the penicillins are narrow-spectrum antibiotics, which work only on Gram-positive bacteria. Ampicillin also works on some Gram-negative bacteria.

The problem of development of resistance was discussed earlier and the penicillins are a classic case. Ampicillin is among those penicillins that are deactivated by a penicillinase enzyme (see Table 18.1 and Figure 18.6). Other penicillins including methicillin, oxacillin, cloxacillin, and nafcillin are resistant to the action of penicillinase, but they are narrow-spectrum antibiotics and are, therefore, less prescribed.

Figure 18.6 Resistance to the penicillins normally occurs due to production by bacteria of one of two enzymes, penicillinase or penicillin amidase. Penicillinase production is by far the more common form of resistance.

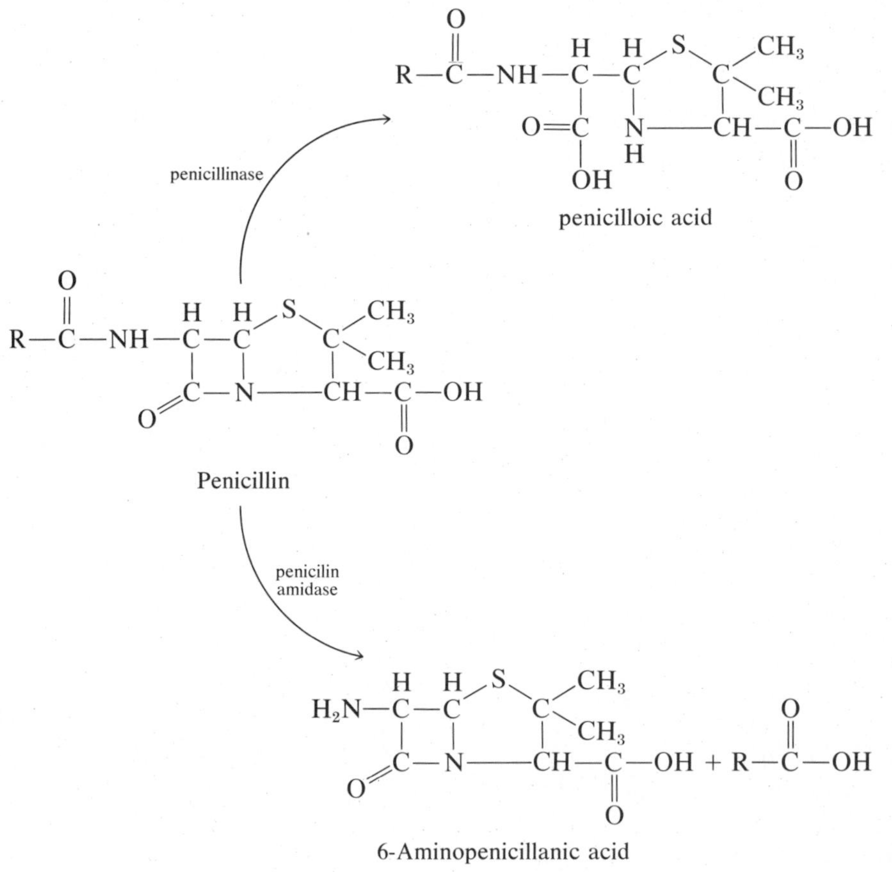

The history of the penicillins is one of the most interesting in all of science.
The first observations of the antibiotic activity of penicillin are a classic example
of an accidental discovery, which was made by Alexander Fleming at the University of London in 1928. Fleming was a bacteriologist who was experimenting with
a staphylococcus microorganism. His samples accidently became contaminated
with a mold known as *Penicillium notatum,* which not only grew on the nutrient
that he had provided for his bacteria, but also inhibited the growth of the staphylococcus organism in the regions where the mold began to grow. Fleming published
his findings and named the unknown antibiotic penicillin because of its source.
Unfortunately, Fleming encountered technical problems in isolating the penicillin.
Spurred on by the early successes of the sulfa drugs, a group headed by Howard
Florey and Ernst Chain picked up the work at Oxford University in 1939, but their
efforts were hindered by the war in Europe. The first clinical trial with a crude
penicillin preparation was carried out in 1941. The treatment was a tremendous
success but the patient died (due to the lack of an adequate supply to sustain the
treatment). Nevertheless, the potential for success was clear and the work moved
to the United States to avoid problems precipitated by the war.

During the next few years, work was aimed toward production of large
quantities of penicillin, and the effort was helped out by discovery of a new strain
of *Penicillium* mold, which was obtained from the surface of a cantaloupe in a supermarket in Peoria, Illinois in 1942. This new strain was mutated with x-rays and
was found to be unusually efficient in producing penicillin (actually penicillin G)
and contributed to the development of the techniques for production of penicillin
in large quantities. By 1946, the conditions for use had been determined and the
supplies were adequate for civilian as well as military needs.

In the years following, the chemical structure of penicillin was determined
and it was found that the highest degree of antibiotic activity occurred in penicillin
G, also known as benzyl penicillin. This and other compounds have since been

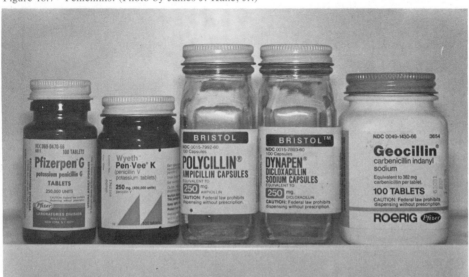

Figure 18.7   Penicillins. (Photo by James J. Kane, Jr.)

CHEMO-
THERAPY
FOR THE
TREATMENT
OF DISEASE

373

synthesized by purely chemical methods, but the most economical approaches have employed microorganisms for all or part of the synthesis process and, with proper control, many different penicillins can be synthesized.

As an example, the synthesis of penicillin G by the mold is favored by addition of phenylacetic acid to the nutrient medium provided for the mold. This

phenylacetic acid

benzyl group

favors incorporation of the benzyl group (see Table 18.1) into the penicillin molecule. Since the mold carries out all of the chemical reactions in the synthesis, penicillin G may be called a *biosynthetic* penicillin.

This same approach would seem to have unlimited potential for incorporation of other groups into the penicillin molecule, but many of the substances that would have to be included in the growth medium are either not tolerated by the mold or are acted upon by the mold in other ways rather than being incorporated into penicillin.

Even greater versatility is achieved by the synthesis of *semisynthetic* penicillins. As noted earlier, a semisynthetic antibiotic is one synthesized partly by a microorganism and partly by chemists. In the case of the penicillins, the crucial compound was 6-aminopenicillanic acid, 6-APA, which is the penicillin molecule

6-aminopenicillanic acid
6-APA

with no side-chain group. It can be isolated by interrupting the normal fermentation process, after which a variety of side-chain groups may be attached.

One of the most useful of the semisynthetic penicillins is phenoxymethyl penicillin, better known as penicillin V, which was the first true oral penicillin because it is stable in the presence of stomach acid. Most of the other penicillins, including ampicillin, which are listed in Table 18.1, are semisynthetic penicillins.

The penicillins are unusually safe except to a small percentage of the population which happens to be allergic to them and may suffer severe side reactions if they are administered.

The mode of action of the penicillins is bacteriocidal. They are lethal to various microorganisms because they interfere with the formation of the cell walls in these bacteria, thus causing the contents of the cell to spill out. Animal cells do not have cell walls so the effect is restricted to bacterial cells. This explains the unusually low toxicity of the penicillins.

Figure 18.8　Merck & Co.'s penicillin facilities at Danville, Pa.

OTHER ANTIBIOTICS

As the potential of antibiotic chemotherapy became increasingly evident, efforts to isolate other antibiotic-producing microbes were greatly increased. Biosynthetic, semisynthetic, and purely synthetic compounds were produced and tested for antibiotic activity. The list of successes is long—too long to consider in any detail here—but let us look at some of the generalizations that can be made regarding the mode of action of most of the antibiotics, in addition to some of their other properties.

Although a penicillin is often the first choice for antibiotic therapy, problems of resistance, allergy, and spectrum of activity often force the use of other antibiotics. Streptomycin, erythromycin, chloramphenicol, the tetracyclines, the cephalosporins, and several others can often be substituted effectively. The structures of some of the antibiotics follow. As usual, it is not necessary to know these structures in any detail, and yet it is interesting to note the great diversity of structure that can lead to antibiotic activity.

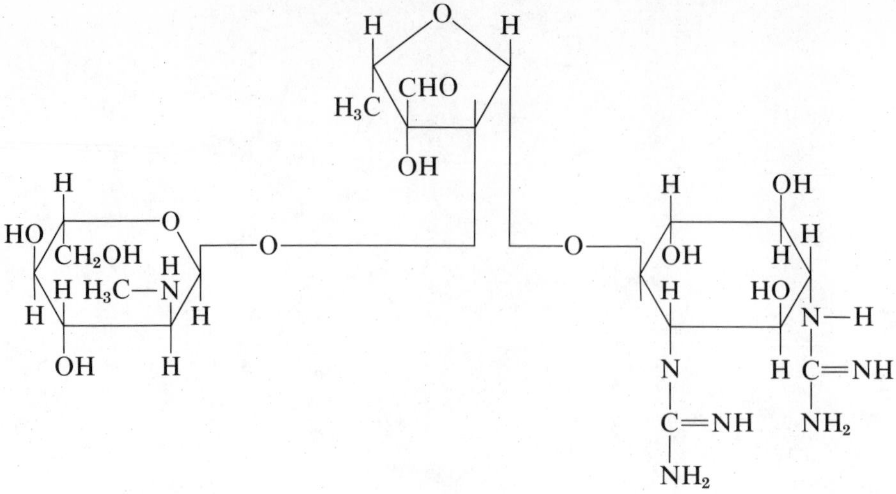

streptomycin

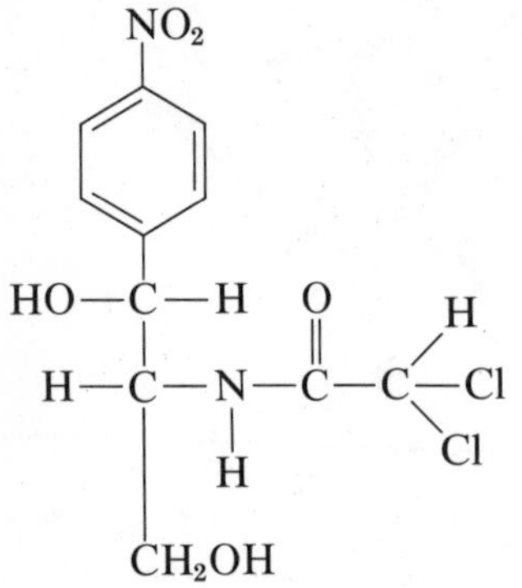

chloramphenicol

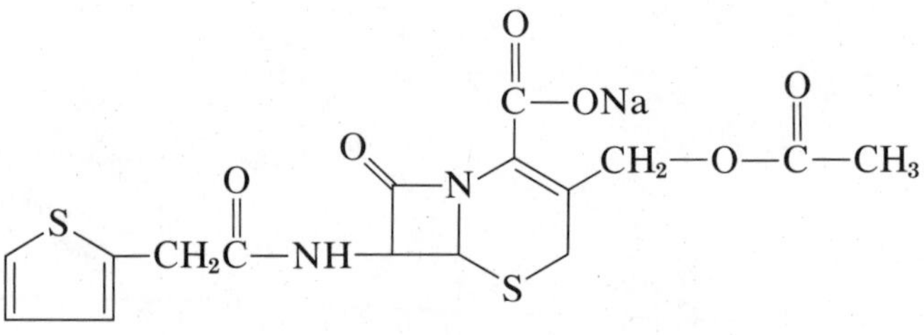

a cephalosporin
cephalothin (sodium)
Keflin®

So far, we have seen the use of antibiotics as antimetabolites (the sulfa drugs) and as inhibitors of cell wall synthesis (the penicillins). The cephalosporins, bacitracin, and others also interfere with the synthesis of cell walls. The third, and most common, mode of action of antibiotics is inhibition of protein synthesis. As both microorganisms and animal cells grow and multiply, proteins play a key role in both the structure and the function of the organism. Some antibiotics are able to interfere with protein synthesis. Since there are differences in the chemical structure of the compounds used for protein synthesis in microbes and in animal cells, the latter are affected much less, although side effects resulting from inhibition of protein synthesis in human cells are not uncommon.

The tetracyclines are an important group of antibiotics, which interfere with protein synthesis in microorganisms. They are bacteriostatic because of this interference, which inhibits the growth of bacterial cells. The first of the tetracyclines was chlortetracycline (Aureomycin®). It was introduced in 1948. As the name suggests, there are four rings in the tetracycline molecules, which differ in the variety of groups attached to the ring structure. Simple tetracycline (Achro-

CHEMO-
THERAPY
FOR THE
TREATMENT
OF DISEASE

377

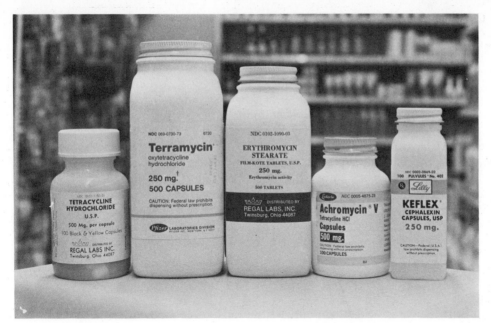

Figure 18.9 Antibiotics. (Photo by James J. Kane, Jr.)

mycin®, Tetracin®) and oxytetracycline (Terramycin®) are other important tetracyclines.

Tetracycline

Chlortetracycline   Oxytetracycline

The tetracyclines are broad-spectrum antibacterial agents. Some are semisynthetic modifications of the basic tetracycline structure that is produced by several microorganisms, many of which have been found by large-scale screening of
microbes obtained from soil samples. Widespread use of the tetracyclines has decreased their effectiveness due to development of resistant microorganisms.

Streptomycin, chloramphenicol (Chloromycetin®) and erythromycin are other antibiotics that were originally isolated from soil samples and have seen considerable use over a period of many years. Each one has a broad spectrum of activity, although each is subject to development of resistant organisms. They are all bacteriostatic agents that act by inhibition of protein synthesis.

Streptomycin was first isolated in 1944. It was the first useful antibiotic against a broad spectrum of Gram-negative bacteria.

Erythromycin is often the antibiotic of choice for treatment of individuals with an allergy to penicillin.

Chloramphenicol has had a very controversial history. It was isolated in 1947 as a fermentation product of a soil microorganism. It is marketed under the brand name, Chloromycetin®, by Parke, Davis and Co. It is one of the few antibiotics that is more economical to produce synthetically. It has a rather broad spectrum of activity and has, therefore, seen heavy use, but it is no longer considered to be the drug of choice for any illness except typhoid fever. The problem is that there is a rare occurrence of fatal aplastic anemia as a side effect—a fact known as early as 1950.

Antiviral Chemotherapy

Due to the parasitic nature of viruses, the success of chemotherapy has been very limited. Since virus particles invade and take over normal cells (see Figure 17.6), many drugs with the potential to interfere with the development of viruses also tend to interfere with the function of normal cells. In addition, clinical symptoms of a viral disease often follow the time when the infection is at a peak. Therefore, the treatments that are available are only really effective for prevention.

Amantidine (Symmetrel®) is one of the better-known antiviral agents, although it is reportedly only effective against the Asian flu, which began its

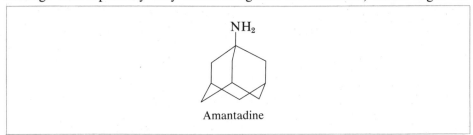

NH$_2$

Amantadine

spread in 1957. Symmetrel was approved for use in 1967, but by that time the Asian flu was almost nonexistent. In the following year the Hong Kong flu began its attack on the world. The effectiveness of Symmetrel against the Hong Kong variety of influenza has been the subject of a major controversy, in which some scientists have been accused of trying to suppress FDA approval of the drug for use against the Hong Kong flu in order to protect their interest in the production of vaccines. It has been noted that the success of antibiotics largely curtailed research on vaccines against bacterial diseases. Some individuals who have opposed the marketing of Symmetrel for use against the Hong Kong flu have been accused of trying to prevent successful development of antiviral chemotherapeutic agents in order to protect their own research efforts, which are directed toward production of vaccines. In any case, the Hong Kong flu is no longer of any significance since a large percentage of the world's population now has antibodies against this disease. The more interesting question is whether Symmetrel or other potential antiviral agents will be subjected to the same hurdles before gaining approval for use against other virus diseases. Vaccines are regarded as quite safe for prevention of viral diseases, whereas an antiviral agent must undergo very thorough and time-consuming testing, which may, in some instances, not be completed until a serious epidemic has passed, at which time the drug may be too late to be of any real significance. This is not to suggest that such thorough testing is undesirable, but it does help to explain why successes in the development of antiviral agents are almost nonexistent. Very often, it is a major pharmaceutical manufacturer that finances the research on a drug, but the instance previously described eliminates the appeal for research in this area because the chances for making a profit are quite slim and, therefore, not worth the risk.

CHEMO-
THERAPY
FOR THE
TREATMENT
OF DISEASE

379

The generally accepted value of Symmetrel lies in its effectiveness (50–70%) in prevention of illness in persons exposed to Asian flu (9). It acts by preventing the penetration of normal cells by virus particles.

Another well-known antiviral agent is methisazone. This drug inhibits the synthesis of proteins under the influence of viruses. It is quite effective against a

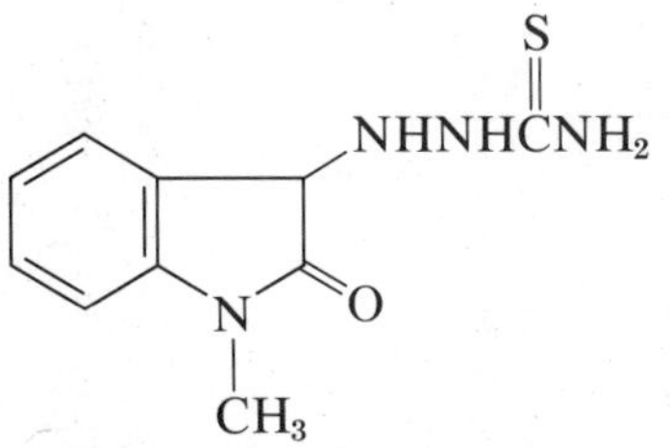

methisazone

major type of smallpox virus, but since smallpox has become such a rare disease, the manufacturers "have been left with the proverbial cure looking for a disease (10)."

Idoxuridine, abbreviated IUdR, was actually the first antiviral agent. It has been shown effective against a virus known as herpes keratitis, a severe eye infec-

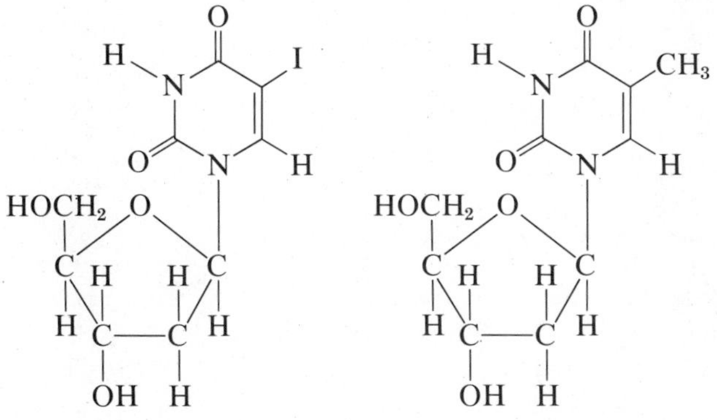

IUdR   Thymidine

tion that is a cause of blindness in many persons who are infected by the virus. IUdR apparently exhibits its antiviral activity as an antimetabolite due to its similarity to thymidine (T), which is one of the building blocks of DNA (see Chapter Two).

A number of other antiviral agents are under investigation, but prospects are not too bright. One intriguing approach is the stimulation of the body's own defenses. Even without having any immunity, the body has a defense mechanism in the form of substances called *interferons*. **Interferons** are carbohydrate-containing proteins. They are synthesized and released by cells as a result of a virus infection and inhibit the multiplication of viruses in other cells. One possible approach to combating a virus infection would be to administer interferon, but there are problems that make this impractical. For one thing, although interferons are effective against a wide variety of viruses, if an interferon is to be used in humans, it must be isolated from humans. This makes large-scale production very costly and difficult and limits the amount available. In addition, being a protein, an interferon could not be given orally since it would be destroyed (metabolized) like any other protein by the digestive enzymes.

This problem is analogous to the use of insulin by a diabetic. Insulin is also a protein and, therefore, cannot be administered orally. There are some substances that are incorrectly called "oral insulins"; they are not insulins at all. An example is tolbutamide (Orinase®), which tends to cause a lowering of the high level of blood sugar typically found in diabetics. In this respect, the result is the same as that achieved by injecting insulin, which is routinely isolated from the

Figure 18.10   The oral insulins. (Photo by James J. Kane, Jr.)

$$CH_3 - \langle \text{benzene ring} \rangle - SO_2NHCNHCH_2CH_2CH_2CH_3$$

$$\overset{O}{\overset{\|}{}}$$

tolbutamide
Orinase®

CHEMO-
THERAPY
FOR THE
TREATMENT
OF DISEASE

381

pancreas of cattle and swine. In fact, some of the oral drugs used by diabetics are thought to stimulate the normal release of insulin by the pancreas.

Here again, the analogy between interferon and insulin may apply since research is underway in an attempt to find drugs that will promote the release of interferon when administered. Unfortunately, even though interferon has been known since 1957, there has not been any success in developing interferon-inducing agents that can be safely administered.

Anticancer Therapy

In Chapter Three, ionizing radiation was discussed as a tool for treatment of cancer. A cancer cell is a normal cell that has become unresponsive to the usual chemical signals that control the activity of the cell. Since more rapidly-dividing cells are more susceptible to ionizing radiation, the radiation is somewhat selective, although there is always some damage to normal cells. This is a kind of side effect that accompanies the treatment.

In using chemotherapy, the strategy is much the same. Most antineoplastic* drugs that are in use interfere with the normal reproduction of DNA, and since DNA (the chromosomes) controls all cellular activity and is more rapidly reproduced in cancer cells, these cells are most susceptible.

On the other hand, cancer cells actually grow more slowly than some normal cells found in the body, such as the cells of the bone marrow, hair, and the gastrointestinal tract. Therefore, radiation or chemotherapy must also be selective in where it attacks in order to minimize adverse side effects on normal cells. In employing radiation therapy, an external source of radiation may be focused on the area of the tumor cells or, in the case of internal radiation therapy using a radioactive drug, i.e., a radiopharmaceutical, the drug that is selected should be one that will concentrate (localize) as much as possible in the tissues under attack by cancer cells in order to minimize side effects.

Simple chemotherapy, as opposed to the use of radiopharmaceuticals, should also be done using drugs that are specific for the tissues invaded by cancer cells. Unfortunately, there are two very analogous problems that make cancer therapy of all types a very serious proposition. The first of these is the phenomenon of *metastasis,* which is the spread of cancer to other tissues in the body. When it occurs, the spread may cause invasion of tissues neighboring directly on the original site of the cancer, or cancer cells may break away from their origin and lodge in other areas of the body. The other problem with therapy is that the effective anticancer drugs do not specifically concentrate in the problem areas. In other words, cancer cells and anticancer agents may affect many areas of the body. This means that many different types of normal cells may be invaded by

---

* A *neoplasm* is an abnormal growth or tumor.

cancer or may be adversely affected during cancer chemotherapy. In short, chemotherapy alone is not often effective, although it has been successfully used in combination with surgery and radiation therapy.

382 Another Use of Antibiotics

Antibiotics have been employed as feed additives for animals for almost 25 years. During this time, many studies have clearly confirmed the favorable effect of the antibiotics, which lead to faster growth with production of more meat, eggs, and milk for the same amount of feed consumption by the animals. As a result, the level of antibiotic consumption by farm animals is comparable to that by humans—reportedly 88.4 million dollars worth in 1973 alone (11, 12).

There are two theories as to why antibiotics exert their effects on growth. It has been suggested that antibiotics may control the population of microorganisms in the intestine in a way that benefits animals. Another theory proposes that antibiotics prevent disease so that the full metabolic capability of the animal is geared to maximum growth rather than to fighting disease. For example, the synthesis of antibody proteins (the immunoglobulins) is an energy (ATP) consuming process and the energy must be provided by metabolism of carbohydrates, etc. If an animal does not have to utilize its own defenses against disease, it may be expected to grow faster. On the other hand, the fact that there are two theories to account for the observed facts suggests that it is not known how antibiotics stimulate growth.

The use of antibiotics in feed has the obvious advantage of increasing supplies of meat, eggs, and milk so that the cost to the consumer is potentially lower, but the practice has also been severely criticized. There are two reasons why this application of antibiotics could have adverse effects. The reasons are resistance and allergy. As noted earlier, the indiscriminate use of antibiotics tends to encourage the development of microorganisms that are resistant to these antibiotics. The more often antibiotics are used, the more likely it is that resistant microbes will appear.

The problem of allergy is potentially even more serious. Some individuals are allergic to antibiotics and may exhibit very severe symptoms when these drugs are consumed. Antibiotics are not given to animals for several days before slaughter to provide time for complete metabolism of the drugs. This would seem to prevent direct exposure of humans to antibiotics assuming that the metabolism is complete, although farm animals could still serve as a breeding ground for resistant microbes. In any case, there is no conclusive evidence of the occurrence of any large number of infections or epidemics in animals or humans due to the use of antibiotics in animal feeds.

SUMMARY

Drugs are usually identified by either the generic name or a trade name (brand name). A company may hold a patent on a drug for 17 years, during which time it has a monopoly on the drug that allows the company to make a reasonable profit and offset the tremendous costs of research that often go into the development and testing of a drug. In fact, the drug industry is one of the most research-oriented of all industries. At the same time, certain sectors of the industry also spend heavily in promotion of their products and include in this effort an attempt to convince physicians and consumers that brand-name drugs are generally superior to the less expensive generic brands, although the evidence suggests that they are equivalent in most cases. There is the possibility that different brands of a drug might not be equivalent, because of the variety of additives that are generally present in tablets and capsules, but production and marketing methods usually eliminate differences between the drugs.

CHEMO-
THERAPY
FOR THE
TREATMENT
OF DISEASE

383

There are many problems that have to be overcome in developing drugs. Sometimes it is necessary to market a drug in a prodrug form, which is converted to the active form in the body.

Antibiotics are one of the major types of drugs that help man to combat infectious diseases. The beginnings of antibiotic therapy date back to the early part of this century, but this form of chemotherapy received its biggest boost from the development of the sulfa drugs and the penicillins in the 1930s and 1940s. Among the best-known antibiotics are those with the suffix "-mycin," which signifies that they were originally isolated from microorganisms found in soil samples.

Since antibiotics are usually very complex molecules, which are very difficult to synthesize, microorganisms are normally employed to produce them even when they can be synthesized in the laboratory. These antibiotics are called biosynthetics. Semisynthetic antibiotics are produced by letting a microorganism do part of the job of synthesis, after which the synthesis is completed by the chemist.

The choice of an antibiotic by a physician is sometimes a very crucial decision. Laboratory analysis of the sensitivity of a disease-causing microorganism to various antibiotics may guide the physician, but this requires that the microbe be isolated and grown (cultured) before determining its sensitivity. It routinely requires a few days to carry out this procedure and, if a disease is life-threatening or is accompanied by severe symptoms, it may not be practical to wait for the results of the analysis. Therefore, the physician will often choose an antibiotic with a broad spectrum of activity or, perhaps, a combination of antibiotics so as to increase the chances of using an effective antibiotic. Although such an approach may be best for the individual patient, it is analogous to the use of insectides in fighting insects, since favorable microorganisms may also be killed and dangerous resistant organisms may be encouraged to flourish.

Antibiotics function as either bacteriostatic or bacteriocidal agents. Bacteriostatic agents interfere with a chemical process that is required for normal growth of microorganisms, e.g., protein synthesis, whereas bacteriocidal agents will actually kill susceptible microbes.

New research on antibiotics is often geared to finding drugs with an improved spectrum of activity and to combating microbes that have become resistant to available antibiotics.

Chemotherapy against viruses and cancer has had very limited success due to the nature of these diseases, although the latter can sometimes be treated with drugs in combination with surgery or radiation therapy.

Antibiotics are also used as supplements in animal feeds to increase growth and production of meat, eggs, and milk. The reason for success is not known and the practice has been criticized as contributing to the development of resistant microbes.

PROBLEMS

1. Give a definition or example of each of the following:

   a) bacteriostatic
   b) bacteriocidal
   c) generic name
   d) brand name or trade name
   e) prodrug
   f) antibiotic
   g) broad-spectrum antibiotic

   h) penicillinase
   i) "magic bullet"
   j) antimetabolite
   k) "miracle drugs"
   l) culturing
   m) semisynthetic antibiotic
   n) metastasis

2. How can two chemically-equivalent drugs not be bioequivalent?

3. What are some of the major drug additives?

4. What is the function of starch paste in the production of drug tablets?

5. Why are some drugs coated?

6. What is the "peak and valley effect" for drugs?

7. What are sulfa drugs? How do they work?

8. What does the suffix "-mycin" signify?

9. How can scientists cause changes in the synthesis of antibiotics?

10. Why is diarrhea a frequent side effect of antibiotic therapy?

11. Why does Bristol Laboratories market ampicillin under the brand name Polycillin®?

12. Why is penicillin V better than penicillin G in the treatment of disease?

13. Why do penicillins have unusually low toxicity?

14. What are the advantages and disadvantages of broad-spectrum antibiotics?

15. Why has it been difficult to produce a drug that is effective against viruses?

16. What are interferons? What are the problems involved in the production and administration of interferons?

17. It is not strictly correct to classify aspirin as a prodrug even though it is frequently done. Explain.

REFERENCES

1. "How To Pay Less For Prescription Drugs." *Consumer Reports,* January 1975, p. 48.

2. Silverman, M., and Lee, P.R. *Pills, Profits, and Politics.* Berkeley, California: University of California Press, 1974, Chapter 3.

3. Burack R., and Fox, F.J. *The New Handbook of Prescription Drugs.* New York: Ballantine Books, 1975, p. 3.

4. *Ibid.,* Chapters 1–3.

5. "On New Antibiotics." *Chemistry 48* (11) (1975): 14.

6. Pratt, W.B. *Fundamentals of Chemotherapy.* Toronto: Oxford University Press, 1973, pp. 100–101.

7. Buyske, D.A. "Drugs From Nature." *Chemtech,* June 1975, p. 361.

8. Burack and Fox, *Ibid.,* pp. 426–8.

9. *AMA Drug Evaluations.* 2nd ed. AMA Department of Drugs. Acton, Mass.: Publishing Sciences Group, 1973, p. 899.

10. Maugh, T.H. II, "Chemotherapy: Antiviral Agents Come of Age." *Science 192* (1976): 128.
11. Majtenyi, J.Z. "Antibiotics—Drugs From The Soil." *Chemistry 48* (3) (1975): 15.
12. Finland, M., and Barnes, M.W. "Salmonellosis and Shigellosis at Boston City Hospital." *Journal of the American Medical Association 229* (9) (1974): 1183.
13. Levinson, A.S. "The Structure of Salvarsan and the Arsenic-Arsenic Double Bond." *Journal of Chemical Education 54* (2) (1977): 98.

CHEMO-
THERAPY
FOR THE
TREATMENT
OF DISEASE

385

C HEMOTHERAPY is generally taken to mean the use of drugs to bring about a cure for a disease condition, and yet treatment of symptoms is often equally important. In the case of viral diseases, treatment of symptoms is really the only course of action. In this chapter, we will consider some of the kinds of drugs that are used for treatment of symptoms and some of the related problems of drug abuse.

ANALGESICS

Pain is often the first indication of a malfunction or injury, and its treatment is usually the first item of concern. There is an almost incredible array of analgesics (pain relievers) available, both over the counter and by prescription.

Aspirin heads the list of over-the-counter drugs. It is available alone (plus binder and filler) and in combination with other analgesics or other drugs. Sales of aspirin in all forms are estimated at about 200 tablets of 5 grains (324 milligrams) each per person per year.* About half of the aspirin is consumed in the form of combination pain relievers and cold remedies, which often contain a second analgesic, a decongestant, an antihistamine, or an antacid.

Aspirin is the generic name for the compound, acetylsalicylic acid. It was first synthesized in 1853, but it was not marketed until 1899. It was patented in the United States by the Bayer Corporation until 1917.

As noted earlier, salicylic acid is also an effective pain reliever but addition of the acetyl group greatly reduces the corrosive effect of the molecule on the stomach. Unfortunately, the effect is not totally eliminated and most people do suffer some damage to the lining of the stomach resulting in an insignificant (normally 0.5–2.0 milliliters) loss of blood when they consume aspirin (1). For a very small percentage of the population, the effect of aspirin is more severe. For these

* Children's aspirin contain only 1¼ grains of aspirin per tablet.

CHEMOTHERAPY FOR THE TREATMENT OF SYMPTOMS AND DRUG ABUSE

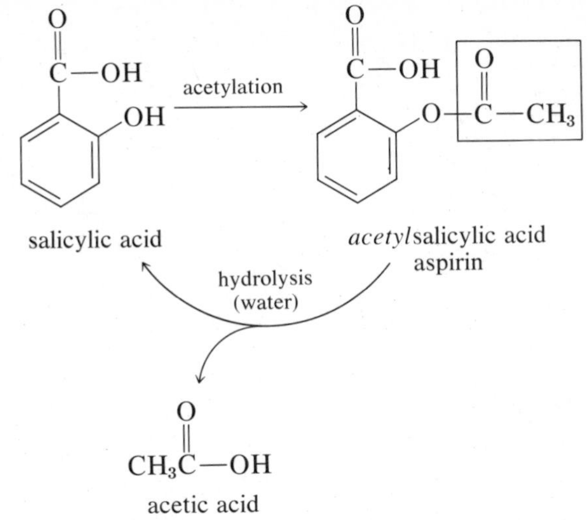

O
‖
C—OH
   acetylation ⟶
OH

salicylic acid

O
‖
C—OH
   O
   ‖
O—C—CH₃

*acetyl*salicylic acid
aspirin

hydrolysis
(water)

O
‖
CH₃C—OH
acetic acid

individuals, acetaminophen (Datril®, Tylenol®, Tempra®, Liquiprin®) is a satisfactory alternative, although it is more expensive.

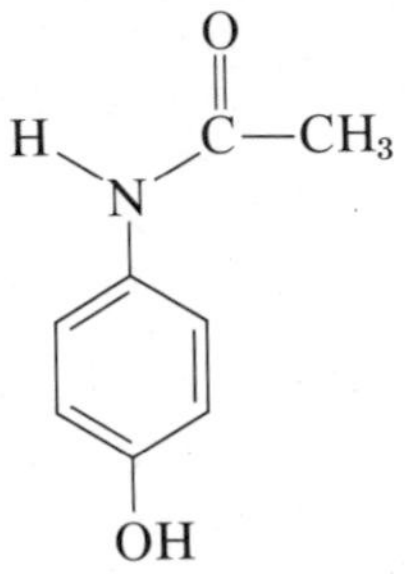

O
‖
C—CH₃
H
N

acetaminophen
Datril®, Tylenol®, Tempra®, Liquiprin®,
Trilium®

OH

Aspirin may undergo a reaction with water to produce salicylic acid and acetic acid. This process is essentially the reverse of the reaction used to make the aspirin from salicylic acid as shown in the scheme on page 361. The acetic acid can be readily detected as the odor of vinegar (5% acetic acid) upon opening an old bottle of aspirin that has had repeated exposure to humidity in the air. Although the smell of acetic acid may be very strong, the amount of acetic acid present is normally very small. In other words, the odor does not usually indicate any significant decomposition of the aspirin.

In addition to its analgesic properties, aspirin is also an effective antipyretic and antiinflammatory agent, i.e., it reduces fever and swelling, respectively. Since swelling often occurs in injured tissues and rheumatic joints, the antiinflammatory property is often very important for the relief of pain, particularly in the treatment of arthritis. Acetaminophen also acts against pain and fever but does not reduce inflammation.

Over-the-counter remedies for treatment of pain from arthritis, such as Arthritis Pain Formula® and Arthritis Strength Bufferin®, contain $7\frac{1}{2}$ grains of aspirin in each tablet compared to the usual 5 grains. A simple calculation suggests that 3 regular aspirin would do the same as 2 of these arthritis-strength tablets,

*"We **treat** illness here, Miss Rothbart. If you insist on being **cured**, you'll have to go to some quack."*

Handelsman, Copyright © 1976 Punch/Rothco. Reprinted by permission.

although the latter usually contain an antacid, which may minimize stomach irritation.

The number of combination remedies available to the consumer is almost overwhelming, although the variations from one product to the next are not very significant. In fact, a thorough analysis of these combinations shows that the major differences between plain aspirin and most combinations are the tremendous promotion of the latter and much higher prices. The promotional effort by the pharmaceutical industry has been estimated at about 70 million dollars annually (2). All of the combinations contain an analgesic—aspirin, acetaminophen, or both (e.g., Excedrin®). Salicylamide is a modification of the salicylic acid mol-

salicylamide

ecule. It also has analgesic and antipyretic properties, although it is regarded as too weak and unreliable to be useful (3). Nevertheless, salicylamide is included in some combination remedies. An extensive list of combination remedies appears in Table 19.1.

TABLE 19.1 SOME WELL-KNOWN COMBINATION REMEDIES

| | INGREDIENTS | | | | |
PRODUCT NAME	ANALGESIC	DECON-GESTANT	ANTI-HISTAMINE	ANTACID	OTHER
Alka-Seltzer (Blue)	aspirin			$NaHCO_3$	citric acid*
Alka-Seltzer Plus	aspirin	phenylpropanol-amine	chlorphenir-amine	$NaHCO_3$	citric acid*
Allerest		phenylpropanol-amine	pyrilamine, methapyrilene		
Anacin	aspirin				caffeine
APC	aspirin phenacetin				caffeine
Arthritis Pain Formula	aspirin			$Al(OH)_3$, $Mg(OH)_2$	
Bufferin	aspirin			$MgCO_3$, aluminum glycinate	
Contac		phenylpropanol-amine	chlorphenir-amine		belladonna alkaloids
Cope	aspirin		methapyrilene	$Al(OH)_3$, $Mg(OH)_2$	caffeine
Coricidin	aspirin		chlorphenir-amine		
Coricidin D	aspirin	phenylpropanol-amine	chlorphenir-amine		
Dristan	aspirin	phenylephrine	phenindamine	$Al(OH)_3$, $Mg(OH)_2$	caffeine
Empirin	aspirin phenacetin				caffeine
Excedrin	aspirin salicylamide phenacetin		methapyrilene†		caffeine
NyQuil	acetaminophen	ephedrine	doxylamine		dextro-methorphan‡, alcohol (25%)
Day Care	acetaminophen	phenylpropanol-amine			dextro-methorphan‡, alcohol (7%)
Romilar	acetaminophen	phenylephrine	chlorphenir-amine		dextro-methorphan‡
Vanquish	aspirin acetaminophen			$Al(OH)_3$, $Mg(OH)_2$	caffeine

* causes an effervescent action
† found in Excedrin PM only
‡ a cough suppressant

Many of the remedies also contain a decongestant, and several contain caffeine. A few contain an antacid or an antihistamine. APC tablets are the old standard combination remedy, which has often been prescribed by doctors for relief of symptoms of cold, flu, etc. The *A* signifies aspirin, *P* is phenacetin, and *C* is caffeine. Phenacetin is also an effective analgesic and antipyretic, but it is not as effective as aspirin in reducing inflammation. Phenacetin has some analgesic activity of its own, although part of this property is due to its metabolism to acetaminophen.

phenacetin   acetaminophen

Recent studies have implicated phenacetin as a cause of some side effects—kidney damage and some types of anemia—although the evidence leading to the conclusion in these studies has been criticized. It was once a frequent ingredient of over-the-counter remedies, but most manufacturers have discontinued its use.

Caffeine is a vasoconstrictor. This means that it causes contraction of blood vessels. Since migraine headaches are attributed to dilation (expansion) of blood vessels in the head, caffeine has potential for treatment, although the amount present in APC formulations is considered insignificant. Perhaps this is fortunate, since caffeine is a stimulant and continual consumption of it might cause some sleepless nights at times when sleep might be the best medicine available. For the sake of comparison, some cold remedies contain 15–30 milligrams of caffeine per tablet, whereas a No-Doz tablet or a cup of coffee contain about 100 milligrams each.

caffeine

The same facts apply when decongestants are included in a combination remedy. Phenylephrine is the most commonly used decongestant. It is supposed to reduce swelling of the nasal passages. It is effective as a nasal spray (Neo-Synephrine® and other brands) for the relief of stuffy nose, but in the amounts normally found in combination remedies, it is ineffective (2). Another common decongestant is phenylpropanolamine.

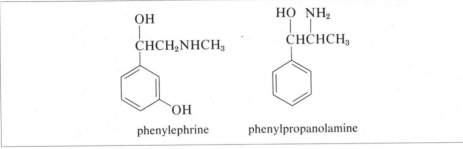

phenylephrine   phenylpropanolamine

CHEMO-
THERAPY
FOR THE
TREATMENT
OF SYMPTOMS
AND DRUG
ABUSE

391

Antihistamines are also frequent ingredients of combination remedies. They are clearly effective for relief of allergies, which stimulate the release of histamine (see Chapter Seventeen for details), but as a remedy for runny nose due to a common cold, they are reported to be ineffective (2). Regardless of their effectiveness, they do cause drowsiness as a side effect, so that they are potentially hazardous to anyone who drives an automobile after consumption.

Finally, an antacid is sometimes included in combination remedies or with aspirin alone. One example is buffered aspirin (Bufferin®).* It is often claimed that the antacid speeds absorption of the aspirin, but the evidence does not confirm this. An antacid is an alkaline substance that reacts with stomach acid. The strategy behind the use of a combination of aspirin and an antacid is that the acid in the stomach will immediately react with the antacid in the tablet, thus causing a more rapid disintegration of the tablet and release of the analgesic. One need only place an aspirin tablet in water to see how quickly it disintegrates, even without any acid or antacid being present.

There is also the claim that the antacid reduces stomach irritation. There is no clear evidence *for* or *against* this claim since it is virtually impossible to measure the effect. It was noted previously that aspirin can irritate the stomach, but an antacid will not eliminate this simply by reacting with some of the HCl in the stomach; so the claim is apparently untrue.

There is one example of an antacid-aspirin combination that may help reduce stomach irritation caused by the aspirin. It is the product sold under the brand name Alka-Seltzer®, which contains the antacid, sodium bicarbonate ($NaHCO_3$).† The effervescent action (the "Fizz Fizz"), which occurs when the Alka-Seltzer tablet comes in contact with water, provides a mixing which helps to dissolve the aspirin in solution. Since the analgesic is in solution when the Alka-Seltzer is consumed, it passes through the stomach more quickly. This would seem ideal, although regular use of antacids is not recommended. Further discussion of antacids appears in a later section.

Treatment of Severe Pain

When pain is more severe, a physician usually enters the picture and prescribes an arsenal of more potent prescription analgesics, most of which are addic-

* A *buffer* is a substance that resists a change of pH when an acid or base is added. This definition does not apply to so-called buffered aspirin, in which an alkaline substance is present and will cause an increase in the pH of the stomach.
† Alka-Seltzer is actually marketed in two forms. One is known as Alka-Seltzer Blue and the other as Alka-Seltzer Gold. The latter does not contain aspirin.

tive. Among them, propoxyphene is the most often used prescription analgesic. It was the third most often prescribed drug in 1973 (4). It is available in a variety of forms under the brand name, Darvon®, and under the generic name, propoxy-phene.*

In spite of its popularity in the United States (about 77 are consumed per capita each year), it is reported that Darvon® in its usual dose is not as effective as aspirin or acetaminophen, and most studies show it is no more effective than a placebo (5). The point is: pain is often a very subjective thing for which treatment is often impossible to evaluate. Therefore, an expensive analgesic like Darvon® may seem to be a better choice than simple aspirin. It is the old story of "you get what you pay for." At least the pharmaceutical industry would have us think so and many individuals suffering from pain are willing, and even eager, to believe that it is true, whether it is or not.

Although the potential for addiction does exist, it exists to a lesser degree with Darvon than with many other drugs, so that it is not subject to narcotic regula-tions of the Controlled Substances Act. Addiction and certain other effects of the drug may be classified as side effects, although it has been suggested that the most important side effect of Darvon® is its price (5).

The other major class of prescription pain killers is the opiates. These in-clude naturally-occurring substances that can be extracted from opium (morphine, codeine), a major semisynthetic (heroin), and pure synthetics (Demerol®, metha-done). The last of these are not derived in any way from opium, but their action and potential for addiction is comparable to the true opiates. The opiates are dis-cussed more fully in the section on drug abuse.

As a pain killer, morphine is about 50 times as potent as aspirin. On the other hand, aspirin is not addictive. Morphine was widely used in the Civil War as a pain killer, but more than 100,000 soldiers became addicts—a problem that be-came known as "soldiers' disease." Codeine is only about one-sixth as potent as morphine, but codeine can be taken orally, whereas morphine must be injected.

Meperidine (Demerol®) is the primary synthetic substance used to treat severe pain. It is between morphine and codeine in potency and is also addictive.

---

* Like a great many drugs, propoxyphene is not marketed as the neutral compound. Compounds used as drugs can often be converted into salts by reaction with acids or bases. Propoxyphene is a typical example of a drug that can be converted into a hydrochloride salt by reaction with HCl due to the pres-ence of a basic nitrogen atom like the one found in ammonia, NH_3, which is converted to ammonium

$$NH_3 + HCl \longrightarrow NH_4^+ + Cl^-$$

chloride when it reacts with HCl. Although the reaction looks more complicated in the case of propox-yphene, it is also a simple neutralization reaction in which an acid and a base react to produce a salt. After conversion to a salt, the stability and ease of handling are often improved.

CHEMO-
THERAPY
FOR THE
TREATMENT
OF SYMPTOMS
AND DRUG
ABUSE

393

meperidine
Demerol®

DRUGS USED IN SURGERY

A well-planned sequence of drugs normally accompanies major surgery, beginning the night before surgery and extending through to the treatment of severe pain in the operating room and in the immediate post-operative period.

The night before surgery, the patient is usually given a tranquilizer to relax him and assure a good night's sleep. During surgery, a combination of drugs is administered to achieve anesthesia, analgesia, and muscle relaxation. Following surgery, morphine or Demerol® may be given for relief of severe pain.

Of these drugs, the most interesting are the ones used during surgery. The old standby is ether—actually diethyl ether. It was first used in dentistry in 1847. It is an effective anesthetic and analgesic and also promotes good muscle relaxation. In addition, there is a wide gap between the amount that causes unconsciousness and the lethal dose, so that it is extremely safe to administer. Like several of the common anesthetics, ether is easily vaporized and administered to a patient in a stream of oxygen. The stream usually contains 10% to 30% ether, while inducing anesthesia and 5% to 15% to maintain the condition. A number of side effects may accompany the use of ether. In addition, it is extremely flammable and explosive, so that its use has been greatly curtailed.

$$CH_3CH_2—O—CH_2CH_3 \qquad \text{diethyl ether}$$

Chloroform, $CHCl_3$, was first used in 1847. It is nonflammable, but it often causes liver damage, and there is a narrow gap between the level of anesthesia and the lethal dose. It is seldom used for surgery.

At the present time, the primary drugs used during surgery are Sodium Pentothal®, nitrous oxide (N_2O), enflurane and halothane. Sodium Pentothal® is a potent barbiturate-hypnotic, which rapidly induces unconsciousness, but does not provide sufficient analgesia and muscle relaxation to be used alone. Enflurane and halothane are the two most popular anesthetics. They are ether-like molecules,

Sodium Pentothal®   halothane   enflurane

which are highly substituted with halogen (fluorine, chlorine, bromine) atoms that make the compounds nonflammable. They are safe and potent substances, which may be used to maintain the anesthesia that has been induced by Sodium Pentothal®.

Enflurane and halothane can also be used to induce anesthesia in place of the Sodium Pentothal®, but most people prefer not to be put to sleep with a mask over the face, whereas once a patient is unconscious, enflurane or halothane may be administered through a mask in a stream of nitrous oxide and oxygen gases.

Normal, dry air contains about 78% nitrogen (N_2) and 21% oxygen (O_2), plus trace amounts of many other gases (e.g., argon, CO_2, neon, and helium.) The stream delivered to a patient in surgery contains about 75% N_2O and 25% O_2 with about 1–2% halothane or enflurane added. Enflurane is the preferred anesthetic of the two because it is also a good analgesic and muscle relaxant. The nitrous oxide is also an excellent anesthetic and analgesic on its own. It is a colorless, sweet-tasting gas, sometimes known as "laughing gas."

Local Anesthetics

For very minor surgery, it is frequently possible to use a local anesthetic, such as procaine (Novocaine®) or lidocaine (Xylocaine®). Either may be injected into the site of the surgery or applied topically (to the surface) to the mucous membranes of the eye, nose, throat, or urethra.

lidocaine (Xylocaine®)

procaine (Novocaine®)

Ethyl chloride is a colorless liquid, which boils at 12°C (54°F), so it is a gas at room temperature, but it can be liquified under pressure in a sealed container. When the liquid is sprayed on the surface of the skin, it quickly evaporates. When a liquid evaporates from a surface, the surface is cooled. This is a

$$CH_3CH_2Cl \quad \text{ethyl chloride}$$

familiar phenomenon, which is noticed when one steps out of water after a shower or a swim. Some of the moisture evaporates and causes a cold sensation. The heat lost during evaporation is called the *heat of vaporization*. Looking at it another way, the heat of vaporization is the energy needed to cause vaporization. Water has a very high heat of vaporization. This is one reason why water is an excellent body fluid, since evaporation of perspiration from the body surface provides an extremely efficient cooling mechanism for the body.

The same strategy applies to the use of ethyl chloride as a local anesthetic. Due to its low boiling temperature, it evaporates very rapidly and cools the surface so quickly that it freezes tissues near the surface, making them insensitive to pain. It is often sprayed on a wound or other injured tissue to deaden pain, e.g., in sporting events. It can also be used for very minor surgery such as draining carbuncles.

CHEMO-
THERAPY
FOR THE
TREATMENT
OF SYMPTOMS
AND DRUG
ABUSE

395

ANTACIDS

Next to treating pain, one of the most frequent concerns is the treatment of indigestion caused by an overproduction of stomach acid (HCl). As in the case of the mild analgesics, the number of over-the-counter antacids is almost overwhelming, although the variation of ingredients is small. The most common antacids contain sodium bicarbonate ($NaHCO_3$), calcium carbonate ($CaCO_3$), magnesium carbonate ($MgCO_3$), aluminum hydroxide [$Al(OH)_3$], magnesium hydroxide [$Mg(OH)_2$], and magnesium trisilicate ($Mg_2Si_3O_8$). The antacid action of each of these and some of the less common antacids is illustrated by the following reactions in which HCl is consumed. The antacid ingredients of several of the commercial products are given in Table 19.2.

$$HCO_3^- \; + HCl \; \longrightarrow H_2CO_3 \; + Cl^-$$

bicarbonate (unstable)
as $NaHCO_3$
or $KHCO_3$

$$\downarrow$$

$$H_2O + CO_2$$

$$OH^- \; + HCl \; \longrightarrow H_2O + Cl^-$$

hydroxide
as $Mg(OH)_2$
or $Al(OH)_3$
or $NaAl(OH)_2CO_3$

$$CO_3^{-2} \; + 2 \, HCl \longrightarrow H_2CO_3 \; + 2 \, Cl^-$$

carbonate (unstable)
as $CaCO_3$
or $MgCO_3$
or $NaAl(OH)_2CO_3$

$$Si_3O_8^{-4} \; + 4 \, HCl \longrightarrow 3 \, SiO_2 + 2 \, H_2O + 4 \, Cl^-$$

trisilicate
as $Mg_2Si_3O_8$

$$C_6H_5O_7^{-3} \; + 3 \, HCl \longrightarrow H_3C_6H_5O_7 + 3 \, Cl^-$$

citrate
as $Na_3C_6H_5O_7$

$$C_4H_4O_6^{-2} \; + 2 \, HCl \longrightarrow H_2C_4H_4O_6 + 2 \, Cl^-$$

tartrate
as $Na_2C_4H_4O_6$

TABLE 19.2   MAJOR COMMERCIAL ANTACID PRODUCTS

PRODUCT NAME	ANTACID INGREDIENTS
Alka-Seltzer Blue	$NaHCO_3$ + citric acid*
Alka-Seltzer Gold	$NaHCO_3$ + citric acid
Bromo-Seltzer	$NaHCO_3$ + citric acid†
Brioschi	$NaHCO_3$ + tataric acid
Milk of Magnesia	$Mg(OH)_2$
Tums	$CaCO_3$
Alka 2	$CaCO_3$
Titrilac	$CaCO_3$
Amitone	$CaCO_3$
Maalox	$Mg(OH)_2$ + $Al(OH)_3$
Mylanta	$Mg(OH)_2$ + $Al(OH)_3$‡
Creamalin	$Mg(OH)_2$ + $Al(OH)_3$
Di-Gel	$Mg(OH)_2$ + $Al(OH)_3$‡
Camalox	$Mg(OH)_2$ + $Al(OH)_3$ + $CaCO_3$
Bisodol	$NaHCO_3$ + $MgCO_3$
Bisodol (low sodium)	$Mg(OH)_2$ + $CaCO_3$
Gelusil	$Mg_2Si_3O_8$ + $Al(OH)_3$
Rolaids	$NaAl(OH)_2CO_3$
Eno	$Na_2C_4H_4O_6$ + $Na_3C_6H_5O_7$

* contains aspirin
† contains acetaminophen
‡ contains a silicone defoaming agent

Different antacids have different advantages and disadvantages. Many individuals prefer effervescent antacids. These products are discussed in the following section.

Magnesium salts may have a laxative effect depending on the dose. In fact, $Mg(OH)_2$, milk of magnesia, is also sold for that purpose. Aluminum salts may be constipating, again depending on the dose. Several antacids are formulated with combinations of magnesium and aluminum salts to prevent side effects. Some ant-

Figure 19.1   Antacids. (Photo by James J. Kane, Jr.)

CHEMO-
THERAPY
FOR THE
TREATMENT
OF SYMPTOMS
AND DRUG
ABUSE

397

acids contain a pain reliever. Those antacids that contain sodium should be avoided by individuals on low-sodium diets.

Effervescent Antacids

Those products that contain sodium bicarbonate and citric acid (or tartaric acid) give an effervescent action when dissolved in water because of the evolution of carbon dioxide gas, which forms via the following reaction.

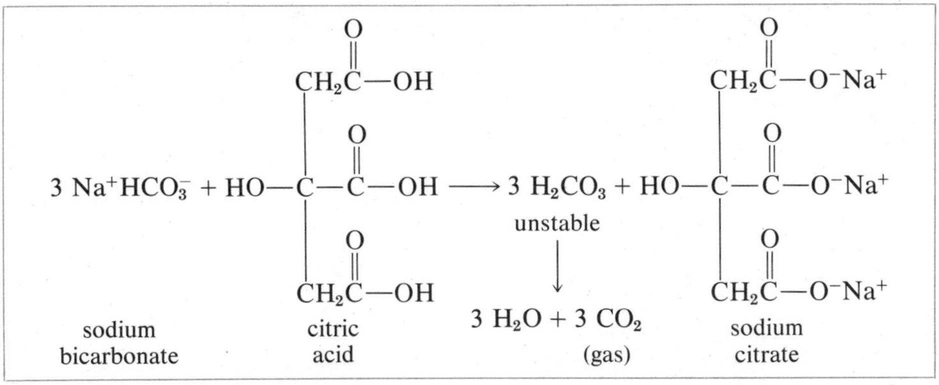

$$
3\ Na^+HCO_3^- + \underset{\substack{\text{citric}\\\text{acid}}}{HO-C\!-\!C-OH} \longrightarrow 3\ H_2CO_3 + \underset{\substack{\text{sodium}\\\text{citrate}}}{HO-C\!-\!C-O^-Na^+}
$$

sodium
bicarbonate

unstable

$$3\ H_2O + 3\ CO_2 \text{ (gas)}$$

Alka-Seltzer® comes sealed in a foil package which prevents contact with moisture in the air prior to opening. The previous reaction does not occur until the ingredients dissolve in water, but even at normal humidity, there is plenty of moisture in the air for the antacid and the acid to combine and react. After standing in contact with air, an Alka-Seltzer will appear "flat" when added to water, much like a carbonated soda which has been left opened and has lost its carbonation.

It has been suggested that the release of CO_2 by $NaHCO_3$ may provide relief from the discomfort of overeating by inducing belching, which aids in expulsion of swallowed air (6).

The sodium citrate, which is produced in the reaction of sodium bicarbonate with citric acid, may also accept hydrogen ions (H^+) and revert back to citric acid. Therefore, even the sodium citrate, which is produced when Alka-Seltzer and other effervescent antacids are dissolved in water, will act as an antacid. In fact, all of the citric acid in an Alka-Seltzer tablet is neutralized to sodium (or potassium) citrate; more than 85% of the bicarbonate is consumed by the effervescent action.

Sodium tartrate is formed from the reaction of tartaric acid and sodium bicarbonate when the antacid, Brioschi®, is dissolved in water. It acts in the same

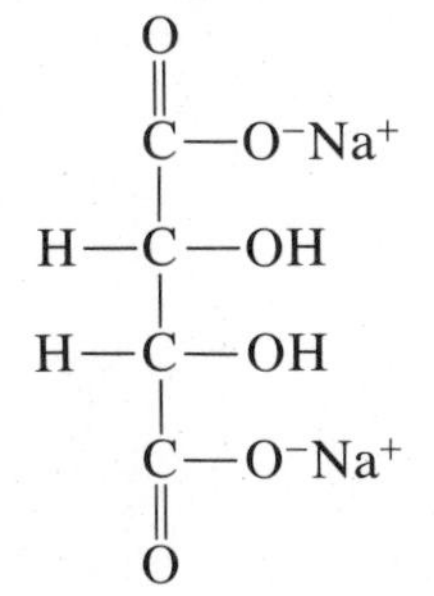

sodium tartrate

way as sodium citrate to accept hydrogen ions from stomach acid. A commercial product known as ENO® reportedly contains both sodium citrate and sodium tartrate, although it is advertised as an effervescent antacid, which means that at least some $NaHCO_3$ is present in the product as it is packaged.

TRANQUILIZERS

Diazepam (Valium®) and chlordiazepoxide (Librium®) were the number 1 and number 4 most prescribed drugs in 1974 (4). They are tranquilizers, or more specifically "minor tranquilizers"—a description meant to distinguish them from drugs that are used in the treatment of severe mental disorders. Valium® and Librium® together total over 25 million prescriptions each year or about one for every ten Americans (7). Miltown® and Equanil® are brand names for another important minor tranquilizer, which is also available under the generic name, meprobamate.

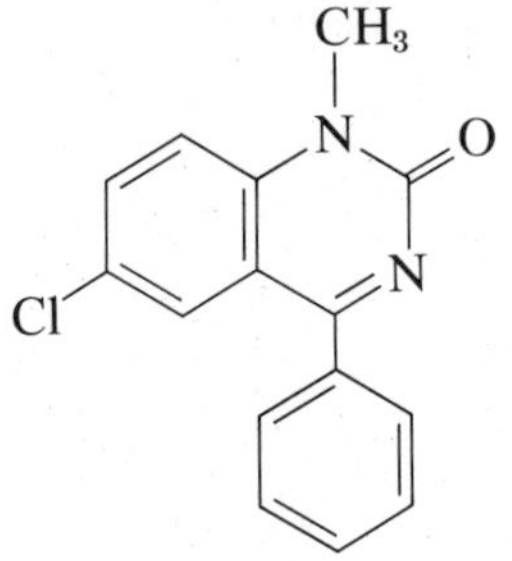

diazepam
Valium®

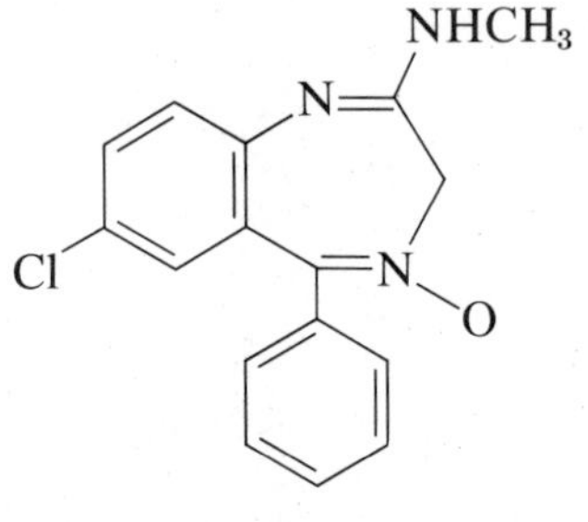

chlordiazepoxide
Librium®

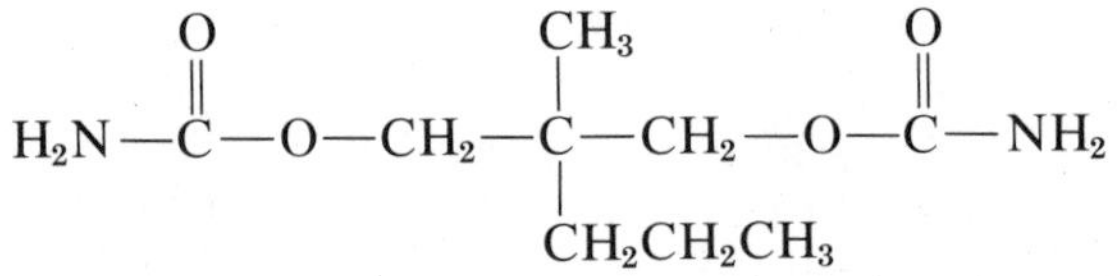

meprobamate
Miltown®, Equanil®

Tranquilizers are drugs which cause sedation without inducing sleep. They are used to relieve anxiety, excitement, and restlessness. Meprobamate was first used as a muscle relaxant.

Chlorpromazine, sold under the brand name Thorazine®, was the first and

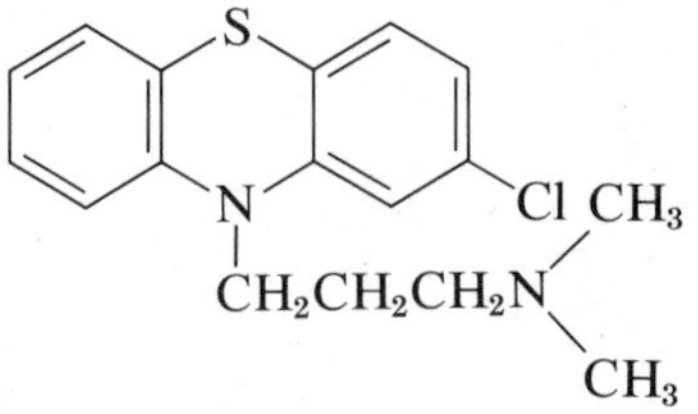

chlorpromazine (Thorazine®)

best known of the major tranquilizers. More than a dozen others are now available. Chlorpromazine was first tried on psychotic patients in 1952. It has been estimated that at least 50 million patients throughout the world have received chlorpromazine therapy. The success of this and related drugs (e.g., reserpine) is shown in Figure 19.2.

The minor, as well as the major tranquilizers are available only by prescription. Over-the-counter remedies that claim to be tranquilizers are generally mild sedatives or antihistamines, which may cause drowsiness as a side effect. Automobile driving should be avoided when taking any of these drugs. The antihistamine most often used is methapyrilene. It is a fast-acting but short-duration antihistamine that is the major ingredient in such well-known products as Sleep-Eze®, Nytol®, Sominex®, Cope®, Compoz®, and Excedrin PM®. Scopolamine and pyrilamine are other compounds that cause drowsiness as a side effect.

CHEMO-
THERAPY
FOR THE
TREATMENT
OF SYMPTOMS
AND DRUG
ABUSE

399

Figure 19.2   Impact of reserpine and chlorpromazine on resident patients in State and County Mental Hospitals. Source: United States Department of Health, Education, and Welfare, National Institute of Mental Health, Biometry Branch. Adapted from *Pills, Profits, and Politics* by Milton Silverman and Philip R. Lee, Copyright © 1974, The Regents of the University of California, by permission of the University of California Press.

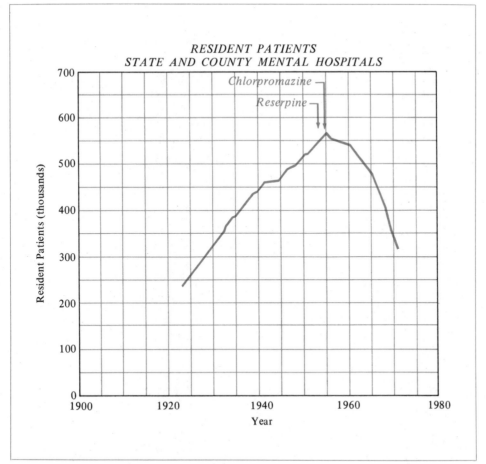

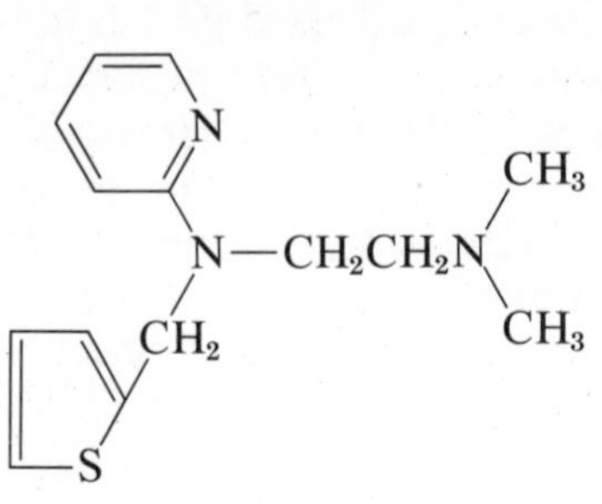

400

methapyrilene

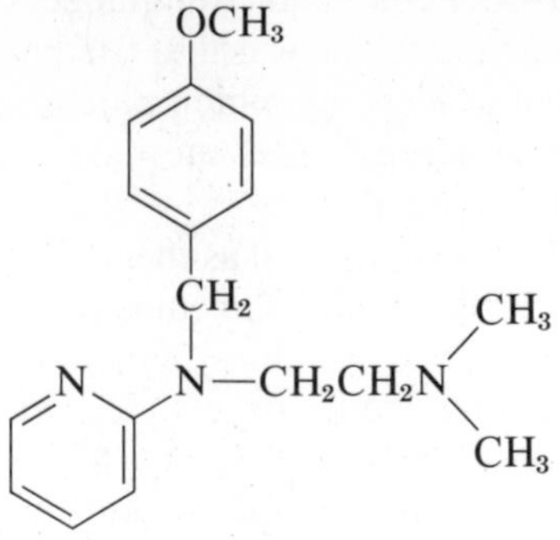

pyrilamine

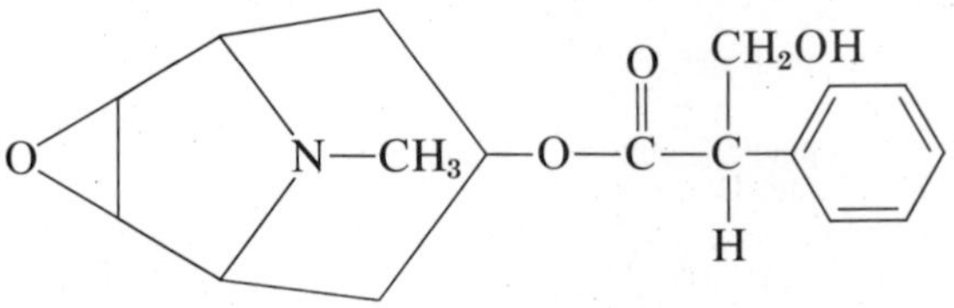

scopolamine

ANTIDEPRESSANTS

Although many individuals seek the peace and tranquility that they think may be obtained by popping some sort of pill, some individuals need the opposite. For this reason, the tricyclic antidepressants known as Elavil® and Tofranil® are often prescribed by physicians. Elavil® was the 29th most prescribed drug in 1974. Tofranil® was number 92.

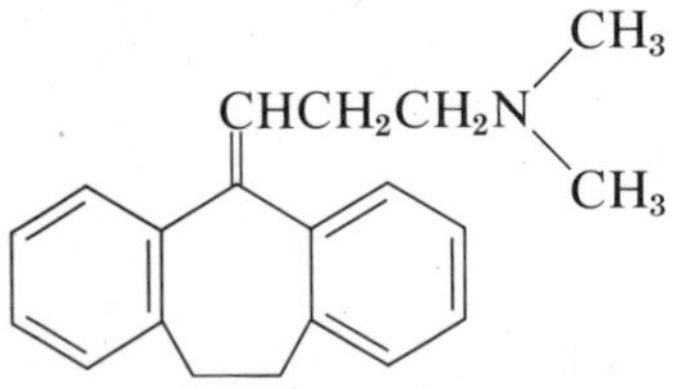

Elavil®

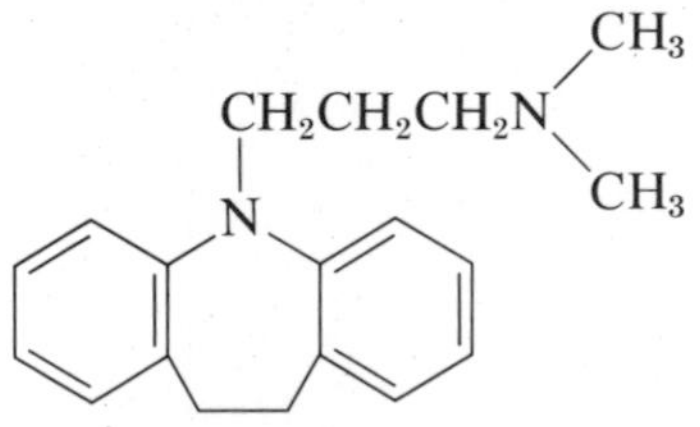

Tofranil®

The amphetamines also cause a stimulation of the central nervous system and were once thought to be safe and nonhabit-forming. Since this has proven untrue, they are seldom used for this purpose. The amphetamines are discussed in greater detail in the section on drug abuse.

DRUG ABUSE

Perhaps no subject is of more widespread concern or is more controversial than drug abuse. The major issues associated with the problem are its causes, its consequences and the methods and problems of dealing with drug addiction. Unfortunately, even the various parts of the problem cannot always be dealt with separately, since methods for treating drug addiction must also recognize the original causes of drug use and the factors that encourage continued use.

CHEMO-
THERAPY
FOR THE
TREATMENT
OF SYMPTOMS
AND DRUG
ABUSE

401

As for the causes of drug abuse, most experts would point to undesirable social and economic aspects of life among the very poor—especially in the inner cities, where drug abuse and its results are most severe. But the problem has spread far beyond the inner cities and the causes are often much more subtle. No doubt, a part of the story is suggested by the following quotations:

> There is the expanding and carefully nurtured tendency of the public—and much of the medical profession as well—to depend on a pill for the solution of every problem of mankind, physical or mental or social. There is the inclination of a patient to request a specific drug, or even to demand it, from his physician. There is the obviously related trend to utilize such drugs as alcohol, marijuana, LSD, and heroin as escape routes from the uncomfortable realities of life.*
>
> So someone is anxious, or hurts a little, or can't sleep, or isn't excited? So what? Everybody is anxious, or hurts a little, or can't sleep, or is depressed from time to time. Why complicate the issue with psychoactive substances? We must learn to cope with, and groove on, reality, without mind- and mood-altering chemicals. In the area of hypnotic-sedative-analgesic-tranquilizer drugs, it is time for the American medical profession to stop arguing over which drug, what dosage, and generic versus trade names, and start asking itself: why should I prescribe a drug at all?
>
> Once a person has a positive psychopharmacologic drug experience, that person may become prone to desire a similar drug experience as a solution when confronted with a similar situation. Psychopharmacologic agents can produce chemical holidays. All drugs are habit-forming (10).

Clearly, the fact that Valium®, Librium®, meprobamate (generic), and Equanil® were the 1st, 4th, 70th and 73rd most prescribed drugs, respectively, in 1974 (4), lends support to the ideas suggested in the preceding quotations.

DRUG ADDICTION

In order to examine the subject of drug abuse, one has to appreciate the meaning of drug addiction, which is normally described in terms of two phenomena: *dependence* and *tolerance*. The term *dependence* signifies a situation in which a user experiences withdrawal symptoms if he stops using a drug to which he has become addicted. The term *tolerance* not only describes an increased ability of the body to tolerate a drug without ill effects, but it also signifies the situation in which the user requires increasing amounts of a drug in order to experience the same effects.

Dependence may be physical or psychological or both. If a physical dependence develops, withdrawal may lead to very severe symptoms and may even be fatal. Withdrawal should always be carried out with medical supervision. Psychic dependence, sometimes called *habituation,* is a more vague condition but a very real one. An individual with a psychological dependence simply finds life unbearable unless he is experiencing the psychological effects of a drug. Many individuals who are critical of addicts seem unable to accept the realities of psychic dependence. Many of these same individuals should also realize the message

* From *Pills, Profits, and Politics,* by Milton Silverman and Philip R. Lee, Copyright © 1974 by The Regents of the University of California, reprinted by permission of the University of California Press (9).

that comes from the fact that the most prescribed drug (Valium®) is a tranquilizer and while a physical dependence on Valium® may develop, it is a psychological need that initiates the use in the first place. In other words, we must face up to the fact that we are not talking about a small group of individuals who make up a "drug culture," we are dealing with a very large percentage of the population that pops pills of all kinds to deal with all sorts of problems, both real and imaginary. Furthermore, far and away the most serious drug problem is alcoholism.

Table 19.3 is a very detailed compilation of the properties of the common drugs of abuse. In the following sections, we will consider a portion of this information with emphasis on the chemistry involved.

Alcohol

The most serious drug problem in America results from the abuse of substances that are depressants of the central nervous system. These include the analgesic narcotics (heroin, morphine, etc.), barbiturates, tranquilizers, and alcohol. Of these, alcohol is the most frequent cause for concern. It is the drug most commonly used legally by adults and illegally by young people. Addiction to alcohol, or simply alcoholism, is the greatest medical problem resulting from the use of drugs, with estimates of the number of alcoholics ranging as high as 10 million in

Copyright © 1964, Cochran. Reprinted by permission.

"No pot for me, thanks. I'll just get drunk like the good Lord intended me to."

the United States alone. It has been estimated that there are about 40 times as many alcoholics in the United States as there are heroin addicts.

CHEMO-
THERAPY
FOR THE
TREATMENT
OF SYMPTOMS
AND DRUG
ABUSE

403

On the other hand, alcohol addiction is not a condition that arises out of short-term use of alcohol, whereas addiction to heroin and other drugs may develop quickly. It is for this reason that alcoholism is more often a disease of middle age, whereas other forms of drug addiction are most prominent among young people.

Withdrawal from alcohol can be quite severe for an alcoholic, more serious than withdrawal from heroin and several drugs.

Crime is a frequent consequence of drug abuse. Drug addicts often steal to support their habit, whereas alcoholics and those who abuse alcohol to a lesser degree may commit even more serious crimes, both intentional and accidental (e.g., while driving), when they are under the influence of alcohol.

Throughout the previous paragraphs, the word alcohol was used several times. To a chemist, the word alcohol is a bit vague, since an alcohol is a compound with the —OH functional group. Organic acids also contain the —OH group, but they are readily distinguished from alcohols due to the presence of the carbonyl ($-\overset{\overset{\textstyle O}{\|}}{C}-$) functional group as shown in the following. The specific compound, simply referred to as "alcohol," is ethyl alcohol. It has very different structure and properties from those of acetic acid even though both substances have two carbon atoms.

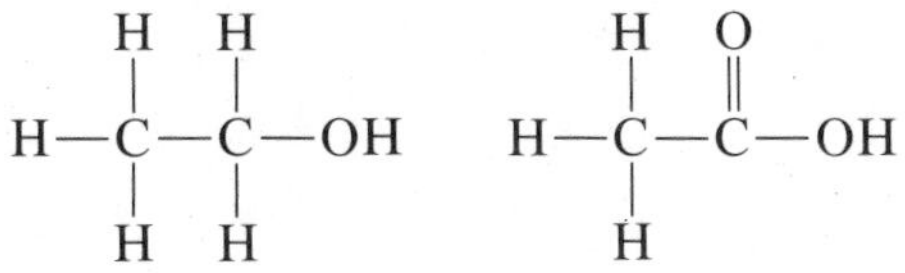

ethyl alcohol   acetic acid

The production of alcohol via fermentation was a major subject considered in Chapter Eleven. An industrial process accounts for more than 75% of all of the ethyl alcohol produced. It is outlined in the following reaction starting with ethylene.

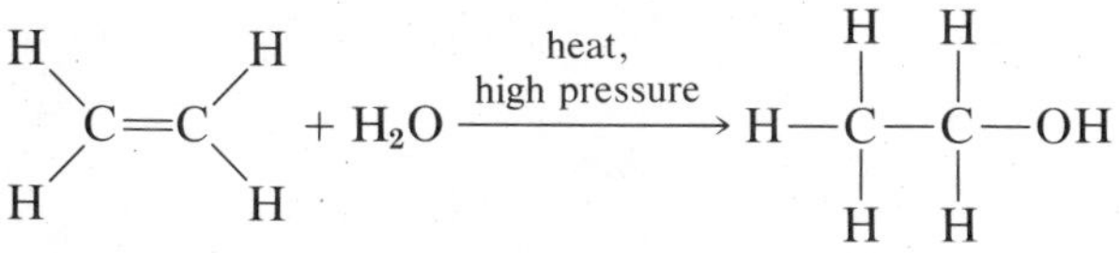

ethylene

METABOLISM

Since alcohol does not have to be digested in the stomach or intestine, it can be absorbed into the bloodstream from either location. Carbon dioxide has the effect of relaxing the pyloric valve, which serves as a gate between the stomach and the

TABLE 19.3 REFERENCE GUIDE TO SOME DRUGS THAT ARE SUBJECT TO ABUSE

	EXAMPLES	SLANG NAMES	PHARMACOLOGICAL CLASSIFICATION	MEDICAL USES
NARCOTICS (Analgesics)	(OPIUM DERIVATIVES) Heroin Morphine Codeine (SYNTHETIC) Methadone Demerol	"H", "Horse", "Smack" "M", "Mary", "Schoolboy"	Central Nervous System Depressants	None To relieve Pain To relieve Pain and Coughing To relieve Pain To relieve Pain
	Cocaine (Coca Leaves)	"Snow", "coke"	Central Nervous System Stimulant	Local Anesthetic
HALLUCINOGENS (Psychedelics)	(Synthetic) LSD DMT DET STP (DOM) LBJ	"Acid", "Cubes", "Trips"	Central Nervous System Stimulants and/or Depressants	None
	Marihuana Hashish (Derivatives of the Hemp Plant— Cannabis Sativa)	"Pot", "Grass", "Mary Jane", "Hash"	Central Nervous System Stimulants and/or Depressants	None
STIMULANTS (Pep Pills)	(Amphetamines) Benzedrine Dexedrine Methedrine Desoxyn	"Ups", "Co-pilots", "Pep Pills", "Bennies", "Dexies", "Hearts", "Meth", "Crystal", "Speed"	Central Nervous System Stimulants	To relieve mild depression and fatigue To reduce appetite To treat narcolepsy (a disease characterized by an almost overwhelming desire to sleep)
DEPRESSANTS (Sedatives and Hypnotics)	(Barbiturates) Amytal Seconal Nembutal Tuinal	"Downs", "Barbs", "Goofballs" "Blue Heavens", "Red Birds", "Red Devils" "Yellow Jackets", "Nemmies" "Rainbows"	Central Nervous System Depressants	To treat insomnia, anxiety, nervous tension and epilepsy; they are also used in the treatment of mental disorders

HOW TAKEN WHEN ABUSED	PHYSICAL SYMPTOMS & BEHAVIOR PATTERNS	MAJOR DANGERS
By injection or sniffed By injection or orally Orally (usually as a cough syrup) By injection or orally By injection or orally	"High" feeling (euphoria) followed by depression; impaired coordination; pinpoint pupils; watery eyes, running nose, chills, sweating, loss of appetite, weight; drowsiness, sleepiness, stupor. The need for narcotics has driven many users to crime.	A very strong physical and psychological dependence may develop. General physical deterioration, possible social deterioration. Painful withdrawal symptoms. Infections, abscesses, tetanus, hepatitis from non-sterile injections. Death from overdose due to respiratory depression.
By injection or sniffed	Excitation, dilated pupils, restlessness, tremors (especially of the hands), and hallucinations.	The stimulation is more pronounced than with amphetamines (stimulants). A very strong psychic dependence may develop. Mental confusion and dizziness, depression, convulsions, death from overdose.
Orally or by injection	Effects vary greatly with dose and individual— may cause: restlessness, exhilaration or depression, dilated pupils; illusions, delusions, hallucinations; distortion or intensification of sensory perceptions; nausea, vomiting, decreased ability to discriminate between fact and fantasy. Unpredictable behaviour, psychotic reactions, acute anxiety.	Permanent personality changes may occur. Bizarre mental effects. Unpredictable behaviour. Exhibit dangerous acts of invulnerability. Suicidal or homicidal tendencies, "Flashbacks" or same type of reactions may occur months later, after drug discontinued.
Smoking or orally Smoking or orally (Hashish 5 times as potent as Marihuana)	Effects vary with the method of ingestion (whether smoked or eaten). Can produce reddening of eyes, dry mouth-throat, coughing spells, talkativeness, laughter. "High" feeling, possible hallucinations, altered perception of time and space; visual distortions. Exaggerated sensory perceptions: can precipitate psychotic acts.	Can lead to aggressive and antisocial behaviour. Moderate psychological dependence liability. Distortion in sense perceptions may lead to accidents. Can lead to more serious drug abuse through contact with "pushers" of other drugs as well as contact with persons using more dangerous drugs.
Orally or by injection	Dilated pupils, loss of appetite, excitation, talkativeness, jumpiness and irritability. Dry nose, lips and mouth, bad breath, extreme fatigue, sleeplessness. With large doses intravenously: delusions, hostility, dangerously aggressive behaviour, hallucinations, induced psychosis with panic.	Can develop high blood pressure or heart attacks. Potential for brain damage, malnutrition, exhaustion, pneumonia. A very strong psychological dependence develops quickly. Engenders reckless behaviour. Can cause coma and death—"Speed Kills."
Orally	Constricted pupils, drunk appearance, slurred speech, incoherency, depression, drowsiness, dullness. Overdose produces unconsciousness, coma, pinpoint pupils, respiratory paralysis and death.	High psychological dependence liability and physical dependence development with continued use. Hazards from faulty judgment and coordination. Painful withdrawal symptoms. Some indication of kidney damage, possible brain and liver damage. Death from overdose (alone or from combination with alcohol and barbiturates).

TABLE 19.3   (*Continued*)

	EXAMPLES	SLANG NAMES	PHARMACOLOGICAL CLASSIFICATION	MEDICAL USES
TRANQUILLIZERS (Minor)	Miltown Equanil Valium Librium	"Downs"	Central Nervous System Depressants	To counteract tension or anxiety and to overcome insomnia without depressing the central nervous system to the extent that barbiturates do. Used also as muscle relaxants

ORGANIC SOLVENTS (Deliriants)	(Toluene)
	Airplane glue
	Plastic Cement
	(Acetone)
	Nail Polish
	   Remover
	(Carbon Tetra-
	   chloride)
	Dry Cleaner Fluid
	Gasoline
	Paint Thinners
	Lighter Fuels
		None	Central Nervous System Depressants	None
				(Too toxic for use)

ALCOHOL (Ethyl)	Wine Beer Whiskies of various types	"Booze" "Sauce"	Central Nervous System Depressant	To sedate, promote sleep, a vaso-dilator (dilates the blood vessels), and food source for energy
NICOTINE (Tobacco)	Cigarettes Cigars Pipe Tobacco Chewing Tobacco Snuff	"Cancer Sticks", "Coffin Nails", "Weeds"	Central Nervous System Stimulant and/or Depressant	None

Drug Reference Chart, Council On Drug Abuse, Toronto, Canada. Reprinted with permission.

HOW TAKEN WHEN ABUSED	PHYSICAL SYMPTOMS & BEHAVIOR PATTERNS	MAJOR DANGERS
Orally	Similar to hypnotics in biological activity. In some people produce unusual feeling of cheerfulness and well-being. Sweating, skin rash, depression, mental sluggishness, urinary retention, anger, anxiety, tension, agitation, excitability, slurred speech.	As with hypnotics, but less so. Less liable to produce psychological and physical dependence. Have an augmented effect with alcohol, barbiturates and opiates (narcotics). Visual disturbances, dizziness, drowsiness. Withdrawal can produce agitation, nausea, depression and sometimes convulsions.
Inhalation (sniffing)	Depending upon exposure, the reactions may last from about 5 minutes to half an hour. Effects include enlarged pupils, confusion, slurred speech, dizziness and a "high" feeling. Distortion of sights and sounds and hallucinations are also reported. Excessive oral secretions (running nose, watering eyes) and poor muscular control may be noted. Drunk appearance. Irritable, drowsiness, unconsciousness.	Moderate psychological dependence liability. Hazards from impaired judgment. Can induce aggressive behaviour, antisocial acts. Possible permanent damage to the brain, liver and kidneys. Accidental death from overdose (choking or suffocation).
Orally	The effects of a given amount of alcohol depend on the rate of drinking, the emotional state of the drinker, the size of the drinker, etc. Heartburn, gastritis, nausea, vomiting, increased urinary flow, malnutrition, various mood states, various diseases, anger, anxiety, tension, fear, belligerence.	Potential for physical and psychological dependence. Hazards from faulty judgment and coordination, emotional liability and increased aggressiveness; accidental death from overdose (alone or in combination with other depressants, e.g. barbiturates). Social and personal deterioration. Antisocial acts. Irreversible damage to brain, liver and kidneys.
Chewed, snuffed or smoked	Tars and smoke irritate the tissues, increasing saliva and bronchiolar secretions. Increased blood pressure and heart rate, and enlarged pupils. With increasing doses causes tremors, vomiting, accelerated respiration, slowing of water excretion by kidneys, paralysis of respiration and convulsions.	High psychological dependence. With chronic use cancer of lungs, larynx and mouth; irritative respiratory syndrome, chronic bronchitis and pulmonary emphysema, damage to heart, blood vessels; impaired vision. An overdose of nicotine can cause convulsions, respiratory failure, and death.

small intestine, so that champagne, sparkling wines, and whiskey mixed with carbonated beverages have a faster effect. Once it has been ingested, alcohol is oxidized (mostly in the liver) according to the following sequence.

A heavy drinker builds up a tolerance to increasing amounts of ethyl alcohol due to the production of increasing levels of certain enzymes that are present in the liver and are responsible for the complete metabolism of alcohol.

If we recall the processes used by the body for regeneration of active niacin (NAD) and active riboflavin (FAD), we can confirm that ethyl alcohol has a relatively high caloric value of 7.1 calories per gram, since a large amount of useful energy (ATP) is derived from the metabolism of alcohol. The steps, called *oxidative phosphorylation,* for formation of ATP are outlined in the following.

It is during the time prior to the metabolism of the alcohol that it exerts its effects on the body. As noted earlier, ethyl alcohol acts as a depressant of the nervous system, although it often appears to act as a stimulant by depressing certain inhibitory effects of the brain. The full range of effects of alcohol are described in Table 19.4.

TREATMENT

Since both physical and psychic dependence on alcohol may develop, withdrawal should be carried out carefully and under medical supervision. Chlordiazepoxide (Librium®) is sometimes used in the treatment of withdrawal symptoms, which

* This is a multistep process known as the Krebs cycle.

CHEMO-
THERAPY
FOR THE
TREATMENT
OF SYMPTOMS
AND DRUG
ABUSE

409

TABLE 19.4 SYMPTOMS OF ALCOHOL CONSUMPTION

BLOOD ALCOHOL (%)	EFFECTS
Up to 0.03	Sedation and tranquility
0.03 to 0.05	Changes in reflex response, reaction time, and some skills (e.g., automobile driving and athletic activity)
0.05 to 0.15	Loss of coordination
0.15 to 0.20	Obvious intoxication, release of inhibition leading to talka-tiveness and boisterous behavior
0.20 to 0.30	blurred vision, dilation of pupils, slurred speech, staggering gait
0.30 to 0.40	depression, stupor, unconsciousness
over 0.05	death

may be quite severe and even fatal. In fact, withdrawal symptoms from alcohol-ism are generally much more severe than those resulting from withdrawal from narcotics such as heroin.

Once an individual has been safely withdrawn, a drug with the generic name disulfiram and the brand name Antabuse® may be administered to help keep a patient "on the wagon." This drug prevents (inhibits) the step in the

$$CH_3CH_2 \quad \overset{S}{\underset{\|}{C}} \qquad \overset{S}{\underset{\|}{C}} \quad CH_2CH_3$$
$$N-C-S-S-C-N$$
$$CH_3CH_2 \qquad\qquad CH_2CH_3$$

disulfiram
Antabuse®

metabolism of alcohol in which acetaldehyde is converted to acetic acid. When this happens, acetaldehyde builds up and causes extreme discomfort. Since the acetaldehyde only builds up as a consequence of alcohol ingestion, an individual taking disulfiram is discouraged from further intake of alcohol once he has experi-enced the effects of the disulfiram.

OTHER ALCOHOLS

Some other common alcohols are listed in Table 19.5. Methyl alcohol (wood al-cohol) may be obtained by destructive distillation of wood. It is quite toxic, but causes blindness in doses far below lethal amounts.

Isopropyl alcohol has a high bacteriocidal activity and finds consid-erable use as rubbing alcohol. It is also quite toxic. Ethylene glycol is about one-half as toxic as ethyl alcohol. Glycerin is less toxic than ethyl alcohol.

The Opiates

The drugs most often associated with drug abuse and drug addiction are the opiates—heroin, morphine and codeine—so-called, because morphine, and to a lesser extent codeine, are derived from opium. Opium is the residue resulting from evaporation of the juice of the poppy grown in the Orient and the Middle East. It contains about 10% morphine and about 0.5% codeine. The chemical structures of the opiates are very similar. Morphine is the simplest, codeine has a methyl group on one of the alcohol groups, and heroin is diacetylmorphine.

TABLE 19.5 SOME IMPORTANT ALCOHOLS

FORMULA	COMMON NAME	IUPAC NAME
H | H—C—OH | H	Methyl alcohol (wood alcohol)	Methanol
H H | | H—C—C—OH | | H H	Ethyl alcohol	Ethanol
H H H | | | H—C—C—C—OH | | | H H H	n-Propyl alcohol	1-Propanol

H H H
| | |
H—C—C—C—H
| | |
H O H
   |
   H
	Isopropyl alcohol
	(rubbing alcohol)
		2-Propanol

H | H—C—OH | H—C—OH | H	Ethylene glycol (antifreeze)	1,2-Ethanediol
H | H—C—OH | H—C—OH | H—C—OH | H	Glycerin (glycerol)	1,2,3-Propanetriol

Heroin is the most potent of the three substances, which are classified as *narcotics* to signify that they cause sedation (narcosis) and relieve pain (analgesia). The term *narcotic* may also be used to describe a drug that has the potential for causing dependence.

Opium has a very long history of use as a pain killer and was legally and conveniently available throughout most of the 19th century through physicians, drug stores, grocery stores and general stores, as well as through the mail in many patent medicines. Morphine (from Morpheus: the god of sleep) was widely used as a pain killer during the Civil War and was produced in several of the Confederate States.

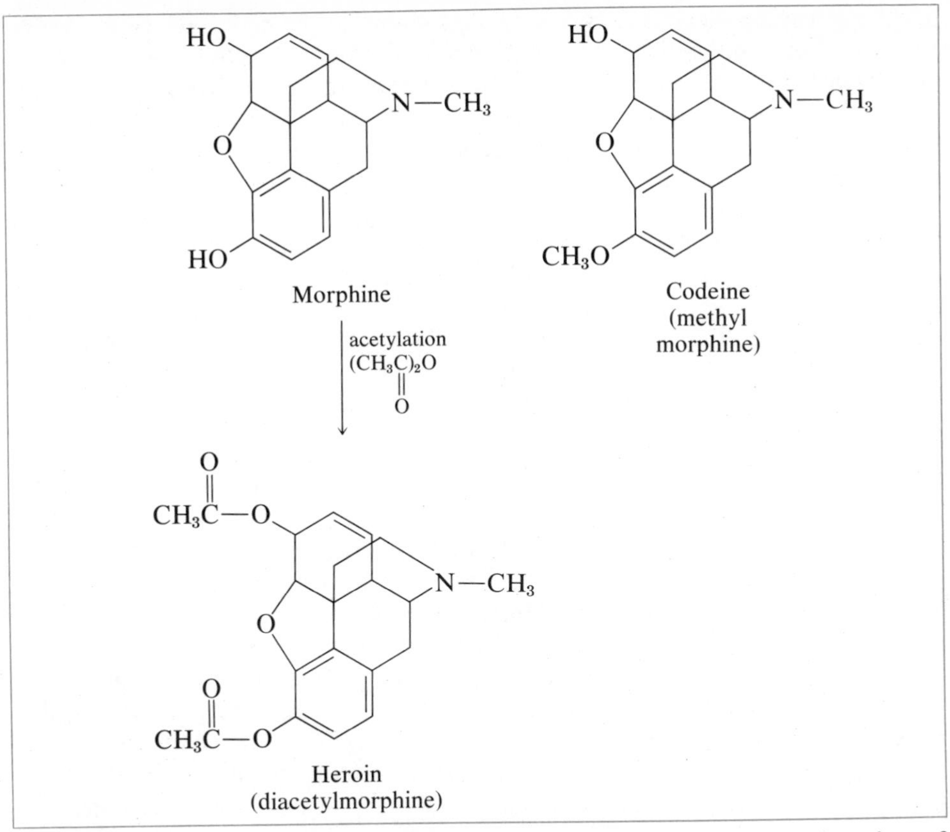

Morphine

Codeine
(methyl
morphine)

acetylation
$(CH_3C)_2O$ ‖ O

Heroin
(diacetylmorphine)

CHEMO-
THERAPY
FOR THE
TREATMENT
OF SYMPTOMS
AND DRUG
ABUSE

411

In addition to narcosis and analgesia, the opiates cause an elevation of mood, euphoria, and a general feeling of peace and tranquility. Constipation is a common side effect, which is often used to advantage for the treatment of severe cases of diarrhea by administering opium as a camphorated solution containing alcohol, known as paregoric. The camphor gives the mixture a bitter taste so as to discourage abuse of the drug.

Codeine is the least potent of the three opiates. It has seen frequent use as a cough suppressant. Although it is naturally present in opium, it is usually synthesized from the more abundant morphine.

Heroin has the same physical and psychological effects as morphine and codeine, but it is a far more potent narcotic and is much more addictive. In fact, the unusually high addiction potential explains why it has been rejected for use by the medical profession and, at the same time, finds such great popularity among drug pushers, who want to keep their customers eager to buy again and again.

Heroin is a synthetic opiate, which is produced by acetylation of morphine. Drug addicts most often take heroin because that is what smugglers smuggle. Since heroin is so potent, it is far more economical to transport and deliver than the less potent morphine and other drugs. The potency of a heroin sample is variable, since the narcotic is commonly "cut" (diluted) several times before it reaches the user by addition of lactose or some other substance.

In the summer of 1972, the average daily cost of a drug habit (usually heroin) was $47.71, compared to $34.68 four years earlier (13). At either level, the price was astronomical. A major factor in supporting the high price is the law of supply and demand, which is often affected by efficient law enforcement activity and has been the subject of great criticism by those who argue that the legalization of many drugs would be an effective measure in curbing their use (14). The point is that prohibition of drugs, as well as alcohol, does not work, since the effect is to raise prices and generate more black market activity. Laws cannot keep drugs away from users. Their primary effect is an increase in the price, which has at least two indirect effects. For one thing, high prices encourage crime among users who need the funds to support their habits. At the same time, high prices force the user to administer the drug in the manner that is most effective. Invariably, this means taking the drug by intravenous injection—a practice called "mainlining." Infectious hepatitis, due to injections given under unsanitary conditions, is a common occurrence among drug users and is a frequent cause of death.

TREATMENT

Treating problems of opiate abuse may be considered as two separate problems—treatment of overdose episodes and treatment of addiction. Both problems are of great significance. The first is an example of short-term emergency treatment, while the second is the very controversial matter of how to deal with drug addiction in general and heroin addiction in particular.

Death due to a heroin overdose is a very rare occurrence. In most cases, several hours pass between administration of the heroin and death, and during this time, the effects of the drug may be countered by use of certain *narcotic antagonists*. The best known narcotic antagonists are nalorphine and naloxone. Both substances are thought to work by attaching themselves to sites in the central nervous

nalorphine   naloxone

system, known as morphine receptors, because the antagonists have a greater affinity for the receptors than the narcotic drugs do. The latter are, therefore, prevented from reaching the nervous system and their effects are blocked.

There is a phenomenon often termed "heroin overdose," which is a frequent cause of death in heroin addicts, but there is good evidence to suggest that it is not the result of an overdose at all. This unusual type of death has been termed *syndrome X*. Its cause is unknown, although it has been speculated that it may be caused by quinine, which is often used as an adulterant, or it may be the result of a combined action of alcohol and heroin (15).

Long-term therapy for heroin addicts focuses on the realities of addiction and its consequences. The reality of addiction is that "once an addict, always an addict." The consequences of heroin addiction are many and varied, but the day-to-day routine of a heroin addict is centered around the need to maintain the supply of the drug in order to avoid withdrawal symptoms. At the same time, normal employment, which might provide the funds to support the drug habit, is impossible due to the narcosis (drowsiness) and euphoria caused by the drug.

CHEMO-
THERAPY
FOR THE
TREATMENT
OF SYMPTOMS
AND DRUG
ABUSE

413

It is because of the picture described in the previous paragraph that the well-known *methadone maintenance* programs have come into prominence for treatment of heroin addiction. Methadone is the generic name for a drug whose brand name is Dolophine® (and others). It was first synthesized during World War II by German chemists. It resembles the opiates in depressing the central nervous

methadone
Dolophine®

system and is a very potent analgesic. On the other hand, it does not cause drowsiness or euphoria (unless mainlined) like heroin or morphine, so an addict who switches to methadone is able to hold down a productive job and return to a more normal way of life.

Another advantage of methadone is the fact that it can be taken orally so that the dangers of infection due to injection are eliminated. Methadone is an addictive drug—a fact which may seem undesirable except that it increases the likelihood that the user will keep coming back for more, rather than reverting back to the use of heroin. In 1963, it was discovered that a dependence on heroin could be transferred to methadone. In other words, methadone can be substituted for heroin and the addict does not experience any withdrawal symptoms as long as he takes a regular dose of methadone, which is usually given daily dissolved in orange juice or other beverage. It has been reported that the mean apparent half-life of methadone in the body is about 15 hours—a fact which explains why a daily dose is sufficient (16).

By 1970, the use of methadone exceeded the use of heroin, and in 1972, there were 65,000 addicts receiving daily doses of methadone in some 450 methadone maintenance programs in the United States. Since the methadone is usually given without charge, the user does not have to resort to stealing in order to support his habit. The legality of methadone is one of its greatest virtues and is a factor that tends to suggest that the legalization of heroin might be an acceptable alternative, as is the case in England. Most critics cite the euphoria and sedation caused by heroin as serious drawbacks to restoring an addict as a normal member of society.

It is frequently claimed that methadone also blocks the euphoric rush that is experienced when heroin is mainlined, but this blockade is probably due only to the amount of methadone that is given. Since the doses of methadone are usually quite high, smaller doses of heroin in addition to the methadone do not cause much of an effect, whereas large doses of heroin will cause the euphoria some addicts may crave.

Bristol Laboratories has marketed an interesting drug named Methenex®, which is a combination of methadone and the narcotic antagonist, naloxone. If an addict tries to inject this form of methadone in order to experience euphoria, the antagonist will interfere with the methadone and prevent the effect. On the other hand, when taken orally, the antagonist has no effect and the methadone works normally. The use of soft drinks and fruit drinks and fruit juices as a means of dispensing methadone is another way of discouraging abuse by injection.

Unfortunately, methadone maintenance is not without drawbacks, since a number of physical side effects and psychological factors must be considered when evaluating the success of this approach. A number of excellent discussions of the issue have appeared (18, 20, 21, 28).

As for the physical problems of methadone maintenance, methadone users reportedly perspire more profusely than normal, they are often constipated, and sexual impotence is common, especially in older men. This last fact, coupled with the knowledge that methadone is addictive, has aroused the suspicions of black and Hispanic minority leaders, who have fears of genocide and enslavement.

There are also some risks associated with methadone maintenance programs. Normal methadone doses are often lethal for nonaddicts. Since much of the heroin sold on the streets is so weak that individuals who use it regularly may not actually become addicted, regardless of what they think, this can be a real danger. In addition, there have been many reports of fatal poisonings of children who accidently took a parent's methadone that had been premixed with fruit juice.

Withdrawal from methadone is generally regarded as being at least as difficult as withdrawal from heroin. The rather large doses of methadone contribute to this problem.

Besides the purely physical problems of methadone maintenance, there are certain psychological factors that argue against this approach. The following quotations sum up this view quite effectively.

> . . . methadone may well contribute to the problem rather than to the solution: one need only consider that the methadone ''solution'' must surely reinforce the popular illusion that a drug can be a fast, cheap, and magical answer to complex human and social problems (18).

> . . . the decision to use methadone on a large scale supports and reinforces for the community at large a drug-oriented approach to the solution of social and personal problems (19).*

> Methadone maintenance, essentially a chemotherapeutic ''fix'' for heroin addiction, has become the predominant means of dealing with the poorly understood problems of the addict and the social consequences of addiction (20).

> Methadone is a good example of useful, incomplete, social problem-solving by means of what can be thought of as ''technology.'' It is a treatment in which a technological approach replaces the ''human touch (21).''*

> The advocates of methadone often derogate the work of therapeutic communities such as Phoenix House or Daytop in New York City. When some successes of such programs are grudgingly admitted, the admission is often accompanied with the criticism that the rehabilitated addict has developed a dependency on the program.

* Copyright © 1973, Smithsonian Institution, from *Smithsonian Magazine,* April 1973.

It is difficult to understand how advocates of methadone maintenance can be critical of programs in which persons become dependent upon human environments instead of drugs. Why should they prefer that persons become dependent upon chemical agents, such as methadone, rather than upon supportive human contacts? Such a preference shows a lack of appreciation of the very factors that initially propel individuals into taking drugs (19).*

In the final analysis, the false model is always the more costly one (19).*

In short, the methadone maintenance approach can convert the dope fiend into a drug addict but, while this is a major accomplishment, it is hardly a satisfactory final answer. At the same time, the abstinence approaches, such as those practiced in self-help organizations such as Synanon, Daytop Village, Gateway Houses, and Phoenix House, do not work for some addicts. For these individuals, methadone is the only viable alternative. Clearly, methadone does not treat the original motivation for abuse, but it does treat the eventual and inevitable need to avoid withdrawal symptoms and permits an addict to achieve a life style that is more consistent with normal members of society.

Amphetamines

Among the other drugs that have been subject to abuse are the amphetamines and barbiturates, which are often referred to as "uppers" and "downers," respectively. Barbiturates are considered in the following section, although they deserve mention in this section since amphetamine users often have to resort to the use of barbiturates in order to be able to sleep.

The chemical formulas of the common amphetamines follow. The first two structures shown are *isomers* of the same compound, which differ in the three-dimensional arrangement of the molecule.

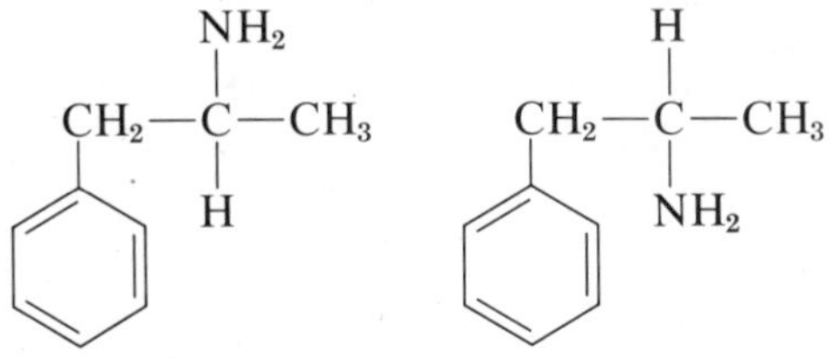

dextroamphetamine
Dexedrine®

amphetamine
Benzedrine®

Amphetamine is not a single compound, but is a mixture of the two isomers, which has been marketed under the name, Benzedrine®. Dextroamphetamine (Dexedrine®) is the name given to the isomer of amphetamine that is shown on the left. Dexedrine alone is a more potent substance than Benzedrine®.

Amphetamines are stimulants of the central nervous system—a fact that is consistent with the chemical structure that is similar to epinephrine, better known as Adrenalin®. Amphetamines induce the same effect as Adrenalin® but the action of amphetamines is more sustained. They increase the heart rate and blood

epinephrine
Adrenalin®

methamphetamine
"speed"

pressure, they increase wakefulness, postpone fatigue, and increase drive and energy. They produce a temporary elevation of mood, but this is usually followed by increased fatigue, irritability, and depression.

Amphetamines are sometimes known as pep pills. Students who are studying for exams and truck drivers who need to stay awake are often cited as users of amphetamines. The latter are said to use slang names such as "coast-to-coasts" and "copilots" to describe amphetamines.

Technically, amphetamines are not physically addictive, but psychic dependence (habituation) and tolerance normally develop very quickly.

The amphetamines were once thought to be safe and nonhabit-forming. They have been used extensively as diet pills, since they do cause anorexia (loss of appetite) and, at the same time, cause an elevation of mood, which tends to encourage physical effort, which helps in losing weight. Benzedrine® was once used as an inhaler for relief of nasal congestion.

Among the more potent amphetamines is methamphetamine (Methedrine®), also known as "speed." Amphetamines can be taken orally with less dramatic effects, but the "speed freak" is an individual who injects Methedrine® intravenously for several days at a stretch (called a "run") and then sleeps for an extended period, usually with the help of barbiturates or heroin. In fact, most speed freaks eventually become heroin addicts. The speed freak is subject to constant wakefulness, incessant babbling, and tremendous nervous energy.

Barbiturates

Alcohol and barbiturates can be largely equated in terms of their effects on the body. One can get drunk on barbiturates and become addicted, both physically and psychologically, to barbiturates. In fact, withdrawal symptoms are among the most severe encountered by drug addicts—usually more severe than those encountered by heroin addicts. Convulsions, *delirium tremens,* and even death, typically result from abrupt withdrawal. Only barbiturates themselves are considered suitable for use in barbiturate withdrawal, since many anticonvulsant drugs are not effective in treating these withdrawal symptoms.

The barbiturates are a family of chemically-related compounds. Each has a structure that is a slight alteration, called a *derivative,* of barbituric acid. Barbituric acid was first synthesized in 1864 from urea and malonic acid.

Barbital was the first barbiturate used in medicine. It was introduced under the name Veronal® in 1903. Phenobarbital, sold also as Luminal®, was the sec-

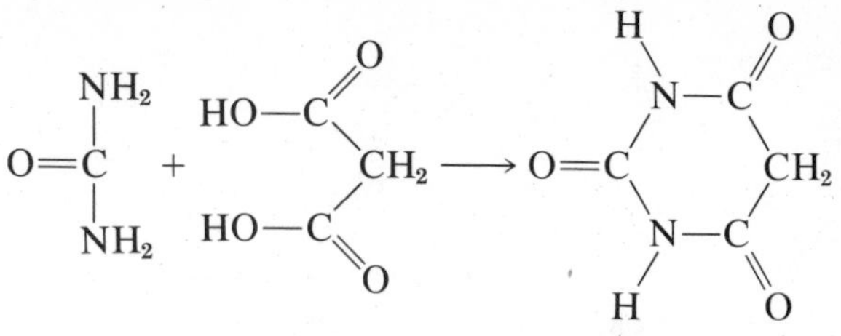

CHEMO-
THERAPY
FOR THE
TREATMENT
OF SYMPTOMS
AND DRUG
ABUSE

417

Urea   Malonic acid   Barbituric acid

ond barbiturate to be used. These and several of the other common barbiturates, including Pentothal® and its sodium salt, follow.

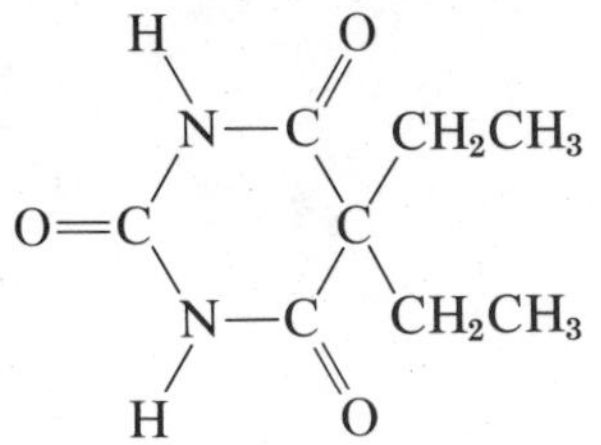

barbital
Veronal®

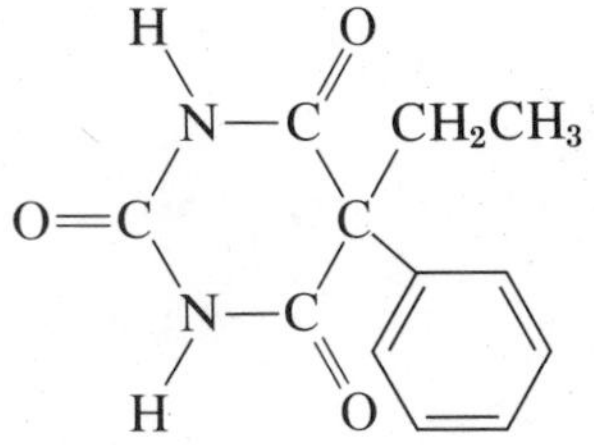

phenobarbital
Luminal®

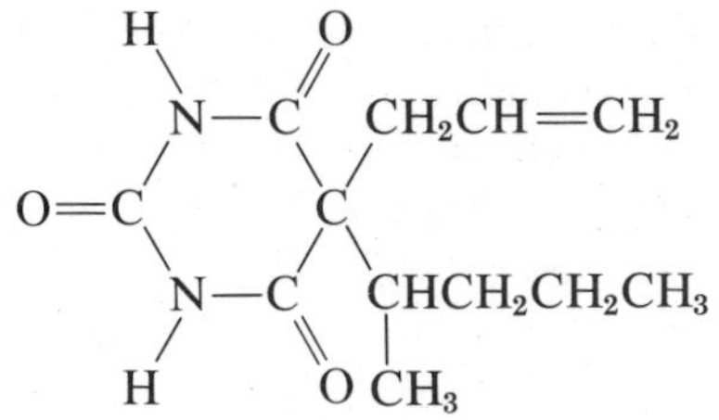

secobarbital
Seconal®

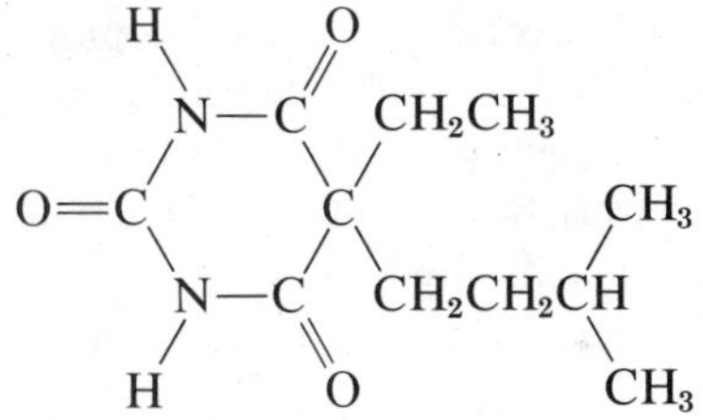

amobarbital
Amytal®

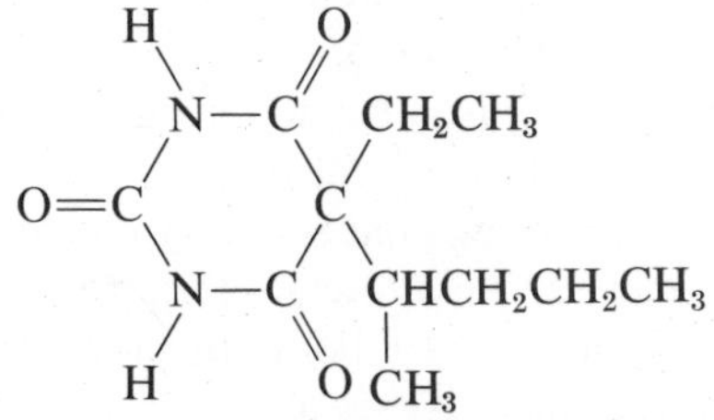

pentobarbital
Nembutal®

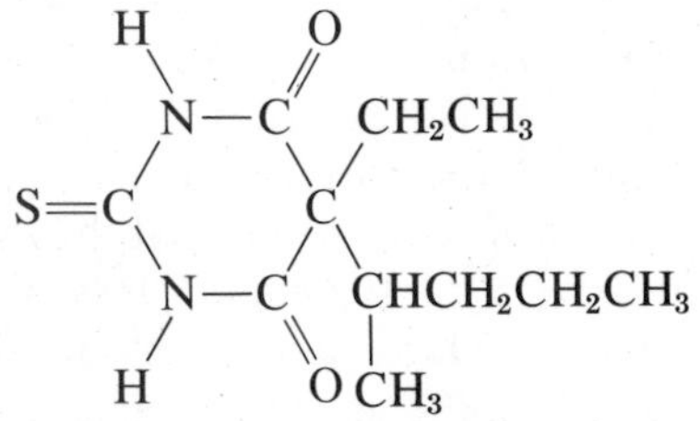

thiopental
Pentothal®

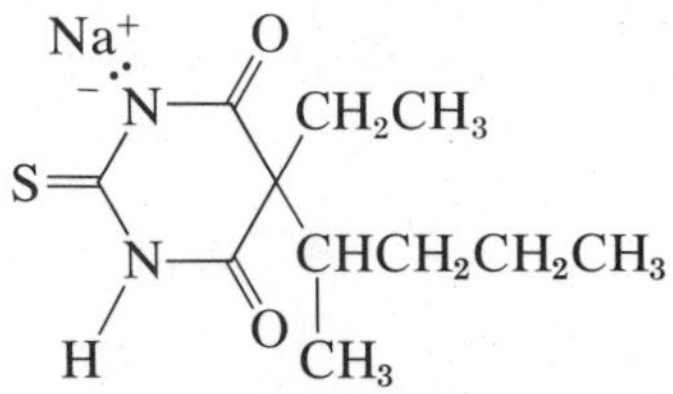

Sodium Pentothal®

Figure 19.3 Barbiturates. (Photo by James J. Kane, Jr.)

Like the opiates, barbiturates are central nervous system depressants ("downers"). The symptoms of barbiturate intoxication are slowed reflexes, impaired coordination, slurred speech, and drowsiness. Medically, barbiturates are used in several ways: as sleeping pills (sedatives), anticonvulsants, anesthetics (Sodium Pentothal®) and preanesthetics, and in psychiatry.

Other than the differences in chemical structure, the major identifying characteristics of each of the barbiturates is the speed and duration of action. Each is classified as long-acting (e.g., Veronal® and phenobarbital), intermediate-acting (e.g., Amytal®), short-acting (e.g., Nembutal® and Seconal®) or ultra-short-acting (e.g., Sodium Pentothal®).

On the drug scene, "nemmies" (Nembutal®), "seccies" (Seconal®), and "tuies" (Amytal® plus Seconal®, called Tuinal®) are slang names for some of the barbiturates, as are the names "yellow jackets" (Nembutal®), "red devils" (Seconal®), and "blue heavens" (Amytal®). The latter set of names describes the colors of the capsules used for each of the drugs. This particular group of barbiturates is the most popular among users because they are all relatively short-acting drugs. This means that they take effect within a few minutes and the effects last for only a few hours or less. Long-acting barbiturates typically require up to an hour to exhibit any effects but the effects may last for six hours or more. For medical use, the ultra-short action of sodium pentothal makes it a frequent choice for inducing anesthesia in surgery. At the other end of the spectrum, phenobarbital (Luminal®) is a long-acting barbiturate, which is often used in treating epilepsy and other brain disorders. Seconal® and Nembutal® are often employed as sleeping pills since they are relatively short-acting. These same two drugs also see frequent use as preanesthetic sedatives, which are given to patients 30–60 minutes before surgery.

Cocaine

"Coke" and "snow" are two slang names for another frequently abused drug. The use of the word snow to describe cocaine stems from the fact that it is a fluffy white powder. Cocaine is a substance obtained from the leaves of the coca plant, *Erythroxylon coca*. It was a drug of abuse around the turn of the century and it has recently reappeared on the drug scene in the United States. It is grown primarily in the cool, mountainous regions of South America, particularly Peru and Bolivia. Cocaine is an expensive drug—more expensive than heroin—and has been used mostly by the "jet set."

CHEMO-
THERAPY
FOR THE
TREATMENT
OF SYMPTOMS
AND DRUG
ABUSE

419

cocaine

Medically, cocaine is an anesthetic, but its use has been restricted to topical (surface) applications to the mucous membranes (nose, throat, etc.). It cannot be safely injected for use as a general anesthetic, because it causes a general constriction of blood vessels and the resulting increase in blood pressure can be dangerous. At the present time, it sees no use as an anesthetic. Other local anesthetics, such as Novocaine® and Xylocaine®, are normally used.

When used on the drug scene, cocaine is usually sniffed—a practice referred to as "snorting." The drug is a strong stimulant of the central nervous system. It causes exhilaration and lasting euphoria, quickens reflexes, and reduces fatigue without any signs of intoxication. It also causes nausea, weight loss, insomnia and other unpleasant side effects.

Cocaine has a strong addiction potential, which takes the form of a psychic dependence. No physical dependence (no withdrawal symptoms) is normally observed. Heavy or repeated use often causes hallucinations and paranoia, sometimes leading to violent behavior.

Hallucinogens

Lysergic acid diethylamide, LSD, is the best known and most popular of a group of drugs known as hallucinogens.

lysergic acid diethylamide LSD

Hallucinogens may cause stimulation, depression, or neither, but their primary effects are to cause changes in perception, especially visual, with vivid hallucinations that are often brightly colored. No physical dependence develops, but bizarre mental effects and permanent personality changes sometimes occur, often long after use of the hallucinogenic drug.

A drug called mescaline is a component of the Peyote cactus, which grows in Mexico and the southwestern United States. It has been used in religious ceremonies by the Indians for many centuries. Psilocybin is another hallucinogenic drug. It is derived from Mexican mushrooms, *Psilocybe mexicana*.

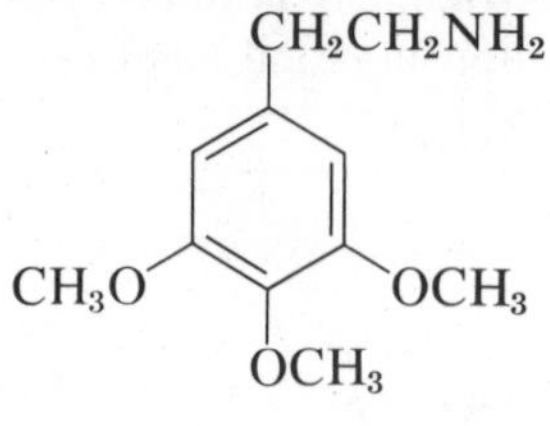

420

Mescaline

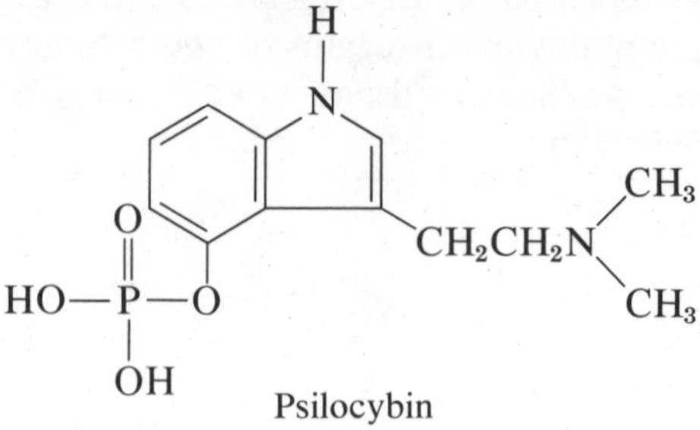

Psilocybin

Marijuana

Perhaps the most controversial of the abused drugs is marijuana, also spelled marihuana. Questions of safety, potential for addiction, and possible legalization are the key issues.

Marijuana is composed of the dried leaves, flowering tops, stems, and seeds of the Indian hemp plant, *Cannabis sativa*. The stems yield tough fibers that are used for making rope. Marijuana is often identified by the name cannabis, although the slang terms "grass," "pot," and "weed" are commonly used.

The active component of cannabis is tetrahydrocannabinol or simply THC. The potency of marijuana is related to the THC content. Typical marijuana available in the United States has had about a 1% THC content. Hashish is a more potent form of cannabis. It is the dried resin obtained from the flowering parts of the

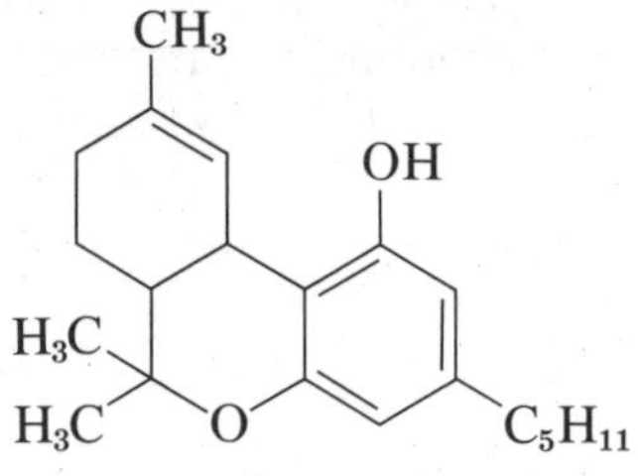

tetrahydrocannabinol
THC

plant. The THC content of hashish is normally in the range of 5%–10%, but may range as high as 20%. There is recent evidence that much of the marijuana now available in the United States contains 2%–3% THC, whereas much of the hashish is down around 4%–5%. In other words, the distinction between marijuana and hashish is not as great as it is generally thought to be.

Powdered marijuana or hashish may be consumed orally, but the effects are felt more rapidly when cannabis is smoked in pipes or cigarettes. The effects typically last for two to three hours. Cannabis may act either as a stimulant or depressant and is sometimes classified as hallucinogenic. Some experts place marijuana in a class by itself, because of the absence of cross tolerance with all other categories of drugs. In other words, when an individual develops a tolerance to a stimulant or depressant, he normally acquires a tolerance to other drugs of the same type. Marijuana does not lead to any cross tolerance, even to drugs normally classified as hallucinogens (11).

CHEMO-
THERAPY
FOR THE
TREATMENT
OF SYMPTOMS
AND DRUG
ABUSE

421

Users of marijuana may exhibit a sort of negative tolerance. Breakdown products remain in the body for many days, and THC persists in the bloodstream for more than three days after a dose, so that a regular user of marijuana may get high on a smaller dose than that needed by the occasional user or even a beginner (22).

The effects of cannabis intoxication are feelings of well-being, excitement, disturbance of associations, alterations in appreciation of time and space, emotional upheaval, illusions, and hallucinations. Well-being sometimes alternates with depression.

As for the dangers of cannabis, the subject has been debated and studied extensively, and while it would be nice to be able to examine a number of hard facts, very few have been clearly established. There is no convincing evidence that casual, infrequent use of marijuana produces any ill effects (23), and the evidence regarding heavy use is not much better. A part of the problem stems from the fact that some very subjective conclusions must be drawn from clinical observations of self-confessed cannabis users. We must remember that marijuana is an illegal drug, so that many users are reluctant to admit that they use it at all. In addition, there is the ethical problem of how to study a drug that may be harmful to humans. Consequently, the safety of cannabis may be a controversy for many years, although there are some negative side effects that do seem to appear frequently in heavy users. These are discussed in the following paragraphs.

Marijuana is not regarded as addictive (tolerance and withdrawal symptoms do not appear) but it can produce a psychic dependence, resulting in restlessness, anxiety, irritability, insomnia, or other symptoms.

There is little doubt that cannabis has a number of short-term effects on the brain. It could not exert its psychoactive effects if it did not. There is evidence that long-term users of marijuana are typically apathetic and sluggish both physically and mentally—a condition that may give the false impression of calm and well-being. This is usually accompanied by the user's belief that he has developed emotional maturity and insight aided by marijuana, although a careful examination of such an individual proves otherwise. Furthermore, the effects may persist for a long time following cessation of cannabis use. This suggests that irreversible brain damage may occur.

The more direct problems of marijuana intoxication are often overlooked. Driving performance, work performance, and relationships with other people are often adversely affected by marijuana use.

It has been reported that regular use of cannabis suppresses the body's immune response, which could increase susceptibility to disease (24, 25). There is also evidence that components of cannabis may cause hormonal changes. The most significant change is a decreased fertility in males, which is apparently attributable to a decrease in concentration of the male hormone, testosterone. In at least some cases, the sexual function is restored when cannabis use is stopped. Another common hormonal effect is the development of female-like breasts in men.

Marijuana cigarettes are also known to contain about 50% more tar than commercial cigarettes and the tar causes skin tumors in rats. Heavy use seems to have adverse effects on the bronchial tract and lungs (25).

Finally let us examine the often mentioned link between the use of marijuana and other drugs, especially heroin. There are some logical ones. Users of marijuana may crave for a better or different sort of high, and thus turn to other drugs. Since marijuana was made illegal in 1937, the traffic in marijuana is by way of the black market and this means contact with pushers, who may also deal in the potent narcotics and who would benefit by getting their customers addicted to other drugs. Since the same supply channels are used for marijuana and heroin, law enforcement activities sometimes dry up the supply of marijuana and increase the temptation to experiment with heroin.

On the other hand, while it is true that most heroin users once used marijuana, it is also true that most addicts started with alcohol, not marijuana. In fact, there is a clear progression from liquor to heroin in which the use of liquor by adolescents may sometimes be "the first experience with breaking the law in connection with drug use (11)."

Caffeine and Nicotine

Surely the section on drug abuse is hardly the place to discuss substances like caffeine and nicotine. The next thing you know, we will be criticizing "mom's apple pie." And yet, caffeine and nicotine are two of the most abused drugs, and each is a very interesting case.

Caffeine   Nicotine

Caffeine is a substance that causes stimulation of the central nervous system. That should sound familiar. Caffeine is commonly consumed in coffee, tea, cocoa, and cola drinks. It produces a more rapid and clear flow of thought and decreases drowsiness and fatigue. After taking caffeine, it is possible to carry out a greater sustained intellectual effort, although physical coordination and timing may be adversely affected (26). The "effective dose" of caffeine is about 200 milligrams. No-Doz® (100 milligrams per tablet) and Vivarin® (200 milligrams per tablet) are commercial products in which caffeine is the major ingredient.

There is no evidence of physical dependence, although some psychic dependence is indicated, for example, in individuals who "cannot face the day without their morning cup of coffee."

As for nicotine, the problem is much less amusing since the consequence of smoking may be much more severe. There are those who smoke for enjoyment or to alleviate stress and there are those who are addicted to nicotine. This latter fact cannot be underestimated, since there are some experts who feel that it is much easier to become dependent on cigarettes than on alcohol or barbiturates. The dependence takes the form of a psychological dependence only.

There are a number of interesting facts that indicate that many individuals smoke strictly for the dose of nicotine. Studies have shown that an addict will take fewer puffs on a high-nicotine cigarette.

CHEMO-
THERAPY
FOR THE
TREATMENT
OF SYMPTOMS
AND DRUG
ABUSE

423

Figure 19.4   In the interest of cancer research, this rabbit at the Tbilisi Institute of Oncology in Russia learned to smoke. Consuming up to nine cigarettes a day, all of the rabbits in the experiment had chronic pneumonia and emphysema by the end of the fourth year. This did not affect their enthusiasm for smoking, however. Photo by G. Kikvadze (Fotokhronika TASS). Reprinted by permission of Sovfoto.

Cigarette smoking is the greatest cause of fatal cancer of any factor in the environment. It causes 15%–20% of all cancer deaths, almost 70,000 per year from lung cancer alone. Per capita consumption of cigarettes fell in the sixties, but is now showing a sharp increase, especially in young people.

Normally, it is the products resulting from the incomplete combustion of tobacco and tobacco paper that are potentially carcinogenic, and not the nicotine. An example is 3,4-benzpyrene, which forms from cellulose by burning cigarette

3,4-benzpyrene

paper. In other words, the nicotine attracts the smoker but it is other components of the smoke that are harmful. On the other hand, it has also been reported that, although nicotine itself is not a carcinogen, it can increase the incidence of tumors in mice due to other carcinogens that may be present in cigarette smoke (27).

Overall, nicotine is generally rated second only to alcohol as a serious drug problem. Some experts have drawn an analogy to methadone as a way of dealing with the problem by suggesting that very high nicotine cigarettes would cut down consumption and minimize risks. In other words, the strategy would be to counteract addiction by supplying a drug—in this case the drug that causes the addiction—or a substitute like methadone. Unfortunately, the idea of using nicotine implies that the nicotine is safe and this may not be true.

SUMMARY

The number one over-the-counter drug is aspirin. It is available alone or in combination with other analgesics or other ingredients such as antacids, decongestants, antihistamines, and caffeine. Acetaminophen is the generic name for an aspirin substitute, which is available under a variety of brand names (Datril®, Tylenol®, and others). Like aspirin, it is an analgesic and antipyretic, but does not have the antiinflammatory action of aspirin.

Propoxyphene (Darvon®) is the number one prescription pain killer. The opiates (usually codeine or morphine) and meperidine (Demerol®) are more potent pain killers, but all have addiction potential. A sequence of drugs is used in surgery in order to induce anesthesia (unconsciousness), analgesia, and muscle relaxation. Enflurane and halothane are ether-like substances that are commonly administered in a stream of oxygen and nitrous oxide (N_2O) after anesthesia has been induced by Sodium Pentothal®. Novocaine® and Xylocaine® are chemicals often used as local anesthetics for very minor surgery. Ethyl chloride can be used locally to freeze small areas of surface tissue and make them insensitive to pain.

Antacids come in a variety of forms, with each containing an alkaline substance, which neutralizes stomach acid (HCl). Effervescent antacids are quite popular.

The most prescribed drug in 1974 was diazepam (Valium®), followed closely by chlordiazepoxide (Librium®) at number four. Both of these drugs are classified as minor tranquilizers. Together, they account for over 25 million prescriptions each year. Antidepressants are sometimes used for the opposite purpose.

The wholesale use of tranquilizers may be a contributing, or at least accompanying, factor in drug abuse. People of all ages find it necessary to use all kinds of drugs including alcohol, the opiates (usually heroin), amphetamines (uppers), barbiturates (downers), cocaine, hallucinogens (e.g., LSD), marijuana, caffeine and nicotine, some of which have serious addiction potential.

Treatment of heroin addicts is a very controversial subject. Methadone is finding heavy use for this purpose, although the practice has been widely criticized.

PROBLEMS

1. Give a definition or example of each of the following:
   a) chemotherapy
   b) analgesic
   c) antipyretic
   d) vasoconstrictor
   e) dependence
   f) tolerance
   g) habituation
   h) narcotic

CHEMO-
THERAPY
FOR THE
TREATMENT
OF SYMPTOMS
AND DRUG
ABUSE

425

    i) "mainlining"   n) hashish
    j) "uppers"   o) "nemmies," "seccies," "tuies"
    k) "downers"   p) "yellow jackets," "blue
    l) "speed freak"   heavens," "red devils"
    m) "snow"   q) snorting

2. What is the side effect of antihistamines that makes them dangerous?

3. What class of drugs is generally used for treatment of very severe pain?

4. How does morphine compare to aspirin as a pain killer?

5. What was "soldiers' disease" in the Civil War?

6. What are the drugs used in surgery, beginning the night before and carrying through the post-operative period?

7. Discuss ether—its effectiveness, its safety, and its side effects.

8. Why is chloroform a poor anesthetic?

9. When and how are halothane and enflurane administered?

10. What is a local anesthetic and how is it used?

11. What is the function and mode of action of ethyl chloride when it is used at athletic events?

12. What is the cause of the effervescent action of some major antacids?

13. What is the antacid(s) in Alka-Seltzer?

14. What was the number one prescription drug in the United States for 1974 and what is its function? What does this say about our society as a whole?

15. What is the most serious drug problem in the United States today?

16. Why is alcoholism most often a disease of middle age, whereas drug addiction seems to be a problem of the young?

17. Describe the action of Antabuse® (disulfiram) on the alcoholic.

18. Why do sparkling wines or whiskey mixed with carbonated beverages have a faster effect than the same amount of alcohol alone?

19. What is an addictive narcotic?

20. What are the advantages and disadvantages of methadone maintenance programs?

21. How do amphetamines work?

22. What kind of dependence is observed in cocaine addiction?

23. What are some side effects of marijuana?

24. Why is marijuana regarded as a drug in a class by itself?

25. What is the greatest cause of fatal cancer?

REFERENCES

1. "Why The Stomach Does Not Dissolve Itself." *Chemistry 46* (5) (1973): 20.

2. "Cold Remedies (What Helps and What Doesn't)." *Consumer Reports,* January 1974, p. 67.

3. Burack, R. and Fox, F.J. *The New Handbook of Prescription Drugs.* New York: Ballantine Books, 1975, p. 65.

4. *Ibid.,* pp. 426–8.

5. *Ibid.,* p. 178.

6. Hem, S.L. "Physiochemical Properties of Antacids." *Journal of Chemical Education 52* (6) (1975): 383.

7. Buyske, D.A. "Drugs From Nature." *Chemtech,* June 1975, p. 361.

8. Silverman, M. and Lee, P.R. *Pills, Profits and Politics.* Berkeley, California: University of California Press, 1974, p. 13.

9. *Ibid.,* p. 2.

10. Lundberg, G.D. "Barbiturates: A Great American Problem." *The Journal of the American Medical Association 224* (11) (1973): 1531.

11. Harms, E. *Drugs and Youth: The Challenge of Today.* New York: Pergamon Press, 1973, Chapter 1.

12. Council On Drug Abuse. "Drug Reference Chart." Toronto, September 1976.

13. "Young Addicts Now Tend To Start With 'Hard' Drugs." *The Journal of the American Medical Association 224* (11) (1973): 1481.

14. Brecher, E.M., and The Editors of Consumer Reports. *Licit and Illicit Drugs.* Boston: Little, Brown and Co., 1972.

15. *Ibid.,* Chapter 12.

16. Inturrisi, C.E., and Verebely, K. "Disposition of Methadone in Man After Single Oral Dose." *Clinical Pharmacology and Therapeutics 13* (1972): 923.

17. "The First Methadone-Naloxone Combination." *Chemical and Engineering News,* August 26, 1974, p. 16.

18. Lennard, H.L., Epstein, L.J., and Rosenthal, M.S. "The Methadone Illusion." *Science 176* (1972): 881.

19. Lennard, H.L., Epstein, L.J., and Rosenthal, M.S. "The Cure Becomes A Problem." *Smithsonian 4* (1) (1973): 51,

20. Nelkin, D. *Methadone Maintenance: A Technological Fix.* New York: George Braziller, 1973.

21. Etzioni, A. "Methadone: Best Hope For Now." *Smithsonian 4* (1) (1973): 50.

22. Hankins, W., and Haukins, M. *Introduction To Chemistry,* St. Louis: C. V. Mosby, 1974, pp. 442–444.

23. Maugh, T.H. Jr. "Marihuana: New Support For Immune and Reproductive Hazards." *Science 190* (1975): 865.

24. "Marihuana: A Weapon Against Cancer." *Chemical and Engineering News,* August 26, 1974, p. 6.

25. Maugh, T.H. Jr. "Marihuana: The Grass May No Longer Be Greener." *Science 185* (1974): 682.

26. Brecher, E. M., and The Editors of Consumer Reports. *Licit and Illicit Drugs.* Boston: Little, Brown and Co., 1972, Chapter 21.

27. "Nicotine's Role in Carcinogenesis." *Chemical and Engineering News,* 17 May 1976, p. 23.

28. Chappel, J.N. "Methadone and Chemotherapy in Drug Addiction." *The Journal of the American Medical Association 228* (6) (1974): 725.

T HE goal of this chapter is to provide an understanding of the strategies involved in the use of oral contraceptives, better known as "the pill." In addition, we will examine the pros and cons of chemical birth control and
some of the available alternatives.

STEROIDS

The study of the chemistry of steroids has been one of the major efforts of modern
chemistry. Since the discovery of cholesterol, research and development in this
field has led to significant changes in medical technology. While almost everyone
has heard of the link between cholesterol and heart disease, it was the success in
synthesizing cortisone that brought the steroids to public awareness. Even greater
publicity has been given to the steroid hormones, whose development and utilization as oral contraceptives have revolutionized birth control technology.

In order to understand the mode of action of the oral contraceptives, we
must first look at the basic chemistry of steroids in general and steroid hormones
in particular. One of the most striking lessons to be learned is that very subtle
changes in the structures of the steroid molecules often lead to very dramatic
changes in their biochemical activity.

The steroids, or sterols, as they are sometimes called, are some of the major
components of a larger class of compounds called *lipids*. Lipids also include the
fat-soluble vitamins, fatty acids, fats and oils, and waxes, to name but a few. Their
occurrence in nature is widespread, being found in all plant and animal tissues, as
well as in the cells of microorganisms. Steroids exist in the free state (i.e., not
combined with other chemicals) or they may be bound to other compounds such
as fatty acids or carbohydrates in the human body. They are principally found in
nerve tissue (including the brain), in cell membranes, in reproductive organs and
glands, in bile, and in the blood.

In order to appreciate the relationships among the various types of natural
and synthetic steroids, let us first examine the basic chemical structure of the
steroid molecules. All steroids have a common structure referred to as the *steroid
nucleus,* which is pictured in two ways in Figure 20.1. The basic building block is
the four-ring system shown on the left of Figure 20.1, which is joined in a way that
makes it relatively rigid when compared to other chemical compounds. The rings
are lettered A, B, C, and D.

The carbon atoms in the ring structure are numbered from 1–17, as shown

STEROIDS
AND BIRTH CONTROL

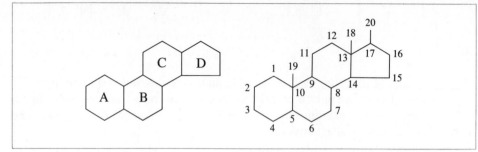

Figure 20.1   The steroid nucleus.

on the right of Figure 20.1, for ease in locating various functional groups that might be attached to the steroid nucleus. Since methyl groups are usually found attached to the steroid at positions 13 and 10, the carbon atoms of these methyl groups are numbered 18 and 19, respectively. Each methyl group is represented by a solid line, using the type of abbreviation introduced in Chapter Six. In Figure 20.1, a methyl group also appears attached to carbon 17, although a simple methyl group seldom appears at this position. The methyl group was added to this figure in order to indicate the system used to number the atoms in a steroid molecule.

Because of the way in which the rings are joined together, the steroid nucleus is nearly planar. All of the atoms attached to the ring system from below are said to be in the alpha configuration, and all those that are attached from above are in the beta configuration. Atoms in the alpha position are designated by a broken line to signify that the bond is projected below the plane of the ring system (see Figure 20.2 for examples) and atoms in the beta position are designated by a solid line to indicate that the bond projects above the plane of the ring system. The biochemical activity of any steroid is very dependent on both *where* (the carbon number) and *how* (above or below) each group is attached to the steroid nucleus. The methyl groups located at positions 10 and 13 are known to chemists as *angular methyl groups* and lie above the plane of the molecule. It is the angular methyl groups, along with various functional groups that may be attached to the parent nucleus, which determine the properties of the different steroid molecules that are found in the body.

We can differentiate among the various kinds of steroids by examining the location and kind of groups attached to the steroid nucleus, including the structure of the side chain located at carbon 17. For example, cholesterol is identified by the beta hydroxyl group at carbon 3, a hydrocarbon chain of 8 carbons attached to carbon 17, and a double bond in ring B between carbons 5 and 6.

HO

Cholesterol

Cholesterol is a white crystalline solid, which was first discovered in gall-stones, from which it got its name (Greek: *chole,* bile; *steros,* solid). Cholesterol has received much notoriety as a chemical in the diet that contributes to heart disease. Much of this story was discussed in Chapter Fifteen (The Cholesterol Controversy), but one major fact needs to be considered here. It is often overlooked that cholesterol is the substance used by the body for synthesis of many steroids, e.g., the steroid hormones, which are required by the body. Consequently, cholesterol is essential to life, although it is not classified as a vitamin because the body can synthesize its own cholesterol. As noted in Chapter Fifteen, the amount synthesized by the body is determined, to some extent, by the amount consumed in the diet. In effect, cholesterol plays the role of a *feedback inhibitor* by shutting down the pathway of reactions that are used in the body for synthesis of cholesterol. The phenomenon of feedback inhibition is a common mechanism by which a product formed in a series of reactions prevents its own synthesis by interfering with some step earlier in the series. We shall see several examples of feedback inhibition later in this chapter.

Once cholesterol is formed, it is converted into other products that are necessary in maintaining life. The major pathway that uses most (about 80%) of the cholesterol is that which results in formation of the bile acids. Smaller amounts of cholesterol are converted into the steroid hormones, vitamin D, and other body chemicals. In addition, cholesterol is a major constituent of the fibers in the brain and nervous tissue.

The Bile Acids

In order for the body to absorb cholesterol and other lipids from the intestine into the bloodstream, another type of steroid molecule is needed that is made from

Figure 20.2   The bile acids.

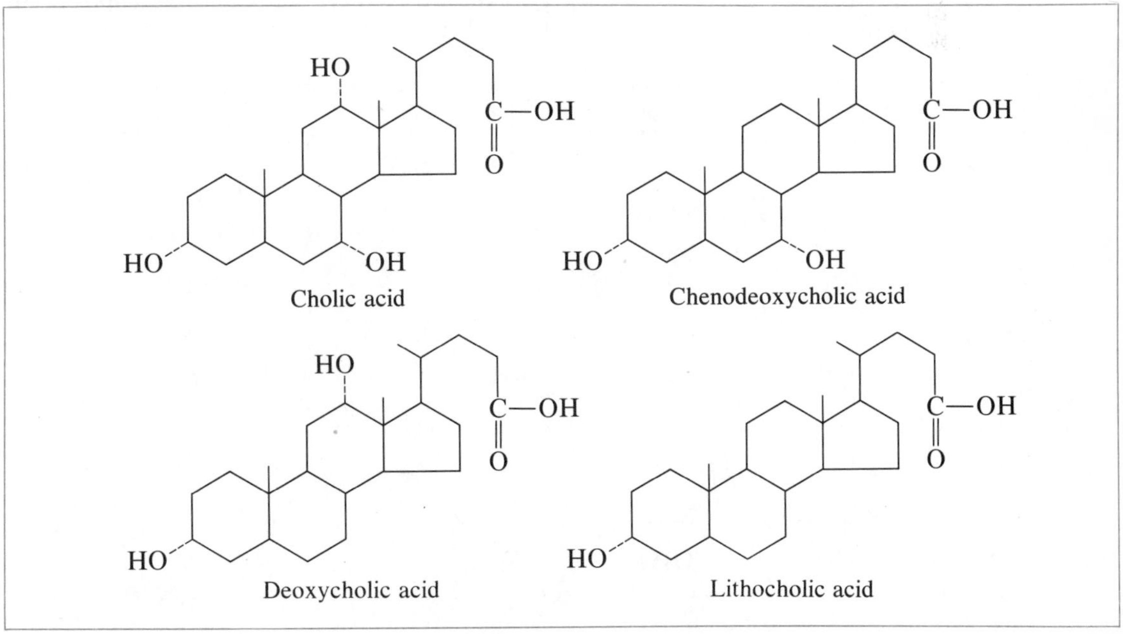

Cholic acid

Chenodeoxycholic acid

Deoxycholic acid

Lithocholic acid

cholesterol—the bile acids. These steroids are derived from the cholesterol by removal of the three terminal carbon atoms from the carbon-17 side chain and saturation (hydrogenation) of the double bond in ring B. Thus, the bile acids are identified by the presence of 24 carbon atoms and the oxidation (insertion of oxygen) of the 3, 7, and 12 positions of the steroid nucleus. The four bile acids are shown in Figure 20.2.

Cholic acid and chenodeoxycholic acid are the major bile acids. They are synthesized in the liver. Deoxycholic acid and lithocholic acid are synthesized in the intestine from cholesterol by bacteria. The bile acids are stored in the gallbladder and are released into the intestine in order to aid in the absorption of dietary lipids, cholesterol included. Thus, the most important breakdown product of cholesterol is also used to assist the body in digesting cholesterol.

The action of the bile acids on dietary lipids is that of an emulsifier. The steroid nucleus itself is very hydrophobic, but the introduction of very hydrophilic groups at several positions of the molecule creates a structure analogous to that found in soaps, detergents, and other emulsifying agents, which assist hydrophobic (nonpolar) lipids in entering the bloodstream (mostly water) so that the lipids may be delivered to the tissues for use as a fuel or for other purposes.

Digitalis

Let us digress briefly from a consideration of the major steroids found in the body to examine the structures and properties of some steroids found in certain plants. Digitoxin and digoxin are two important drugs used in the treatment of heart disease, including congestive heart failure. Digitalis is the dried leaves of a plant named *Digitalis purpurea* or purple foxglove. The odd name is derived from the purple colored flowers that have the shape of a finger. Digitalis has been used with dramatic success since the late eighteenth century in treatment of heart disorders. The leaves contain a variety of substances, most notably the combination sugar and steroid compound known as digitoxin, which is the major active component of digitalis.

Figure 20.3   Digoxin (Lanoxin®) is a drug extracted from a plant and used in the treatment of heart disease. Digitoxin is a related substance that lacks the OH at carbon 12 of the steroid nucleus.

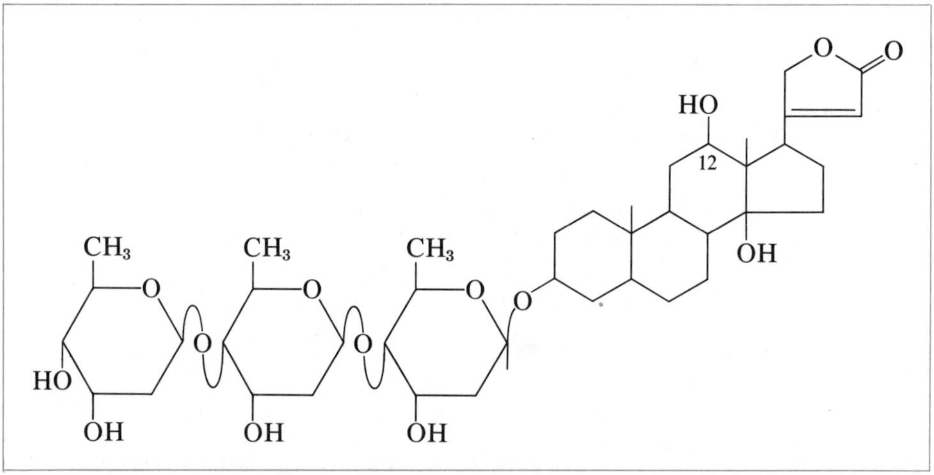

Digoxin is a related compound that has become the number one drug for the treatment of heart disorders. The brand name form, Lanoxin®, and the generic digoxin were the number 14 and 61 most prescribed drugs in 1974 (1). Digoxin may be obtained by extraction and treatment of the extract from *Digitalis lanata*, also known as white foxglove.

Cortisone

The adrenal glands (located on top of the kidneys) utilize cholesterol for synthesis of a series of hormones, of which the best known is cortisone. In the bile acids, the side chain at carbon 17 has been shortened from 8 carbons to 5. In going on to the hormones synthesized by the adrenal gland, the side chain is shortened from 8 to 3 carbons. Other changes in functional groups also occur on the steroid nucleus.

In 1938, E. C. Kendall* at the Mayo Clinic first isolated cortisone, which was later shown to produce dramatic relief in patients with rheumatoid arthritis. The antiinflammatory action, which is responsible for the relief of arthritis symptoms, is just one of many effects caused by this and related hormones. The hormones in this group are collectively known as *adrenal corticosteroids*. Several are shown in the following.

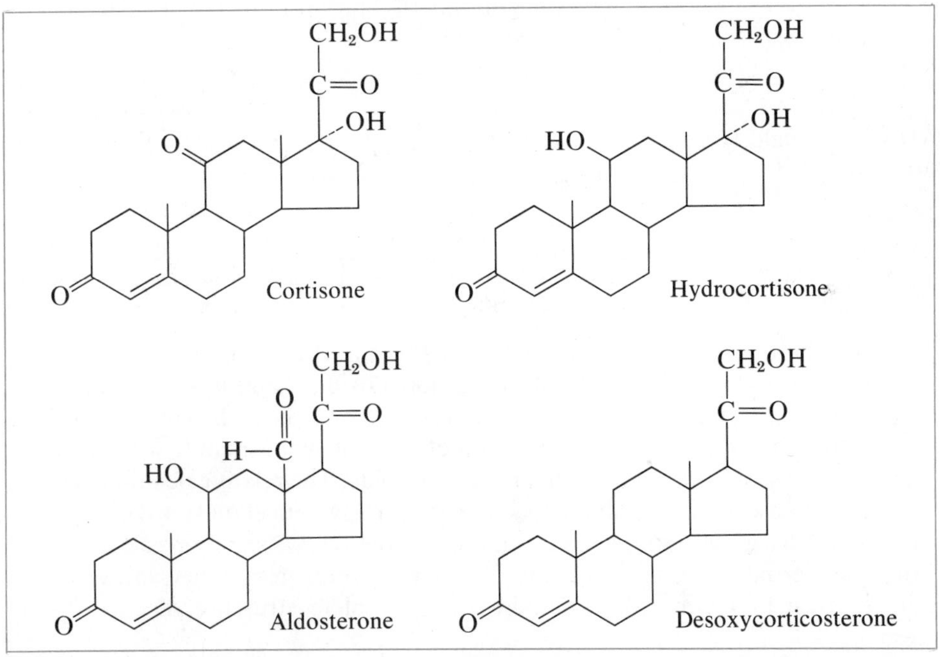

Cortisone

Hydrocortisone

Aldosterone

Desoxycorticosterone

Hydrocortisone, also known as cortisol, is often administered in place of cortisone due to greater potency, although it is less suited for injection because it is less soluble in body fluids than cortisone.

The Sex Hormones

Up to this point, we have seen two types of steroid molecules, the bile acids and the adrenal corticosteroids, which are synthesized in the body from cholesterol.

* Kendall received the Nobel Prize in 1950 for his work on the adrenal hormones.

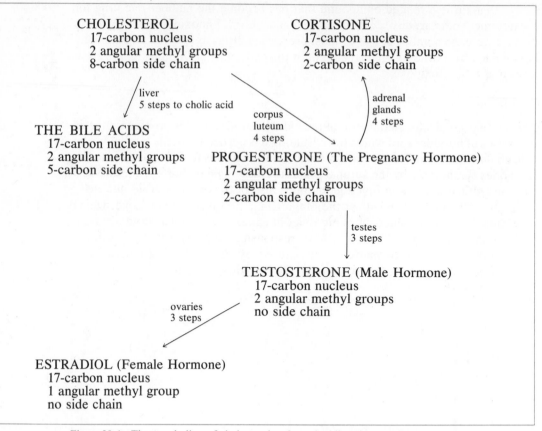

Figure 20.4 The metabolism of cholesterol to form the bile acids, cortisone, and the sex hormones.

Although there are some very important changes in the functional groups on the steroid nucleus and in the length of the carbon-17 side chain, it is also interesting to observe the change in the number of carbon atoms as each type of steroid is formed from its precursor. The discussion of steroids will conclude with a consideration of the sex hormones, which are also steroids. The relative position of these steroids in the series of metabolic reactions beginning with cholesterol is shown in Figure 20.4 in a greatly simplified form. Most of the arrows signifying a conversion from one steroid to the next actually represent several steps. The same information is given in Figure 20.5, including the complete structure of each of the steroids.

The number of carbon atoms in the steroid nucleus remains the same throughout the pathway of reactions, whereas the side chain at carbon 17 and the angular methyl groups are changed by this scheme of reactions, as indicated in Figure 20.5.

Adjacent to each of the arrows in Figure 20.4, is the name of the organ or gland in which each type of steroid is an end product. For example, progesterone is a precursor to cortisone, testosterone (a male hormone), and estradiol (a female hormone), but, except during pregnancy, a gland called the corpus luteum (yellow body) converts cholesterol only as far as progesterone. On the other hand, the

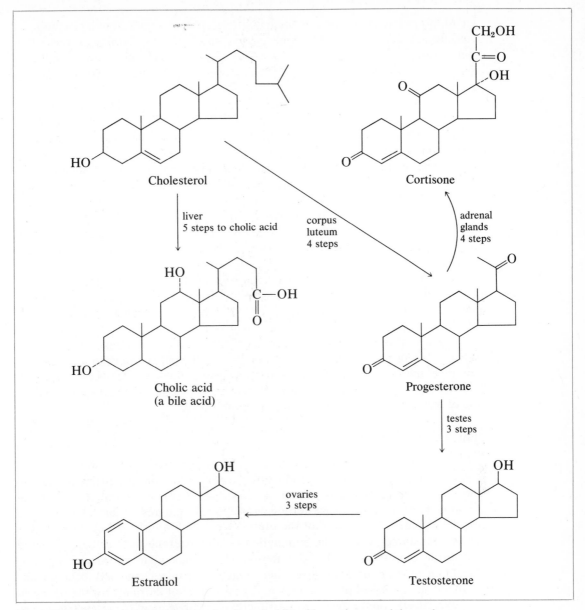

Figure 20.5   The metabolism of cholesterol to form the bile acids, cortisone, and the sex hormones.

testes carry the synthetic sequence all the way from cholesterol to testosterone, whereas in the ovaries, the pathway goes all the way from cholesterol to estradiol.

Although we will concentrate on the three sex hormones, each is actually just one member of the classes of hormones known as *progestogens* (pregnancy hormones), *androgens* (male hormones), and *estrogens* (female hormones). Each member of each class of steroids differs from the others in the kinds and positions of functional groups that are present.

Up to this point, the term *hormone* has been used many times, but let us stop and consider exactly what a hormone is. Chemically, most hormones are either steroids or proteins. The thyroid hormones, adrenalin and others are exceptions. We have already seen some of the steroid hormones. Insulin and glucagon are examples of protein hormones.

All types of hormones function in much the same way. Hormones are produced by various glands in the body and then released into the bloodstream. Once in the circulation, they are able to move to other parts of the body and exert their effects on target tissues.

The scientific field involving the study of hormones is called **endocrinology.** The parts of the body that produce and release hormones are called *endocrine glands*. Hormones affect a target tissue by regulating the activity of enzymes within that tissue. The term *hormone* comes from the Greek, *horman*, to excite.

In several cases, a hormone released from one endocrine gland stimulates or suppresses the production of a hormone by another gland. For example, the pituitary gland (located at the base of the brain) produces protein hormones called *gonadotropic hormones* or simply *gonadotropins*. The gonads are the sex glands. The word *gonadotropic* indicates the ability to act on the sex glands. Thus, we find that the protein hormones released by the pituitary stimulate the production of steroids by the sex glands, i.e., the ovaries and testes.

The entire scheme of interactions of the various hormones is a fascinating subject that provides the basis for the strategy behind the development of oral contraceptives; but before we get into that aspect of the story, let us consider each of the major types of sex hormones. We will do so by considering the hormones in the order in which they are synthesized in the body.

The Progestogens

Although there are several progestogens, the most active is progesterone. It is a steroid synthesized as an end product by the corpus luteum, which is generated in the ovaries. Progesterone is often described as the pregnancy hormone for several reasons. For one thing, progesterone helps to prevent a miscarriage by promoting the development of the wall of the uterus into which a fertilized egg (ovum) becomes implanted. Under the stimulation of progesterone (and estradiol), blood vessels develop in the wall of the uterus, which thickens in preparation for possible implantation of a fertilized egg. Once a fertilized egg has become implanted, the increased blood supply acts as a source of nutrients for the developing fetus. In addition, progesterone also prevents subsequent ovulation (release of an ovum), which could result in a second fertilization. Consequently, progesterone is a built-in birth control mechanism for prevention of a second pregnancy. As we shall see later, progesterone-like compounds are major components of some of the oral contraceptives for exactly this reason.

The Androgens

The male hormones are known as *androgens*. Like the other sex hormones, they are produced from cholesterol, but they are the end products of the metabolism of cholesterol in the testes. The most active of the androgen steroids is testosterone.

"... and now, Dr. Preston, reporting on some interesting side effects of moose hormone experimentation."

Copyright © 1965, Cartoon Features Syndicate. Reprinted by permission.

The androgens promote certain characteristics normally associated with masculinity, such as muscle strength, deep voice, and facial hair.

The Estrogens

The estrogens are the female or feminine hormones, which stimulate the growth of female tissues such as the breasts and the lining of the uterus. In the latter function, they act in collaboration with progesterone. Estradiol is the principal estrogen, although a steroid known as estrone also is quite active. It is important to note that cholesterol is converted to testosterone and then on to estradiol in the ovaries. When there is a breakdown in the formation of estradiol, the result may be masculinization of the female, due to a buildup of testosterone.

THE BIOLOGY OF THE REPRODUCTIVE PROCESS

In order to understand the role that hormones play in reproduction, we must first have an understanding of the principal steps involved in the process. Much of the story is portrayed in Figures 20.6 and 20.7. Let us trace the course of events as they would occur in the event of a successful fertilization leading to pregnancy.

The first event is the development of a mature ovum (egg), which occurs in a cavity of the ovary called a *Graafian follicle*. The ovum and the surrounding follicle mature in about 12 to 14 days to a point when the follicle ruptures and a mature ovum is released from the ovary. This is called *ovulation*. Following release of the ovum, the follicle changes into an endocrine gland, called the corpus luteum, which takes on the job of producing progesterone. The entire process up to this point is shown in Figure 20.6.

Moving on to Figure 20.7 we see the reproductive system in more detail, including all of the events leading up to and following ovulation. As indicated in this illustration, only one side of the reproductive system is normally active in any cycle. In humans, the cycle alternates approximately every 28 days from one side to the other, terminating in menstruation at the end of each cycle, except in the event of a successful fertilization leading to pregnancy. Following release from the ovary, the mature (but unfertilized) ovum passes into the fallopian tube through which it eventually travels to the uterus. Fertilization is only possible in the fallopian tubes and only for a short period after the ovum is first released from the ovary. Sperm released into the female reproductive tract must make their way up the fallopian tubes in order to cause fertilization. If the timing is not right, the unfertilized ovum will travel down the fallopian tube into the uterus, where it cannot be fertilized. In addition, the sperm that enter the system survive only a short time, since the supply of nutrients is not sufficient to sustain them.

Once fertilization is successful, the fertilized egg moves down the fallopian tube into the uterus and becomes implanted into the wall of the uterus, which has become saturated with blood vessels carrying a supply of nutrients to sustain the embryo. The development of the uterine wall into this state begins shortly after the time when the follicle starts to develop in the ovary. In other words, the devel-

Figure 20.6 Growth and rupture of the ovarian (Graafian) follicle and formation of the corpus luteum. (Adapted from *Birth Control*, by Philip Rhodes, Oxford University Press, 1971.)

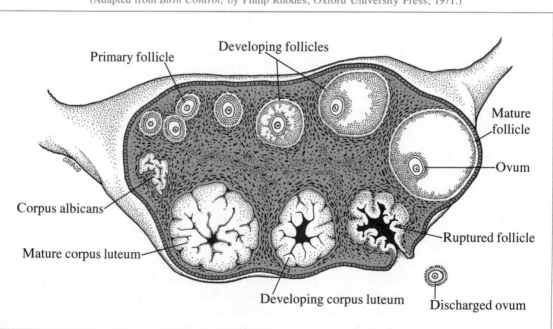

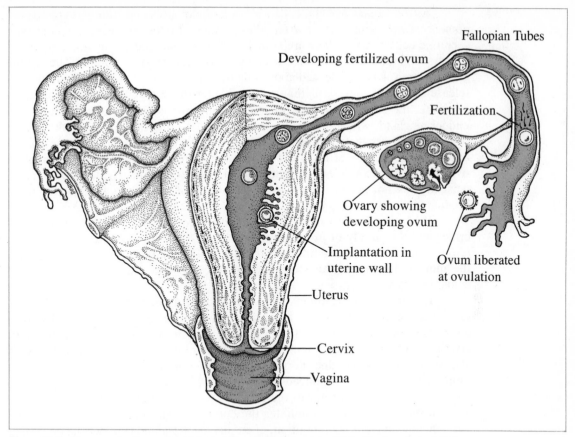

Figure 20.7 The reproductive cycle. (Adapted from *Birth Control,* by Philip Rhodes, Oxford University Press, 1971.)

opment of the uterine wall begins prior to fertilization and does not depend on a successful fertilization. If fertilization does occur and the fertilized egg becomes implanted, the uterine wall remains intact and the embryo (fetus) develops. If no fertilization occurs, the uterine wall is sloughed off and menstrual flow begins. An unfertilized ovum does not become implanted in the uterine wall.

THE CHEMISTRY OF THE REPRODUCTIVE PROCESS

Now let us look at the role that hormones play in the progress of the cycle, which is sometimes referred to as *the ovarian cycle.* The entire process is triggered by the release of hormones by the hypothalamus region of the brain. These hormones then travel to the pituitary gland and stimulate the release of pituitary hormones into the bloodstream. The two principal hormones released by the pituitary are called *follicle stimulating hormone,* symbolized FSH, and *luteinizing hormone,* symbolized LH. Both of these pituitary hormones are proteins. They are the gonadotropic hormones mentioned earlier, which stimulate the sex glands (the gonads).

As the name suggests, FSH travels to the ovaries and stimulates the development of an ovum within a Graafian follicle. In the process, the hormone-secreting cells of the follicle are stimulated to produce increasing amounts of estrogens while the ovum matures to the state necessary for ovulation. The second gonadotropin, LH, is released about this time and it causes the ripened follicle to release the ovum, i.e., ovulation occurs. Under the stimulation of the luteinizing hormone, the empty follicle is transformed into the corpus luteum, a temporary endocrine gland, which produces progesterone. As the level of the sex hormones (estrogen and progesterone) builds up, the pituitary gland recognizes the change and stops the release of FSH and LH so that there is no development of another ovum. This is an example of a negative feedback mechanism in which the end products, the sex hormones, suppress the repetition of the events that caused these products to form.

When fertilization and implantation do not occur, the corpus luteum degenerates and the supply of progesterone is cut off. The drop in circulating progesterone is the signal for the pituitary to release FSH and begin a new cycle. Since the lining of the uterus is also dependent on progesterone and estrogen, it dies and is lost in menstruation.

If fertilization and implantation are successful, the corpus luteum still eventually degenerates; however, it functions until the placenta takes up the job of producing progesterone at levels high enough to sustain the inhibition of the pituitary and prevent further ovulation until birth. The continued output of progesterone is also stimulated by one final hormone, called human chorionic gonadotropin, or simply HCG, which is produced by the developing fetus.

It is common to refer to the pituitary as the master gland because the release of pituitary hormones initiates the complex series of chemical events, and yet, once the cycle is set in motion, even this master gland is under the control of chemicals produced in other glands.

In addition, the progesterone causes the glands of the cervix to produce a sticky secretion. This makes it more difficult for sperm to penetrate the cervix and reinforces the effect of the progesterone on the pituitary in preventing a second pregnancy. As we will see in the next section, this phenomenon can also contribute to the effectiveness of oral contraceptives.

Before looking at the strategies employed in the use of oral contraceptives, let us summarize the reproductive process and view the steps as shown in Figure 20.8, in which the variation in hormone levels is indicated on the same time scale as the events that occur in the ovary and the uterus.

Day 1 (same as day 29) of the cycle is normally taken as the beginning of menstruation (menses), at which time the lining of the uterus begins to slough off. This is the time in the cycle when the level of progesterone has dropped very low. Recall that progesterone, in cooperation with estrogen, is responsible for maintaining the uterine lining in the event of successful fertilization and implantation. The situation described in Figure 20.8 is one in which no fertilization occurs, so that the corpus luteum degenerates and the supply of progesterone decreases. If pregnancy had occurred, the progesterone production by the corpus luteum would have continued until about the fifth week of pregnancy, when the placenta normally takes over the production of progesterone.

It is during the early days of the cycle that the levels of the gonadotropic hormones (FSH and LH) are elevated as they are stimulating the development of

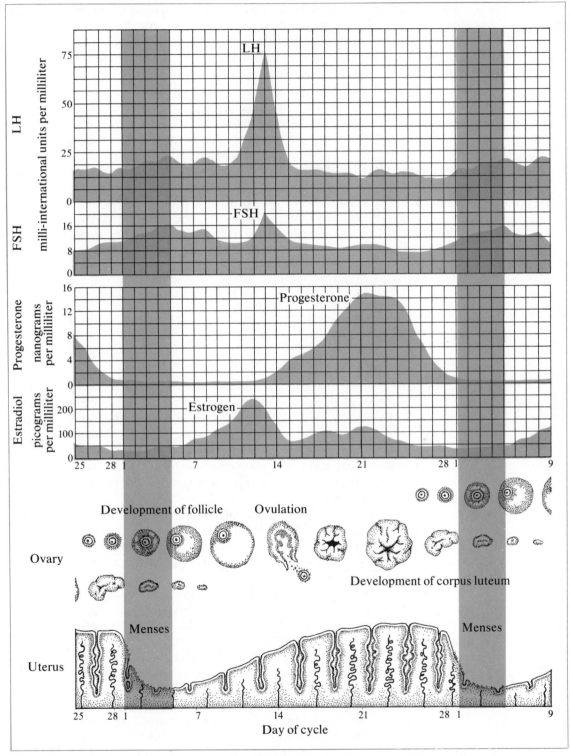

Figure 20.8 Ovulatory and menstrual cycles. (Adapted from ''The Physiology of Human Reproduction,'' S. J. Segal, Scientific American, Inc., 1974.)

a new follicle. Ovulation occurs around day 14, in response to a sharp rise in LH, after which the progesterone level rises and suppresses further release of FSH and LH. It can be seen from Figure 20.8 that the peak of progesterone coincides with the time when the corpus luteum is fully developed. As the corpus luteum begins to degenerate, the progesterone level decreases. Around day 26, several interesting events are occurring. At this time, as the progesterone level is dropping sharply, the follicle stimulating hormone is on the rise and a new follicle is beginning to develop in the other ovary. Therefore, the development of a new follicle has already started by day 1, if pregnancy has not occurred.

In short, we see a sort of inverse relationship between the protein hormones (FSH and LH) and the steroid hormones. When one set of hormones is high, the other set is depressed. When the first set of hormones subsides, the second set increases in level. When the estrogen level drops following a peak at day 12, the gonadotropins rise sharply and trigger ovulation, during which the FSH and LH levels drop and the progesterone begins to rise. This pattern is the phenomenon of feedback inhibition operating at a very high degree of sophistication.

THE STRATEGY OF "THE PILL"

Although oral contraceptives are commonly known as "the pill," there are actually three types of oral contraceptives. These will be discussed in the order in which they were developed.

The First-Generation Birth Control Pill

With an understanding of the sequence of events that occur in the female cycle and the involvement of chemicals in that sequence, we can now proceed to a consideration of the mode of action of the first type of oral contraceptive, which is often described as the conventional approach or the "combination pill."

Contraceptive technology has improved with a better knowledge of the female cycle. It has been known for some time that disorders of the cycle could be controlled by administering steroid hormones to patients, but it was not until the development of synthetic steroids in the early 1950s that it became practical to use chemicals to prevent conception. The combination pill has a mixture of two synthetic steroids, an estrogen and a progestin. The term *progestin* is commonly used for steroids that are synthetic modifications of the progesterone molecule but with the same hormonal properties. The term *estrogen* is used to describe both the natural and synthetic estrogens.

Each pill in the first-generation oral contraceptive contains a constant amount of estrogen and progestin and is taken every day beginning on day 5 and continuing for a period of 21 days.* At the end of the 21 days, i.e., after day 25, the routine is stopped and menstruation can begin. The same steps are followed in each subsequent month beginning on day 5. In each month, day 1 is defined as the day when menstrual flow begins. Some preparations on the market contain pills that are taken beyond day 25. In any case, estrogen and progestin are discontinued after 21 days and the rest of the pills may contain a placebo or, perhaps, an iron supplement to counteract the loss of iron that occurs during menstrual flow. In other words, some of the events of the normal cycle do take place, e.g., the

---

* Some products are taken for 20 days.

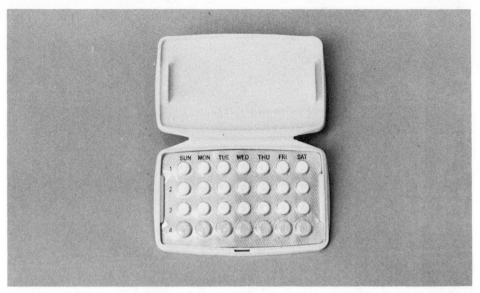

Figure 20.9   Birth control pills. Demulen-28 is a combination (first-generation) oral contraceptive that consists of 21 pills with progestin and estrogen plus 7 pills that are placebos to help the user keep an accurate count of the days of the cycle. Some similar products contain an iron supplement in the 7 placebos in order to counteract the loss of iron in menstruation. (Courtesy of G. D. Searle & Co.)

development of the lining of the uterus, which occurs under the influence of the synthetic steroids. This normal functioning helps to assure that pregnancy can happen at any time that a woman decides she is ready to have children. If the routine of building up the uterine lining followed by menstruation did not occur regularly, the tissues might become inactive and nonfunctional even when the use of an oral contraceptive is discontinued.

On the other hand, while part of the cycle occurs normally, both ovulation and fertilization are prevented while using the combination pills. The constant high level of synthetic steroids in the blood causes a negative feedback inhibition of the release of follicle stimulating hormone and luteinizing hormone, which are necessary for development of the ovum and ovulation. It is the estrogen that inhibits FSH release and the progestin that inhibits LH release.

At the same time, in the presence of the progestin, the glands of the cervix produce the sticky secretion that normally develops in response to the natural progesterone. This occurrence enhances the contraceptive effect of the pill by making it more difficult for sperm to penetrate the cervix. In short, the sperm are unable to reach an ovum which is not there anyway.

The Second-Generation Birth Control Pill

The second type of oral contraceptive is often referred to as the "sequential pill." As a result of the appearance of a number of undesirable side effects, the sequential approach was developed in an effort to more closely mimic the natural levels of the ovarian hormones found in the body. With these pills, the normal routine is to begin taking the pills containing only estrogen on day 5 for a period of about 16 days, after which a combination of estrogen and progestin is taken for about an-

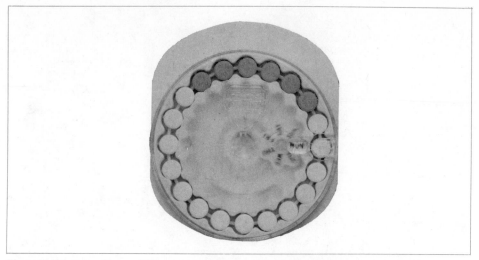

Figure 20.10   Ortho-Novum SQ is a sequential (second-generation) oral contraceptive, which consists of 14 pills containing only estrogen and 6 pills containing a combination of estrogen and progestin. (Courtesy of Ortho Pharmaceutical Corp.)

other 5 days.* In this way, the level of progestin is elevated at about the same time in the cycle when progesterone levels are normally elevated.

In contrast to the mechanism of the conventional or combination pills, the sequential pills suppress the release of FSH due to the presence of the estrogen, but do not suppress the release of LH, due to the absence of a progestogen prior to day 21. Whether or not the LH is suppressed is of no concern, since a mature ovum cannot develop in the absence of FSH.

As with the conventional pill, the buildup of the lining of the uterus and menstruation occur normally during each cycle. On the other hand, the absence of the progestin in the early stages of the cycle causes the fluid in the cervix to remain clear so that sperm can penetrate normally, during the time when ovulation would ordinarily occur. In short, the sperm are now able to reach an ovum, which is not there.

The Third-Generation Birth Control Pill

The most recent development in the technology of chemical birth control is the so-called "minipill." Once again, in an effort to further minimize side effects, an approach was developed that omits estrogen altogether. The only chemical agent is the synthetic progestogen, which is taken every day of the year. The amount of progestin is also lower than that found in conventional pills, which explains the use of such terms as "minipill" and "minidose." Because of the low dose of progestin and the absence of estrogen, the release of FSH and LH is not blocked and ovulation occurs normally. However, as in the first- and second-generation pills, the progestin stimulates the cervix into producing sticky secretions, which block passage of the sperm. In short, in this case, the sperm are unable to reach the ovum, which is available for fertilization.

* The routine varies slightly from one product to the next.

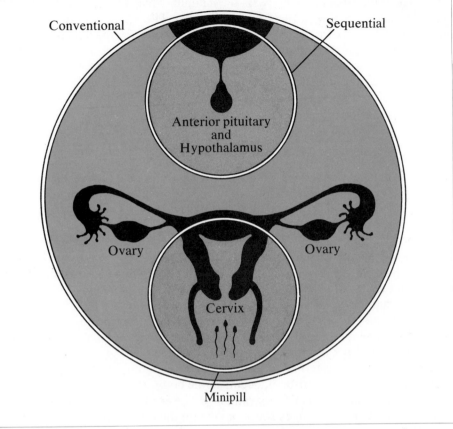

Figure 20.11 The modes of action of oral contraceptives. Circles indicate areas affected by different contraceptive tablets. (1) Large circle includes both the brain-pituitary network, through which ovulation is prevented, and the cervix, whose normal secretions are made to form a barrier to sperm. These are the areas affected by the "first generation" tablets. (2) Upper circle includes only the brain-pituitary network, which is primarily affected by the sequential pills to prevent ovulation. (3) Lower circle includes only the cervix, which is the primary area of influence of the "minipill." This approach permits normal ovulation but causes development of a barrier to sperm. (Courtesy of Syntex Laboratories, Inc.)

Summary Of Oral Contraceptives

Since about 1960, there have been three approaches available for chemical birth control. The first one was the conventional or combination pill, which contains both estrogen and progestin. It is given in a fixed dose, usually for 21 days of the month. The strategy of this approach is prevention of both ovulation and fertilization. The latter may be enhanced by altering the fluid in the cervix to form a barrier to penetration by sperm.

The second oral contraceptive to be developed was the sequential type—estrogen only (15 days), followed by estrogen plus progestin (5 days). The strategy here is only to prevent ovulation. This approach was developed in hopes of mimicking natural hormone levels more closely, thereby reducing side effects to a minimum.

The third oral contraceptive is the minipill, which is taken every day of the year. It contains a low dose of progestin only. The strategy here is to permit normal ovulation but interfere with fertilization by altering the fluid in the cervix. Once again, one of the major goals in developing the minipill was elimination of side effects.

The areas of influence of the three types of contraceptives within the body are compared in Figure 20.11.

Synthetic Hormones Used In Oral Contraceptives

In the previous sections, there has been frequent reference to synthetic estrogens and synthetic progestogens, called progestins. At this point, we should consider the kinds of changes that were made at the molecular level in order to produce effective synthetic steroids.

The problems that have to be overcome in producing synthetic steroids are much the same as those encountered with most drugs and include such things as side effects, cost, effective dosage, and effectiveness when consumed orally. Let us consider each of these in more detail.

The matter of side effects is always of great concern, and, in the case of oral contraceptives, has received much notoriety. This topic is discussed in depth in a following section, but let us recall one lesson that we have already learned about the natural sex hormones. As outlined in Figures 20.4 and 20.5, the metabolic route from progesterone to testosterone to estradiol, with branches leading to other steroids (e.g., cortisone), occurs in the body. In a similar fashion, there is the possibility that a synthetic progestin might be converted to an androgen (male hormone) or, perhaps, to an estrogen *in vivo* (in the body). In fact, it is even possible that enzymes in the body might convert a steroid with the properties of a sex hormone into one with entirely different properties. Given all these possibilities, plus the knowledge that steroids can cause dramatic feedback effects on the brain and pituitary, it is not surprising that side effects are a particular concern with oral contraceptives.

In addition, there is the fact that these agents are administered to healthy persons. Undesirable side effects are often considered an acceptable risk when the drug that causes them cures a disease or counteracts some severe symptoms, but when a person is healthy in the first place, side effects are less likely to be considered acceptable.

A second matter for concern is the cost and effective dosage of a drug. These may seem unrelated, but the potency of a drug determines how much is needed and is, therefore, a factor in determining the cost of the drug and whether or not it will be profitable to market the drug. As with many other drugs, the synthetic substance is often more potent than the natural one, and if the potency is directed to the target tissue, the lower dosage may result in fewer side effects.

Finally, there was the obvious need to develop a contraceptive that could be taken orally, since regular injections would be very undesirable. Seven steroids have come into prominence for use in oral contraceptives—two estrogens and five progestins. An additional progestin, known as dimethisterone, has been discontinued as a suspected carcinogen. The structures of the synthetic steroids are shown in Table 20.1. Both of the synthetic estrogens differ very little from the natural estradiol. In fact, the only significant change is the *ethinyl group* (—C≡C—H) at carbon-17, which is located alpha, as signified by the dotted line. The ethinyl group is a key feature of all the synthetic sex hormones. It enhances two properties of the steroids—potency and stability in the gastrointestinal tract. The latter is particularly important because it makes it possible to take them orally. In contrast, the natural hormones are largely inactivated during digestion.

As for the progestins, there are some more dramatic differences between the natural hormone, progesterone, and the synthetics. Once again, the ethinyl group is present for the reasons cited above, but in addition, the methyl group (carbon 19) is missing from the number 10 position of the steroid nucleus. This change is reflected in the use of the prefix "nor" to indicate that there is no carbon 19 in steroids such as 17-α-ethinyl-19-nortestosterone, also known as norethindrone (see Table 20.1).

Note also that norethindrone and all of the other progestins have a chemical structure similar to testosterone, as reflected in the name, because of the absence of a carbon-containing chain in the 17-beta position and the presence of the hydroxy group at that site. In spite of this, the progestins have progestogenic

Testosterone

properties because of the absence of carbon 19. In fact, the chemical structure is very similar to the estrogens except for the multiple unsaturation in ring A.

A listing of some of the products that have been marketed appears in Table 20.2.

Side Effects

The adverse reactions or side effects resulting from oral contraceptives have been a point of much public concern ever since the introduction of these drugs in the 1960s. It is not surprising that side effects have shown up since steroids exert control over so many body functions. The estrogen and progestogen compounds used in the oral contraceptives are highly potent hormones, which are meant to almost completely suppress the natural synthesis of certain steroids in the body, as well as other body chemicals, e.g., the gonadotropins. In addition, women in the reproductive age differ greatly in the amount of natural hormone that they synthesize. Consequently, the administration of oral contraceptives with fixed amounts of synthetic hormones is almost certain to cause undesirable side effects if the natural balance of hormone levels is not met by a particular pill. Of course, a woman and her physician have the option of shopping around for different products with different combinations of synthetic steroids—either different steroids or different amounts (see Table 20.2), but this puts the woman into the role of a guinea pig while the search is on. And even a proper balance of steroids does not assure the

TABLE 20.1   THE NATURAL AND SYNTHETIC FEMALE SEX HORMONES

ESTROGENS	PROGESTOGENS
Natural	*Natural*

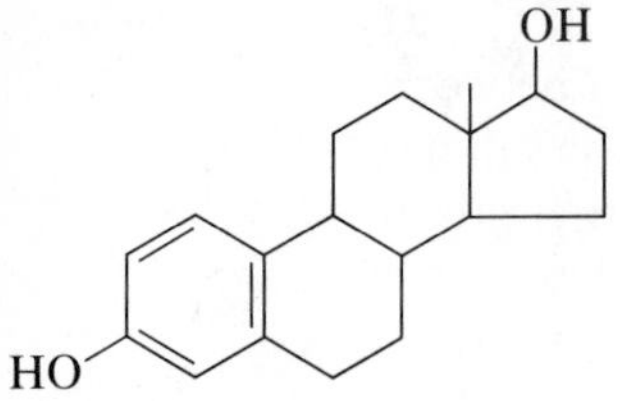

estradiol

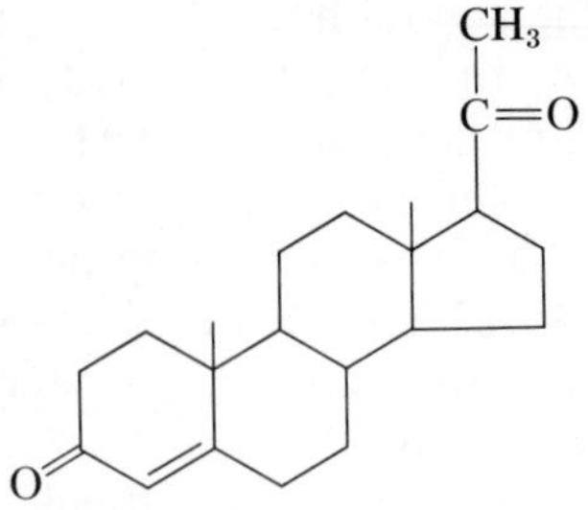

progesterone

Synthetic

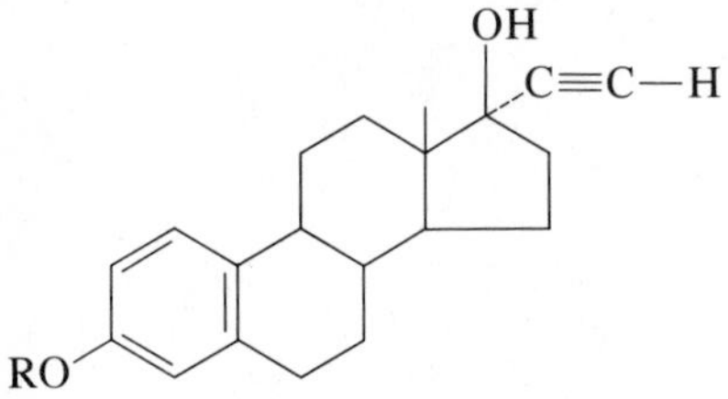

mestranol (R=CH₃)
ethinyl estradiol (R=H)

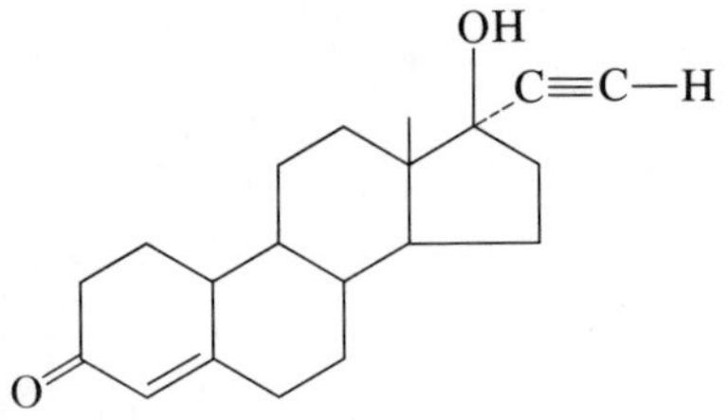

Synthetic

norethindrone

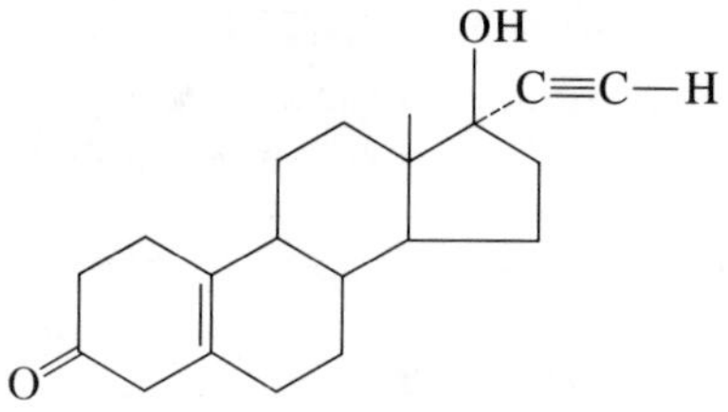

norethynodrel

elimination of all side effects, since the synthetic steroids do have chemical structures that are slightly different from the natural hormones, and may, therefore, have different effects on certain body processes. Some of the side effects are symptoms that are often encountered during pregnancy. This is not surprising since the pill creates a pseudopregnancy condition in the body by artificial elevation of the steroid hormones.

When the pill was introduced in the United States in 1960, a great deal of controversy arose over the adverse reactions to the drug. This led to extensive research in both Great Britain and the United States and these studies have shown a statistical probability that certain risks are associated with these drugs. The evidence suggests an increased risk of blood clots, hypertension (high blood pres-

TABLE 20.1   THE NATURAL AND SYNTHETIC FEMALE SEX HORMONES

ESTROGENS	PROGESTOGENS

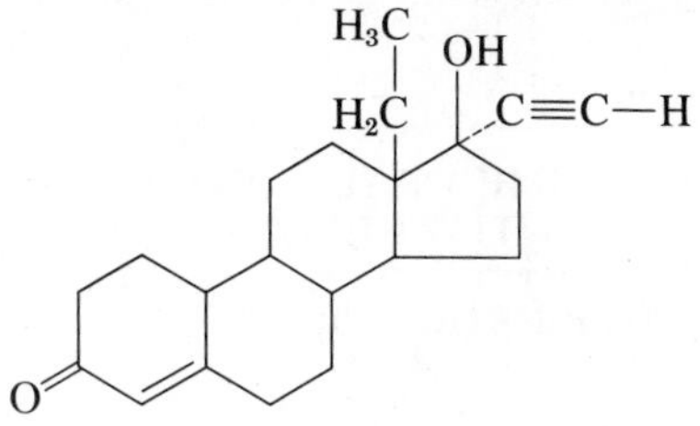

norgestrel

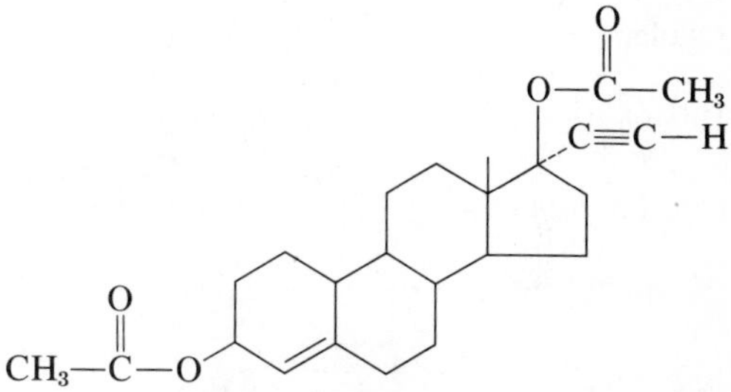

ethynodiol
diacetate

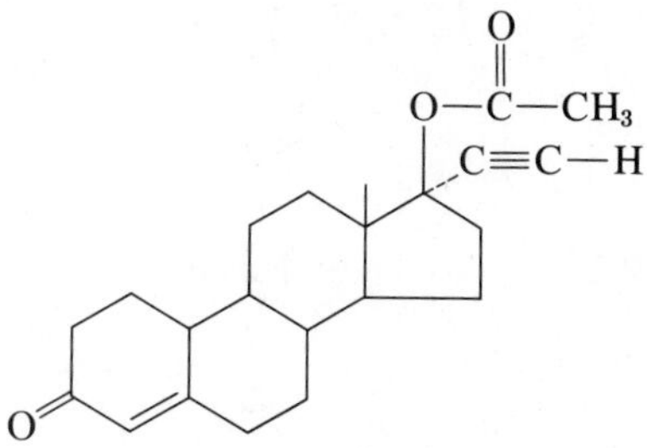

norethindrone
acetate

sure), depression and psychiatric disturbances, gallbladder disease, benign liver tumors, urinary tract infections, and other risks. Initially, there was great apprehension that birth control pills may increase the risk of breast or genital cancer, but studies have produced no definitive evidence to support these fears. In fact, it now appears that the estrogen and progestogen compounds protect against the development of benign breast tumors and ovarian cysts.

The use of oral contraceptives also produces profound changes in the metabolism of proteins, lipids, and carbohydrates, which are undoubtedly related to alterations in the levels of hormones released by the thyroid and adrenal glands.

A study in New York during the period from 1968 to 1973 showed an increased risk of birth defects from breakthrough pregnancy while using oral contraceptives or from supportive hormone therapy during pregnancy. It is also apparent that the defects are sex specific, since most of the affected babies are male.

In most cases, the increased risk of adverse side effects has been linked to the estrogen component in the pill. In response to this finding, oral contraceptives

TABLE 20.2   SOME ORAL CONTRACEPTIVES THAT HAVE BEEN AVAILABLE IN THE UNITED STATES

| COMBINATION PILLS | | | TABLET COMPOSITION | | |
TRADE NAMES	MANU-FACTURER	PROGESTIN	DOSE (mg)	ESTROGEN	DOSE (mg)
Enovid-E, 5	Searle	norethynodrel	2.5–5.0	mestranol	0.075–0.1
Ovulen	Searle	ethynodiol diacetate	1.0	mestranol	0.1
Demulen	Searle	ethynodiol diacetate	1.0	ethinyl estradiol	0.05
Ovral & Lo/Ovral	Wyeth	norgestrel	0.3–0.5	ethinyl estradiol	0.03–0.05
Ovcon-35, -50	Mead-Johnson	norethindrone	0.4–1.0	ethinyl estradiol	0.035–0.05
Brevicon	Syntex	norethindrone	0.5	ethinyl estradiol	0.035
Norinyl 2, 1 + 50, 1 + 80†	Syntex	norethindrone	1.0–2.0	mestranol	0.05–0.1
Ortho-Novum 2, $\frac{1}{50}$, $\frac{1}{80}$†	Ortho	norethindrone	1.0–2.0	mestranol	0.05–0.1
Norlestrin $\frac{1}{50}$, 2.5 mg	Parke-Davis	norethindrone acetate	1.0–2.5	ethinyl estradiol	0.05
Norlestrin Fe‡ 1, 2.5 mg	Parke-Davis	norethindrone acetate	1.0–2.5	ethinyl estradiol	0.05
Loestrin $\frac{1}{20}$, $\frac{15}{30}$	Parke-Davis	norethindrone acetate	1.0–1.5	ethinyl estradiol	0.02–0.03
Zorane $\frac{1}{20}$, $\frac{1}{50}$, $\frac{15}{30}$	Lederle	norethindrone acetate	1.0–1.5	ethinyl estradiol	0.02–0.05
SEQUENTIAL PILLS*					
Oracon	Mead-Johnson	dimethisterone	25	ethinyl estradiol	0.1
Norquen	Syntex	norethindrone	2.0	mestranol	0.08
Ortho-Novum SQ	Ortho	norethindrone	2.0	mestranol	0.08
MINIPILL*					
Micronor	Ortho	norethindrone	0.35	none	——
Nor QD	Syntex	norethindrone	0.35	none	——
Ovrette	Wyeth	norgestrel	0.075	none	——

* Removed from the United States market.
† Symbolism such as 1 + 80 or $\frac{1}{80}$ indicates that the product contains 1 milligram of progestogen and 80 micrograms (0.080 mg) of estrogen.
‡ Contains iron.

have been manufactured that have a lower estrogen content, and the minipill was produced with no estrogen. Studies also indicate that the sequential pills carry a higher risk of side effects even though this approach was developed to more closely mimic natural levels of the sex hormones during the reproductive cycle. Consequently, the first-generation combination pills are preferred over the sequential type.

With the increased use of the progestin-containing minipill, liver disorders are found to occur with a higher incidence. This suggests that these side effects are due to the progestin and not estrogen.

So, it is obvious that the use of oral contraceptives is associated with an increased risk of some serious medical problems. From the evidence presented in the scientific literature, the absolute risk of developing any of these problems is quite small, particularly for young women, but it is advisable to use other forms of contraception for women with pre-existing symptoms that argue against the use of the pills. In the jargon of the medical profession, the use of oral contraceptives is contraindicated in some cases, although the risk of death associated with the use of the pill is much lower than the risk of death in childbirth for all women except those over forty (2).

ALTERNATIVE METHODS OF BIRTH CONTROL

A highly effective alternative to oral contraceptives is the intrauterine device or simply IUD. The most important advantage of the IUD over other methods is that once the device is inserted by a physician, the user no longer has to worry about it or adhere to a pill-taking schedule. The device can be safely left in place for some time with no decrease in effectiveness. Upon removal of the device, fertility returns within a few days. In addition to the lack of motivation needed by the user to assure effectiveness, the cost is low.

The earliest report of an entirely intrauterine device was in 1909 (3). At that time, the IUDs were made of silkworm gut and secured with a piece of fine wire. The devices were used in conjunction with glass discs and silver rings until about 1930, and then virtually ceased to exist for the next thirty years. With increased public attention on birth control, a revival of interest led to the development of a new series of IUDs, which were introduced on the market at about the same time as the oral contraceptives.

Several of the IUDs that have been used are pictured in Figure 20.12. From observations in animals and experience of women, it was known that the presence of a foreign body in the uterus prevents pregnancy, but the exact mode of action of an IUD is not clearly understood. An IUD does not suppress ovulation or completely block the sperm, as do other mechanical methods. The mechanism of action appears to be to prevent implantation of a fertilized egg by creating a hostile environment in the uterus. Observations in animals indicate that there is an inflammation of the uterine lining as a result of the contact with the IUD.

The most common adverse reactions to the IUDs are expulsion, perforation of the uterus, irregular bleeding, uterine cramps, and pelvic pain. One factor that seems to be important here is the overall size and shape of the device in rela-

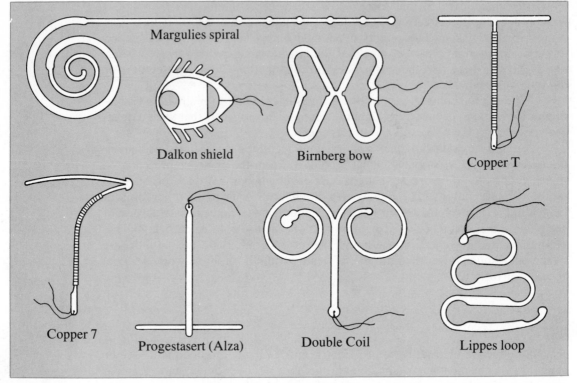

Margulies spiral

Dalkon shield   Birnberg bow   Copper T

Copper 7   Progestasert (Alza)   Double Coil   Lippes loop

Figure 20.12   Intrauterine devices (IUDs) are among the most effective methods of birth control. Several examples are pictured here. The Margulies spiral, the Dalkon shield, and the Birnberg bow have been removed from the market.

tion to the uterine cavity. The degree to which the IUD becomes embedded in the uterine wall may account for most of the problems with these devices.

Recent improvements in the shapes, sizes, and materials used for IUDs have helped to reduce most of the adverse reactions. In addition, the use of copper metal and these progestins or progesterone in combination with these devices has increased the contraceptive effectiveness. In each type, the enhanced anti-implanation activity is believed to be the result of a continuous release of small amounts of copper or progestogen. Although the precise mechanism of action of these modified devices is not known, it is thought to be similar to that of the minipill (4).

A recent center of controversy was the Dalkon shield, which is pictured in Figure 20.12. It was taken off the market by the manufacturer after 14 deaths and almost 300 spontaneous abortions due to blood poisoning in users who were unknowingly pregnant.

The copper T is another intrauterine device that has been used. A new application of the copper T as a "morning after" birth control device has been reported (5). This new emergency birth control method may prevent implantation of a fertilized egg as late as five days after intercourse. It is known that the egg takes approximately five days to nest in the uterine wall.

Other birth control methods are available that utilize chemical substances.

Spermicidal (sperm-killing) agents, in the form of jellies and foams, prove to be less effective because of the high motivation necessary for their effective use. Mechanical methods, such as the condom and diaphragm, do not require a chemical agent since the mechanism of action is simply to block passage of the sperm. The most drastic methods of contraception are the sterilization methods—tubal ligation (cutting the fallopian tubes) in the female and vasectomy in the male. These two surgical methods are essentially irreversible because of the poor success rate in reversing the procedures.

FUTURE METHODS OF BIRTH CONTROL

When oral contraceptives and intrauterine devices became widespread in the 1960s, there was great hope that new, more effective, chemical birth control compounds would soon be marketed. The much publicized side effects led to more stringent screening of new agents, which has resulted in a failure to produce any new chemical substances that can meet the rigid standards for safety. Consequently, the new methods of contraception for women involve new ways of delivering the existing hormonal agents.

Among the newer techniques being tested are weekly, monthly, or yearly administration of synthetic hormones that would have minimal concentration so as to reduce adverse side effects. Another approach is oral administration of these hormones during the phase following ovulation in order to prevent or interrupt im-

Figure 20.13   The use of a spermicidal foam for contraception.

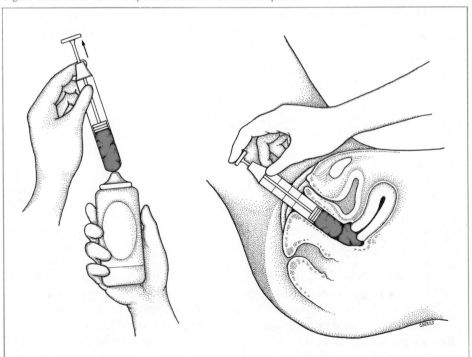

plantation through specific inhibition of progesterone synthesis by the corpus luteum. The goal is to develop compounds that would suppress the corpus luteum, and thus bring about the menstrual phase by eliminating the progesterone support of the uterine lining needed to maintain implantation. Some progestin compounds have been found to render the cervical mucous impenetrable to sperm rapidly enough to be administered up to a few hours before coitus. Initial studies have indicated that only a few hours are required to establish changes in the cervical glands, uterine wall, and possibly the corpus luteum (6).

Postcoital agents, or "morning after pills," have been developed, but harmful side effects have been associated with continued use. Oral administration of large amounts of estrogens, namely diethylstilbesterol (DES), for five days

Diethylstilbesterol

beginning within 72 hours after insemination, has been used for many years. In addition, DES was once used in low doses to prevent miscarriages. Recent studies have strongly implicated DES as a cause of vaginal or cervical cancer in daughters of mothers who received it during pregnancy. As a result, the Food and Drug Administration has approved the use of DES only for emergency situations such as rape or incest (7).

The search for suitable postcoital agents has led to the prostaglandins, which are long-chain fatty acids with biological activity in the induction of uterine contractions that stimulate menstruation. Two prostaglandins, designated PGF_2

PGF_2

PGE_2

and PGE_2, have been used to induce labor. No orally active agents have been found. The prostaglandins are administered by injection or in suppository form. The high cost of these compounds makes their use impractical at this time.

Injectable birth control drugs have also been introduced as new methods of delivery of progestin compounds. Intramuscular injections of progestins, lasting 90 days, have been tested and found to suppress ovulation by inhibiting the midcycle surge of LH. Some new progestins appear to be safe and effective. Spacing of the injections is critical, since pregnancy has resulted after only 107 days from the last injection in some cases (8).

The discovery that continuous low doses of progestins were effective in fertility control opened up the possibility of subdermal (under the skin) implants. Progestins are encapsulated in polymeric compounds or in biodegradable polymers, which continuously release minute quantities of the hormone into the bloodstream. These implants would be effective for up to one year before another capsule must be inserted.

Immunization is another potential method for contraception in women. A contraceptive vaccine has been under study in Australia. The strategy is to induce the production of antibodies to the hormone, human chorionic gonadotropin (HCG), which is released into the bloodstream soon after pregnancy and is necessary for sustaining pregnancy. The method would involve inoculation with the hormone or a similar substance that would stimulate antibody formation. Consequently, natural production of the hormone, as a result of pregnancy, would evoke a memory response resulting in efficient antibody production and destruction of the hormone. The result would be an inability to sustain a pregnancy due to interference with the natural hormones (9).

CHEMICAL METHODS OF MALE CONTRACEPTION

The development of the diaphragm and the pill as effective methods of fertility control gave the responsibility to women. Sensing the loss of control, there is a revival of interest in contraceptive methods that involve the male. Given the role of the sperm in fertilization, there are three strategies available for male contraception—interference with sperm production (spermatogenesis), inhibition of sperm development after spermatogenesis, and blockage of the transport of sperm to the female. Existing methods involve the last of these strategies by employing physical means, such as the condom and coitus interruptus, and by surgical sterilization (vasectomy).

Progress is very slow in male contraception, because very little is known about the male reproductive system when compared to the female system. Several synthetic compounds have been developed in an attempt to suppress the gonadotropin release from the pituitary via the negative feedback mechanism. In this way, the spermatogenesis in the testes is inhibited, but so is the male sex drive. The difficulty that arises in this method is the fact that the suppression of gonadotropin with progestins also suppresses the release of luteinizing hormone, which is responsible for the production of testosterone. To combat this side effect, the progestins are administered both alone and in combination with low doses of androgens. The androgens are given to maintain normal secondary sex characteristics because, in the male cycle, there is no surge of LH or FSH, but a continuous flow of these gonadotropins. Consequently, any suppression of these hormones will have serious side effects on the sex drive of the male. Testing is currently underway in an attempt to use progestins and testosterone-like compounds to suppress gonadotropin secretion in men, and thus abolish spermatogenesis. A male pill method, a male injection method, or a male implant method may result if these studies prove successful.

Research into the possible killing of sperm that have already been produced is another method that has proved difficult. Sperm development is directly dependent on the male hormone. The danger in all of the methods affecting sperm is the possibility of changes in a sperm that might later fertilize an ovum, resulting in birth defects.

SUMMARY

The 17-carbon steroid nucleus is the primary characteristic of all steroids. Two methyl groups, called angular methyl groups and located at carbon 10 and carbon 13, plus several other functional groups determine the biochemical properties of each steroid. In drawing steroid molecules, solid lines represent bonds to groups located above (beta) the steroid nucleus and dotted lines represent bonds to groups located below (alpha) the steroid nucleus. A major steroid is cholesterol, which has a beta hydroxy group at carbon 3, a double bond in ring B, and an 8-carbon side chain at carbon 17, attached beta.

Cholesterol is a precursor for many other major steroids. The bile acids are produced from cholesterol in the liver and assist in the digestion of lipids by detergent action. Cortisone and related steroids are produced in the adrenal glands from cholesterol.

The sex hormones are also synthesized from cholesterol in the sex glands (ovaries and testes), called gonads. There are three types of sex hormones—progestogens (pregnancy hormones), androgens (male hormones) and estrogens (female hormones).

A hormone is a substance (usually steroid or protein) that is released into the bloodstream from one location in the body and exerts its effects elsewhere in the body. The release of the steroid sex hormones is controlled by protein hormones released by the pituitary gland. The pituitary hormones are called gonadotropic hormones because they stimulate the gonads to become active and release the sex hormones.

Progesterone (a progestogen) and estradiol (an estrogen) are required in order to prepare the lining of the uterus for implantation of a fertilized egg (ovum). Progesterone also prevents ovulation (release of an ovum) in the event of a successful fertilization and, thus, acts as a natural birth control chemical to prevent a second pregnancy. Progesterone-like compounds are a major ingredient in all birth control pills.

The interplay between the gonadotropic hormones and the sex hormones is the basis for the action of most oral contraceptives on the market. In general, the gonadotropic hormones by a mechanism called feedback inhibition and, thus, during and following menstruation, and peak sharply at about mid cycle. The sharp rise triggers ovulation, after which the levels of the sex hormones become elevated. The elevation of the sex hormones suppresses further release of the gonadotropic hormones by a mechanism called feedback inhibition and, thus, prevents further ovulation. The female cycle ends in either pregnancy or menstruation. If menstruation occurs, the sex hormones drop to low levels and the gonadotropic hormones begin to rise in preparation for another cycle and another ovulation. In the event of pregnancy, the level of the sex hormones, particularly progesterone, remains high and suppresses the gonadotropic hormones.

In addition to suppression of the gonadotropic hormones, progesterone also prevents a second pregnancy by causing the glands of the cervix to produce a sticky secretion, which blocks the passage of sperm and, thus, prevents fertilization. In effect, progesterone has a double-barreled action in preventing a second pregnancy. It prevents fertilization of an ovum that is not released in the first place.

Since the early 1960s, there have been three strategies employed in birth control pills. The first- and second-generation pills employ synthetic sex hormones, which act by suppressing ovulation by feedback inhibition of the gonadotropic hormones.

The first-generation pills also affect the secretions in the cervix, whereas the second-generation pills do not. The third-generation pills do not suppress ovulation. They act only by producing the sticky secretions in the cervix.

The second- and third-generation pills were developed because of side effects that sometimes accompany the use of the first-generation pills, but unfortunately, the second- and third-generation pills also cause problems for some users. Likewise, other methods of contraception are available or are in the developing stages. The intrauterine device (IUD) is a popular approach to birth control. A number of other methods are under investigation, including some involving the male.

PROBLEMS

1. Give a definition or example of each of the following:
   a) steroid nucleus
   b) angular methyl group
   c) bile acid
   d) hormone
   e) progestogens
   f) androgens
   h) estrogens
   i) gonadotropins
   j) feedback inhibition
   k) Graafian follicle
   l) corpus luteum
   m) FSH
   n) LH
   o) HCG
   p) combination pill
   q) sequential pill
   r) minipill
   s) IUD

2. Draw the steroid nucleus and label the rings and carbon atoms.

3. How does one specify the location and configuration of functional groups attached to the steroid nucleus?

4. Since the basic chemical structure is the same for all steroids, what causes the various classes to function in different ways?

5. What substance is essential for the synthesis of many steroids required by the body?

6. What is the function of the bile acids?

7. Which steroid is the common precursor to those produced by the sex glands and adrenal glands?

8. Draw the chemical structure for estradiol and testosterone. What is the major structural difference between these steroid hormones?

9. Draw the chemical structures for a synthetic estrogen and a progestin. What

modifications of these molecules have caused the enhanced chemical activity that makes it possible to take them orally?

10. Explain the strategy of each of the following, including the intended advantages of the later ones.
   a) 1st-generation oral contraceptives
   b) 2nd-generation oral contraceptives
   c) 3rd-generation oral contraceptives

11. How does an IUD work and what are its advantages and disadvantages?

REFERENCES

1. Burack, R. *The New Handbook Of Prescription Drugs*. New York: Ballantine Books, 1975, pp. 426–8.
2. "The Controversy Over 'The Pill'." *Executive Health XIII* (1), October 1976.
3. Pan, E.L. "Contraception With Intrauterine Devices." *BioScience 23* (5) (1973): 281.
4. Kistner, R.W. "The Pill and IUD: not perfect, but still the best we have." *Modern Medicine,* 11 November 1974, pp. 36–44.
5. "Copper IUD For The Morning After." *Medical World News 19* (10) (1975): 34.
6. Segal, S.J. "The Physiology Of Human Reproduction." *Scientific American 231* (9) (1974): 53–62.
7. 'DES (Diethylstilbesterol)." *Good Housekeeping,* January 1975, p. 122.
8. "90-day Birth Control Drug Tests Out Safe." *Chemical and Engineering News,* 17 February 1975, p. 23.
9. "A Contraceptive Vaccine Effective For A Year," *Chemical and Engineering News,* 15 December 1975, p. 19.

SOMEWHERE THE SUN IS SHINING

PART SIX

I N this chapter, we turn our attention to one of man's most abundant consumer products, pollution. In a time when there seems to be a shortage, real or otherwise, of almost everything, it is pleasant to imagine the possibility that we may someday have a shortage of some major pollutants.

Unfortunately, the event is not imminent. Perhaps Tom Lehrer said it best in his song, Pollution:

"Pollution, pollution, we got smog and sewage and mud,
Turn on your tap and get hot and cold running crud."

Or, in another verse from the same song:

"Pollution, pollution, you can use the latest toothpaste,
And then rinse your mouth with industrial waste (1)."

Many of us view pollution in these same terms, often rightly so. Pollution is a very emotional topic. Everyone knows pollution is bad, but not everyone knows what pollution is and what it is not. We will consider two aspects of pollution, air pollution and water pollution. These are the two most common types and we will consider them together so that comparisons can be made to increase your understanding of both.

In order to define pollution, we must first define unpolluted air and water. This is not always easy. Just ask the question, what are the characteristics of unpolluted air (or water)? Unpolluted by whom, or what? By virtue of its existence, air and water will carry impurities. Whether these are to be regarded as pollutants will depend on our definition of pollution. The dictionary is no help. It defines polluting as making or rendering unclean, defiling, desecrating, profaning. One man's joy may be another's pollution. Someone building a cabin and dock on a pristine lake may be able to enjoy the lake more fully, but may cause other people to claim he has destroyed the native beauty of the lake. Such situations will not be within our scope of coverage. We will restrict ourselves to substances found in air and water.

CLEAN AIR

Air is a mixture of various molecules. The definition of clean air must include a judgment on what air is or was before any undesirable substances were added to

AIR AND
WATER POLLUTION

"You know as well as I do, a man is measured by the amount of pollution he creates."

Copyright © 1977, S. Harris. Reprinted by permission.

it. Some of the components of air are regarded as pollutants because there is now more present than the amount defined for clean air. The components of clean air are listed in Table 21.1.

The total mass of air in the earth's atmosphere is approximately 5×10^{15} metric tons. The metric ton is also known as the *long ton* and is equal to 2200 pounds. Seventy-five percent of the mass of air is located inside a seven-mile thick layer beginning at the earth's surface.

The components in Table 21.1 are listed according to the *percent by volume,* because this is the common way of describing concentrations of air components. Whenever anyone says nitrogen makes up approximately 80 percent of the earth's atmosphere, volume percent is being used. The other common percentage is weight percent. Volume percent is calculated as follows.

$$\text{Volume \%} = \frac{\text{volume of component}}{\text{total volume of air}} \times 100$$

When the volume percent is small, this unit becomes awkward and we change to *parts per million (ppm),* which is calculated as follows.

TABLE 21.1   COMPONENTS OF CLEAN DRY AIR (2, 3)

COMPONENT	PERCENT BY VOLUME	PPM BY VOLUME	TOTAL MASS IN THE ATMOSPHERE*
Nitrogen, N_2	78.084		3900
Oxygen, O_2	20.946		1200
Argon, Ar	0.934		67
Carbon Dioxide, CO_2	0.033		2.5
Neon, Ne	0.001818	18.18	0.065
Helium, He		5.24	0.004
Methane, CH_4		2	0.004
Krypton, Kr		1.14	0.017
Hydrogen, H_2		0.5	0.0002
Nitrous Oxide, N_2O		0.5	0.002
Carbon Monoxide, CO		0.1	0.0006
Xenon, Xe		0.08	0.002
Ozone, O_3		0.02	0.003
Ammonia, NH_3		0.006	0.00002
Nitrogen Dioxide, NO_2		0.006	0.000013
Nitric Oxide, NO		0.0006	0.000005
Sulfur Dioxide, SO_2		0.0002	0.000002
Hydrogen Sulfide, H_2S		0.0002	0.000001

* In trillions of metric tons.

$$\text{Parts per million (ppm)} = \frac{\text{volume of component}}{1,000,000 \text{ volumes of air}}$$

Thus, the concentration in parts per million is essentially the number of parts of a component in one million parts of air, where the number of parts are determined by volume measurements.* The total mass of each component is also listed in Table 21.1. The mass is not necessarily in the same order as the volume, because the density will vary for each component.

POLLUTED AIR

Of the components listed in Table 21.1, the following are known as air pollutants: carbon dioxide, methane, nitrous oxide, carbon monoxide, ozone, ammonia, nitrogen dioxide, nitric oxide, sulfur dioxide, and hydrogen sulfide. In addition, we have not provided an exhaustive list of air components in Table 21.1. There are a number of components present in even smaller amounts. However, the entries in the table will be sufficient to illustrate several things.

We can use the values in Table 21.1 as acceptable minimum levels for each of these pollutants, but we need information about maximum levels in order to

* Parts per million can also be given using weight. This is normally done when describing the concentration of a solute that is a solid or a liquid.

realize the range for each pollutant. What kind of maximum are we to define? The ultimate maximum is, of course, the lethal limit for man, but we need to define a lower, more workable limit. This kind of definition is not simple.

Carbon Monoxide

Carbon monoxide is a well-known poison to living systems as well as a well-known air pollutant. For human beings, if the concentration of carbon monoxide is 400-500 ppm and it is inhaled for one hour, there is no appreciable effect. If the concentration is 600–700 ppm, inhalation for one hour gives barely detectable effects. A concentration of 1000–2000 ppm is classified as dangerous, while 4000 ppm or higher is fatal in less than one hour (4). These figures apply to normal, healthy human beings. The data do not include differences in tolerance to carbon monoxide for various individuals, i.e., the ill and aged. Neither do these data suggest the effects on living systems, other than humans. For example, what is the effect on growing crops, on trees, on cattle, on hogs, or other species essential to human life? All these factors must be taken into account when setting standards for maximum levels of the various pollutants. One additional difficulty is that the necessary data are not always available for making the appropriate decisions and the best estimates must be used until the data are available. Therefore, we find frequent revisions in pollution standards. The air quality standards for 1975 as established by the Federal Government are listed in Table 21.2.

Sulfur Oxides

In order to understand some of the entries in Table 21.2, we need to consider some of the chemistry involved. The use of a figure for total sulfur oxides is justified by the fact that sulfur dioxide, SO_2, is normally formed when sulfur-containing compounds are burned. The sulfur dioxide is then slowly oxidized in air to sulfur trioxide, SO_3.

$$2 \ SO_2 \ (gas) + O_2 \ (gas) \rightarrow 2 \ SO_3 \ (gas)$$

Once sulfur trioxide is formed, it reacts quickly with water to produce sulfuric acid.

TABLE 21.2   UNITED STATES GOVERNMENT POLLUTION
STANDARDS (5)

COMPONENT	FEDERAL STANDARD	NATURAL ABUNDANCE (ppm)
Carbon Monoxide	9 ppm average for 8 hours 35 ppm average for 1 hour	0.1
Total Sulfur Oxides	0.03 ppm average for 1 year 0.14 ppm average for 24 hours	0.0002 for SO_2 alone
Nitrogen Oxides	0.05 ppm average for 1 year	0.0016 (NO + NO_2)
Hydrocarbons	0.24 ppm average for 3 hours*	

* Not including methane or other volatile hydrocarbons.

$$SO_3 \text{ (gas)} + H_2O \text{ (liquid)} \rightarrow H_2SO_4 \text{ (liquid)}$$

The sulfuric acid in polluted air causes extensive damage when present in large amounts close to the earth's surface.

Marble is rock that was originally deposited as limestone, $CaCO_3$, and later changed into marble due to heat and pressure, which cause changes in the size of the calcium carbonate crystals. Recalling that $CaCO_3$ is often used as an antacid (Table 19.2), we expect that marble will be particularly susceptible to the effects of sulfuric acid, which results from SO_2 pollution. The problem is explained by the following chemical reaction.

$$CaCO_3 + H_2SO_4 \longrightarrow CaSO_4 + H_2CO_3$$

marble   unstable

$$\downarrow$$

$$H_2O + CO_2$$

When exposed to sulfuric acid, the surface of the marble is changed to calcium sulfate. That would not be so bad except that $CaSO_4$ is more soluble in water than $CaCO_3$, so that the surface coating is washed away more readily by rainfall

Figure 21.1 The impact of 50 years of air pollution is evident in photos of this Renaissance Italian fresco of the Madonna. The right photo, taken in 1970, shows the erosion that has occurred since the left photo was taken in 1920. (Courtesy of the Italian Art and Landscape Foundation, Inc.)

SOMEWHERE
THE SUN
IS SHINING

464

and a fresh surface of $CaCO_3$ is exposed. Even the $CaSO_4$ is not very soluble, so the effects are long-term, although the more SO_2 pollution, the faster the erosion of the marble.

The priceless statuary and buildings of Venice, Italy are being severely eroded by the sulfuric acid formed from SO_2 produced by the extensive industry along the coast outside Venice. The famous lions outside the New York City Public Library are suffering the same fate.

The longer SO_2 remains in the atmosphere, the more likely it is to be converted to sulfuric acid. If the SO_2 remains within the first few thousand feet of the atmosphere, it will have the maximum opportunity to affect living systems (e.g., the lungs), either as SO_2 gas or as sulfuric acid in reactable concentrations. If the SO_2 is kept at altitudes in excess of ten thousand feet, the conversion to SO_3 is virtually complete, the sulfuric acid formed has an opportunity to become greatly diluted, and the harmful effects are considerably decreased. Thus, smokestacks that emit SO_2 are now required to be sufficiently high so their emissions will have a long residence time in the higher atmosphere. In any event, the amount of SO_2 and SO_3 combined is the potential SO_3. Therefore, it is convenient to speak of combined sulfur oxides, which indicates the sum of the SO_2 and SO_3 concentrations.

Nitrogen Oxides

Just as SO_2 may be oxidized to SO_3, a similar reaction occurs with nitric oxide, NO, to form nitrogen dioxide, NO_2. In fact, this reaction occurs much more easily

$$2 \text{ NO (gas)} + O_2 \text{ (gas)} \longrightarrow 2 \text{ NO}_2 \text{ (gas)}$$
$$\text{nitric} \qquad\qquad\qquad\qquad \text{nitrogen}$$
$$\text{oxide} \qquad\qquad\qquad\qquad \text{dioxide}$$

than SO_2 oxidation. When NO_2 reacts with water, a solution of nitric acid (HNO_3) is produced. The natural presence of NO and NO_2 in the atmosphere is partly due to lightning. The electric discharge, which is lightning, causes the nitrogen and ox-

$$N_2 \text{ (gas)} + O_2 \text{ (gas)} \xrightarrow[\text{discharge}]{\text{electric}} 2 \text{ NO (gas)}$$

ygen to react to form nitric oxide. The NO then reacts with oxygen to form NO_2, which, in turn, reacts with water to form nitric acid. However, no matter what the source of NO or NO_2, the easy conversion of NO to NO_2 makes it convenient to speak of combined nitrogen oxides, which usually means NO and NO_2.

Ozone

Some components of air are essential at one altitude and pollutants at another. Ozone, O_3, is essential at altitudes between ten and twelve miles, called the *stratosphere,* in order to filter out dangerous ultraviolet rays from the sun. These ultraviolet rays would quickly cause skin cancer, at least, and quickly kill most species. It is speculated that life began in places that were shielded from this radiation, such as in water. However, if ozone exists very near the earth's surface, it is a dangerous, poisonous gas with a chlorine-like odor, which causes respiratory

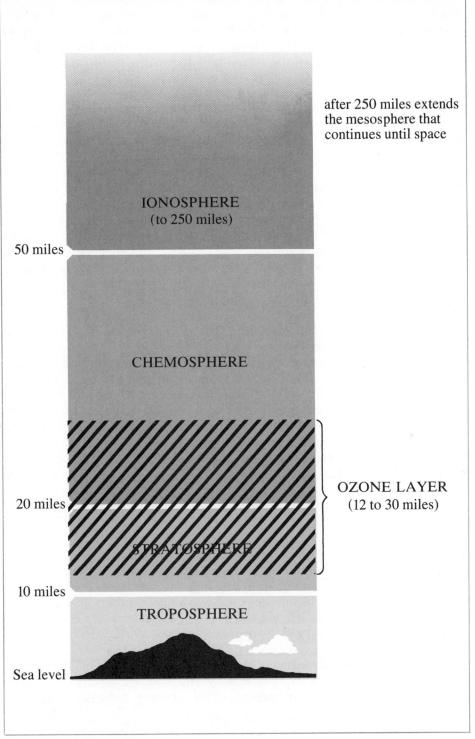

after 250 miles extends
the mesosphere that
continues until space

IONOSPHERE
(to 250 miles)

50 miles

CHEMOSPHERE

OZONE LAYER
(12 to 30 miles)

20 miles

STRATOSPHERE

10 miles

TROPOSPHERE

Sea level

Figure 21.2   The atmosphere.

problems, reacts readily with other molecules in the air to form equally dangerous species, and severely injures most plants. If an ozone concentration is given in a pollution report, it does not refer to the layer of ozone protecting us in the stratosphere; it indicates the level of ozone in the air at ground level, the *troposphere*.

Krypton

In a somewhat different situation, krypton (Kr) is an inert gas—inert in the sense that it is chemically unreactive. It is present in small amounts in the atmosphere and will not be added to or subtracted from by normal chemical processes. However, nuclear power plants produce an isotope of krypton, ^{85}Kr, which is radioactive with a half-life of 9.4 years. The mode of radioactive decay is beta decay and gamma ray emission. It is possible that ^{85}Kr will build up in the atmosphere as a result of nuclear power generation and nuclear fuel reprocessing. There will be no chemical effects due to the krypton, nor will the krypton be removed by any chemical action, but the electrons emitted from the beta decay could cause disturbances in the atmosphere and, thereby, change cycles affecting cloud formation and rainfall. It is still too early to say whether or not this will indeed occur.

New Findings on Ozone

We can see how new data can alter the predictions that were based on initial thinking if we return to the ozone story.

Due to the release of nitrogen oxides, the supersonic transport (SST) airplanes were expected to severely deplete the ozone layer in the stratosphere. Based on a specific number of SSTs flying for a specified number of hours each day, a depletion of twelve percent in the ozone layer was predicted for as long as the airplanes are in use. More recent data has caused a revision of the initial predictions to suggest the depletion would be only two percent, rather than twelve. The reduction in the predicted depletion is due to increased understanding of the chemistry of the pollutants in the ozone layer. It is known that chlorine atoms and nitric oxide, NO, will react with ozone to destroy the ozone and regenerate the chlorine atoms and nitric oxide according to the following equations.*

$$NO + O_3 \rightarrow NO_2 + O_2 \qquad Cl + O_3 \rightarrow ClO + O_2$$
$$NO_2 + O \rightarrow NO + O_2 \qquad ClO + O \rightarrow Cl + O_2$$

However, new findings indicate that NO_2 and ClO react appreciably with one another to form chlorine nitrate, $ClONO_2$, which has a low reactivity with ozone (6). Removal of the NO_2 and ClO prevents the second step in each of the sequences above so that NO and Cl are not regenerated for repeated reaction with O_3.

This is an excellent example of the need for more research into these chemical reaction systems before we know exactly what is happening and what effects man will have on them. Fortunately, in this situation, increased knowledge

* NO may be present due to lightning or as a pollutant from a variety of sources, including the automobile. Chlorine atoms may be present due to the breakdown of chlorofluorocarbons (freons) in the atmosphere. Oxygen atoms are abundant in the upper atmosphere.

showed man's effect to be less than initially anticipated. It is entirely possible for the opposite to occur. A revised estimate, based on increased knowledge, could show a greater effect than initially anticipated.

Carbon Dioxide: The Greenhouse Effect

While there is much that is not known about pollution effects in the atmosphere, we do know about the potential warming effect of excess carbon dioxide, which is produced by the burning of fossil fuels in many situations.

Carbon dioxide is a compound that permits the passage of high energy radiation in the ultraviolet and visible regions of the spectrum, such as is contained in sunlight. On the other hand, carbon dioxide efficiently absorbs some of the lower energy radiation in the infrared region of the spectrum, such as heat. Consequently, as the concentration of atmospheric CO_2 builds up, we would expect no change in the amount of energy from the sun, but a decrease in the amount of heat radiated away from the earth and a warming trend. This phenomenon is commonly termed "the greenhouse effect," since the glass covering a greenhouse behaves in the same way as CO_2.*

The greenhouse effect could cause melting of part of the polar ice caps and flooding of coastal cities, but, while we are reasonably certain of the qualitative effect, i.e., that it will indeed occur, we are much less certain of the extent to which it will occur. The lowest estimate on the increase in temperature due to the greenhouse effect differs from the highest estimate by a factor of two. It is estimated that a doubling of the concentration of CO_2 in the atmosphere would cause an increase in the average world temperature of 3.8°C (6.8°F). During the period from 1900 to 1969, the atmospheric CO_2 increased by 7.4% (8).

In addition, if more heat is prevented from escaping from the troposphere, this could lead to cooling of the stratosphere and, therefore, a cooling of the ozone layer. If the ozone is cooler, the reactions causing its depletion will go slower. The ozone layer is always being depleted by natural reactions. It is this natural depletion, plus the tendency of the ultraviolet radiation to generate more ozone, which

$$3\ O_2 \xrightarrow[\text{radiation}]{\text{ultraviolet}} 2\ O_3$$

leads to the balance that is present today. It has been estimated that a doubling of the CO_2 concentration could lead to an increase of three percent in the ozone concentration. Thus, if we double the CO_2 concentration, and the present model for depletion of ozone due to the impact of the SST is correct, the effects almost cancel each other out. But take little comfort from this, for it all depends on the accuracy of the models that we have been discussing and whether there are any other effects that we have not considered.

The types of interaction that we have seen occurring between various systems makes prediction of effects even more difficult. The need for more research is very evident even from these few examples.

---

* The temperature on Venus is reportedly about 500°C. This is believed to be due to the very dense atmosphere, including CO_2, which reduces the loss of heat from the surface of the planet (7).

Particulates

Our comments so far have dealt with gases as pollutants. The atmosphere carries vast amounts of *particulate matter*. These particles are extremely complex in nature. They may be solids or liquids, organic or inorganic. The size of these particles varies from too small to be settled by gravity (up to about one micron)* to beach sand and rain (about 5000 microns). Particulates over ten microns settle rapidly to the earth, usually within a day or two. The smaller particles, which do not settle due to gravity, usually combine with other particles to form aggregates of larger size, which then settle to the ground.

The variety of particulates in the atmosphere includes viruses, smoke, automobile exhaust, bacteria, fly ash, coal dust, cement dust, and pollen, to name a few. The particle that causes the greatest lung damage is in the range of 0.1 micron to 3 microns and encompasses some smoke, bacteria, coal dust, cement dust, and fly ash. Some of these particles can cause scar tissue in the lungs and eventually lead to emphysema and death. Small particles of asbestos are especially prone to cause this scar tissue. This is one reason that there are increasing restrictions on the use of asbestos materials in construction.

Smog

The combination of particulate matter plus the reactive nitrogen oxides, sulfur oxides, and ozone, as well as unreacted hydrocarbons, constitute the material known as *smog*. The particles provide catalytic surfaces for the conversion of SO_2 to SO_3 and the reaction of ozone with hydrocarbons. The products from these catalyzed reactions are also components of smog and are usually as dangerous as the initial air pollutants. The reaction of smog components is also initiated by ultraviolet light. The combination of ultraviolet light, the catalysis by the particle surfaces, and the pollutants abundantly provided by transportation vehicle exhaust and industrial emissions in large cities provide the ideal conditions for smog formation. For this reason, smog is normally found only in the atmosphere of large metropolitan areas.

Acid Rain: Air Pollution Sometimes Equals Water Pollution

A report was recently published that dealt with the rising acidity of rain and snow in the United States and Europe (9). The pH of snow before the onslaught of industrial pollution can be determined by obtaining samples from deep inside glaciers, where the pH is greater than 5. For the past twenty years, the pH of rain and snow has been decreasing, which means the acidity has been increasing; pH levels as low as 3 have been recorded. The reasons for this decrease in pH are not clear. Correlations have been made with the amount of fossil fuel burned. Fossil fuels often contain sulfur and the sulfur burns to produce sulfur dioxide. In Norway, the presence of sulfate (from H_2SO_4) in rainwater has increased in a fashion similar to the acidity, but not enough to account for all the increase in acidity of rain and snow. In other words, it is not really known why the acidity is increasing to the extent that it is.

* 1 micron = 10^{-6} centimeters.

Figure 21.3   Photographs taken of Los Angeles before and after smog has rolled in. (Courtesy Air Pollution District of the City of Los Angeles.)

The increased acidity of rain and snow does correlate well with the lower pH of fresh water lakes, and there is a direct relationship between the decrease in pH and the decrease in fish population of these lakes. The acid-forming species, which cause the pH change and the fish population decrease, originate in the atmosphere as air pollutants, but become water pollutants when rain occurs. Therefore, correction of these effects must begin with a study of the air pollutants, which cause the acid species. This again requires additional research before we will know precisely how to handle the problem.

WATER POLLUTION

In earlier chapters, we examined several examples of water pollution. The major one was the problem of eutrophication promoted by the presence of phosphate builders in detergents. We also considered the problems associated with insecticides, fertilizers, and the potential for pollution by wastes from nuclear power plants.

In this section, we will examine some general matters relating to water pollution and some additional examples.

CLEAN WATER

The definition of clean water is even more difficult than that of clean air. As a simple example, clean water from the ocean is not the same as clean water from a fresh water lake. Ocean water contains various salts such as sodium chloride and magnesium chloride.

Pure water is easy to define; it contains *only* H_2O. However, natural water is never pure. The mere act of running along or through the earth's surface provides ample opportunity for the water to dissolve some impurities. Not all of these impurities are objectionable—some are even beneficial. We can classify various types of natural water according to the amount of dissolved minerals, as shown in Table 21.3.

The table does not contain the category, hard water, which usually applies when the concentration of dissolved minerals is in the range 120–250 ppm. The phrase *hard water* is only used to indicate the presence of certain minerals. This subject will be discussed shortly.

Water from the oceans averages 35,000 ppm of dissolved minerals, which places it in the salty category as expected. Some of the elements found as components of minerals dissolved in the oceans are shown in Table 21.4.

TABLE 21.3 CLASSIFICATIONS OF NATURAL WATERS (10)

NATURAL WATER	DISSOLVED MINERALS (ppm)
Fresh	0–1000
Brackish	1000–10,000
Salty	10,000–100,000
Brine	>100,000

"*The Chairman is meticulous in checking the
environment for possible pollution.*"

Taylor, Copyright © 1975, *Punch*/Rothco. Reprinted by permission.

Many other elements have been found in seawater, ranging down to radon
at concentrations of 9×10^{-15} ppm. We should point out that, at their natural concentrations, magnesium, bromine, and iodine have been mined from the sea, and
sodium and chlorine are derived from seawater as sodium chloride for use as
common table salt. These are only the better-known materials mined from the
oceans. There has, of course, been discussion of extraction of gold, at 4×10^{-6}
ppm, and silver, at 3×10^{-4} ppm, from the seas. The low concentrations of these
precious metals mask the vast absolute amount actually present in the oceans.

Neither Table 21.3 or Table 21.4 includes organic impurities from natural
sources. Such sources are abundant and range from materials dissolved out of
leaves to water-soluble compounds formed by decaying vegetation in swamps.
Ocean water contains organic materials in concentrations of about 0.5 ppm.
Various swamps and brine pools have organic impurities up to 100 ppm. The dissolved organic compounds include amino acids, sugars, alcohols, aldehydes,
organic acids, amines, and some hydrocarbons. If concentrations are significantly
above natural levels, a pollution problem exists. However, it must be recognized
that even natural levels of impurities may make the water unfit for human use. As
an example, seawater is unfit for drinking.

TABLE 21.4   THE ELEMENTS FOUND IN SEAWATER (11)

ELEMENT	CONCENTRATION (ppm)
Chlorine	19,000
Sodium	10,000
Magnesium	1,300
Sulfur	900
Calcium	400
Potassium	380
Bromine	65
Carbon	28
Oxygen	8
Strontium	8
Boron	4.8
Silicon	3.0
Fluorine	1.3
Nitrogen	0.8
Argon	0.6
Lithium	0.2
Rubidium	0.12
Phosphorus	0.07
Iodine	0.05

In addition to dissolved matter, water contains suspended particulate matter called *sediment* or *silt*. These sediment loads vary from 100 ppm to 100,000 ppm.

The Hydrologic Cycle

The water on earth is distributed in various places and forms. This distribution is shown in Table 21.5. Water is not lodged forever in each form. The water from one form is converted to another form through the *hydrologic cycle*. Within this cycle, water is carried from the oceans into the atmosphere by evaporation and deposits on land as rain or snow. The remainder falls back into the oceans. Some

TABLE 21.5   THE DISTRIBUTION OF WATER ON EARTH (12)

SOURCE OF WATER	PERCENT CONTAINED
Oceans	94
Ice and snow	4.2
Underground	1.2
Surface and soil	0.4
Atmosphere	0.001
Living species	0.00003

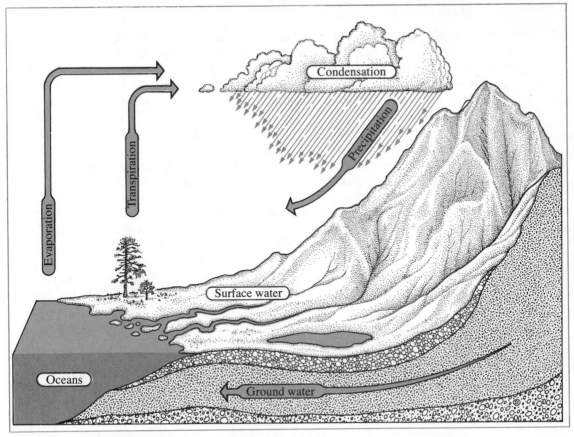

Figure 21.4   The hydrologic cycle.

water on the land evaporates into the atmosphere as well. Once the water falls onto the land, some is used, some is absorbed, and some runs off. This cycle is shown in Figure 21.4.

All of the water on the earth is not in motion at any one time, but all the water is in the cycle. It is this cycle that allows renewal of lakes and rivers from rain, even as the fresh water runs off into the oceans. In effect, this is a natural cycle that provides fresh water for the earth on a continuous basis. Every day, the sun provides well over one trillion metric tons of water by evaporation from the oceans. Some of this water will fall on the land masses as fresh water. From this rainfall, the United States finds almost five billion metric tons of water flowing through its rivers each day.

Unfortunately, the demands for this fresh water are great and increasing at an accelerated rate. It has been estimated that the demand for fresh water will equal the supply shortly after the year 2000. This means that water must be recycled or the demand decreased. Only an extremely small percentage of this demand is for drinking. The vast majority of the water is used for agriculture and industry. If the demand is to be decreased, it must be done in these areas. If the demand cannot be decreased, the water used will have to be recycled in a condition fit for reuse.

Hard Water

We mentioned hard water earlier when discussing the various types of natural water according to the concentration of dissolved minerals. Hard water is characterized by the presence of soap-precipitating agents, the deposition of scale in boilers and teapots, the inability to properly clean clothes in the presence of sufficient soap, and the inability to raise suds from soap. All of these are related to the presence of cations such as calcium and magnesium. Ca^{+2} and Mg^{+2} form scums, i.e., insoluble compounds, with a wide variety of soaps. (These scums are very slippery as solids and are often used as lubricants.) The source of the hardness is the land over which the water flows. Limestone ($CaCO_3$) and dolomite ($CaCO_3$ plus $MgCO_3$) formations provide excellent sources of calcium and magnesium ions, upon which much of the hard water chemistry is based.

The solubility of calcium carbonate in water is only about 9 ppm. Yet we characterized hard water as having dissolved minerals in the range of 120–250 ppm, most of which is $CaCO_3$. Thus, there must be some mechanism for increasing the concentration of $CaCO_3$ in water. This mechanism is related to the presence of CO_2 in the atmosphere. Carbon dioxide dissolves in water to form an acid solution known as carbonic acid and carbonic acid reacts with $CaCO_3$.

$$CO_2 + H_2O \rightarrow H_2CO_3 \text{ (carbonic acid)}$$
$$CaCO_3 + H_2CO_3 \rightarrow Ca^{+2} + 2\ HCO_3^-$$

The amount of CO_2 that dissolves in water is sufficient to cause a pH of about 5 to 5.5. in natural water. This is a slightly acid pH. The pH of a saturated $CaCO_3$ solution is very close to 10, a basic pH. Thus, the solubility of $CaCO_3$ is increased due to an acid–base reaction in which $CaCO_3$ is converted to the soluble calcium bicarbonate, $Ca(HCO_3)_2$. The reaction of $CaCO_3$ with H_2CO_3 is a typical neutralization reaction.

The product $Ca(HCO_3)_2$ is easily decomposed, even in solution. If the water is boiled, or even heated, the reaction is reversed, the CO_2 is driven off and

Figure 21.5 Actual pipe sections filled with scale from continued use in a hard water supply. (Reprinted from R.S. Drago, *Principles of Chemistry with Practical Perspectives, Second Edition*, Boston: Allyn and Bacon, Inc., 1977.)

the $CaCO_3$ precipitates out of the water. This is the reason a hard crust forms in boilers and steam irons: the heat drives off the carbon dioxide and precipitates the limestone ($CaCO_3$). This condition in water is known as *temporary hardness* because the hardness is so easily removed. The same reaction sequence is used for $MgCO_3$.

If $CaCO_3$ and $MgCO_3$ were the only compounds responsible for water hardness, the problems of hard water would be solved easily. A simple heating would precipitate most of the hardness-causing materials. What did precipitate as a result of the heating would be easily removed by treating with strong acid, which would react as follows.

$$CaCO_3 + 2\,H^+ \rightarrow Ca^{+2} + CO_2 + H_2O$$

However, this does not occur exactly as we have described it. The $CaCO_3$ does indeed precipitate, and if acid is added to the crust, it does bubble and give off CO_2 gas, but not all the scum reacts, nor does all the hardness precipitate on heating. In addition to the carbonates, natural waters pick up calcium sulfate, $CaSO_4$, an abundant species that also causes hardness. Unlike $CaCO_3$, $CaSO_4$ does not precipitate on heating, nor does it dissolve in strong acid easily. The $CaSO_4$ is a primary cause of so-called *permanent hardness*.

The precipitation of soap by hardness cations such as Ca^{+2} and Mg^{+2} and the subsequent difficulty caused to washing machines and clothing has led to the use of water softening apparatus. Water softeners act by binding the hardness cations onto another material, called an *ion exchange resin,* and exchanging them for sodium ions, Na^+ (see discussion in Figure 21.6 for details). Sodium ions do not react with soaps or precipitate onto boilers or cause laundry problems. However, more recent research has shown that hard water is beneficial to the health and the availability of excess sodium is somewhat detrimental.

Thus, while we solve one problem we may create another. Instead, the use of detergents has come into widespread use. The chemistry of soaps and detergents has been discussed in Chapter Sixteen. The reader is referred to this section for more information on this topic.

Drinking Water

The water we drink must have one overriding characteristic; it must be safe! This is not a well defined term. The history of mankind is replete with plagues and epidemics in which unsafe drinking water played its part. Cholera is spread through drinking water by being too close to sewage runoff. The definition of unsafe water is the presence of disease-carrying or producing species, frequently as a result of mixing water removal and drinking sources. Today, we recognize the necessity of eliminating these biologically detrimental species. Drinking water is tested and treated to destroy disease-carrying species. The usual treatment is sterilization by chlorination or oxygenation.

The chlorine (Cl_2) that we add to water supplies reacts with these germs and disease-producing species to destroy them. Oxygenation works similarly. The water is exposed to air as a spray to allow maximum opportunity for oxygen to

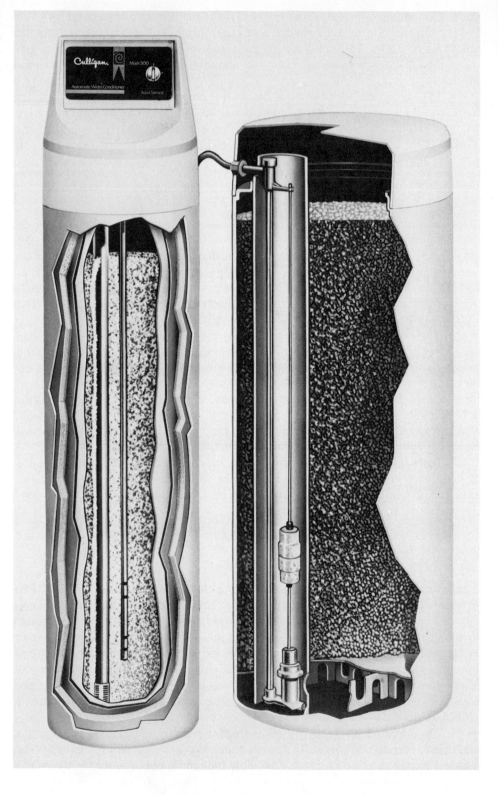

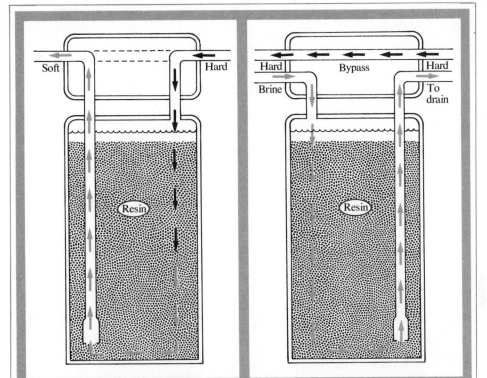

Figure 21.6 Home water softener. Water may be "softened" by the use of ion exchange resins, which efficiently remove calcium and magnesium ions from water. A home water softener is pictured in cutaway view (page 476). The taller tank contains the ion exchange resin that removes the hardness cations from the water as it passes through. The shorter tank contains a large supply of brine (NaCl), which is periodically required for recharging the ion exchange resin. The schematic diagram (Figure above) shows the operation of the system during the softening cycle (left) and the recharging cycle (right). During the recharge cycle, water is passed through the tank of brine, picks up a high concentration of NaCl, and then passes through the ion exchange resin. The excess NaCl is then washed off the resin and the softening cycle is resumed. The entire process can be automated and timed to conform to the degree of hardness of the water supply and the amount of water consumed in each household. The details of the chemistry of the process are discussed in Figure 21.7. (Courtesy of Culligan Water Institute.)

dissolve. Many undesirable organisms are anaerobic and cannot tolerate oxygen. Not only does the oxygen aid in destroying these species but it improves the taste of the water as well. Frequently, chlorination and oxygenation are both used in treating water supplies. This treatment removes the biological health threat. However, we have now discovered that some of the products of chlorination may be dangerous. The chemical reaction can generate polychlorinated species in small but significant amounts when certain pollutants are present in the water. This is by no means the primary source of the chlorinated compounds in our water systems. Other sources are pesticides, such as aldrin, dieldrin, and DDT. In addition, a wide variety of industries use substances called polychlorinated biphenyls, PCBs, in a number of applications. For example, PCBs are often used to increase the flexibility of polymers, such as polyvinyl chloride (PVC), which is a hard plastic in

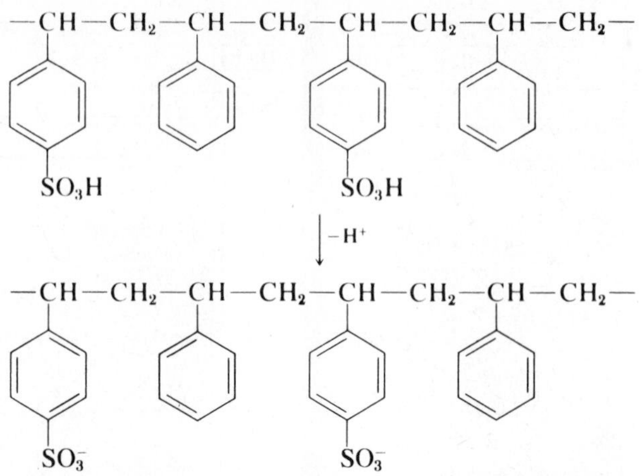

Figure 21.7   The chemistry of water softening.

Ion exchange resins are complex substances that may have a variety of chemical structures and find use in a variety of applications. For water softening, a class of ion exchange resins called *zeolites* were often used for softening water, but in the 1950s, the zeolites were largely replaced by a type of ion exchange resin called a *sulfonated polystyrene*. Only the latter will be discussed.

Polystyrene is an important polymer that is often used as insulating material, e.g., under the brand name Styrofoam. A portion of a polystyrene resin may be represented as shown with sulfonic acid groups attached and pictured in both the neutral and ionic forms.

Not every one of the rings contains a sulfonic acid group or the corresponding anion, but each of the negative sites in the resin provides a point for binding of cations.

When the ion exchange resin is ready to be used for softening, sodium ions (Na^+) are bound to the anionic sites on the resin. As water containing hardness cations passes through the resin, the calcium and magnesium ions, Ca^{+2} and Mg^{+2}, displace the sodium ions from the resin and bind to the anionic sites. This is the phenomenon of *ion exchange,* or more specifically *cation exchange.* The hardness cations are able to replace the sodium ions due to the greater positive charge of Ca^{+2} and Mg^{+2}, which causes a tighter binding to the anionic sites on the resin.

Quite obviously, an increased concentration of sodium ions appears in the water that comes out of the water softener, but Na^+ is not a hardness ion and does not cause precipitation of soap, formation of boiler scale, etc.

Needless to say, a point is eventually reached when all of the sodium ions in the resin have been replaced by hardness ions so that hard water will be unaffected. At this point, it is necessary to recharge the resin by stripping off the hardness cations. This is done by flushing the tank with concentrated NaCl solution, which is obtained by directing a flow of water through the second tank containing NaCl (brine) and then through the resin. Although Ca^{+2} and Mg^{+2} do bind more tightly than Na^+ to the resin, when the concentration of Na^+ in the flow is high enough, the Na^+ will effectively compete for the negative sites on the polystyrene and the resin will again be in the sodium form and, therefore, an effective water softening agent.

the absence of a plasticizer. With plasticizer present, PVC can be used for making plastic raincoats, garden hose, and auto seat covers. In any case, the PCBs have also shown up as water pollutants, and the level of such pollutants has sometimes been found to increase as a result of attempts to purify water by chlorination. In essence, as the biological problem has been reduced, the chemical problems have emerged.

Other species are now coming into prominence as toxic agents in drinking water. Nitrate (NO_3^-) is known to be particularly harmful to newborn babies, causing the so-called blue baby syndrome, in which nitrite (NO_2^-), which is produced from nitrate in the water supplies, interferes with the ability of hemoglobin

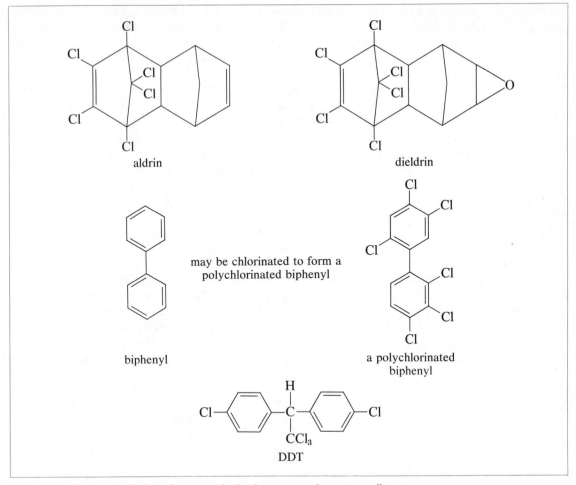

Figure 21.8   Some polychlorinated compounds that have emerged as water pollutants.

to carry oxygen. The quantity of nitrates in natural water has increased considerably in recent years due to the extensive use of nitrogen fertilizers. Even the manure generated by cattle feed lots is the source of quantities of nitrate, since the manure is not spread on fields as fertilizer due to high transportation costs from the feed lots. Instead, the manure is piled outside the cattle areas and the soluble components, of which nitrate is one, are leached out by rain. The rain runoff then becomes part of our natural waterways and may enter our drinking water. The nitrogen fertilizers used are readily soluble in water. Those not used quickly are washed away by succeeding rainfall, again entering the local waterways.

Nitrates are not the only problem associated with drinking water. Most heavy metals, e.g., cadmium, mercury, and lead, are poisonous to mankind. The ones that have received the widest attention lately are mercury and cadmium. As a general rule, an excess of most heavy metals is undesirable. In small amounts, however, some are essential to human life. The list of essential elements is increasing as research shows the function of these in the body. A very recent example is selenium. There is evidence to suggest selenium is a significant factor in combating a variety of cancers in the body (13). The levels examined are approximately 0.3 ppm in the blood. Selenium (Se) is also known as a deadly poison, but in significantly larger quantities. The desirable quantity of selenium is absorbed from food and water. Therefore, this research suggests it is not proper to completely eliminate selenium from the diet, but rather to keep the levels in the useful range.

This same research suggests zinc interferes with the cancer-fighting abilities of selenium. Yet zinc is on the list of essential elements for human life. Such interactions are one of the great unknown areas of biochemistry. It is not enough to know the effects of an element alone; its interaction with other elements and species present must also be known. Before we can establish detailed specifications for foods and water supplies, these interactions must be known. In lieu of this knowledge it is better to be conservative in our specifications.

Dissolved Oxygen

It was mentioned earlier that oxygen dissolves in water. When water is saturated with oxygen, the concentration is 10 ppm. The very small amount plays a crucial role in sustaining life in natural waters. Fish are known to require 4 to 5 ppm of oxygen for life. Oxygen is used to remove a variety of objectionable materials from water. We have already mentioned its role in the treatment of drinking water. As an example with an undesirable effect, oxygen in natural waters reacts with iron pyrites, FeS_2, which is frequently exposed by mining, to produce sulfuric acid and insoluble iron compounds according to the following sequence of reactions.

$$4\ FeS_2 + 14\ O_2 + 4\ H_2O \rightarrow 4\ Fe^{+2} + 8\ SO_4^{-2} + 8\ H^+$$
$$4\ Fe^{+2} + O_2 + 4\ H^+ \rightarrow 4\ Fe^{+3} + 2\ H_2O$$
$$4\ Fe^{+3} + 12\ H_2O \rightarrow 4\ Fe(OH)_3 + 12\ H^+$$
$$\text{``yellow boy''}$$

The result is acid mine drainage, which pollutes streams due to the acidity and the deposits of "yellow boy," the insoluble iron compound, which also forms in the process.

Oxygen is depleted from natural waters when an overabundance of oxidizable material is present. This material can be anything from iron pyrites to sewage to industrial wastes. Most of the material with which oxygen reacts is biochemical in nature. The oxygen required to react with this biochemical waste is known as the *biochemical oxygen demand* or *BOD*. The BOD is formally defined as the amount of oxygen used during a five day period at 20°C. The paper used to make this book generates organic waste with a BOD of 16,000 to 25,000 ppm, while the brewing of beer generates waste with a BOD of 500 to 1200 ppm. (Some individuals may suggest it is more environmentally sound to get drunk than to read this book. The author, however, does not subscribe to this view.)

Certainly these BOD's far exceed the oxygen available at saturation. As the oxygen is used, it is renewed by dissolving additional oxygen from the atmosphere. However, different types of natural water dissolve additional oxygen with greater or lesser ease. Swiftly moving streams with rapids and falls expose more water to the air and replenish their oxygen faster. Slow-moving streams and lakes replenish oxygen far slower. The definition of a dead lake or stream is one where the BOD is so great that no life can be supported other than organisms involved in the oxidation of waste. In deep lakes into which waste is dumped, the oxygen at the bottom is used up very quickly and replaced very slowly. During the present century, the amount of oxygen near the bottom of the northern end of the Baltic Sea has decreased from approximately 3 ppm to virtually zero because of waste disposal in the Baltic (14).

The BOD for organic waste disposal can be estimated quickly and easily from the weight of dry organic matter. The reaction between carbon in the organic waste and oxygen forms CO_2 according to the following equation.

$$C + O_2 \rightarrow CO_2$$

From this chemical reaction, plus a knowledge of the atomic weights of the atoms involved, we know that 12 grams of carbon will react with 32 grams of oxygen, or a ratio of 32 g O_2 per 12 g carbon = 8 g O_2 per 3 g carbon, or in general terms, 8 parts of oxygen for each 3 parts of carbon. This means for every three parts of carbon present, we require eight parts of oxygen. If we assume that the waste material is mostly carbon when dry, we can translate this into eight parts of oxygen for every three parts of dry organic matter. On this basis, a saturated solution of oxygen in water (10 ppm of O_2) can handle 3.75 ppm of organic matter, assuming no additional oxygen enters the water from the atmosphere. This condition is the usual situation for deep, slow-moving rivers or lakes. If the water can accept oxygen readily from the atmosphere, the BOD can be higher without removing all the oxygen.

Untreated sewage has a BOD of 100 to 300 ppm. Sewage treatment does not remove all the organic matter. About twenty percent of the organic materials are normally passed along as soluble components. These soluble parts have a BOD of 20 to 60 ppm, well in excess of the 10 ppm available in saturated water.

CONCLUSION

Pollution is a problem that must be solved. Unfortunately, the solution is not simple. It will require much additional research, frequently in seemingly unrelated areas, in order to make intelligent decisions and set meaningful specifications. Much can be done based on our present knowledge, but the public must be aware that science is a growing discipline and revised specifications merely reflect more recent knowledge. Failure to set specifications because we lack complete data may be tantamount to committing suicide. But no set of specifications can be allowed to be treated as inviolable.

SUMMARY

In order to characterize polluted air, it is necessary to have a definition of clean air, but this is difficult. Clean air contains many components that are hazardous pollutants when they are present in large amounts. Carbon monoxide, sulfur oxides, ozone, krypton, and carbon dioxide are examples. In addition, ozone is a substance that is very useful in the stratosphere, but is hazardous in the troposphere.

One of the most difficult problems encountered when studying pollutants is the possibility of reactions between pollutants or their decomposition products, such as may occur in smog. The phenomenon of acid rain demonstrates the inter-relationship between air and water.

Pure water is easy to define; polluted water is not. Salt water from the ocean is not polluted, but neither is it fit for use as drinking water. Hard water can be categorized according to the degree of temporary or permanent hardness, which is distinguished by the ability of CO_2 to affect the solubility of the dissolved ions (Ca^{+2} and Mg^{+2}). Water softeners can be used to remove these ions.

Chlorine and oxygen are used routinely for treatment of drinking water to remove biological pollutants. Chlorination can affect certain dissolved pollutants and render them more dangerous. In other words, chemical hazards may be created while eliminating the biological hazards. Even dissolved oxygen can contribute to pollution due to reaction with iron pyrites, FeS_2, released during mining. The biochemical oxygen demand (BOD) can be used to evaluate the extent of some pollution problems.

PROBLEMS

1. Give a definition or example of each of the following:
   a) percent by volume
   b) percent by weight
   c) parts per million (ppm)
   d) stratosphere
   e) ozone
   f) troposphere
   g) ionosphere
   h) chemosphere
   i) particulate matter
   j) smog
   k) eutrophication
   l) BOD

2. What factors must be considered when setting standards for defining polluted air?
3. How do sulfur oxides contribute to air pollution and why are they undesirable in air?
4. Why is ozone desirable in the stratosphere but a pollutant in the troposphere?
5. What effect could the supersonic transport (SST) have on the ozone layer in the stratosphere? What might be the consequences?
6. What is "The Greenhouse Effect"? What are its possible consequences in terms of air pollutants?
7. Why is it difficult to define unpolluted water?
8. What is the "hydrologic cycle?"
9. What causes temporary hardness in water and what effect does this hardness have?
10. What causes permanent hardness in water and how can it be removed?
11. How can water be made safe for drinking and what are the drawbacks of these methods?
12. What is acid mine drainage and how does it contribute to water pollution?
13. What is a dead lake or stream?

REFERENCES

1. Lehrer, T. "Pollution." On *That Was The Year That Was,* Reprise Records. Port Washington, N.Y.: Alfred Publishing Co.

2. Giddings, J.C. *Chemistry, Man, and Environmental Change*. San Francisco: Canfield Press, 1973, p. 199.

3. Weast, R.C., ed. *Handbook Of Chemistry and Physics*. 51st edition. Cleveland: The Chemical Rubber Co., 1970, p. F-151.

4. Sax, N.I. *Dangerous Properties Of Industrial Materials*. New York: Van Nostrand Reinhold, 1968, pp. 533–4.

5. Giddings, J.C. *Chemistry, Man, and Environmental Change,* San Francisco: Canfield Press, 1973, p. 229.

6. Davis, D.D., Smith, G., Tesi, G., and Ravishankan, A. "A Photochemical And Kinetics Study Of The Stratospheric Stability Of Chlorine Nitrate." *Abstracts, 172 nd Meeting, American Chemical Society* 1976, Fluo. 21.

7. Johnston, D.O., et al. *Chemistry and the Environment*. Philadelphia: W.B. Saunders, 1973, p. 417.

8. Jones, M.M., et al. *Chemistry, Man and Society*. Philadelphia: W.B. Saunders, 1976, p. 518.

9. Likens, G.E. "Acid Precipitation." *Chemical and Engineering News,* 22 November 1976, p. 29–37.

10. Giddings, J.C. *Chemistry, Man, and Environmental Change*. San Francisco: Canfield Press, 1973, p. 315.

11. Pryde, L.T. *Environmental Chemistry*. San Francisco: Canfield Press, 1973, p. 213.

12. Ward, R.C. *Principles of Hydrology*. New York: McGraw-Hill, 1975, p. 5.

13. "Studies Firm Up Some Metals' Role In Cancer." *Chemical and Engineering News,* 17 January 1977, p. 35.

14. Giddings, J.C. *Chemistry, Man, and Environmental Change*. San Francisco: Canfield Press, 1973, p. 295.

CHAPTER TWENTY-TWO

T HE emergence of man as a distinct species on the planet Earth has been accompanied by an increasing desire to experience a controlled interaction with the environment. In its earliest stages, this interaction was a matter of minimal survival, requiring only protection from a variety of external hostile forces and a need to continually silence the reoccurring internal hunger signals. The resulting struggle to participate in a parade of progress, that which Eisely called "The Immense Journey" and Bronowski defined as "The Ascent of Man," has been filled with brilliant achievements and monumental failures.

In a paper written while he was a student at Cal Tech, movie director Frank Capra proposed an intriguing "equation of the mind,"

$$P = KI^2$$

where *progress equals knowledge times intelligence squared* (1). Based on the premise that intelligence is a constant, while knowledge is the "variable in the progress of man," the equation predicts that each step in the accumulation of knowledge produces a large increase in progress.

Observations of nature and the results of experimentation have been collected for millions of years. Since the publication of the first scientific journal some 300 years ago, the rate of compilation has been steadily increasing. In recent years, this rate has reached alarming proportions, as the volume of scientific literature published between 1660 and 1960 was doubled from 1960 to 1973. Yet Capra's premise suggests that the intelligence of the first cave man is no different than that of the modern scientist and that we are still working with the same basic tool for the interpretation and application of a massive body of facts that the cave man used to analyze the first entries in the scientific ledger. (Ironically, we still have little understanding of the function and operation of the human mind—a topic that some have called the last great frontier for investigation.)

If we equate progress with growth, Capra's equation appears to give an acceptable result. Unfortunately, P might represent problems just as well as progress. There appears to be no end to the increasing complexity of the difficulties accompanying each step forward. The Madison Avenue slogan "Better Living Through Chemistry" seems questionable as we face the disposal of wastes from the very industrial and nuclear processes that are supposedly providing that "better living" we are searching for. In other words, progress often carries a very high price tag.

RESEARCH: NO DEPOSIT, NO RETURN

Thus, we are confronted with the same basic problems that our early ancestors struggled with. Our main concerns are still protection from the unpredictable forces of nature and sustaining life. Energy production, food requirements, the treatment of disease, and the interference of man with the balance of nature are some of the most critical problems that have to be dealt with. While ancient man searched for answers, we are re-searching for different answers to similar questions.

The nature of this research, the direction it should take, and the support it should receive have been widely debated issues for some time. Unfortunately, a negative attitude toward research is not uncommon. For example, Sir Humphrey Davy had suggested as early as 1800 that nitrous oxide had pain-killing properties that might be of value in surgery. However, painless operations were considered so impossible that "laughing gas" and ether were used only for amusement until 1846, when William Morton successfully demonstrated the anesthetic effects of ether in an operation at Massachusetts General Hospital. While the medical profession was then quick to adopt anesthesia, the surgical mortality rate remained at about 80% for decades, because of infection. The less dramatic antiseptic techniques formulated by Joseph Lister in Scotland required over thirty years for acceptance as a common practice.

Around 1843, a commissioner of the United States Patent Office supposedly resigned his position and stated that most discoveries had already been made and there would be little need for the issuance of additional patents. In spite of such pronouncements, there were many individuals on both sides of the Atlantic who continued to search for the key to successful manned flight. By the beginning of 1903, glider flights had become commonplace, and Dr. Samuel P. Langley, a respected scientist and head of the Smithsonian Institution in Washington, had flown many types of powered model aircraft with wingspans up to thirteen feet. On October 7, 1903, Charles M. Manley was launched from a houseboat in the Potomac River in a large Langly-built and designed, full-scale, gasoline-powered machine with the hope of becoming the first human to make a sustained flight. According to the *Washington Post,* the machine "simply slid into the water like a handful of mortar." After some modification, a second attempt was made on December 8, but the results were the same.

A *New York Times* editorialist was not enthusiastic about the time, effort, and money spent on such foolishness. On October 9, 1903, he wrote that "the ridiculous fiasco which attended the attempt at aerial navigation in the Langley flying machine was not unexpected. The flying machine which will really fly might be evolved by the combined and continuous efforts of mathematicians and mechanicians in from one to ten million years—provided we can meanwhile eliminate such little drawbacks and embarrassments as the existing relation between weight and strength of materials." Langley's second failure brought an additional comment on the same editorial page, where on December 10, the *Times* indicated that the eminent scientist "is capable of services to humanity incomparably greater than can be expected to result from trying to fly." In Dayton, Ohio, just one week later, on December 17, Bishop Milton Wright received a telegram from his son Orville, which began "Success four flights Thursday morning."

While there should be no place in scientific research for resistance to discovery, the history of science is filled with similar examples that have resulted in

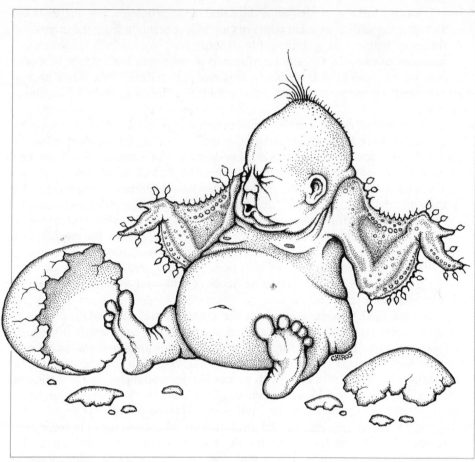

If man were meant to fly . . .

the retardation of scientific thought and knowledge. And it does lead to a very interesting question—just how are scientific discoveries made? How many doctors had participated in "ether frolics" without seeing what is now an obvious application? How many biologists had thrown away molding culture dishes and thus missed the immortality acquired by Fleming due to his discovery of penicillin? How many chemists had discarded glass reaction vessels because they contained polymeric messes that were considered undesirable by-products? How many engineers overlooked the warping control mechanism developed by the Wrights?

In his book *The Flash of Genius,* Alfred B. Garrett has cited case histories of scientific discoveries and placed them into three categories: trial and error, planned research, and accident (2). It is not easy to place a discovery into one of these categories, and, indeed, elements of all three may be involved in a single discovery. The success of the trial and error approach depends on just doing enough experiments; the larger the number of experiments, the greater the chance of a successful result. The massive screening program conducted by The Drug Evaluation Branch of the National Cancer Institute is typical. In this program, any chemical compound that can be obtained is routinely tested in a carefully-designed procedure to determine if the substance might have potential use as an anticancer agent.

Planned research is often carried out in response to an immediate need. In 1942, there was a need to develop a super weapon, and the Manhattan Project led to the production of the first atomic bomb. In 1957, there was the push to develop a space exploration program because of a small object called Sputnik circling the Earth, and men walked on the moon. In the chemical industry, there is a need to satisfy the stockholders, and a variety of consumer products are produced as a result. In all three cases, however, the overall results far surpass the initial objectives, and the benefits that accrue are not part of the initial planning or predictions. Peaceful uses of atomic energy and their application to medicine and energy were not important in 1942. Miniaturization of electronic components for calculators and watches and accurate weather forecasting were not the primary goals in 1957. The discovery of Teflon by Dr. Roy Plunkett of Dupont was strictly an accident that no one could have anticipated would result from some simple experiments with tetrafluoroethylene.

Discovery by accident is difficult to categorize. Are accidental discoveries actually made by simply mixing things at random and seeing what happens? Not necessarily. In the "scientific method," the role of observation is extremely important. Not everyone will observe a particular event in the same way, as we have already seen. One must know how to take advantage of observation, and this requires being informed, organized, objective, flexible, and creative. In Dr. Zhivago, Boris Pasternak writes that "the fabulous is never anything but the commonplace touched by the hand of genius." Francis Bacon felt that "there remains simple experience, which if taken as it comes is called accident, if sought for, experiment." A frequently quoted statement belongs to Louis Pasteur: "In fields of observation, chance favors only minds that are prepared."

Thus, accidental discoveries are not necessarily pure accident. And Garrett includes the item of serendipity as important to the trigger-tripping process of discovery by accident. *Serendipity,* a term coined by Horace Walpole in 1754, stems from an old fable that the princes of Ceylon (or Serendip) had the faculty of making happy and unexpected discoveries by accident. The term has been applied to those that experience these happy coincidences, while recently it has also been used to describe the trait of recognizing interesting and exciting phenomena in everyday activities.

With these ideas in mind, whether creativity, serendipity, a prepared mind, or genius are operating, let us look at some examples of discoveries where the accident, in whatever form it may take, suggests something to the experimenter that usually speeds progress towards a satisfactory conclusion.

Collodion, a solution of nitrocellulose in alcohol and ether, was a common item in many laboratories. Not only was it used to seal labels on laboratory bottles, but it was frequently used to cover minor cuts and scratches because of its ability to form a protective surface coating. Many users who occasionally spilled a bottle of collodion found an annoying, sticky mess. But to three scientists, this annoyance led to three different and important discoveries.

Alfred Nobel was a self-educated man whose interest in chemistry coupled with his father's business of manufacturing gunpowder led to his building a plant for the production of nitroglycerin near Stockholm in 1865. This highly explosive material was so sensitive to shock that trains and boats used in transporting it frequently exploded. When Nobel noticed that the nitroglycerin leaking from a container on a loading dock was readily absorbed by the kieselguhr packing material used to cushion the cans, he recognized the possibility of solving the shipping and handling problems. The result was dynamite, a product that brought him immense wealth and fame. But Nobel was not satisfied, and after eleven years he was still trying to find a better solidifier for his nitroglycerin. While experimenting with this problem, he accidentally cut his finger in his laboratory. Following a common practice of the day, he coated the cut with collodion, which after evaporation of the solvent left a "new skin" residue of flexible nitrocellulose. It then occurred to him that if he dissolved solid nitrocellulose in nitroglycerine, he might even obtain a double explosive. The idea worked, and the resulting blasting gelatin proved to be insensitive to shock and more adaptable to the larger guns that were being made at the time.

At about the same time in the United States, a totally different problem, which eventually involved collodion, was being investigated. Since the natural supply of ivory was rapidly being depleted, the need arose to find a suitable substitute. The company of Phelander and Collander, a major manufacturer of ivory billiard balls, sponsored a contest with a first prize of $10,000 for the best synthetic alternative for ivory to use in their product. A New Jersey printer, John Wesley Hyatt, and his brother prepared batch after batch of formulations that were unsuccessful. While making a potential billiard ball from sawdust and paper bonded together with glue, Hyatt, like Nobel, cut his finger. When he went for a bottle of collodion, he found that the bottle had been upset, and the collodion that had run out of the bottle had formed a tough sheet on the shelf. Hyatt immediately recognized that this substance might be a better binder than the glue he was using. A process of heating nitrocellulose and camphor under pressure was developed for a satisfactory billiard ball, and subsequent improvements led to a new plastic material called "celluloid."

For the third part of the collodion story, we turn to France, where in 1865, the silk industry was being threatened by an epidemic that was killing silkworms. It was Louis Pasteur who, after a four-year investigation, was able to locate the source of the disease that was threatening the silkworm and eradicate it. Pasteur was assisted by a young chemist named Hilarie de Chardonnet, and during his four years of intense work, de Chardonnet realized that the uncertainties in natural silk production would make an artificial substitute a highly desirable commercial item. One day, while working in a darkroom developing photographic plates, de Chardonnet accidentally spilled a bottle of collodion. He did not attempt to clean up the spilled material until his work with the plates was completed, and by this time, it had become quite sticky. When a cloth was wiped over the spilled area, long strands adhered to both the rag and the table, and their resemblance to silk fibers led de Chardonnet to reason that he had a possible starting point for his silk substitute. Six years later, he had produced a synthetic fiber that was comparable to silk, drawing the filaments from a solution of mulberry leaves in collodion. In 1891, commercial production of "artificial silk" began, and in 1924, the name was changed to "rayon."

Blasting gelatin, celluloid, and rayon—all from bottles of collodion. But none of these could have been written up in advance as a research proposal to be submitted to a government agency or a vice president in charge of research for approval and funding.

There are several examples in which frustration on the part of an experimenter led to fame and wealth and the formation of another new chemical industry. Let us look at one example. In 1892, Thomas L. Wilson, an engineer from North Carolina, was trying to obtain metallic calcium by fusing lime (CaO) and coal tar in an electric furnace. Instead of the expected metal, he obtained a dark-colored mass in a crucible. In disgust, Wilson disposed of the product in a nearby stream. When the solid hit the water, a large volume of gas was liberated and immediately ignited, producing a bright yellow sooty flame. Realizing that the gas was a hydrocarbon because of the characteristics of the flame it produced, Wilson repeated the experiment, and by careful analysis was able to show that the material produced in the crucible was calcium carbide, CaC_2, and the gas was acetylene, which formed according to the following equation:

$$CaC_2 + 2\ H_2O \rightarrow H{-}C{\equiv}C{-}H + Ca(OH)_2$$

Wilson eventually formed his own company to manufacture calcium carbide and to utilize the acetylene it produced in lighting fixtures such as miners lamps and automobile headlamps.

Photographic plates have been involved in two important accidental discoveries. Henri Becquerel's experience with sealed and unused plates that became clouded when placed next to a sample of uranium ore is the well-known beginning of the discovery of natural radioactivity. However, a similar experience that happened to the German physicist Wilhelm Roentgen is less well-known. In 1895, Roentgen was performing experiments with glass tubes called cathode ray tubes, or Crookes tubes. These glass containers had electrodes sealed at each end and were used to study the effect of the passage of electricity (cathode rays) through gases. Roentgen's laboratory has been described as unconventional. In reality, it was quite a mess, with every available space littered with books, papers, and equipment.

When breaking for lunch one day, Roentgen disconnected the Crookes tube he had been working with and placed it on top of a nearby book. Returning from lunch, he picked up his camera and several photographic plates that were lying under the book and went outside to take some pictures. When these plates were developed, one was found to have the image of a key. Suspecting that a coworker was playing a practical joke, Roentgen immediately questioned his staff, but had no success in determining the origin of the unexpected picture. He then reconstructed the arrangement of his laboratory on the previous day, placing a fresh plate under the book. The cathode ray tube was connected and allowed to glow for an identical time. The tube was disconnected and placed on the same book for the same time period. When the plate was developed, the image of the key appeared on the new plate as well. Roentgen then realized that he often used various objects as book marks, and soon found a key inside the book. The known properties of the cathode rays would not explain the development of a plate hidden underneath a book, and it took two more years of intense and planned research before Roentgen was ready to announce that he had discovered a new ray that he named the x-ray, after the algebraic symbol x for unknown.

There are many other examples of discoveries that have an accidental nature, and most of them occurred in the last century or the early part of this century. One might argue that since chemistry has become so highly developed both experimentally and theoretically, there is really no further room for an accident to be of value in research. Certainly the amount of preparedness needed today in terms of formal education far surpasses that of Nobel, Wilson, or Roentgen. Yet accidents still happen.

During the summer of 1938, Dr. Roy Plunkett of Dupont was studying the synthesis of various fluorochlorohydrocarbons. When he needed about one hundred pounds of tetrafluoroethylene for his research, he found that only very small quantities had ever been prepared. Plunkett constructed a small pilot plant for preparing the desired amount and stored the samples in steel cylinders cooled with dry ice. Since this compound was the starting material for the synthesis of other compounds, the experimental arrangement involved passing the gaseous fluorocarbon from one of the storage cylinders through a flow meter for monitoring purposes and then into a reaction chamber. One day, a small cylinder containing two pounds of the gas was attached to the system and the valve was opened to start the flow of gas. A few minutes later, an assistant noticed that the gas flow had stopped, although the weight of the cylinder indicated that an appreciable quantity of gas remained inside. The valve was opened completely, but no gas escaped. A small wire was run through the valve opening, but to no avail. Finally, the valve was removed, and a white powder poured from the cylinder opening. Out of desperation, Plunkett obtained a hacksaw and cut the cylinder in half. The gas had disappeared and in its place were large quantities of a white powder. Further investigation of the white powder culminated with the marketing of the polymer Teflon.

Perhaps the most recent example of a prepared mind taking advantage of a chemical observation occurred in 1962 when Professor Neil Bartlett of British Columbia destroyed a chemical principle that was one of the most sacred in the long history of chemistry. Bartlett was reacting fluorine with various materials, including platinum. In one experiment, he heated some platinum wire surrounded by fluorine gas and obtained a red solid that he identified as containing the elements platinum, fluorine, and oxygen. In this compound (O_2^+ PtF_6^-), the oxygen came from the glass vessel in which the reaction was carried out. This in itself was not planned, but the really surprising thing was the fact that the oxygen had undergone a loss of electrons during the reaction. This loss of electrons from an oxygen atom is just the opposite of what normally occurs, as oxygen atoms are usually found to gain electrons to form the inert gas structure of neon.

In a subsequent reaction using the compound platinum hexafluoride, PtF_6, oxygen in the air was found to participate in the same reaction. While doing so, the PtF_6 functions as an oxidizing agent in removing an electron from oxygen and Bartlett reasoned that PtF_6 was the strongest oxidizing agent then known. If this were true, then were there other compounds that could lose electrons to PtF_6 when they would not normally do so? As the ultimate test of this hypothesis, Bartlett picked the chemical virgins known as the inert gases. These elements, found in group VIIIA in the Periodic Table, include helium, neon, argon, krypton, and xenon. They were indeed inert and no one had been able to pry an electron from any one of these elements to form a compound containing an inert gas cation.

Bartlett constructed an all-glass apparatus in which a bulb containing PtF_6 was separated by a glass septum from another bulb containing xenon gas. When Bartlett was ready to perform the experiment, it was late on a Friday afternoon and the laboratories were deserted. His graduate students, showing no optimism, had already left for the weekend. When the glass septum separating the red gas PtF_6 from the colorless gas Xe was broken and the gases mixed, there was an immediate reaction producing a yellow-orange solid ($XePtF_6$). Bartlett knew that he had his result—the once in a lifetime experiment that shattered chemical dogma—by reacting the unreactive. Textbooks would be rewritten and a whole new branch of chemistry was opened up.

In the preceding examples, we have looked at several important discoveries that have been termed accidental. It is interesting to speculate what factors were important in the recognition of a particular event as significant and its subsequent utilization for future developments. Of all the possibilities, the "prepared mind" is probably the most critical.

Nobel Laureate Paul J. Flory has emphasized that "there is much more to invention than . . . a bolt out of the blue. Knowledge in depth and in breadth are virtual prerequisites. Unless the mind is thoroughly charged beforehand, the proverbial spark of genius, if it should manifest itself, probably will find nothing to ignite. At the present level of technological sophistication, creative invention, without a firm grasp of underlying principles, becomes increasingly rare (3)."

Obviously, there must be a strong educational commitment at all levels, from the grade schools to the postdoctoral level, to identify, encourage, and nurture those strongly motivated individuals who are truly creative. In addition to natural intelligence, they should be unusually curious, flexible, and independent in thought and action. Like every other research worker before them, they will be in the unique position of being able to evaluate all of the existing concepts and results in order to plan and implement the synthesis of additional stepping stones in the pathway of knowledge. In short, they will find themselves with Sir Isaac Newton, "standing on the shoulders of giants in order to see farther."

Where will all this activity take place? The day of the single investigator working with only basic tools has largely passed, for it is doubtful that many major discoveries today will be made in isolated laboratories. Instead, much of the interaction between scientific specialties now occurs in academic, governmental, or industrial laboratories, where interdisciplinary expertise can be utilized efficiently.

What is the form of this activity? Generally, research is considered as either basic or applied. The phrase *basic research,* also referred to as *pure research,* generally describes research that is aimed toward development of a better understanding of certain substances or phenomena, rather than an immediate consumer application. In other words, the research is either "pure" with no immediate application, or oriented towards commercialization. However, this does not mean that basic and applied research are entirely separate entities. The results of basic research, both anticipated and unexpected, provide the raw material for technology (applied science).

According to Professor Flory, "basic research has a more pervasive mission in advancing knowledge, in providing incisive concepts, and in sharpening insights. These are the ingredients that nurture enduring innovations of the broadest scope. They do not enjoy the visibility and the attention-arresting qualities of radi-cal discoveries, but . . . these less tangible contributions of basic research are of greater importance than the latter.''

What is the source of financial support for these activities? In his presiden-tial address to the Royal Society in 1943, Sir Henry Dale stressed that ''care for the practical fruits of the tree of knowledge was never so urgent as today; but the tree will wither unless we take care that the roots have nourishment and room for spreading.'' Yet in recent years, the financial commitment to basic research has declined steadily. It is not reasonable to expect substantial support from industry when the bottom line must justify the investment. The burden, therefore, falls on the Federal Government.

In a period when the federal budget increased by 100%, allocations for re-search decreased by 20%. Since most of this support is provided for special pro-jects, there is precious little remaining for basic research. When a large inflation rate over the same period is included, the results can be disastrous. Does a ''tech-nology gap'' now exist? Are the raw materials *from* basic research being depleted? While these questions are not easily answered, the possibilities are real. For ex-ample, there is already some indication that products of American industry are less competitive than they were on the international market.

The situation in the People's Republic of China emphasizes this problem even more strongly. For three decades, Chinese scientists, under strict political orders, have virtually ignored basic research while concentrating only on solving important national problems such as agricultural production and health care. Ac-cording to Frederick Seitz, president of Rockefeller University, the result is that ''they're doing basic and applied research to the best of their limited ability, but are still dependent for their innovations on the scientific community of the rest of the world.'' The new Chinese leaders have only recently begun efforts to bring their science and technology up to that of the rest of the world (4).

Senator Charles Mathias, a strong advocate of increased support for basic research, has used the energy crisis as a typical example of the need for Congress to reverse the funding trend of recent years: ''We cannot hope to meet this and fu-ture problems by buying knowledge on a crisis-by-crisis basis. We cannot expect to push a button and have science supply answers instantaneously (5).''

Scientists are often called upon to solve current problems spanning a wide range of disciplines including energy production, agriculture, health care, and many more. But what is often overlooked is the fact that the ability to deal with these problems is based on a foundation of long-term research, much of which has no clear application at the time it is carried out but, nevertheless, contributes to an understanding of various phenomena.

Many of our needs are global, and scientific research around the world must continue with the support and respect it so rightfully deserves. Otherwise, the day may come when we will join the fans of the immortal Casey who approached the plate as the hero who could make all their dreams of victory come true. And after the ''air had been shattered by the force of Casey's blow,'' it was reported that

. . . somewhere in this favored land the sun is shining bright,
The band is playing somewhere, and somewhere hearts are light;
And somewhere men are laughing, and somewhere children shout,
But there is no joy in Mudville . . .

REFERENCES

1. Capra, F., *The Name Above The Title,* New York: Macmillan, 1971, p. 183.
2. Garrett, A.B., *The Flash of Genius,* Princeton, N.J.: D. Van Nostrand, 1963, p. 1.
3. "Paul Flory on Basic Research," *Chemical and Engineering News,* February 28, 1977, p. 4.
4. "The Stalled Leap Forward," *Time,* August 1, 1977, p. 75.
5. Mathias, C.M., Jr., "Wanted, More Support for Basic Research," *Chemical and Engineering News,* February 14, 1977, p. 2.

S CIENTIFIC measurements and calculations often require the use of very large and very small numbers. Since both can be very unwieldy to write, a shorthand expression is often used. This expression is called the exponential notation and takes the following form:

$$N \times 10^{\text{exponent}}$$

where N is usually a number between 1 and 10 and the exponent is a whole number. The exponent may be either positive or negative.

The notation is summarized in Table A.1 and is easily understood once it is recognized that the exponent does nothing more than describe the location of the decimal point. For example, the number one million can be written in exponential notation by moving the decimal point six places *to the left* and using the exponent $+6$.

$$1 \text{ million} = 1,000,000 = 1.0 \times 10^6$$

move decimal
6 places to
left

One one-millionth can also be expressed in exponential notation by moving the decimal point *to the right* by six places and using the exponent -6.

$$1 \text{ one-millionth} = 0.000001 = 1.0 \times 10^{-6}$$

move decimal
6 places to
right

In other words, *when converting a number into the shorthand exponential notation, a positive exponent signifies that the decimal has been moved to the left, whereas a negative exponent indicates that the decimal has been moved to the right.*

On the other hand, when converting a number expressed in exponential no-

EXPONENTIAL NOTATION

TABLE A.1 EXPONENTIAL NOTATION

NUMBER	EXPONENTIAL NOTATION
1,000,000,000	1.0×10^9
1,000,000	1.0×10^6
10,000	1.0×10^4
1,000	1.0×10^3
100	1.0×10^2
10	1.0×10^1
1	1.0×10^0
0.1	1.0×10^{-1}
0.01	1.0×10^{-2}
0.001	1.0×10^{-3}
0.0001	1.0×10^{-4}
0.000001	1.0×10^{-6}
0.000000001	1.0×10^{-9}

tation into its complete form, the opposite is true. That is, *to write the full number, a positive exponent signifies that you must move the decimal to the right, whereas a negative exponent indicates that the decimal should be moved to the left.*

One could commit these rules to memory or simply recall that positive exponents are used for large numbers and negative exponents are used for small numbers. This information serves as a guide when making the conversion in either direction. Examples of the different conversions are given in the following four problems. Additional examples follow.

Problem A.1:

Express 62,000 in exponential notation.

Answer: Since the number N in the exponential is normally given between 1 and 10, it is necessary to move the decimal 4 places to the left, and use the exponent $+4$. Therefore, the correct answer is 6.2×10^4. Notations such as 62×10^3 or 0.62×10^5 are also correct, but are usually not used, since N is not between 1 and 10 in either notation.

Problem A.2:

Express 0.000029 in exponential notation.

Answer: To give N between 1 and 10, it is necessary to move the decimal 5 places to the right, which gives the exponent -5. Therefore, the correct answer is 2.9×10^{-5}.

Problem A.3:

Express 2.36×10^2 as a common number.

Answer: 236

Problem A.4:

Express 8.6×10^{-6} as a common number.

Answer: 0.0000086

ADDITIONAL EXAMPLES

Problem A.5:
Express 0.02 as an exponential number.
Problem A.6:
Express 3,600,000 as an exponential number.
Problem A.7:
Express 907 as an exponential number.
Problem A.8:
Express 0.00083 as an exponential number.
Problem A.9:
Express 9.6×10^{-3} as a common number.
Problem A.10:
Express 4.13×10^{-1} as a common number.
Problem A.11:
Express 1.85×10^4 as a common number.
Problem A.12:
Express 6.02×10^{23} as a common number.
Answers: A.5: 2×10^{-2}; A.6: 3.6×10^6; A.7: 9.07×10^2; A.8: 8.3×10^{-4}; A.9: 0.0096; A.10: 0.413; A.11: 18,500; A.12: 602,000,000,000,000,000,000,000.

A MERICANS are familiar with the English system of measurement, which uses the foot as the unit of length, the pound as the unit of weight, and the gallon as the unit of volume. Elsewhere in the world, including Great Britain, and in all fields of science, the metric system is used exclusively.

The popularity of the metric system arises from the fact that it is a decimal system in which all units may be interconverted by multiplying or dividing by $10^{exponent}$.* In other words, one need only move the decimal point in order to convert from one unit of length to another. In contrast, one must divide by 12 in order to convert inches to feet, or divide by 3 to convert feet to yards, or divide by 5280 to convert feet into miles.

TABLE B.1   PREFIXES USED IN THE METRIC SYSTEM

MULTIPLE	PREFIX
10^9	giga
10^6	mega
10^3	kilo
10^2	hecto
10^1	deka
10^{-1}	deci
10^{-2}	centi
10^{-3}	milli
10^{-6}	micro
10^{-9}	nano

The unit of length in the metric system is the *meter,* which is slightly longer than a yard. The units of mass and volume are the *gram* and the *liter.* Large and small weights of a substance are commonly expressed in kilograms and milligrams, respectively. Large and small lengths are commonly expressed in kilometers and millimeters, respectively. These and other common prefixes are summarized in Table B.1. The common metric units of length, mass, and volume are shown in Tables B.2, B.3, and B.4, respectively. Some common English-metric conversions appear in Table B.5.

* See Appendix A.

THE METRIC SYSTEM

TABLE B.2   COMMON METRIC UNITS OF LENGTH

1 kilometer (km)	= 1000 meters (m)
1 *meter* (m)	= 100 centimeters (cm)
1 centimeter (cm)	= 10 millimeters (mm)
1 millimeter (mm)	= 1000 micrometers (μm)

TABLE B.3   COMMON METRIC UNITS OF MASS

1 kilogram (kg)	= 1000 grams (g)
1 *gram* (g)	= 1000 milligrams (mg)
1 milligram (mg)	= 1000 micrograms (μg)

TABLE B.4   COMMON METRIC UNITS OF VOLUME

1 *liter* (l)	= 1000 milliliters (ml)
1 milliliter (ml)	= 1000 microliters (μl)

TABLE B.5   SOME ENGLISH-METRIC CONVERSIONS

LENGTH

1 mile (mi)	= 1.61 kilometers (km)
1 yard (yd)	= 0.914 meter (m)
1 inch (in)	= 2.54 centimeters (cm)

MASS

1 pound (lb)	= 454 grams (g)
1 pound (lb)	= 0.454 kilogram (kg)
1 ounce (oz)	= 28.4 grams (g)

VOLUME

1 gallon (gal)	= 3.78 liters (l)
1 quart (qt)	= 0.946 liter (l)
1 pint (pt)	= 0.473 liter (l)
1 fluid ounce (fl oz)	= 29.6 milliliters (ml)

THE INTERNATIONAL SYSTEM, SI

A modified form of the metric system is coming into increasing use. It is called the International System of Units, officially abbreviated SI.

In order to consider the differences between the standard metric system and the SI, it would be necessary to consider many other units of measurement such as force and energy. We have considered the metric units of mass, length, and volume and, of these, only the units of volume are different in the International System of Units, and even then in a very subtle way.

The basic unit of volume in the metric system is the liter (l). In the SI, the basic unit of volume is the cubic meter (m^3). Since the cubic meter is a very large volume, chemists continue to express volumes in liters or milliliters. One milliliter (ml) is equivalent to 1 cubic centimeter (cm^3 or cc).

acid: A substance that turns litmus red; its solutions in water have a pH less than 7; a proton donor (Brønsted definition)

active niacin (NAD): A form of the B vitamin (niacin), which is often required for the oxidation of organic compounds, e.g., in the fermentation of glucose

addition polymer: A high molecular weight compound formed when unsaturated monomers combine by simple addition without the involvement of any functional groups other than the unsaturation

aerobic: Functioning in the presence of oxygen

albumin: Water soluble proteins that occur in blood plasma or serum, muscle, the whites of eggs, milk, and other animal substances and in many plant tissues and fluids

alkali metals: Elements in group IA in the periodic table

alkaline: Having a pH greater than 7; basic

alkaline earth metals: Elements found in group IIA of the periodic table

allomone: A chemical substance released by a plant or animal that provides an advantage to the producer, e.g., a defensive secretion

alloy: A solution of solids

amalgam: A solution of solids containing mercury

amino acids: The repeating unit in proteins; this unit is characterized by having an amino group ($-NH_2$) and an acid group ($-COOH$)

amorphous: Having no definite shape

anaerobic: Functioning in the absence of oxygen

analgesia: Relief from pain

androgens: Male sex hormones

anion: A negatively charged particle

antibiotic: A drug that interferes with the growth of microorganisms

antibody: A protein molecule that provides immunity by inhibiting the growth of microorganisms

antigen: A substance that stimulates the release of antibodies

antimetabolite: A substance that interferes with normal metabolism

antioxidant: A substance that prevents oxidation

antipyretic: A drug that reduces fever

antitoxin: A substance that acts against poisons (toxins)

aqueous: Relating to water

atherosclerosis: Hardening of the arteries

atom: The smallest unit of an element, which possesses all the properties of that element

atomic number: The number of protons in the nucleus

atomic weight: The average relative weight of an element compared to a standard of carbon with an atomic weight of 12 atomic mass units (amu)

autoxidation: Oxidation by atmospheric oxygen

bacteriocidal: Able to kill bacteria

bacteriostatic: Able to inhibit the growth of bacteria

base: A substance that turns litmus blue; its solutions in water have a pH greater than 7; a proton acceptor (Brønsted definition)

basic: Having a pH greater than 7; alkaline

bifunctional: A substance containing two functional groups

biochemical oxygen demand (BOD): The amount of oxygen consumed by a sample during a five-day period at 20°C

biodegradable: Able to be broken down by organisms in the biosphere

bioequivalent: Having the same biological effect

biosphere: The part of the world in which life can exist

boiling point: The temperature at which a liquid is transformed into a vapor

bottled gas: A mixture of propane and butane that is removed from natural gas

brachytherapy: The technique of implanting radioactive material at a disease site

breeder reactor: A type of nuclear reactor that produces fissionable fuel

brine: A water solution containing a very high concentration ($> 100,000$ ppm) of dissolved minerals

broad spectrum antibiotic: A drug capable of interfering with the growth of a wide variety of microorganisms

builder: A substance added to detergents to tie up "hardness" ions

Cannabis sativa: The Indian hemp plant that is the source of marijuana

caramelization: The process of heating sugar to produce a brown amorphous substance

carbohydrates: Any of various neutral compounds of carbon, hydrogen, and oxygen (as sugars, starch, cellulose), most of which are formed by green plants and which constitute a major class of animal food

carbonation: The addition of carbon dioxide, usually done under pressure

caries: Progressive destruction of bones or teeth

casein: A group of proteins found in milk

catalyst: A substance that increases the speed of a reaction and is not consumed by the reaction

cation: Positively charged ion

cellulose: A polysaccharide consisting of glucose units joined by beta linkages

chemotherapy: Treatment of disease with drugs (chemicals)

chlorination: Treatment with chlorine

cis: A form of isomerism of unsaturated organic compounds in which two designated atoms or groups attached to the two carbon atoms of the double bond are located on the same side of the molecule

coenzyme: A substance that acts as a catalyst in cooperation with an enzyme

compost: New topsoil produced by the decay of organic matter

compound: A substance formed by combining elements

condensation polymer: A high molecular weight compound formed by reactions involving functional groups found in the monomers

congener: A substance produced as a side product of fermentation

corpus luteum: An endocrine gland formed in the ovaries following the release of an ovum. It releases the hormone progesterone, which aids in the development of a fertilized ovum

cosmic radiation: Radiation that penetrates the earth's atmosphere from outer space

covalent bond: A nonionic chemical bond formed by shared electrons

critical: A condition in a nuclear reactor in which an average of one product neutron becomes a bombarding neutron

critical mass: The weight of the isotope required for the reactor to achieve the supercritical condition in which more than one product neutron becomes a bombarding neutron

cross section: Probability of capture by a nucleus

culture: Growth of an organism in a nutrient media

DNA — deoxyribonucleic acid: A polymeric substance that is located in cell nuclei and contains the information to direct the functioning of the cell

demineralization: The breakdown of minerals, e.g., hydroxyapatite (tooth enamel)

density: The weight of a given volume of a substance

deuterium: The isotope of hydrogen that contains one neutron in the nucleus

dialysis: The separation of substances in solution by means of their unequal diffusion through semipermeable membranes

disaccharide: A carbohydrate consisting of two monosaccharide units bonded together covalently

divalent: Having the capacity to combine with two other atoms

effervescent: Releasing bubbles of gas

electron: An elementary particle with a negative charge numerically equal to that of a proton and a mass of 9.107×10^{-28}g or 1/1837 of a proton

electronegativity: The attraction of an atom for electrons in a molecule

element: A substance that contains only one kind of atom

emulsifying agent: A substance that stabilizes a suspension of two immiscible liquids, e.g., oil and water

emulsion: A suspension of fine particles of one liquid in another liquid

endocrinology: The study of glandular secretions

enteric drug: A substance treated to pass through the stomach unchanged and to disintegrate in the intestines

entomology: The study of insects

enzymes: A protein that catalyzes chemical reactions in living organisms

estrogens: Female sex hormones

eutrophication: The change of a body of water into a condition stimulating the growth of algae to the detriment of animal life

exponential notation: A shorthand notation used in writing very small and very large numbers, e.g., $1 \times 10^6 =$ one million

fast reactor: A reactor that utilizes unmoderated bombarding particles

fermentation: The breakdown of carbohydrates by microorganisms, such as the conversion of sugar into alcohol and carbon dioxide by the action of enzymes from yeast

fertile isotope: An isotope that can be used to breed a fissionable isotope

fission: The splitting of an atom as a result of bombardment of the nucleus by neutrons

fortification: Addition of alcohol

functional group: An atom or group of atoms other than carbon or hydrogen that are often responsible for the chemical and physical properties of the compound

fusion: The combination of two small nuclei to produce one larger nucleus

gene: A segment of a DNA molecule that is responsible for directing the synthesis of one protein

generic name: A technical, unsystematic type of name often used when describing drugs

genetic code: A three-letter code that is formulated from the chemical building blocks of DNA and directs the placement of the 20 common amino acids into their proper sequence in proteins

genetics: A branch of biology that deals with the heredity and variation of organisms

geometrical isomers: Molecules which differ in the three-dimensional arrangement of atoms, e.g., *cis-trans* isomers

glycolysis: The metabolic breakdown of glucose

gonads: Sex glands

gram negative: A type of bacteria to which certain dyes will not adhere

gram positive: A type of bacteria to which certain dyes will adhere

group: A vertical row of elements in the periodic table

habituation: Psychological dependence on a drug

half-life: The length of time required for one-half of a sample of a radioactive isotope to decay

hallucinogen: A substance that produces perceptions that do not exist

hard water: Water that contains ions, e.g., Ca^{++}, Mg^{++}, which form insoluble compounds with soap

hard wheat: Wheat that has a high content of the proteins that go into the making of high gluten doughs for baking

heat of vaporization: The amount of energy (in calories) required to change one gram of liquid to vapor at its boiling point

heavy water (^2H_2O or D_2O): Water in which the element hydrogen is present as the isotope deuterium with one neutron in the nucleus

herbicides: Weed killers

homogenization: A process for reducing the size of fat particles of an emulsion to uniform size

hormone: A substance secreted by endocrine glands, carried by the blood and used for the regulation of other tissues in the body

hybrid: An offspring of two animals or plants of different species

hydrocarbons: Organic compounds containing only hydrogen and carbon

hydrogenation: Addition of hydrogen

hydronium ion: H_3O^+

hydrophilic: Having an attraction for water

hydrophobic: Repelled by water

hydroxyapatite: Tooth enamel; a crystalline substance containing calcium, hydroxide, and phosphate ions; symbolized $Ca_5(PO_4)_3OH$

hypothalamus: A portion of the brain that exerts control over the function of many organs in the body and is involved in many sensations, e.g., thirst, hunger, blood pressure, body temperature

IUPAC: The International Union of Pure and Applied Chemistry; an organization that formulates rules for naming chemical compounds

immune response: The release of antibodies in response to invasion by an antigen

immunity: The ability to resist development of a pathogenic organism due to the acquired ability to produce antibodies that act against such an organism

immunochemistry: The branch of chemistry dealing with immunity

immunology: The branch of science dealing with immunity

inactive niacin (NADH): A form of the B vitamin niacin, which is sometimes utilized directly in metabolism but is usually converted to active niacin (NAD)

inert gas: A member of group VIIIA of the periodic table; the noble gases that have a valence level that is filled to capacity with electrons

influenza: A virus disease; flu

invert sugar: An equal mixture of glucose and fructose usually obtained by cleavage of sucrose

in vitro: In a test tube or other artificial environment outside the body

in vivo: In the living body

iodine value: The number of grams of iodine (I_2) consumed by 100 grams of a fat

ion: An atom existing as a charged particle

ionizing radiation: Any particulate or electromagnetic radiation that will cause a target substance to become charged

isomer: One of two or more compounds having the same formula but a different arrangement of the atoms within the molecule

isotopes: Atoms of the same element differing only in the number of neutrons in the nucleus

kairomone: A chemical substance released by a plant or animal, which provides an advantage to the recipient, e.g., repellents from toxic substances

lactic acid fermentation: The metabolism of glucose under anaerobic conditions to produce lactic acid

leavening: The rising of baked goods due to the release of CO_2, NH_3, or other gases trapped in the product during and prior to baking

legumes: Plants that are inhabited by microorganisms that carry out nitrogen fixation

light water (H_2O): Water containing only the isotope of hydrogen with no neutrons in the nucleus

lime: Calcium oxide (CaO); also known as quicklime

limestone: Calcium carbonate ($CaCO_3$); also known as "agricultural lime"

lye: Sodium hydroxide (NaOH)

mass: The amount of material contained in a substance

mass number: The sum of the number of protons and neutrons in the nucleus

metabolism: The chemical changes occurring in living cells by which energy is provided for vital processes and activities

metastasis: A secondary growth of a malignant tumor in another part of the body

micelle: A cluster of molecules in solution, which are oriented so as to permit maximum interaction of water with the hydrophilic portion of the molecules and minimum interaction of water with the hydrophobic portion

microbiology: The study of microorganisms

micron: 1×10^{-6} meter

mineralization: Formation of inorganic ions from organic compounds; the combination of inorganic ions

miscible: Mutually soluble

moderator: A substance used to slow neutrons down so they can come under the influence of nuclear forces and cause reactions to occur

mole: An amount of a substance equal to its molecular weight

molecular biology: The field of science that is an outgrowth of the study of genetics and biochemistry

molecular weight: The sum of the weights of all the atoms in a molecule, expressed in atomic mass units

molecule: The smallest unit of a compound that possesses all the properties of that compound

monomers: The chemical building blocks of polymers

monosaccharide: A sugar that cannot be cleaved into simpler sugars

monovalent: Having the capacity to combine with one other atom

must: The juice from grapes prior to fermentation

mutation: A change in the amino acid sequence in a protein causing the appearance of a hybrid form of the organism

narcotic: A substance that causes sedation (narcosis) and relieves pain

narrow spectrum antibiotic: A drug capable of interfering with the growth of only a very few microorganisms

neoplasm: An abnormal growth or tumor

neutralization: The reaction of an acid and a base to produce a salt

neutrino: An uncharged particle with zero mass, which is released during beta decay

neutron: An uncharged particle with a mass nearly equal to that of a proton and present in all known atomic nuclei except the hydrogen atom

noble gases: Any of the inert or nonreactive gases such as helium, neon, argon, krypton, xenon, and radon that appear in group VIIIA of the periodic table

nuclear fission: The splitting of an atomic nucleus resulting in the release of large amounts of energy

nuclear fusion: The union of atomic nuclei to form heavier nuclei resulting in the release of large quantities of energy when certain light elements unite

octet rule: Any element is most stable when it has a set of eight valence electrons

osmosis: The flow of solvent molecules through a semipermeable membrane in order to bring two solutions on either side of the membrane to equal concentration

ovulation   The release of an egg from the ovaries

oxidation: The combination of a substance with oxygen; loss of electrons; dehydrogenation

oxidative phosphorylation: A combination of the processes by which active niacin is regenerated and ATP is formed: phosphate

$+ \text{NADH} + \text{O}_2 + \text{ADP} \rightarrow \text{NAD} + \text{H}_2\text{O} + \text{ATP}$

oxygenation: Treatment with oxygen

pandemic: Global epidemic

parts per million (ppm): The number of parts of a component of a sample per one million parts of the total sample

passive immunity: A type of temporary immunity that is acquired by transferring preformed antibodies to an individual to provide protection against a particular infection

pathogenic: Causing disease

period: A horizontal row of elements in the periodic table

pH: A numerical scale used to describe the level of acidity or basicity of a water solution

pharmacology: The study of the action of drugs on the body

pheromone: A chemical compound that is used for communication among animals, e.g., a sex attractant

photosynthesis: The process by which chlorophyll-containing plants use energy from the sun to convert carbon dioxide and water into oxygen and sugar

plaque: A collection of microorganisms on teeth

plasticizer: A substance that may be added to a rigid polymeric material in order to impart flexibility

plastic range: The temperature range in which a fat product retains its shape but is readily deformed

polymer: A substance of high molecular weight, synthesized from repeating units called monomers

polyunsaturated fat: Triglycerides containing a high level of unsaturated fatty acids

potash: Potassium oxide (K_2O)

precursor: A compound that serves as a source of another compound, e.g., glucose as a precursor to alcohol via fermentation

prodrug: An inactive form of the drug that is converted to the active drug after it is consumed

protein: A polymeric substance in which amino acids are the repeating units

proteolytic: Protein-cleaving

psychedelic: Intensely pleasureful sensual perception

quicklime: (CaO) calcium oxide

rad: The unit of absorbed dose of radiation

radical: The form of a compound resulting from the removal of a hydrogen atom

radiopharmaceutical: A radioactive drug

reduction: Gain of electrons; hydrogenation

rem: The unit of absorbed dose expressed in terms of the biological effect of radiation, which varies from one type of tissue to another and from one type of radiation to another

remineralization: The formation of tooth enamel (hydroxyapatite)

rendering: Isolation of a fat from an animal source

resin: The polymeric component of a plastic

resistance: The ability to avoid destruction by a substance due to the presence of an enzyme that causes a chemical change in that substance

respiration: The utilization of oxygen by living organisms in order to produce energy

reversion: Spoilage of a fat by oxidation

Roentgen: The unit describing the quantity of radioactivity emitted by a sample (only applicable to X- or gamma radiation)

ruminant: A cud chewing animal that has several stomach compartments, one of which is called the rumen

salometer: A unit used to describe salt concentration; 1% NaCl = 4° salometer

salt: A compound containing a cation and an anion

saturated: Containing only carbon–carbon

single bonds; solution that contains all the solute it can dissolve at a given temperature

sequestering agent: A substance that binds metal ions

smelting: A process for obtaining metals from their ores by a process that includes melting

soft radiation: Radiation of very low energy

soft wheat: Wheat that has a low content of the proteins that go into the making of high gluten doughs for baking

solute: A substance dissolved in a solvent

solution: A homogeneous mixture of two or more substances

solvent: A liquid that dissolves another substance

specific gravity: The weight of a given volume of a substance compared to another substance used as a standard

specific heat: The amount of energy (in calories) needed to raise the temperature of one gram of a substance one degree Celsius

spermicide: A substance capable of killing sperm

starch: A polysaccharide consisting of repeating units of glucose

steady state: The condition under which the rate of formation and the rate of decay of a substance are equal

surfactant: A substance that is both hydrophobic and hydrophilic and is able to bring about mixing of hydrophobic and hydrophilic substances

syndet: Synthetic detergent

synergist: A substance that works in cooperation with another

teletherapy: Exposure to an external source of radiation

teratogen: A substance that causes birth defects

tetravalent: Having the ability to combine with four other atoms

thermal neutrons: Slow neutrons

thermoplastic resin: A polymeric material that can be heated to melting and restored to its original chemical form upon cooling

thermosetting resin: A polymeric material that cannot be heated to melting without destroying its chemical structure; contains cross-links between polymer chains

tolerance: The ability to tolerate increasing amounts of a drug without ill effects; requiring increasing amounts of a drug in order to experience the same effects

toxin: A poison

toxoid: An inactivated toxin

tracers: Substances that can be used to monitor the movement of chemical compounds

trans: A form of isomer of unsaturated organic compounds in which two designated atoms or groups attached to two carbon atoms of a double bond are located on opposite sides of the molecule

transition elements: Metallic elements located in the center portion of the periodic table in the B groups

transmutation: The conversion of one element into another either naturally or artificially

transuranium elements: Elements following uranium in the periodic table

triglyceride: A triester resulting from the combination of glycerin and three fatty acid molecules

tritium: An isotope of hydrogen with two neutrons in the nucleus

trivalent: Having the capacity to combine with three other atoms

troposphere: The layer of atmosphere ranging from ground level to ten miles into space

unsaturation: Double or triple bonds

valence: The combining capacity of an atom

valence electrons: The number of electrons found in the outermost shell of the atom

weight: The force with which a body is attracted toward the earth by gravitation

COMMON SIMPLE IONS

ELEMENT	COMMON SIMPLE ION	NAME OF ELEMENT	NAME OF ION (IF DIFFERENT FROM NEUTRAL ATOM)
H	H^+	hydrogen	proton
Li	Li^+	lithium	
O	O^{-2}	oxygen	oxide
F	F^-	fluorine	fluoride
Na	Na^+	sodium	
Mg	Mg^{+2}	magnesium	
Al	Al^{+3}	aluminum	
S	S^{-2}	sulfur	sulfide
Cl	Cl^-	chlorine	chloride
K	K^+	potassium	
Ca	Ca^{+2}	calcium	
Fe	Fe^{+2}	iron	ferrous or iron (II)
	Fe^{+3}		ferric or iron (III)
Cu	Cu^{+1}	copper	cuprous or copper (I)
	Cu^{+2}		cupric or copper (II)
Zn	Zn^{+2}	zinc	
Br	Br^-	bromine	bromide
Rb	Rb^+	rubidium	
Sr	Sr^{+2}	strontium	
Ag	Ag^+	silver	
Cd	Cd^{+2}	cadmium	
Sn	Sn^{+2}	tin	stannous or tin (II)
	Sn^{+4}		stannic or tin (IV)
I	I^-	iodine	iodide
Cs	Cs^+	cesium	
Ba	Ba^{+2}	barium	
Pb	Pb^{+2}	lead	
Ra	Ra^{+2}	radium	

COMMON COMBINATION IONS

ELEMENT	COMBINATION ION	NAME OF ION
H	OH^-	hydroxide
C	HCO_3^-	hydrogen carbonate or bicarbonate
	CO_3^{-2}	carbonate
N	NH_4^+	ammonium
	NO_3^-	nitrate
	NO_2^-	nitrite
P	$H_2PO_4^-$	dihydrogenphosphate
	HPO_4^{-2}	monohydrogenphosphate
	PO_4^{-3}	phosphate
S	HSO_4^-	hydrogen sulfate or bisulfate
	SO_4^{-2}	sulfate
	HSO_3^-	hydrogen sulfite or bisulfite
	SO_3^{-2}	sulfite
Cl	OCl^-	hypochlorite
Cr	CrO_4^-	chromate
	$Cr_2O_7^{-2}$	dichromate
Mn	MnO_4^-	permanganate
As	AsO_4^{-3}	arsenate
Br	OBr^-	hypobromite
	BrO_3^-	bromate
	BrO_4^-	perbromate